SEI/ASCE 7-02
Second Edition

American Society of Civil Engineers

Minimum Design Loads for Buildings and Other Structures

This second edition incorporates the corrections as shown in the errata found on www.seinstitute.org.

Revision of ASCE 7-98

This document uses both Système International (SI) units and customary units.

Structural Engineering Institute

Published by the American Society of Civil Engineers
1801 Alexander Bell Drive
Reston, Virginia 20191-4400

ABSTRACT

This revision of the ASCE Standard *Minimum Design Loads for Buildings and Other Structures* is a replacement of ASCE 7-98. This Standard provides requirements for dead, live, soil, flood, wind, snow, rain, ice, and earthquake loads, and their combinations that are suitable for inclusion in building codes and other documents.

Substantial changes were made to the wind, snow, earthquake, and ice provisions. In addition, substantial new material was added regarding the determination of flood loads. The structural loading requirements provided by this Standard are intended for use by architects, structural engineers, and those engaged in preparing and administering local building codes.

Library of Congress Cataloging-in-Publication Data

Minimum design loads for buildings and other structures / American Society of Civil Engineers.
    p. cm. — (ASCE standard.)
  "Revision of ASCE 7-98."
  ISBN 0-7844-0624-3
  1. Structural engineering — United States. 2. Standards, Engineering — United States. I. American Society of Civil Engineers.

TH851 .M56 2002
624.1'72'021873 — dc21

                    2002038610

STANDARDS

In April 1980, the Board of Direction approved ASCE Rules for Standards Committees to govern the writing and maintenance of standards developed by the Society. All such standards are developed by a consensus standards process managed by the Codes and Standards Activities Committee (CSAC). The consensus process includes balloting by the balanced standards committee made up of Society members and nonmembers, balloting by the membership of ASCE as a whole, and balloting by the public. All standards are updated or reaffirmed by the same process at intervals not exceeding 5 years.

The following Standards have been issued:

ANSI/ASCE 1-82 N-725 Guideline for Design and Analysis of Nuclear Safety Related Earth Structures

ANSI/ASCE 2-91 Measurement of Oxygen Transfer in Clean Water

ANSI/ASCE 3-91 Standard for the Structural Design of Composite Slabs and ANSI/ASCE 9-91 Standard Practice for the Construction and Inspection of Composite Slabs

ASCE 4-98 Seismic Analysis of Safety-Related Nuclear Structures

Building Code Requirements for Masonry Structures (ACI 530-02/ASCE 5-02/TMS 402-02) and Specifications for Masonry Structures (ACI 530.1-02/ASCE 6-02/TMS 602-02)

SEI/ASCE 7-02 Minimum Design Loads for Buildings and Other Structures

ANSI/ASCE 8-90 Standard Specification for the Design of Cold-Formed Stainless Steel Structural Members

ANSI/ASCE 9-91 listed with ASCE 3-91

ASCE 10-97 Design of Latticed Steel Transmission Structures

SEI/ASCE 11-99 Guideline for Structural Condition Assessment of Existing Buildings

ANSI/ASCE 12-91 Guideline for the Design of Urban Subsurface Drainage

ASCE 13-93 Standard Guidelines for Installation of Urban Subsurface Drainage

ASCE 14-93 Standard Guidelines for Operation and Maintenance of Urban Subsurface Drainage

ASCE 15-98 Standard Practice for Direct Design of Buried Precast Concrete Pipe Using Standard Installations (SIDD)

ASCE 16-95 Standard for Load and Resistance Factor Design (LRFD) of Engineered Wood Construction

ASCE 17-96 Air-Supported Structures

ASCE 18-96 Standard Guidelines for In-Process Oxygen Transfer Testing

ASCE 19-96 Structural Applications of Steel Cables for Buildings

ASCE 20-96 Standard Guidelines for the Design and Installation of Pile Foundations

ASCE 21-96 Automated People Mover Standards — Part 1

ASCE 21-98 Automated People Mover Standards — Part 2

ASCE 21-00 Automated People Mover Standards — Part 3

SEI/ASCE 23-97 Specification for Structural Steel Beams with Web Openings

SEI/ASCE 24-98 Flood Resistant Design and Construction

ASCE 25-97 Earthquake-Actuated Automatic Gas Shut-Off Devices

ASCE 26-97 Standard Practice for Design of Buried Precast Concrete Box Sections

ASCE 27-00 Standard Practice for Direct Design of Precast Concrete Pipe for Jacking in Trenchless Construction

ASCE 28-00 Standard Practice for Direct Design of Precast Concrete Box Sections for Jacking in Trenchless Construction

SEI/ASCE 30-00 Guideline for Condition Assessment of the Building Envelope

SEI/ASCE 32-01 Design and Construction of Frost-Protected Shallow Foundations

EWRI/ASCE 33-01 Comprehensive Transboundary International Water Quality Management Agreement

EWRI/ASCE 34-01 Standard Guidelines for Artificial Recharge of Ground Water

EWRI/ASCE 35-01 Guidelines for Quality Assurance of Installed Fine-Pore Aeration Equipment

CI/ASCE 36-01 Standard Construction Guidelines for Microtunneling

SEI/ASCE 37-02 Design Loads on Structures During Construction

CI/ASCE 38-02 Standard Guideline for the Collection and Depiction of Existing Subsurface Utility Data

FOREWORD

The material presented in this Standard has been prepared in accordance with recognized engineering principles. This Standard should not be used without first securing competent advice with respect to its suitability for any given application. The publication of the material contained herein is not intended as a representation or warranty on the part of the American Society of Civil Engineers, or of any other person named herein, that this information is suitable for any general or particular use or promises freedom from infringement of any patent or patents. Anyone making use of this information assumes all liability from such use.

ACKNOWLEDGEMENTS

The American Society of Civil Engineers (ASCE) acknowledges the work of the Minimum Design Loads on Buildings and Other Structures Standards Committee of the Codes and Standards Activities Division of the Structural Engineering Institute. This group comprises individuals from many backgrounds including: consulting engineering, research, construction industry, education, government, design, and private practice.

This revision of the standard began in 1999 and incorporates information as described in the commentary.

This Standard was prepared through the consensus standards process by balloting in compliance with procedures of ASCE's Codes and Standards Activities Committee. Those individuals who serve on the Standards Committee are:

Kharaiti L. Abrol
Robert E. Bachman
Charles C. Baldwin
Demirtas C. Bayar
John E. Breen
David G. Brinker
Ray A. Bucklin
James R. Cagley
Charles J. Carter
Jack E. Cermak
William L. Coulbourne
Jay H. Crandell
Stanley W. Crawley
Majed Dabdoub
James M. Delahay
Bernard J. Deneke
Bradford K. Douglas
John F. Duntemann
Donald Dusenberry
Bruce R. Ellingwood, Vice-Chair
Edward R. Estes
Mohammed M. Ettouney
David A. Fanella
Lawrence Fischer
Theodore V. Galambos
Satyendra K. Ghosh
Dennis W. Graber
Lawrence G. Griffis
David S. Gromala
Robert D. Hanson
Gilliam S. Harris
James R. Harris, Chair
Joseph P. Hartman
Scott R. Humphreys
Nicholas Isyumov
Christopher P. Jones
John H. Kapmann, Jr.
A. Harry Karabinis
Jon P. Kiland
Randy Kissell
Uno Kula

Edward M. Laatsch
John V. Loscheider
Fred Morello
Ian Mackinlay
Sanjeev Malushte
Bonnie Manley
Harry W. Martin
Rusk Masih
George M. Matsumura
Therese P. McAllister
Robert R. McCluer
Richard McConnell
Kishor C. Mehta
Joseph J. Messersmith, Jr.
Joe N. Nunnery
Michael O'Rourke
George N. Olive
Frederick J. Palmer
Alan B. Peabody
David B. Peraza
Clarkson W. Pinkham
Robert D. Prince
Robert T. Ratay
Mayasandra K. Ravindra
Lawrence D. Reaveley
Abraham J. Rokach
James A. Rossberg, Secretary
Herbert S. Saffir
Phillip J Samblanet
Suresh C. Satsangi
Andrew Scanlon
Jeffery C. Sciaudone
William L. Shoemaker
Emil Simiu
Thomas D. Skaggs
Thomas L. Smith
James G. Soules
Theodore Stathopoulos
Donald R. Strand
Harry B. Thomas
Wayne N. Tobiasson

Brian E. Trimble
David P. Tyree
Thomas R. Tyson
Joseph W. Vellozzi
Francis J. Walter, Jr.
Yi Kwei Wen
Peter J.G. Willse
Lyle L. Wilson
Joseph A. Wintz, III

Task Committee on General Structural Requirements
James S. Cohen
John F. Duntemann
Donald Dusenberry, Chair
John L. Gross
Anatol Longinow
John V. Loscheider
Robert T. Ratay
James G. Soules
Jason J. Thompson

Task Committee on Strength
Bruce R. Ellingwood, Chair
Theodore V. Galambos
David S. Gromala
James R. Harris
Clarkson Pinkham
Massandra K. Ravindra
Andrew Scanlon
James G. Soules
Yi Kwei Wen

CONTENTS

Commentary

SECTION 1.0
GENERAL

SECTION 1.1
SCOPE

This standard provides minimum load requirements for the design of buildings and other structures that are subject to building code requirements. Loads and appropriate load combinations, which have been developed to be used together, are set forth for strength design and allowable stress design. For design strengths and allowable stress limits, design specifications for conventional structural materials used in buildings and modifications contained in this standard shall be followed.

SECTION 1.2
DEFINITIONS

The following definitions apply to the provisions of the entire standard.

ALLOWABLE STRESS DESIGN. A method of proportioning structural members such that elastically computed stresses produced in the members by nominal loads do not exceed specified allowable stresses (also called working stress design).

AUTHORITY HAVING JURISDICTION. The organization, political subdivision, office, or individual charged with the responsibility of administering and enforcing the provisions of this standard.

BUILDINGS. Structures, usually enclosed by walls and a roof, constructed to provide support or shelter for an intended occupancy.

DESIGN STRENGTH. The product of the nominal strength and a resistance factor.

ESSENTIAL FACILITIES. Buildings and other structures that are intended to remain operational in the event of extreme environmental loading from wind, snow, or earthquakes.

FACTORED LOAD. The product of the nominal load and a load factor.

HAZARDOUS MATERIAL. Chemicals or substances classified as a physical or health hazard whether the chemicals or substances are in a usable or waste condition.

HEALTH HAZARD. Chemicals or substances classified by the authority having jurisdiction as toxic, highly toxic, or corrosive.

LIMIT STATE. A condition beyond which a structure or member becomes unfit for service and is judged either to be no longer useful for its intended function (serviceability limit state) or to be unsafe (strength limit state).

LOAD EFFECTS. Forces and deformations produced in structural members by the applied loads.

LOAD FACTOR. A factor that accounts for deviations of the actual load from the nominal load, for uncertainties in the analysis that transforms the load into a load effect, and for the probability that more than one extreme load will occur simultaneously.

LOADS. Forces or other actions that result from the weight of all building materials, occupants and their possessions, environmental effects, differential movement, and restrained dimensional changes. Permanent loads are those loads in which variations over time are rare or of small magnitude. All other loads are variable loads. (see also nominal loads.)

NOMINAL LOADS. The magnitudes of the loads specified in Sections 3 through 9 (dead, live, soil, wind, snow, rain, flood, and earthquake) of this standard.

NOMINAL STRENGTH. The capacity of a structure or member to resist the effects of loads, as determined by computations using specified material strengths and dimensions and formulas derived from accepted principles of structural mechanics or by field tests or laboratory tests of scaled models, allowing for modeling effects and differences between laboratory and field conditions.

OCCUPANCY. The purpose for which a building or other structure, or part thereof, is used or intended to be used.

OTHER STRUCTURES. Structures, other than buildings, for which loads are specified in this standard.

P-DELTA EFFECT. The second-order effect on shears and moments of frame members induced by axial loads on a laterally displaced building frame.

PHYSICAL HAZARD. Chemicals or substances in a liquid, solid, or gaseous form that are classified by the authority having jurisdiction as combustible, flammable, explosive, oxidizer, pyrophoric, unstable (reactive), or water reactive.

RESISTANCE FACTOR. A factor that accounts for deviations of the actual strength from the nominal strength and

the manner and consequences of failure (also called strength reduction factor).

STRENGTH DESIGN. A method of proportioning structural members such that the computed forces produced in the members by the factored loads do not exceed the member design strength (also called load and resistance factor design).

TEMPORARY FACILITIES. Buildings or other structures that are to be in service for a limited time and have a limited exposure period for environmental loadings.

SECTION 1.3
BASIC REQUIREMENTS

1.3.1 Strength. Buildings and other structures, and all parts thereof, shall be designed and constructed to support safely the factored loads in load combinations defined in this document without exceeding the appropriate strength limit states for the materials of construction. Alternatively, buildings and other structures, and all parts thereof, shall be designed and constructed to support safely the nominal loads in load combinations defined in this document without exceeding the appropriate specified allowable stresses for the materials of construction.

1.3.2 Serviceability. Structural systems, and members thereof, shall be designed to have adequate stiffness to limit deflections, lateral drift, vibration, or any other deformations that adversely affect the intended use and performance of buildings and other structures.

1.3.3 Self-Straining Forces. Provision shall be made for anticipated self-straining forces arising from differential settlements of foundations and from restrained dimensional changes due to temperature, moisture, shrinkage, creep, and similar effects.

1.3.4 Analysis. Load effects on individual structural members shall be determined by methods of structural analysis that take into account equilibrium, general stability, geometric compatibility, and both short- and long-term material properties. Members that tend to accumulate residual deformations under repeated service loads shall have included in their analysis the added eccentricities expected to occur during their service life.

1.3.5 Counteracting Structural Actions. All structural members and systems, and all components and cladding in a building or other structure, shall be designed to resist forces due to earthquake and wind, with consideration of overturning, sliding, and uplift, and continuous load paths shall be provided for transmitting these forces to the foundation. Where sliding is used to isolate the elements, the effects of friction between sliding elements shall be included as a force. Where all or a portion of the resistance to these forces is provided by dead load, the dead load shall be taken as the minimum dead load likely to be in place during the event causing the considered forces. Consideration shall be given to the effects of vertical and horizontal deflections resulting from such forces.

SECTION 1.4
GENERAL STRUCTURAL INTEGRITY

Buildings and other structures shall be designed to sustain local damage with the structural system as a whole remaining stable and not being damaged to an extent disproportionate to the original local damage. This shall be achieved through an arrangement of the structural elements that provides stability to the entire structural system by transferring loads from any locally damaged region to adjacent regions capable of resisting those loads without collapse. This shall be accomplished by providing sufficient continuity, redundancy, or energy-dissipating capacity (ductility), or a combination thereof, in the members of the structure.

SECTION 1.5
CLASSIFICATION OF BUILDINGS AND OTHER STRUCTURES

1.5.1 Nature of Occupancy. Buildings and other structures shall be classified, based on the nature of occupancy, according to Table 1-1 for the purposes of applying flood, wind, snow, earthquake, and ice provisions. The categories range from I to IV, where Category I represents buildings and other structures with a low hazard to human life in the event of failure and Category IV represents essential facilities. Each building or other structure shall be assigned to the highest applicable category or categories. Assignment of the same structure to multiple categories based on use and the type of load condition being evaluated (e.g., wind, seismic, etc.) shall be permissible.

When buildings or other structures have multiple uses (occupancies), the relationship between the uses of various parts of the building or other structure and the independence of the structural systems for those various parts shall be examined. The classification for each independent structural system of a multiple use building or other structure shall be that of the highest usage group in any part of the building or other structure that is dependent on that basic structural system.

1.5.2 Hazardous Materials and Extremely Hazardous Materials. Buildings and other structures containing hazardous materials or extremely hazardous materials are permitted to be classified as Category II structures if it can be demonstrated to the satisfaction of the authority having jurisdiction by a hazard assessment as part of an overall

risk management plan (RMP) that a release of the hazardous material or extremely hazardous material does not pose a threat to the public.

In order to qualify for this reduced classification, the owner or operator of the buildings or other structures containing the hazardous materials or extremely hazardous materials shall have a risk management plan that incorporates three elements as a minimum: a hazard assessment, a prevention program, and an emergency response plan.

As a minimum, the hazard assessment shall include the preparation and reporting of worst-case release scenarios for each structure under consideration, showing the potential effect on the public for each. As a minimum, the worst-case event shall include the complete failure (instantaneous release of entire contents) of a vessel, piping system, or other storage structure. A worst-case event includes (but is not limited to) a release during the design wind or design seismic event. In this assessment, the evaluation of the effectiveness of subsequent measures for accident mitigation shall be based on the assumption that the complete failure of the primary storage structure has occurred. The off-site impact must be defined in terms of population within the potentially affected area. In order to qualify for the reduced classification, the hazard assessment shall demonstrate that a release of the hazardous material from a worst-case event does not pose a threat to the public outside the property boundary of the facility.

As a minimum, the prevention program shall consist of the comprehensive elements of process safety management, which is based on accident prevention through the application of management controls in the key areas of design, construction, operation, and maintenance. Secondary containment of the hazardous materials or extremely hazardous materials (including, but not limited to, double wall tank, dike of sufficient size to contain a spill, or other means to contain a release of the hazardous materials or extremely hazardous material within the property boundary of the facility and prevent release of harmful quantities of contaminants to the air, soil, ground water, or surface water) are permitted to be used to mitigate the risk of release. When secondary containment is provided, it shall be designed for all environmental loads and is not eligible for this reduced classification. In hurricane-prone regions, mandatory practices and procedures that effectively diminish the effects of wind on critical structural elements or that alternatively protect against harmful releases during and after hurricanes may be used to mitigate the risk of release.

As a minimum, the emergency response plan shall address public notification, emergency medical treatment for accidental exposure to humans, and procedures for emergency response to releases that have consequences beyond the property boundary of the facility. The emergency response plan shall address the potential that resources for response could be compromised by the event that has caused the emergency.

SECTION 1.6
ADDITIONS AND ALTERATIONS TO EXISTING STRUCTURES

When an existing building or other structure is enlarged or otherwise altered, structural members affected shall be strengthened if necessary so that the factored loads defined in this document will be supported without exceeding the specified design strength for the materials of construction. When using allowable stress design, strengthening is required when the stresses due to nominal loads exceed the specified allowable stresses for the materials of construction.

SECTION 1.7
LOAD TESTS

A load test of any construction shall be conducted when required by the authority having jurisdiction whenever there is reason to question its safety for the intended occupancy or use.

TABLE 1-1
CLASSIFICATION OF BUILDINGS AND OTHER STRUCTURES FOR FLOOD, WIND, SNOW, EARTHQUAKE, AND ICE LOADS

Nature of Occupancy	Category

Buildings and other structures that represent a low hazard to human life in the event of failure including, but not limited to:
    Agricultural facilities
    Certain temporary facilities
    Minor storage facilities
	I

All buildings and other structures except those listed in Categories I, III, and IV	II

Buildings and other structures that represent a substantial hazard to human life in the event of failure including, but not limited to:
    Buildings and other structures where more than 300 people congregate in one area
    Buildings and other structures with day care facilities with capacity greater than 150
    Buildings and other structures with elementary school or secondary school facilities with capacity greater than 250
    Buildings and other structures with a capacity greater than 500 for colleges or adult education facilities
    Health care facilities with a capacity of 50 or more resident patients but not having surgery or emergency treatment facilities
    Jails and detention facilities
    Power generating stations and other public utility facilities not included in Category IV
 	
Buildings and other structures not included in Category IV (including, but not limited to, facilities that manufacture, process, handle, store, use, or dispose of such substances as hazardous fuels, hazardous chemicals, hazardous waste, or explosives) containing sufficient quantities of hazardous materials to be dangerous to the public if released.
 	
Buildings and other structures containing hazardous materials shall be eligible for classification as Category II structures if it can be demonstrated to the satisfaction of the authority having jurisdiction by a hazard assessment as described in Section 1.5.2 that a release of the hazardous material does not pose a threat to the public.
	III

Buildings and other structures designated as essential facilities including, but not limited to:
    Hospitals and other health care facilities having surgery or emergency treatment facilities
    Fire, rescue, ambulance, and police stations and emergency vehicle garages
    Designated earthquake, hurricane, or other emergency shelters
    Designated emergency preparedness, communication, and operation centers and other facilities required for emergency response
    Power generating stations and other public utility facilities required in an emergency
    Ancillary structures (including, but not limited to, communication towers, fuel storage tanks, cooling towers, electrical substation structures, fire water storage tanks or other structures housing or supporting water, or other fire-suppression material or equipment) required for operation of Category IV structures during an emergency
    Aviation control towers, air traffic control centers, and emergency aircraft hangars
    Water storage facilities and pump structures required to maintain water pressure for fire suppression
    Buildings and other structures having critical national defense functions
 	
Buildings and other structures (including, but not limited to, facilities that manufacture, process, handle, store, use, or dispose of such substances as hazardous fuels, hazardous chemicals, hazardous waste, or explosives) containing extremely hazardous materials where the quantity of the material exceeds a threshold quantity established by the authority having jurisdiction.
 	
Buildings and other structures containing extremely hazardous materials shall be eligible for classification as Category II structures if it can be demonstrated to the satisfaction of the authority having jurisdiction by a hazard assessment as described in Section 1.5.2 that a release of the extremely hazardous material does not pose a threat to the public. This reduced classification shall not be permitted if the buildings or other structures also function as essential facilities.
	IV

SECTION 2.0
COMBINATIONS OF LOADS

SECTION 2.1
GENERAL

Buildings and other structures shall be designed using the provisions of either Section 2.3 or 2.4. Either Section 2.3 or 2.4 shall be used exclusively for proportioning elements of a particular construction material throughout the structure.

SECTION 2.2
SYMBOLS AND NOTATION

D = dead load;
D_i = weight of ice;
E = earthquake load;
F = load due to fluids with well-defined pressures and maximum heights;
F_a = flood load;
H = load due to lateral earth pressure, ground water pressure, or pressure of bulk materials;
L = live load;
L_r = roof live load;
R = rain load;
S = snow load;
T = self-straining force;
W = wind load;
W_i = wind-on-ice determined in accordance with Section 10.

SECTION 2.3
COMBINING FACTORED LOADS USING STRENGTH DESIGN

2.3.1 Applicability. The load combinations and load factors given in Section 2.3.2 shall be used only in those cases in which they are specifically authorized by the applicable material design standard.

2.3.2 Basic Combinations. Structures, components, and foundations shall be designed so that their design strength equals or exceeds the effects of the factored loads in the following combinations:

1. $1.4(D + F)$
2. $1.2(D + F + T) + 1.6(L + H) + 0.5(L_r$ or S or $R)$
3. $1.2D + 1.6(L_r$ or S or $R) + (L$ or $0.8W)$
4. $1.2D + 1.6W + L + 0.5(L_r$ or S or $R)$
5. $1.2D + 1.0E + L + 0.2S$
6. $0.9D + 1.6W + 1.6H$
7. $0.9D + 1.0E + 1.6H$

Exceptions:

1. The load factor on L in combinations (3), (4), and (5) is permitted to equal 0.5 for all occupancies in which L_o in Table 4.1 is less than or equal to 100 psf, with the exception of garages or areas occupied as places of public assembly.

2. The load factor on H shall be set equal to zero in combinations (6) and (7) if the structural action due to H counteracts that due to W or E. Where lateral earth pressure provides resistance to structural actions from other forces, it shall not be included in H but shall be included in the design resistance.

Each relevant strength limit state shall be investigated. Effects of one or more loads not acting shall be investigated. The most unfavorable effects from both wind and earthquake loads shall be investigated, where appropriate, but they need not be considered to act simultaneously. Refer to Section 9.2.2 for specific definition of the earthquake load effect E.[1]

2.3.3 Load Combinations Including Flood Load. When a structure is located in a flood zone (Section 5.3.1), the following load combinations shall be considered:

1. In V-Zones or Coastal A-Zones, $1.6W$ in combinations (4) and (6) shall be replaced by $1.6W + 2.0F_a$.
2. In noncoastal A-Zones, $1.6W$ in combinations (4) and (6) shall be replaced by $0.8W + 1.0F_a$.

2.3.4 Load Combinations Including Atmospheric Ice Loads. When a structure is subjected to atmospheric ice and wind-on-ice loads, the following load combinations shall be considered:

1. 0.5 $(L_r$ or S or $R)$ in combination (2) shall be replaced by $0.2D_i + 0.5S$.
2. $1.6W + 0.5$ $(L_r$ or S or $R)$ in combination (4) shall be replaced by $D_i + W_i + 0.5S$.
3. $1.6W$ in combination (6) shall be replaced by $D_i + W_i$.

---

[1] The same E from Section 9 is used for both Section 2.3.2 and Section 2.4.1. Refer to the Commentary for Section 9.

SECTION 2.4
COMBINING NOMINAL LOADS USING ALLOWABLE STRESS DESIGN

2.4.1 Basic Combinations. Loads listed herein shall be considered to act in the following combinations; whichever produces the most unfavorable effect in the building, foundation, or structural member being considered. Effects of one or more loads not acting shall be considered.

1. $D + F$
2. $D + H + F + L + T$
3. $D + H + F + (L_r$ or S or $R)$
4. $D + H + F + 0.75(L + T) + 0.75(L_r$ or S or $R)$
5. $D + H + F + (W$ or $0.7E)$
6. $D + H + F + 0.75(W$ or $0.7E) + 0.75L + 0.75(L_r$ or S or $R)$
7. $0.6D + W + H$
8. $0.6D + 0.7E + H$

The most unfavorable effects from both wind and earthquake loads shall be considered, where appropriate, but they need not be assumed to act simultaneously. Refer to Section 9.2.2 for the specific definition of the earthquake load effect E.[2]

Increases in allowable stress shall not be used with the loads or load combinations given in this standard unless it can be demonstrated that such an increase is justified by structural behavior caused by rate or duration of load.

---

[2] The same E from Section 9 is used for both Section 2.3.2 and Section 2.4.1. Refer to the Commentary for Section 9.

2.4.2 Load Combinations Including Flood Load. When a structure is located in a flood zone, the following load combinations shall be considered:

1. In V-Zones or Coastal A-Zones (Section 5.3.1), $1.5F_a$ shall be added to other loads in combinations (5), (6), and (7) and E shall be set equal to zero in (5) and (6).
2. In noncoastal A-Zones, $0.75F_a$ shall be added to combinations (5), (6), and (7) and E shall be set equal to zero in (5) and (6).

2.4.3 Load Combinations Including Atmospheric Ice Loads. When a structure is subjected to atmospheric ice and wind-on-ice loads, the following load combinations shall be considered:

1. $0.7D_i$ shall be added to combination (2).
2. $(L_r$ or S or $R)$ in combination (3) shall be replaced by $0.7D_i + 0.7W_i + S$.
3. W in combination (7) shall be replaced by $0.7D_i + 0.7W_i$.

SECTION 2.5
LOAD COMBINATIONS FOR EXTRAORDINARY EVENTS

Where required by the applicable code, standard, or the authority having jurisdiction, strength and stability shall be checked to ensure that structures are capable of withstanding the effects of extraordinary (i.e., low-probability) events such as fires, explosions, and vehicular impact.

SECTION 3.0
DEAD LOADS

SECTION 3.1
DEFINITION

Dead loads consist of the weight of all materials of construction incorporated into the building including but not limited to walls, floors, roofs, ceilings, stairways, built-in partitions, finishes, cladding and other similarly incorporated architectural and structural items, and fixed service equipment including the weight of cranes.

SECTION 3.2
WEIGHTS OF MATERIALS AND CONSTRUCTIONS

In determining dead loads for purposes of design, the actual weights of materials and constructions shall be used provided that in the absence of definite information, values approved by the authority having jurisdiction shall be used.

SECTION 3.3
WEIGHT OF FIXED SERVICE EQUIPMENT

In determining dead loads for purposes of design, the weight of fixed service equipment such as plumbing stacks and risers, electrical feeders, and heating, ventilating, and air conditioning systems shall be included.

SECTION 4.0
LIVE LOADS

SECTION 4.1
DEFINITION

Live loads are those loads produced by the use and occupancy of the building or other structure and do not include construction or environmental loads such as wind load, snow load, rain load, earthquake load, flood load, or dead load. Live loads on a roof are those produced (1) during maintenance by workers, equipment, and materials, and (2) during the life of the structure by movable objects such as planters and by people.

SECTION 4.2
UNIFORMLY DISTRIBUTED LOADS

4.2.1 Required Live Loads. The live loads used in the design of buildings and other structures shall be the maximum loads expected by the intended use or occupancy but shall in no case be less than the minimum uniformly distributed unit loads required by Table 4-1.

4.2.2 Provision for Partitions. In office buildings or other buildings where partitions will be erected or rearranged, provision for partition weight shall be made, whether or not partitions are shown on the plans, unless the specified live load exceeds 80 lb/ft^2 (3.83 kN/m^2).

SECTION 4.3
CONCENTRATED LOADS

Floors and other similar surfaces shall be designed to support safely the uniformly distributed live loads prescribed in Section 4.2 or the concentrated load, in pounds or kilonewtons (kN), given in Table 4-1, whichever produces the greater load effects. Unless otherwise specified, the indicated concentration shall be assumed to be uniformly distributed over an area 2.5 ft square (762 mm square) [6.25 ft^2 (0.58 m^2)] and shall be located so as to produce the maximum load effects in the structural members.

Any single panel point of the lower chord of exposed roof trusses or any point along the primary structural members supporting roofs over manufacturing, commercial storage and warehousing, and commercial garage floors shall be capable of carrying safely a suspended concentrated load of not less than 2000 lb (pound-force) (8.90 kN) in addition to dead load. For all other occupancies, a load of 200 lb (0.89 kN) shall be used instead of 2000 lb (8.90 kN).

SECTION 4.4
LOADS ON HANDRAILS, GUARDRAIL SYSTEMS, GRAB BAR SYSTEMS, VEHICLE BARRIER SYSTEMS, AND FIXED LADDERS

4.4.1 Definitions.

HANDRAIL. A rail grasped by hand for guidance and support. A handrail assembly includes the handrail, supporting attachments, and structures.

FIXED LADDER. A ladder that is permanently attached to a structure, building, or equipment.

GUARDRAIL SYSTEM. A system of building components near open sides of an elevated surface for the purpose of minimizing the possibility of a fall from the elevated surface by people, equipment, or material.

GRAB BAR SYSTEM. A bar provided to support body weight in locations such as toilets, showers, and tub enclosures.

VEHICLE BARRIER SYSTEM. A system of building components near open sides of a garage floor or ramp, or building walls that act as restraints for vehicles.

4.4.2 Loads.

(a) Handrail assemblies and guardrail systems shall be designed to resist a load of 50 lb/ft (pound-force per linear ft) (0.73 kN/m) applied in any direction at the top and to transfer this load through the supports to the structure. For one- and two-family dwellings, the minimum load shall be 20 lb/ft (0.29 kN/m).

Further, all handrail assemblies and guardrail systems shall be able to resist a single concentrated load of 200 lb (0.89 kN) applied in any direction at any point along the top, and have attachment devices and supporting structure to transfer this loading to appropriate structural elements of the building. This load need not be assumed to act concurrently with the loads specified in the preceding paragraph.

Intermediate rails (all those except the handrail), balusters, and panel fillers shall be designed to withstand a horizontally applied normal load of 50 lb (0.22 kN) on an area not to exceed 1 ft^2 (305 mm^2) including openings and space between rails. Reactions due to this loading are not required to be superimposed with those of either preceding paragraph.

(b) Grab bar systems shall be designed to resist a single concentrated load of 250 lb (1.11 kN) applied in any direction at any point.

(c) Vehicle barrier systems for passenger cars shall be designed to resist a single load of 6000 lb (26.70 kN) applied horizontally in any direction to the barrier system, and shall have anchorages or attachments capable of transferring this load to the structure. For design of the system, the load shall be assumed to act at a minimum height of 1 ft, 6 in. (460 mm) above the floor or ramp surface on an area not to exceed 1 ft^2 (305 mm^2), and is not required to be assumed to act concurrently with any handrail or guardrail loadings specified in the preceding paragraphs of Section 4.4.2. Garages accommodating trucks and buses shall be designed in accordance with an approved method, which contains provision for traffic railings.

(d) The minimum design live load on fixed ladders with rungs shall be a single concentrated load of 300 lbs, and shall be applied at any point to produce the maximum load effect on the element being considered. The number and position of additional concentrated live load units shall be a minimum of 1 unit of 300 lbs for every 10 ft of ladder height.

(e) Where rails of fixed ladders extend above a floor or platform at the top of the ladder, each side rail extension shall be designed to resist a concentrated live load of 100 lbs in any direction at any height up to the top of the side rail extension. Ship ladders with treads instead of rungs shall have minimum design loads as stairs, defined in Table 4-1.

SECTION 4.5
LOADS NOT SPECIFIED

For occupancies or uses not designated in Section 4.2. or 4.3, the live load shall be determined in accordance with a method approved by the authority having jurisdiction.

SECTION 4.6
PARTIAL LOADING

The full intensity of the appropriately reduced live load applied only to a portion of a structure or member shall be accounted for if it produces a more unfavorable effect than the same intensity applied over the full structure or member.

SECTION 4.7
IMPACT LOADS

The live loads specified in Sections 4.2.1 and 4.4.2 shall be assumed to include adequate allowance for ordinary impact conditions. Provision shall be made in the structural design for uses and loads that involve unusual vibration and impact forces.

4.7.1 Elevators. All elevator loads shall be increased by 100% for impact and the structural supports shall be designed within the limits of deflection prescribed by Refs. 4-1 and 4-2.

4.7.2 Machinery. For the purpose of design, the weight of machinery and moving loads shall be increased as follows to allow for impact:

(1) elevator machinery, 100%;

(2) light machinery, shaft- or motor-driven, 20%;

(3) reciprocating machinery or power-driven units, 50%;

(4) hangers for floors or balconies, 33%. All percentages shall be increased where specified by the manufacturer.

SECTION 4.8
REDUCTION IN LIVE LOADS

The minimum uniformly distributed live loads, L_o in Table 4-1, may be reduced according to the following provisions.

4.8.1 General. Subject to the limitations of Sections 4.8.2 through 4.8.5, members for which a value of $K_{LL}A_T$ is 400 ft^2 (37.16 m^2) or more are permitted to be designed for a reduced live load in accordance with the following formula:

$$L = L_o \left(0.25 + \frac{15}{\sqrt{K_{LL}A_T}} \right) \qquad \textbf{(Eq. 4-1)}$$

In SI:

$$L = L_o \left(0.25 + \frac{4.57}{\sqrt{K_{LL}A_T}} \right)$$

where

L = reduced design live load per square ft (m) of area supported by the member

L_o = unreduced design live load per square ft (m) of area supported by the member (see Table 4-1)

K_{LL} = live load element factor (see Table 4-2)

A_T = tributary area in ft^2 (m^2)

L shall not be less than $0.50L_o$ for members supporting one floor and L shall not be less than $0.40L_o$ for members supporting two or more floors.

4.8.2 Heavy Live Loads. Live loads that exceed 100 lb/ft^2 (4.79 kN/m^2) shall not be reduced except the live loads for members supporting two or more floors may be reduced by 20%.

4.8.3 Passenger Car Garages. The live loads shall not be reduced in passenger car garages except the live loads for members supporting two or more floors may be reduced by 20%.

4.8.4 Special Occupancies. Live loads of 100 lbs/ft^2 (4.79 kN/m^2) or less shall not be reduced in public assembly occupancies.

4.8.5 Limitations on One-Way Slabs. The tributary area, A_T, for one-way slabs shall not exceed an area defined by the slab span times a width normal to the span of 1.5 times the slab span.

SECTION 4.9
MINIMUM ROOF LIVE LOADS

4.9.1 Flat, Pitched, and Curved Roofs. Ordinary flat, pitched, and curved roofs shall be designed for the live loads specified in Eq. 4-2 or other controlling combinations of loads as discussed in Section 2, whichever produces the greater load. In structures such as greenhouses, where special scaffolding is used as a work surface for workmen and materials during maintenance and repair operations, a lower roof load than specified in Eq. 4-2 shall not be used unless approved by the authority having jurisdiction.

$$L_r = 20R_1R_2 \text{ where } 12 \leq L_r \leq 20 \quad \textbf{(Eq. 4-2)}$$

in SI: $\quad L_r = 0.96R_1R_2$ where $0.58 \leq L_r \leq 0.96$

where $\quad L_r = $ roof live load/ft^2 of horizontal projection in lbs/ ft^2 (kN/m^2).

The reduction factors R_1 and R_2 shall be determined as follows:

$$R_1 = \begin{cases} 1 & \text{for } A_t \leq 200 \text{ ft}^2 (18.58 \text{ m}^2) \\ 1.2 - 0.001A_t & \text{for } 200 \text{ ft}^2 < A_t < 600 \text{ ft}^2 \\ 0.6 & \text{for } A_t \geq 600 \text{ ft}^2 (55.74 \text{ m}^2) \end{cases}$$

in SI:

$$R_1 = \begin{cases} 1 & \text{for } A_t \leq 18.58 \text{ m}^2 \\ 1.2 - 0.01076A_t & \text{for } 18.58 \text{ m}^2 < A_t < 55.74 \text{ m}^2 \\ 0.6 & \text{for } A_t \geq 55.74 \text{ m}^2 \end{cases}$$

where $A_t = $ tributary area in ft^2 (m^2) supported by any structural member and

$$R_2 = \begin{cases} 1 & \text{for } F \leq 4 \\ 1.2 - 0.05\ F & \text{for } 4 < F < 12 \\ 0.6 & \text{for } F \geq 12 \end{cases}$$

where, for a pitched roof, $F = $ number of inches of rise per ft (in SI: $F = 0.12 \times$ slope, with slope expressed in percentage points) and, for an arch or dome, $F = $ rise-to-span ratio multiplied by 32.

4.9.2 Special-Purpose Roofs. Roofs used for promenade purposes shall be designed for a minimum live load of 60 lb/ft^2 (2.87 kN/m^2). Roofs used for roof gardens or assembly purposes shall be designed for a minimum live load of 100 lb/ft^2 (4.79 kN/m^2). Roofs used for other special purposes shall be designed for appropriate loads as approved by the authority having jurisdiction.

4.9.3 Special Structural Elements. Live loads of 100 lbs/ft^2 (4.79 kN/m^2) or less shall not be reduced for roof members except as specified in 4.9.

SECTION 4.10
CRANE LOADS

The crane live load shall be the rated capacity of the crane. Design loads for the runway beams, including connections and support brackets, of moving bridge cranes and monorail cranes shall include the maximum wheel loads of the crane and the vertical impact, lateral, and longitudinal forces induced by the moving crane.

4.10.1 Maximum Wheel Load. The maximum wheel loads shall be the wheel loads produced by the weight of the bridge, as applicable, plus the sum of the rated capacity and the weight of the trolley with the trolley positioned on its runway at the location where the resulting load effect is maximum.

4.10.2 Vertical Impact Force. The maximum wheel loads of the crane shall be increased by the percentages shown below to determine the induced vertical impact or vibration force:

Monorail cranes (powered)	25
Cab-operated or remotely operated bridge cranes (powered)	25
Pendant-operated bridge cranes (powered)	10
Bridge cranes or monorail cranes with hand-geared bridge, trolley, and hoist	0

4.10.3 Lateral Force. The lateral force on crane runway beams with electrically powered trolleys shall be calculated as 20% of the sum of the rated capacity of the crane and the weight of the hoist and trolley. The lateral force shall be assumed to act horizontally at the traction surface of a runway beam, in either direction perpendicular to the beam, and shall be distributed with due regard

to the lateral stiffness of the runway beam and supporting structure.

4.10.4 Longitudinal Force. The longitudinal force on crane runway beams, except for bridge cranes with hand-geared bridges, shall be calculated as 10% of the maximum wheel loads of the crane. The longitudinal force shall be assumed to act horizontally at the traction surface of a runway beam in either direction parallel to the beam.

SECTION 4.11
REFERENCES

Ref. 4-1 ANSI. (1988). "American National Standard Practice for the Inspection of Elevators, Escalators, and Moving Walks (Inspectors' Manual)." *ANSI A17.2.*

Ref. 4-2 ANSI/ASME. (1993). "American National Standard Safety Code for Elevators and Escalators." *ANSI/ASME A17.1.*

TABLE 4-1
MINIMUM UNIFORMLY DISTRIBUTED LIVE LOADS, L_O, AND MINIMUM CONCENTRATED LIVE LOADS

Occupancy or Use	Uniform psf (kN/m^2)	Conc. lbs (kN)
Apartments (see residential)		
Access floor systems Office use Computer use	 50 (2.4) 100 (4.79)	 2000 (8.9) 2000 (8.9)
Armories and drill rooms	150 (7.18)	
Assembly areas and theaters Fixed seats (fastened to floor) Lobbies Movable seats Platforms (assembly) Stage floors	 60 (2.87) 100 (4.79) 100 (4.79) 100 (4.79) 150 (7.18)	
Balconies (exterior) On one- and two-family residences only, and not exceeding 100 ft.^{2} (9.3 m^2)	100 (4.79) 60 (2.87)	
Bowling alleys, poolrooms, and similar recreational areas	75 (3.59)	
Catwalks for maintenance access	40 (1.92)	300 (1.33)
Corridors First floor Other floors, same as occupancy served except as indicated	 100 (4.79)	
Dance halls and ballrooms	100 (4.79)	
Decks (patio and roof) Same as area served, or for the type of occupancy accommodated		
Dining rooms and restaurants	100 (4.79)	
Dwellings (see residential)		
Elevator machine room grating (on area of 4 in.^{2} (2580 mm^2))		300 (1.33)
Finish light floor plate construction (on area of 1 in.^{2} (645 mm^2))		200 (0.89)
Fire escapes On single-family dwellings only	100 (4.79) 40 (1.92)	
Fixed ladders		See Section 4.4
Garages (passenger vehicles only) Trucks and buses	40 (1.92) Note (2)	Note (1)

(continued)

Occupancy or Use	Uniform psf (kN/m²)	Conc. lbs (kN)
Grandstands (see stadium and arena bleachers)		
Gymnasiums, main floors, and balconies	100 (4.79) Note (4)	
Handrails, guardrails, and grab bars	See Section 4.4	
Hospitals		
Operating rooms, laboratories	60 (2.87)	1000 (4.45)
Private rooms	40 (1.92)	1000 (4.45)
Wards	40 (1.92)	1000 (4.45)
Corridors above first floor	80 (3.83)	1000 (4.45)
Hotels (see residential)		
Libraries		
Reading rooms	60 (2.87)	1000 (4.45)
Stack rooms	150 (7.18) Note (3)	1000 (4.45)
Corridors above first floor	80 (3.83)	1000 (4.45)
Manufacturing		
Light	125 (6.00)	2000 (8.90)
Heavy	250 (11.97)	3000 (13.40)
Marquees and canopies	75 (3.59)	
Office buildings		
File and computer rooms shall be designed for heavier loads based on anticipated occupancy		
Lobbies and first floor corridors	100 (4.79)	2000 (8.90)
Offices	50 (2.40)	2000 (8.90)
Corridors above first floor	80 (3.83)	2000 (8.90)
Penal institutions		
Cell blocks	40 (1.92)	
Corridors	100 (4.79)	
Residential		
Dwellings (one- and two-family)		
Uninhabitable attics without storage	10 (0.48)	
Uninhabitable attics with storage	20 (0.96)	
Habitable attics and sleeping areas	30 (1.44)	
All other areas except stairs and balconies	40 (1.92)	
Hotels and multifamily houses		
Private rooms and corridors serving them	40 (1.92)	
Public rooms and corridors serving them	100 (4.79)	
Reviewing stands, grandstands, and bleachers	100 (4.79) Note (4)	
Roofs	See Sections 4.3 and 4.9	

(continued)

TABLE 4-1 — continued
MINIMUM UNIFORMLY DISTRIBUTED LIVE LOADS, L_O, AND MINIMUM CONCENTRATED LIVE LOADS

Occupancy or Use	Uniform psf (kN/m^2)	Conc. lbs (kN)
Schools		
Classrooms	40 (1.92)	1000 (4.45)
Corridors above first floor	80 (3.83)	1000 (4.45)
First floor corridors	100 (4.79)	1000 (4.45)
Scuttles, skylight ribs, and accessible ceilings		200 (9.58)
Sidewalks, vehicular driveways, and yards subject to trucking	250 (11.97) Note (5)	8000 (35.60) Note (6)
Stadiums and arenas		
Bleachers	100 (4.79) Note (4)	
Fixed Seats (fastened to floor)	60 (2.87) Note (4)	
Stairs and exit-ways	100 (4.79)	Note (7)
One- and two-family residences only	40 (1.92)	
Storage areas above ceilings	20 (0.96)	
Storage warehouses (shall be designed for heavier loads if required for anticipated storage)		
Light	125 (6.00)	
Heavy	250 (11.97)	
Stores		
Retail		
First floor	100 (4.79)	1000 (4.45)
Upper floors	75 (3.59)	1000 (4.45)
Wholesale, all floors	125 (6.00)	1000 (4.45)
Vehicle barriers	See Section 4.4	
Walkways and elevated platforms (other than exit-ways)	60 (2.87)	
Yards and terraces, pedestrians	100 (4.79)	

Notes

(1) Floors in garages or portions of building used for the storage of motor vehicles shall be designed for the uniformly distributed live loads of Table 4-1 or the following concentrated load: (1) for garages restricted to passenger vehicles accommodating not more than nine passengers, 3000 lb (13.35 kN) acting on an area of 4.5 in. by 4.5 in. (114 mm by 114 mm, footprint of a jack); (2) for mechanical parking structures without slab or deck which are used for storing passenger car only, 2250 lb (10 kN) per wheel.

(2) Garages accommodating trucks and buses shall be designed in accordance with an approved method, which contains provisions for truck and bus loadings.

(3) The loading applies to stack room floors that support nonmobile, double-faced library bookstacks subject to the following limitations:

    a. The nominal bookstack unit height shall not exceed 90 in. (2290 mm);

    b. The nominal shelf depth shall not exceed 12 in. (305 mm) for each face; and

    c. Parallel rows of double-faced bookstacks shall be separated by aisles not less than 36 in. (914 mm) wide.

(4) In addition to the vertical live loads, the design shall include horizontal swaying forces applied to each row of the seats as follows: 24 lbs/ linear ft of seat applied in a direction parallel to each row of seats and 10 lbs/ linear ft of seat applied in a direction perpendicular to each row of seats. The parallel and perpendicular horizontal swaying forces need not be applied simultaneously.

(5) Other uniform loads in accordance with an approved method, which contains provisions for truck loadings, shall also be considered where appropriate.

(6) The concentrated wheel load shall be applied on an area of 4.5 in. by 4.5 in. (114 mm by 114 mm, footprint of a jack).

(7) Minimum concentrated load on stair treads (on area of 4 in.^{2} (2580 mm^2)) is 300 lbs (1.33 kN).

TABLE 4-2
LIVE LOAD ELEMENT FACTOR, K_{LL}

Element	K_{LL} (Note 1)
Interior Columns	4
Exterior Columns without cantilever slabs	4
Edge Columns with cantilever slabs	3
Corner Columns with cantilever slabs	2
Edge Beams without cantilever slabs	2
Interior Beams	2

All Other Members Not Identified Above including:
 	
    Edge Beams with cantilever slabs
    Cantilever Beams
    One-way Slabs
    Two-way Slabs
    Members without provisions for
       continuous shear transfer normal
       to their span
	1

Note 1. In lieu of the values above, K_{LL} is permitted to be calculated.

SECTION 5.0
SOIL AND HYDROSTATIC PRESSURE AND FLOOD LOADS

SECTION 5.1
PRESSURE ON BASEMENT WALLS

In the design of basement walls and similar approximately vertical structures below grade, provision shall be made for the lateral pressure of adjacent soil. Due allowance shall be made for possible surcharge from fixed or moving loads. When a portion or the whole of the adjacent soil is below a free-water surface, computations shall be based on the weight of the soil diminished by buoyancy, plus full hydrostatic pressure.

Basement walls shall be designed to resist lateral soil loads. Soil loads specified in Table 5-1 shall be used as the minimum design lateral soil loads unless specified otherwise in a soil investigation report approved by the authority having jurisdiction. The lateral pressure from surcharge loads shall be added to the lateral earth pressure load. The lateral pressure shall be increased if soils with expansion potential are present at the site as determined by a geotechnical investigation.

SECTION 5.2
UPLIFT ON FLOORS AND FOUNDATIONS

In the design of basement floors and similar approximately horizontal elements below grade, the upward pressure of water, where applicable, shall be taken as the full hydrostatic pressure applied over the entire area. The hydrostatic head shall be measured from the underside of the construction. Any other upward loads shall be included in the design.

Where expansive soils are present under foundations or slabs-on-ground, the foundations, slabs, and other components shall be designed to tolerate the movement or resist the upward pressures caused by the expansive soils, or the expansive soil shall be removed or stabilized around and beneath the structure.

SECTION 5.3
FLOOD LOADS

The provisions of this section apply to buildings and other structures located in areas prone to flooding as defined on a flood hazard map.

5.3.1 Definitions. The following definitions apply to the provisions of Section 5.3.

APPROVED. Acceptable to the authority having jurisdiction.

BASE FLOOD. The flood having a 1% chance of being equaled or exceeded in any given year.

BASE FLOOD ELEVATION (BFE). The elevation of flooding, including wave height, having a 1% chance of being equaled or exceeded in any given year.

BREAKAWAY WALL. Any type of wall subject to flooding that is not required to provide structural support to a building or other structure, and that is designed and constructed such that, under base flood or lesser flood conditions, it will collapse in such a way that: (1) it allows the free passage of floodwaters, and (2) it does not damage the structure or supporting foundation system.

COASTAL A-ZONE. An area within a Special Flood Hazard Area, landward of a V-Zone or landward of an open coast without mapped V-Zones. To be classified as a Coastal A-Zone, the principal source of flooding must be astronomical tides, storm surges, seiches, or tsunamis, not riverine flooding.

COASTAL HIGH-HAZARD AREA (V-ZONE). An area within a Special Flood Hazard Area, extending from offshore to the inland limit of a primary frontal dune along an open coast, and any other area that is subject to high-velocity wave action from storms or seismic sources. This area is designated on FIRMs as V, VE, VO, or V1-30.

DESIGN FLOOD. The greater of the following two flood events: (1) the Base Flood, affecting those areas identified as Special Flood Hazard Areas on the community's FIRM; or (2) the flood corresponding to the area designated as a Flood Hazard Area on a community's Flood Hazard Map or otherwise legally designated.

DESIGN FLOOD ELEVATION (DFE). The elevation of the Design Flood, including wave height, relative to the datum specified on a community's Flood Hazard Map.

FLOOD HAZARD AREA. The area subject to flooding during the Design Flood.

FLOOD HAZARD MAP. The map delineating Flood Hazard Areas adopted by the authority having jurisdiction.

FLOOD INSURANCE RATE MAP (FIRM). An official map of a community on which the Federal Insurance and Mitigation Administration has delineated both Special

Flood Hazard Areas and the risk premium zones applicable to the community.

SPECIAL FLOOD HAZARD AREA (AREA OF SPECIAL FLOOD HAZARD). The land in the floodplain subject to a 1% or greater chance of flooding in any given year. These areas are delineated on a community's Flood Insurance Rate Map (FIRM) as A-Zones (A, AE, A1-30, A99, AR, AO, or AH) or V-Zones (V, VE, VO, or V1-30).

5.3.2 Design Requirements.

5.3.2.1 Design Loads.
Structural systems of buildings or other structures shall be designed, constructed, connected, and anchored to resist flotation, collapse, and permanent lateral displacement due to action of flood loads associated with the design flood (see Section 5.3.3) and other loads in accordance with the load combinations of Section 2.

5.3.2.2 Erosion and Scour.
The effects of erosion and scour shall be included in the calculation of loads on buildings and other structures in flood hazard areas.

5.3.2.3 Loads on Breakaway Walls.
Walls and partitions required by Ref. 5-1, to break away, including their connections to the structure, shall be designed for the largest of the following loads acting perpendicular to the plane of the wall:

1. The wind load specified in Section 6.
2. The earthquake load specified in Section 9.
3. 10 psf (0.48 kN/m²).

The loading at which breakaway walls are intended to collapse shall not exceed 20 psf (0.96 kN/m²) unless the design meets the following conditions:

1. Breakaway wall collapse is designed to result from a flood load less than that which occurs during the base flood; and
2. The supporting foundation and the elevated portion of the building shall be designed against collapse, permanent lateral displacement, and other structural damage due to the effects of flood loads in combination with other loads as specified in Section 2.

5.3.3 Loads During Flooding.

5.3.3.1 Load Basis.
In flood hazard areas, the structural design shall be based on the design flood.

5.3.3.2 Hydrostatic Loads.
Hydrostatic loads caused by a depth of water to the level of the design flood elevation shall be applied over all surfaces involved, both above and below ground level, except that for surfaces exposed to free water, the design depth shall be increased by 1 ft (0.30 m).

Reduced uplift and lateral loads on surfaces of enclosed spaces below the design flood elevation shall apply only if provision is made for entry and exit of floodwater.

5.3.3.3 Hydrodynamic Loads.
Dynamic effects of moving water shall be determined by a detailed analysis utilizing basic concepts of fluid mechanics.

Exception: Where water velocities do not exceed 10 ft/sec (3.05 m/s), dynamic effects of moving water shall be permitted to be converted into equivalent hydrostatic loads by increasing the design flood elevation for design purposes by an equivalent surcharge depth, d_h, on the headwater side and above the ground level only, equal to:

$$d_h = \frac{aV^2}{2g} \qquad \text{(Eq. 5-1)}$$

where

V = average velocity of water in ft/sec (m/s)
g = acceleration due to gravity, 32.2 ft/sec (9.81 m/s²)
a = coefficient of drag or shape factor (not less than 1.25)

The equivalent surcharge depth shall be added to the design flood elevation design depth and the resultant hydrostatic pressures applied to, and uniformly distributed across, the vertical projected area of the building or structure which is perpendicular to the flow. Surfaces parallel to the flow or surfaces wetted by the tailwater shall be subject to the hydrostatic pressures for depths to the design flood elevation only.

5.3.3.4 Wave Loads.
Wave loads shall be determined by one of the following three methods: (1) using the analytical procedures outlined in this section, (2) by more advanced numerical modeling procedures or, (3) by laboratory test procedures (physical modeling).

Wave loads are those loads that result from water waves propagating over the water surface and striking a building or other structure. Design and construction of buildings and other structures subject to wave loads shall account for the following loads: waves breaking on any portion of the building or structure; uplift forces caused by shoaling waves beneath a building or structure, or portion thereof; wave runup striking any portion of the building or structure; wave-induced drag and inertia forces; wave-induced scour at the base of a building or structure, or its foundation. Wave loads shall be included for both V-Zones and A-Zones. In V-Zones,

waves are 3 ft (0.91 m) high, or higher; in coastal floodplains landward of the V-Zone, waves are less than 3 ft high (0.91 m).

Nonbreaking and broken wave loads shall be calculated using the procedures described in Sections 5.3.3.2 and 5.3.3.3 to calculate hydrostatic and hydrodynamic loads.

Breaking waves loads shall be calculated using the procedures described in Sections 5.3.3.4.1 through 5.3.3.4.4. Breaking wave heights used in the procedures described in Sections 5.3.3.4.1 through 5.3.3.4.4 shall be calculated for V Zones and Coastal A Zones using Eqs. 5-2 and 5-3.

$$H_b = 0.78\,d_s \qquad \textbf{(Eq. 5-2)}$$

where

H_b = breaking wave height in ft (m)
d_s = local stillwater depth in ft (m)

The local stillwater depth shall be calculated using Eq. 5-3, unless more advanced procedures or laboratory tests permitted by this section are used.

$$d_s = 0.65(\text{BFE-G}) \qquad \textbf{(Eq. 5-3)}$$

where

BFE = Base Flood Elevation in ft (m)
G = Ground elevation in ft (m)

5.3.3.4.1 Breaking Wave Loads on Vertical Pilings and Columns.
The net force resulting from a breaking wave acting on a rigid vertical pile or column shall be assumed to act at the stillwater elevation and shall be calculated by the following:

$$F_D = 0.5\gamma_w C_D D H_b^2 \qquad \textbf{(Eq. 5-4)}$$

where

F_D = net wave force, in pounds (kN)
γ_w = unit weight of water, in pounds per cubic ft (kN/m^3), = 62.4 pcf (9.80 kN/m^3) for fresh water and 64.0 pcf (10.05 kN/m^3) for salt water
C_D = coefficient of drag for breaking waves, = 1.75 for round piles or columns, and = 2.25 for square piles or columns
D = pile or column diameter, in ft (m) for circular sections, or for a square pile or column, 1.4 times the width of the pile or column in ft (m).
H_b = breaking wave height, in ft (m)

5.3.3.4.2 Breaking Wave Loads on Vertical Walls.
Maximum pressures and net forces resulting from a normally incident breaking wave (depth-limited in size, with $H_b = 0.78d_s$) acting on a rigid vertical wall shall be calculated by the following:

$$P_{\max} = C_p\gamma_w d_s + 1.2\gamma_w d_s \qquad \textbf{(Eq. 5-5)}$$

and

$$F_t = 1.1 C_p\gamma_w d_s^2 + 2.4\gamma_w d_s^2 \qquad \textbf{(Eq. 5-6)}$$

where

$P_{\max}$ = maximum combined dynamic ($C_p\gamma_w d_s$) and static ($1.2\gamma_w d_s$) wave pressures, also referred to as shock pressures in lbs/ft^2 (kN/m^2)
F_t = net breaking wave force per unit length of structure, also referred to as shock, impulse or wave impact force in lbs/ft (kN/m), acting near the stillwater elevation
C_p = dynamic pressure coefficient ($1.6 < C_p < 3.5$) (see Table 5-2)
γ_w = unit weight of water, in lbs/ft^3 (kN/m^3), = 62.4 pcf (9.80 kN/m^3) for fresh water and 64.0 pcf (10.05 kN/m^3) for salt water
d_s = stillwater depth in ft (m) at base of building or other structure where the wave breaks

This procedure assumes the vertical wall causes a reflected or standing wave against the waterward side of the wall with the crest of the wave at a height of 1.2 d_s above the stillwater level. Thus, the dynamic static and total pressure distributions against the wall are as shown in Figure 5-1.

This procedure also assumes the space behind the vertical wall is dry, with no fluid balancing the static component of the wave force on the outside of the wall. If free water exists behind the wall, a portion of the hydrostatic component of the wave pressure and force disappears (see Figure 5-2) and the net force shall be computed by Eq. 5-7 (the maximum combined wave pressure is still computed with Eq. 5-5).

$$F_t = 1.1\, C_p\gamma_w d_s^2 + 1.9\gamma_w d_s^2 \qquad \textbf{(Eq. 5-7)}$$

where

F_t = net breaking wave force per unit length of structure, also referred to as shock, impulse, or wave impact force in lbs/ft (kN/m), acting near the stillwater elevation
C_p = dynamic pressure coefficient ($1.6 < C_p < 3.5$) (see Table 5-2)
γ_w = unit weight of water, in lbs/ft^3 (kN/m^3), = 62.4 pcf (9.80 kN/m^3) for fresh water and 64.0 pcf (10.05 kN/m^3) for salt water
d_s = stillwater depth in ft (m) at base of building or other structure where the wave breaks

5.3.3.4.3 Breaking Wave Loads on Nonvertical Walls. Breaking wave forces given by Eq. 5-6 and Eq. 5-7 shall be modified in instances where the walls or surfaces upon which the breaking waves act are nonvertical. The horizontal component of breaking wave force shall be given by:

$$F_{nv} = F_t \sin^2 \alpha \qquad \textbf{(Eq. 5-8)}$$

where

F_{nv} = horizontal component of breaking wave force in lbs/ft (kN/m)

F_t = net breaking wave force acting on a vertical surface in lbs/ft (kN/m)

α = vertical angle between nonvertical surface and the horizontal

5.3.3.4.4 Breaking Wave Loads from Obliquely Incident Waves. Breaking wave forces given by Eq. 5-6 and Eq. 5-7 shall be modified in instances where waves are obliquely incident. Breaking wave forces from non-normally incident waves shall be given by:

$$F_{oi} = F_t \sin^2 \alpha \qquad \textbf{(Eq. 5-9)}$$

where

F_{oi} = horizontal component of obliquely incident breaking wave force in lbs/ft (kN/m)

F_t = net breaking wave force (normally incident waves) acting on a vertical surface in lbs/ft (kN/m)

α = horizontal angle between the direction of wave approach and the vertical surface

5.3.3.5 Impact Loads. Impact loads are those that result from debris, ice, and any object transported by floodwaters striking against buildings and structures, or parts thereof. Impact loads shall be determined using a rational approach as concentrated loads acting horizontally at the most critical location at or below the design flood elevation.

**SECTION 5.4
REFERENCE**

Ref. 5-1 ASCE. (1998). "Flood Resistant Design and Construction." *SEI/ASCE 24-98.*

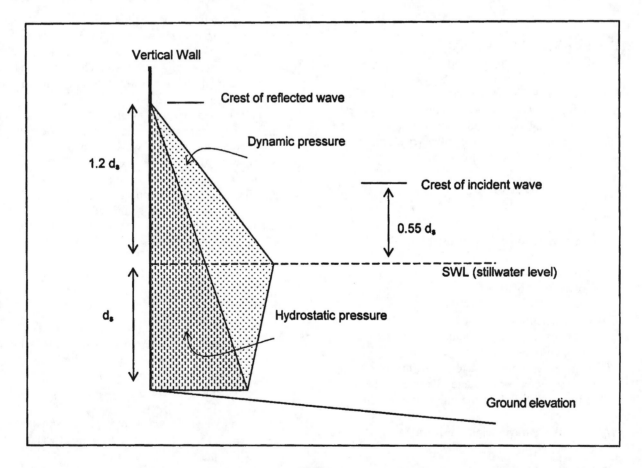

**FIGURE 5.1
NORMALLY INCIDENT BREAKING WAVE PRESSURES AGAINST A VERTICAL WALL (SPACE BEHIND VERTICAL WALL IS DRY)**

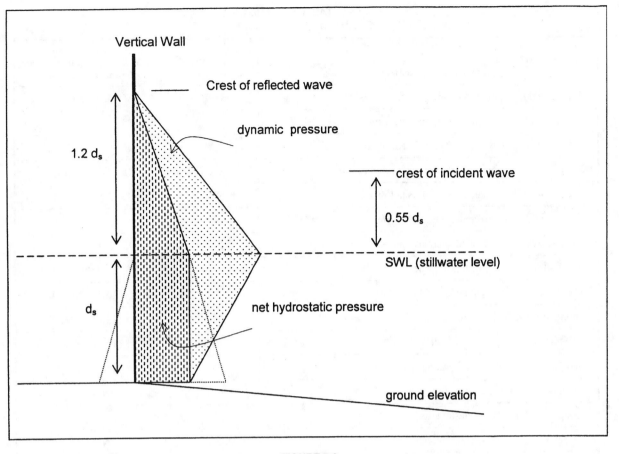

FIGURE 5.2
NORMALLY INCIDENT BREAKING WAVE PRESSURES AGAINST A VERTICAL WALL (STILLWATER LEVEL EQUAL ON BOTH SIDES OF WALL)

TABLE 5-1
DESIGN LATERAL SOIL LOAD

Description of Backfill Material	Unified Soil Classification	Design Lateral Soil Load (Note A) psf/ft of depth (kN/m²/m of depth)
Well-graded, clean gravels; gravel-sand mixes	GW	35 (5.50) Note C
Poorly graded clean gravels; gravel-sand mixes	GP	35 (5.50) Note C
Silty gravels, poorly graded gravel-sand mixes	GM	35 (5.50) Note C
Clayey gravels, poorly graded gravel-and-clay mixes	GC	45 (7.07) Note C
Well-graded, clean sands; gravelly-sand mixes	SW	35 (5.50) Note C
Poorly graded clean sands; sand-gravel mixes	SP	35 (5.50) Note C
Silty sands, poorly graded sand-silt mixes	SM	45 (7.07) Note C
Sand-silt clay mix with plastic fines	SM-SC	85 (13.35) Note D
Clayey sands, poorly graded sand-clay mixes	SC	85 (13.35) Note D
Inorganic silts and clayey silts	ML	85 (13.35) Note D
Mixture of inorganic silt and clay	ML-CL	85 (13.35) Note D
Inorganic clays of low to medium plasticity	CL	100 (15.71)
Organic silts and silt-clays, low plasticity	OL	Note B
Inorganic clayey silts, elastic silts	MH	Note B
Inorganic clays of high plasticity	CH	Note B
Organic clays and silty clays	OH	Note B

Note A. Design lateral soil loads are given for moist conditions for the specified soils at their optimum densities. Actual field conditions shall govern. Submerged or saturated soil pressures shall include the weight of the buoyant soil plus the hydrostatic loads.

Note B. Unsuitable as backfill material.

Note C. For relatively rigid walls, as when braced by floors, the design lateral soil load shall be increased for sand and gravel type soils to 60 psf (9.43 kN/m²) per ft (meter) of depth. Basement walls extending not more than 8 ft (2.44 m) below grade and supporting light floor systems are not considered as being relatively rigid walls.

Note D. For relatively rigid walls, as when braced by floors, the design lateral load shall be increased for silt and clay type oils to 100 psf (15.71 kN/m²) per ft (meter) of depth. Basement walls extending not more than 8 ft (2.44 m) below grade and supporting light floor systems are not considered as being relatively rigid walls.

TABLE 5-2
VALUE OF DYNAMIC PRESSURE
COEFFICIENT, C_p

Building Category	C_p
I	1.6
II	2.8
III	3.2
IV	3.5

SECTION 6.0
WIND LOADS

SECTION 6.1
GENERAL

6.1.1 Scope. Buildings and other structures, including the main wind force-resisting system and all components and cladding thereof, shall be designed and constructed to resist wind loads as specified herein.

6.1.2 Allowed Procedures. The design wind loads for buildings and other structures, including the main wind force-resisting system and component and cladding elements thereof, shall be determined using one of the following procedures: (1) Method 1 — Simplified Procedure as specified in Section 6.4 for buildings meeting the requirements specified therein; (2) Method 2 — Analytical Procedure as specified in Section 6.5 for buildings meeting the requirements specified therein; (3) Method 3 — Wind Tunnel Procedure as specified in Section 6.6.

6.1.3 Wind Pressures Acting on Opposite Faces of Each Building Surface. In the calculation of design wind loads for the main wind force-resisting system and for components and cladding for buildings, the algebraic sum of the pressures acting on opposite faces of each building surface shall be taken into account.

6.1.4 Minimum Design Wind Loading. The design wind load, determined by any one of the procedures specified in Section 6.1.2, shall be not less than specified in this Section.

6.1.4.1 Main Wind Force-Resisting System. The wind load to be used in the design of the main wind force-resisting system for an enclosed or partially enclosed building or other structure shall not be less than 10 lb/ft^2 (0.48 kN/m^2) multiplied by the area of the building or structure projected onto a vertical plane normal to the assumed wind direction. The design wind force for open buildings and other structures shall be not less than 10 lb/ft^2 (0.48 kN/m^2) multiplied by the area A_f.

6.1.4.2 Components and Cladding. The design wind pressure for components and cladding of buildings shall not be less than a net pressure of 10 lb/ft^2 (0.48 kN/m^2) acting in either direction normal to the surface.

SECTION 6.2
DEFINITIONS

The following definitions apply only to the provisions of Section 6:

APPROVED. Acceptable to the authority having jurisdiction.

BASIC WIND SPEED, V. 3-second gust speed at 33 ft (10 m) above the ground in Exposure C (see Section 6.5.6.3) as determined in accordance with Section 6.5.4.

BUILDING, ENCLOSED. A building that does not comply with the requirements for open or partially enclosed buildings.

BUILDING ENVELOPE. Cladding, roofing, exterior walls, glazing, door assemblies, window assemblies, skylight assemblies, and other components enclosing the building.

BUILDING AND OTHER STRUCTURE, FLEXIBLE. Slender buildings and other structures that have a fundamental natural frequency less than 1 Hz.

BUILDING, LOW-RISE. Enclosed or partially enclosed buildings that comply with the following conditions:

1. mean roof height h less than or equal to 60 ft (18 m); and
2. mean roof height h does not exceed least horizontal dimension.

BUILDING, OPEN. A building having each wall at least 80% open. This condition is expressed for each wall by the equation $A_o \geq 0.8 A_g$ where

A_o = total area of openings in a wall that receives positive external pressure, in ft^2 (m^2)
A_g = the gross area of that wall in which A_o is identified, in ft^2 (m^2)

BUILDING, PARTIALLY ENCLOSED. A building that complies with both of the following conditions:

1. the total area of openings in a wall that receives positive external pressure exceeds the sum of the areas of openings in the balance of the building envelope (walls and roof) by more than 10%, and
2. the total area of openings in a wall that receives positive external pressure exceeds 4 ft^2 (0.37 m^2) or

1% of the area of that wall, whichever is smaller, and the percentage of openings in the balance of the building envelope does not exceed 20%.

These conditions are expressed by the following equations:

1. $A_o > 1.10A_{oi}$
2. $A_o > 4\,ft^2$ (0.37 m²) or $> 0.01A_g$, whichever is smaller, and $A_{oi}/A_{gi} \leq 0.20$ where

   A_o, A_g are as defined for Open Building
   A_{oi} = the sum of the areas of openings in the building envelope (walls and roof) not including A_o, in ft² (m²)
   A_{gi} = the sum of the gross surface areas of the building envelope (walls and roof) not including A_g, in ft² (m²)

BUILDING OR OTHER STRUCTURE, REGULAR SHAPED. A building or other structure having no unusual geometrical irregularity in spatial form.

BUILDING OR OTHER STRUCTURES, RIGID. A building or other structure whose fundamental frequency is greater than or equal to 1 Hz.

BUILDING, SIMPLE DIAPHRAGM. An enclosed or partially enclosed building in which wind loads are transmitted through floor and roof diaphragms to the vertical main wind force-resisting system.

COMPONENTS AND CLADDING. Elements of the building envelope that do not qualify as part of the main wind force-resisting system.

DESIGN FORCE, F. Equivalent static force to be used in the determination of wind loads for open buildings and other structures.

DESIGN PRESSURE, p. Equivalent static pressure to be used in the determination of wind loads for buildings.

EFFECTIVE WIND AREA. The area used to determine GC_p. For component and cladding elements, the effective wind area in Figures 6-11 through 6-17 is the span length multiplied by an effective width that need not be less than one-third the span length. For cladding fasteners, the effective wind area shall not be greater than the area that is tributary to an individual fastener.

ESCARPMENT. Also known as scarp, with respect to topographic effects in Section 6.5.7, a cliff or steep slope generally separating two levels or gently sloping areas (see Figure 6-4).

GLAZING. Glass or transparent or translucent plastic sheet used in windows, doors, skylights, or curtain walls.

GLAZING, IMPACT RESISTANT. Glazing that has been shown by testing in accordance with ASTM E 1886 [6-1] and ASTM E 1996 [6-2] or other approved test methods to withstand the impact of wind-borne missiles likely to be generated in wind-borne debris regions during design winds.

HILL. With respect to topographic effects in Section 6.5.7, a land surface characterized by strong relief in any horizontal direction (see Figure 6-4).

HURRICANE-PRONE REGIONS. Areas vulnerable to hurricanes; in the United States and its territories defined as:

1. the U.S. Atlantic Ocean and Gulf of Mexico coasts where the basic wind speed is greater than 90 mph, and
2. Hawaii, Puerto Rico, Guam, Virgin Islands, and American Samoa.

IMPACT-RESISTANT COVERING. A covering designed to protect glazing, which has been shown by testing in accordance with ASTM E 1886 [6-1] and ASTM 1996 [6-2] or other approved test methods to withstand the impact of wind-borne debris missiles likely to be generated in wind-borne debris regions during design winds.

IMPORTANCE FACTOR, I. A factor that accounts for the degree of hazard to human life and damage to property.

MAIN WIND FORCE-RESISTING SYSTEM. An assemblage of structural elements assigned to provide support and stability for the overall structure. The system generally receives wind loading from more than one surface.

MEAN ROOF HEIGHT, h. The average of the roof eave height and the height to the highest point on the roof surface, except that, for roof angles of less than or equal to 10 degrees, the mean roof height shall be the roof eave height.

OPENINGS. Apertures or holes in the building envelope that allow air to flow through the building envelope and that are designed as "open" during design winds as defined by these provisions.

RECOGNIZED LITERATURE. Published research findings and technical papers that are approved.

RIDGE. With respect to topographic effects in Section 6.5.7, an elongated crest of a hill characterized by strong relief in two directions (see Figure 6-4).

WIND-BORNE DEBRIS REGIONS. Areas within hurricane-prone regions located

1. within 1 mile of the coastal mean high water line where the basic wind speed is equal to or greater than 110 mph and in Hawaii, or
2. in areas where the basic wind speed is equal to or greater than 120 mph.

SECTION 6.3
SYMBOLS AND NOTATIONS

The following symbols and notations apply only to the provisions of Section 6:

A = effective wind area, in ft^2 (m^2);

A_f = area of open buildings and other structures either normal to the wind direction or projected on a plane normal to the wind direction, in ft^2 (m^2);

A_g = the gross area of that wall in which A_o is identified, in ft^2 (m^2);

A_{gi} = the sum of the gross surface areas of the building envelope (walls and roof) not including A_g, in ft^2 (m^2);

A_o = total area of openings in a wall that receives positive external pressure, in ft^2 (m^2);

A_{oi} = the sum of the areas of openings in the building envelope (walls and roof) not including A_o, in ft^2 (m^2);

A_{og} = total area of openings in the building envelope ft^2 (m^2);

a = width of pressure coefficient zone, in ft (m);

B = horizontal dimension of building measured normal to wind direction, in ft (m);

$\bar{b}$ = mean hourly wind speed factor in Eq. 6-14 from Table 6-2;

$\hat{b}$ = 3-second gust speed factor from Table 6-2;

C_f = force coefficient to be used in determination of wind loads for other structures;

C_p = external pressure coefficient to be used in determination of wind loads for buildings;

c = turbulence intensity factor in Eq. 6-5 from Table 6-2

D = diameter of a circular structure or member, in ft (m);

D' = depth of protruding elements such as ribs and spoilers, in ft (m);

G = gust effect factor;

G_f = gust effect factor for main wind force-resisting systems of flexible buildings and other structures;

GC_{pn} = combined net pressure coefficient for a parapet

GC_p = product of external pressure coefficient and gust effect factor to be used in determination of wind loads for buildings;

GC_{pf} = product of the equivalent external pressure coefficient and gust effect factor to be used in determination of wind loads for main wind force-resisting system of low-rise buildings;

GC_{pi} = product of internal pressure coefficient and gust effect factor to be used in determination of wind loads for buildings;

g_Q = peak factor for background response in Eqs. 6-4 and 6-8;

g_R = peak factor for resonant response in Eq. 6-8;

g_v = peak factor for wind response in Eqs. 6-4 and 6-8;

H = height of hill or escarpment in Figure 6-4, in ft (m);

h = mean roof height of a building or height of other structure, except that eave height shall be used for roof angle θ of less than or equal to 10 degrees, in ft (m);

I = importance factor;

$I_{\bar{z}}$ = intensity of turbulence from Eq. 6-5;

K_1, K_2, K_3 = multipliers in Figure 6-4 to obtain K_{zt};

K_d = wind directionality factor in Table 6-4;

K_h = velocity pressure exposure coefficient evaluated at height $z = h$;

K_z = velocity pressure exposure coefficient evaluated at height z;

K_{zt} = topographic factor;

L = horizontal dimension of a building measured parallel to the wind direction, in ft (m);

L_h = distance upwind of crest of hill or escarpment in Figure 6-4 to where the difference in ground elevation is half the height of hill or escarpment, in ft (m);

$L_{\bar{z}}$ = integral length scale of turbulence, in ft (m);

ℓ = integral length scale factor from Table 6-2, ft (m);

M = larger dimension of sign, in ft (m);

N = smaller dimension of sign, in ft (m);

N_1 = reduced frequency from Eq. 6-12;

n_1 = building natural frequency, Hz;

p = design pressure to be used in determination of wind loads for buildings, in lb/ft^2 (N/m^2);

p_L = wind pressure acting on leeward face in Figure 6-9;

p_{net30} = net design wind pressure for exposure B at $h = 30$ ft and $I = 1.0$ from Figure 6-3;

p_p = combined net pressure on a parapet from Eq. 6-20;

p_{s30} = simplified design wind pressure for exposure B at $h = 30$ ft and $I = 1.0$ from Figure 6-2;

p_W = wind pressure acting on windward face in Figure 6-9;

Q = background response factor from Eq. 6-6;

q = velocity pressure, in lb/ft^2 (N/m^2);

q_h = velocity pressure evaluated at height $z = h$, in lb/ft^2 (N/m^2);

q_i = velocity pressure for internal pressure determination;

q_p = velocity pressure at top of parapet;

q_z = velocity pressure evaluated at height z above ground, in lb/ft^2 (N/m^2);

R = resonant response factor from Eq. 6-10;

R_B, R_h, R_L = values from Eq. 6-13;

R_i = reduction factor from Eq. 6-16;

R_n = value from Eq. 6-11;

r = rise-to-span ratio for arched roofs;

V = basic wind speed obtained from Figure 6-1, in mph (m/s). The basic wind speed corresponds to a 3-second gust speed at 33 ft (10 m) above ground in Exposure Category C;

V_i = unpartitioned internal volume ft^3 (m^3);

$\overline{V}_{\bar{z}}$ = mean hourly wind speed at height $\bar{z}$, ft/s (m/s);

W = width of building in Figures 6-12, and 6-14A and B and width of span in Figures 6-13 and 6-15, in ft (m);

X = distance to center of pressure from windward edge in Figure 6-18, in ft (m);

x = distance upwind or downwind of crest in Figure 6-4, in ft (m);

z = height above ground level, in ft (m);

$\bar{z}$ = equivalent height of structure, in ft (m);

z_g = nominal height of the atmospheric boundary layer used in this standard. Values appear in Table 6-2;

$z_{\min}$ = exposure constant from Table 6-2;

α = 3-sec gust speed power law exponent from Table 6-2;

$\hat{\alpha}$ = reciprocal of α from Table 6-2;

$\bar{\alpha}$ = mean hourly wind speed power law exponent in Eq. 6-14 from Table 6-2;

β = damping ratio, percent critical for buildings or other structures;

ϵ = ratio of solid area to gross area for open sign, face of a trussed tower, or lattice structure;

λ = adjustment factor for building height and exposure from Figures 6-2 and 6-3;

$\bar{\epsilon}$ = integral length scale power law exponent in Eq. 6-7 from Table 6-2;

η = value used in Eq. 6-13 (see Section 6.5.8.2);

θ = angle of plane of roof from horizontal, in degrees;

υ = height-to-width ratio for solid sign.

SECTION 6.4
METHOD 1 — SIMPLIFIED PROCEDURE

6.4.1 Scope. A building whose design wind loads are determined in accordance with this Section shall meet all the conditions of 6.4.1.1 or 6.4.1.2. If a building qualifies only under 6.4.1.2 for design of its components and cladding, then its main wind force-resisting system shall be designed by Method 2 or Method 3.

6.4.1.1 Main Wind Force-Resisting Systems. For the design of main wind force-resisting systems the building must meet all of the following conditions:

1. the building is a simple diaphragm building as defined in Section 6.2,

2. the building is a low-rise building as defined in Section 6.2,

3. the building is enclosed as defined in Section 6.2 and conforms to the wind-borne debris provisions of Section 6.5.9.3,

4. the building is a regular shaped building or structure as defined in Section 6.2,

5. the building is not classified as a flexible building as defined in Section 6.2,

6. the building does not have response characteristics making it subject to across-wind loading, vortex shedding, instability due to galloping or flutter; and does not have a site location for which channeling effects or buffeting in the wake of upwind obstructions warrant special consideration,

7. the building structure has no expansion joints or separations,

8. the building is not subject to the topographic effects of 6.5.7 (i.e., $K_{zt} = 1.0$),

9. the building has an approximately symmetrical cross section in each direction with either a flat roof, or a gable or hip roof with $\theta \le 45$ degrees.

6.4.1.2 Components and Cladding. For the design of components and cladding the building must meet all the following conditions:

1. the mean roof height $h \le 60$ ft,

2. the building is enclosed as defined in Section 6.2 and conforms to the wind-borne debris provisions of Section 6.5.9.3,

3. the building is a regular shaped building or structure as defined in Section 6.2,

4. the building does not have response characteristics making it subject to across-wind loading, vortex shedding, instability due to galloping or flutter; and does not have a site location for which channeling effects or buffeting in the wake of upwind obstructions warrant special consideration,

5. the building is not subject to the topographic effects of Section 6.5.7 (i.e., $K_{zt} = 1.0$),

6. the building has either a flat roof, or a gable roof with $\theta \leq 45$ degrees, or a hip roof with $\theta \leq 27$ degrees.

6.4.2 Design Procedure.

1. The *basic wind speed V* shall be determined in accordance with Section 6.5.4. The wind shall be assumed to come from any horizontal direction.

2. An *importance factor I* shall be determined in accordance with Section 6.5.5.

3. An *exposure category* shall be determined in accordance with Section 6.5.6.

4. A height and exposure adjustment coefficient, λ, shall be determined from Figure 6-2.

6.4.2.1 Main Wind Force-Resisting System. Simplified design wind pressures, p_s, for the main wind force-resisting systems of low-rise simple diaphragm buildings represent the net pressures (sum of internal and external) to be applied to the horizontal and vertical projections of building surfaces as shown in Figure 6-2. For the horizontal pressures (Zones A, B, C, D), p_s is the combination of the windward and leeward net pressures. p_s shall be determined by the following equation:

$$p_s = \lambda I p_{S30} \qquad \text{(Eq. 6-1)}$$

where

λ = adjustment factor for building height and exposure from Figure 6-2.
I = importance factor as defined in Section 6.2.
p_{S30} = simplified design wind pressure for exposure B, at $h = 30$ ft, and for $I = 1.0$, from Figure 6-2.

6.4.2.1.1 Minimum Pressures. The load effects of the design wind pressures from Section 6.4.2.1 shall not be less than the minimum load case from Section 6.1.4.1 assuming the pressures, p_s, for Zones A, B, C, and D all equal to +10 psf, while assuming Zones E, F, G, and H all equal to 0 psf.

6.4.2.2 Components and Cladding. Net design wind pressures, p_{net}, for the components and cladding of buildings designed using Method 1 represent the net pressures (sum of internal and external) to be applied normal to each building surface as shown in Figure 6-3. p_{net} shall be determined by the following equation:

$$p_{net} = \lambda I p_{net30} \qquad \text{(Eq. 6-2)}$$

where

λ = adjustment factor for building height and exposure from Figure 6-3.

I = importance factor as defined in Section 6.2.
p_{net30} = net design wind pressure for exposure B, at $h = 30$ ft, and for $I = 1.0$, from Figure 6-3.

6.4.2.2.1 Minimum Pressures. The positive design wind pressures, p_{net}, from Section 6.4.2.2 shall not be less than +10 psf, and the negative design wind pressures, p_{net}, from 6.4.2.2 shall not be less than −10 psf.

6.4.3 Air-Permeable Cladding. Design wind loads determined from Figure 6-3 shall be used for all air-permeable cladding unless approved test data or recognized literature demonstrate lower loads for the type of air-permeable cladding being considered.

SECTION 6.5
METHOD 2 — ANALYTICAL PROCEDURE

6.5.1 Scope. A building or other structure whose design wind loads are determined in accordance with this Section shall meet all of the following conditions:

1. The building or other structure is a regular shaped building or structure as defined in Section 6.2, and

2. The building or other structure does not have response characteristics making it subject to across-wind loading, vortex shedding, instability due to galloping or flutter; or does not have a site location for which channeling effects or buffeting in the wake of upwind obstructions warrant special consideration.

6.5.2 Limitations. The provisions of Section 6.5 take into consideration the load magnification effect caused by gusts in resonance with along-wind vibrations of flexible buildings or other structures. Buildings or other structures not meeting the requirements of Section 6.5.1, or having unusual shapes or response characteristics, shall be designed using recognized literature documenting such wind load effects or shall use the wind tunnel procedure specified in Section 6.6.

6.5.2.1 Shielding. There shall be no reductions in velocity pressure due to apparent shielding afforded by buildings and other structures or terrain features.

6.5.2.2 Air-Permeable Cladding. Design wind loads determined from Section 6.5 shall be used for air-permeable cladding unless approved test data or recognized literature demonstrate lower loads for the type of air-permeable cladding being considered.

6.5.3 Design Procedure.

1. *The basic wind speed V and wind directionality factor K_d* shall be determined in accordance with Section 6.5.4.

2. An *importance factor I* shall be determined in accordance with Section 6.5.5.

3. An *exposure category or exposure categories and velocity pressure exposure coefficient K_z or K_h,* as applicable, shall be determined for each wind direction in accordance with Section 6.5.6.

4. A *topographic factor K_{zt}* shall be determined in accordance with Section 6.5.7.

5. A *gust effect factor G or G_f,* as applicable, shall be determined in accordance with Section 6.5.8.

6. An *enclosure classification* shall be determined in accordance with Section 6.5.9.

7. *Internal pressure coefficient GC_{pi}* shall be determined in accordance with Section 6.5.11.1.

8. *External pressure coefficients C_p or GC_{pf}, or force coefficients C_f,* as applicable, shall be determined in accordance with Section 6.5.11.2 or 6.5.11.3, respectively.

9. *Velocity pressure q_z or q_h,* as applicable, shall be determined in accordance with Section 6.5.10.

10. *Design wind load p or F* shall be determined in accordance with Sections 6.5.12 and 6.5.13, as applicable.

6.5.4 Basic Wind Speed. The basic wind speed, V, used in the determination of design wind loads on buildings and other structures shall be as given in Figure 6-1 except as provided in Sections 6.5.4.1 and 6.5.4.2. The wind shall be assumed to come from any horizontal direction.

6.5.4.1 Special Wind Regions. The basic wind speed shall be increased where records or experience indicate that the wind speeds are higher than those reflected in Figure 6-1. Mountainous terrain, gorges, and special regions shown in Figure 6-1 shall be examined for unusual wind conditions. The authority having jurisdiction shall, if necessary, adjust the values given in Figure 6-1 to account for higher local wind speeds. Such adjustment shall be based on meteorological information and an estimate of the basic wind speed obtained in accordance with the provisions of Section 6.5.4.2.

6.5.4.2 Estimation of Basic Wind Speeds from Regional Climatic Data. Regional climatic data shall only be used in lieu of the basic wind speeds given in Figure 6-1 when: (1) approved extreme-value statistical-analysis procedures have been employed in reducing the data; and (2) the length of record, sampling error, averaging time, anemometer height, data quality, and

terrain exposure of the anemometer have been taken into account.

In hurricane-prone regions, wind speeds derived from simulation techniques shall only be used in lieu of the basic wind speeds given in Figure 6-1 when (1) approved simulation or extreme-value statistical-analysis procedures are used (the use of regional wind speed data obtained from anemometers is not permitted to define the hurricane wind speed risk along the Gulf and Atlantic coasts, the Caribbean, or Hawaii) and (2) the design wind speeds resulting from the study shall not be less than the resulting 500-year return period wind speed divided by $\sqrt{1.5}$.

6.5.4.3 Limitation. Tornadoes have not been considered in developing the basic wind-speed distributions.

6.5.4.4 Wind Directionality Factor. The wind directionality factor, K_d, shall be determined from Table 6-4. This factor shall only be applied when used in conjunction with load combinations specified in Sections 2.3 and 2.4.

6.5.5 Importance Factor. An importance factor, I, for the building or other structure shall be determined from Table 6-1 based on building and structure categories listed in Table 1-1.

6.5.6 Exposure. For each wind direction considered, an exposure category that adequately reflects the characteristics of ground roughness and surface irregularities shall be determined for the site at which the building or structure is to be constructed. Account shall be taken of variations in ground surface roughness that arises from natural topography and vegetation as well as constructed features.

6.5.6.1 Wind Directions and Sectors. For each selected wind direction at which the wind loads are to be evaluated, the exposure of the building or structure shall be determined for the two upwind sectors extending 45 degrees either side of the selected wind direction. The exposures in these two sectors shall be determined in accordance with Sections 6.5.6.2 and 6.5.6.3 and the exposure resulting in the highest wind loads shall be used to represent the winds from that direction.

6.5.6.2 Surface Roughness Categories. A ground surface roughness within each 45-degree sector shall be determined for a distance upwind of the site as defined in Section 6.5.6.3 from the categories defined below, for the purpose of assigning an exposure category as defined in Section 6.5.6.3.

  Surface Roughness B: Urban and suburban areas, wooded areas or other terrain with numerous

closely spaced obstructions having the size of single-family dwellings or larger.

Surface Roughness C: Open terrain with scattered obstructions having heights generally less than 30 ft (9.1 m). This category includes flat open country, grasslands, and all water surfaces in hurricane-prone regions.

Surface Roughness D: Flat, unobstructed areas and water surfaces outside hurricane-prone regions. This category includes smooth mud flats, salt flats, and unbroken ice.

6.5.6.3 Exposure Categories.

Exposure B: Exposure B shall apply where the ground surface roughness condition, as defined by Surface Roughness B, prevails in the upwind direction for a distance of at least 2630 ft (800 m) or 10 times the height of the building, whichever is greater.

> **Exception:** For buildings whose mean roof height is less than or equal to 30 ft (9.1 m), the upwind distance may be reduced to 1500 ft (457 m).

Exposure C: Exposure C shall apply for all cases where exposures B or D do not apply.

Exposure D: Exposure D shall apply where the ground surface roughness, as defined by surface roughness D, prevails in the upwind direction for a distance at least 5000 ft (1524 m) or 10 times the building height, whichever is greater. Exposure D shall extend inland from the shoreline for a distance of 660 ft (200 m) or 10 times the height of the building, whichever is greater.

For a site located in the transition zone between exposure categories, the category resulting in the largest wind forces shall be used. Exception: An intermediate exposure between the above categories is permitted in a transition zone provided that it is determined by a rational analysis method defined in the recognized literature.

6.5.6.4 Exposure Category for Main Wind Force-Resisting Systems.

6.5.6.4.1 Buildings and Other Structures. For each wind direction considered, wind loads for the design of the main wind force-resisting system determined from Figure 6-6 shall be based on the exposure categories defined in Section 6.5.6.3.

6.5.6.4.2 Low-Rise Buildings. Wind loads for the design of the main wind force-resisting systems for low-rise buildings shall be determined using a velocity pressure q_h based on the exposure resulting in the highest wind loads for any wind direction at the site when external pressure coefficients GC_{pf} given in Figure 6-10 are used.

6.5.6.5 Exposure Category for Components and Cladding.

6.5.6.5.1 Buildings with Mean Roof Height h Less Than or Equal to 60 ft (18 m). Components and cladding for buildings with a mean roof height h of 60 ft (18 m) or less shall be designed using a velocity pressure q_h based on the exposure resulting in the highest wind loads for any wind direction at the site.

6.5.6.5.2 Buildings with Mean Roof Height h Greater Than 60 ft (18 m) and Other Structures. Components and cladding for buildings with a mean roof height h in excess of 60 ft (18 m) and for other structures shall be designed using the exposure resulting in the highest wind loads for any wind direction at the site.

6.5.6.6 Velocity Pressure Exposure Coefficient. Based on the exposure category determined in Section 6.5.6.3, a velocity pressure exposure coefficient K_z or K_h, as applicable, shall be determined from Table 6-3.

6.5.7 Topographic Effects.

6.5.7.1 Wind Speed-Up over Hills, Ridges, and Escarpments. Wind speed-up effects at isolated hills, ridges, and escarpments constituting abrupt changes in the general topography, located in any exposure category, shall be included in the design when buildings and other site conditions and locations of structures meet all of the following conditions:

1. The hill, ridge, or escarpment is isolated and unobstructed upwind by other similar topographic features of comparable height for 100 times the height of the topographic feature (100 H) or 2 miles (3.22 km), whichever is less. This distance shall be measured horizontally from the point at which the height H of the hill, ridge, or escarpment is determined;

2. The hill, ridge, or escarpment protrudes above the height of upwind terrain features within a 2-mile (3.22-km) radius in any quadrant by a factor of two or more;

3. The structure is located as shown in Figure 6-4 in the upper half of a hill or ridge or near the crest of an escarpment;

4. $H/L_h \geq 0.2$; and

5. H is greater than or equal to 15 ft (4.5 m) for Exposures C and D and 60 ft (18 m) for Exposure B.

6.5.7.2 Topographic Factor.

The wind speed-up effect shall be included in the calculation of design wind loads by using the factor K_{zt}:

$$K_{zt} = (1 + K_1 K_2 K_3)^2 \qquad \textbf{(Eq. 6-3)}$$

where K_1, K_2, and K_3 are given in Figure 6-4.

6.5.8 Gust Effect Factor.

6.5.8.1 Rigid Structures.

For rigid structures as defined in Section 6.2, the gust effect factor shall be taken as 0.85 or calculated by the formula:

$$G = 0.925 \left(\frac{(1 + 1.7 g_Q I_{\bar{z}} Q)}{1 + 1.7 g_v I_{\bar{z}}} \right) \qquad \textbf{(Eq. 6-4)}$$

$$I_{\bar{z}} = c(33/\bar{z})^{1/6} \qquad \textbf{(Eq. 6-5)}$$

where $I_{\bar{z}}$ = the intensity of turbulence at height $\bar{z}$ where $\bar{z}$ = the equivalent height of the structure defined as 0.6 h but not less than z_{min} for all building heights h. z_{min} and c are listed for each exposure in Table 6-2; g_Q and g_v shall be taken as 3.4. The background response Q is given by

$$Q = \sqrt{\frac{1}{1 + 0.63 \left(\dfrac{B + h}{L_{\bar{z}}} \right)^{0.63}}} \qquad \textbf{(Eq. 6-6)}$$

where B, h are defined in Section 6.3; and $L_{\bar{z}}$ = the integral length scale of turbulence at the equivalent height given by

$$L_{\bar{z}} = \ell (\bar{z}/33)^{\bar{\in}} \qquad \textbf{(Eq. 6-7)}$$

in which ℓ and $\bar{\in}$ are constants listed in Table 6-2.

6.5.8.2 Flexible or Dynamically Sensitive Structures.

For flexible or dynamically sensitive structures as defined in Section 6.2, the gust effect factor shall be calculated by:

$$G_f = 0.925 \left(\frac{1 + 1.7 I_{\bar{z}} \sqrt{g_Q^2 Q^2 + g_R^2 R^2}}{1 + 1.7 g_v I_{\bar{z}}} \right) \qquad \textbf{(Eq. 6-8)}$$

g_Q and g_v shall be taken as 3.4 and g_R is given by

$$g_R = \sqrt{2 \ln(3600 \, n_1)} + \frac{0.577}{\sqrt{2 \ln(3600 \, n_1)}} \qquad \textbf{(Eq. 6-9)}$$

R, the resonant response factor, is given by

$$R = \sqrt{\frac{1}{\beta} R_n R_h R_B (0.53 + 0.47 R_L)} \qquad \textbf{(Eq. 6-10)}$$

$$R_n = \frac{7.47 \, N_1}{(1 + 10.3 N_1)^{5/3}} \qquad \textbf{(Eq. 6-11)}$$

$$N_1 = \frac{n_1 L_{\bar{z}}}{\overline{V_{\bar{z}}}} \qquad \textbf{(Eq. 6-12)}$$

$$R_\ell = \frac{1}{\eta} - \frac{1}{2\eta^2}(1 - e^{-2\eta}) \text{ for } \eta > 0 \qquad \textbf{(Eq. 6-13a)}$$

$$R_\ell = 1 \text{ for } \eta = 0 \qquad \textbf{(Eq. 6-13b)}$$

where the subscript ℓ in Eq. 6-13 shall be taken as h, B, and L respectively where h, B, L are defined in Section 6.3.

n_1 = building natural frequency;
$R_\ell = R_h$ setting $\eta = 4.6 n_1 h / \overline{V_{\bar{z}}}$;
$R_\ell = R_B$ setting $\eta = 4.6 n_1 B / \overline{V_{\bar{z}}}$;
$R_\ell = R_L$ setting $\eta = 15.4 n_1 L / \overline{V_{\bar{z}}}$;
β = damping ratio, percent of critical; and
$\overline{V_{\bar{z}}}$ = mean hourly wind speed (ft/sec) at height $\bar{z}$ determined from Eq. 6-14.

$$\overline{V_{\bar{z}}} = \bar{b} \left(\frac{\bar{z}}{33} \right)^{\bar{\alpha}} V \left(\frac{88}{60} \right) \qquad \textbf{(Eq. 6-14)}$$

where $\bar{b}$ and $\bar{\alpha}$ are constants listed in Table 6-2 and V is the basic wind speed in mph.

6.5.8.3 Rational Analysis.

In lieu of the procedure defined in Sections 6.5.8.1 and 6.5.8.2, determination of the gust effect factor by any rational analysis defined in the recognized literature is permitted.

6.5.8.4 Limitations.

Where combined gust effect factors and pressure coefficients (GC_p, GC_{pi}, and GC_{pf}) are given in figures and tables, the gust effect factor shall not be determined separately.

6.5.9 Enclosure Classifications.

6.5.9.1 General.

For the purpose of determining internal pressure coefficients, all buildings shall be classified as enclosed, partially enclosed, or open as defined in Section 6.2.

6.5.9.2 Openings.

A determination shall be made of the amount of openings in the building envelope in order to determine the enclosure classification as defined in Section 6.5.9.1.

6.5.9.3 Wind-Borne Debris. Glazing in buildings classified as Category II, III, or IV (Note 1) located in wind-borne debris regions shall be protected with an impact-resistant covering or be impact-resistant glazing according to the requirements (Note 2) specified in Ref. 6-1 and Ref. 6-2 referenced therein or other approved test methods and performance criteria.

Notes:

1. In Category II, III, or IV buildings, glazing located over 60 ft (18.3 m) above the ground and over 30 ft (9.2 m) above aggregate surface roof debris located within 1500 ft (458 m) of the building shall be permitted to be unprotected.

   Exceptions: In Category II and III buildings (other than health care, jail, and detention facilities, power generating and other public utility facilities), unprotected glazing shall be permitted, provided that unprotected glazing that receives positive external pressure is assumed to be an opening in determining the buildings' enclosure classification.

2. The levels of impact resistance shall be a function of Missile Levels and Wind Zones specified in Ref. 6-2.

6.5.9.4 Multiple Classifications. If a building by definition complies with both the "open" and "partially enclosed" definitions, it shall be classified as an "open" building. A building that does not comply with either the "open" or "partially enclosed" definitions shall be classified as an "enclosed" building.

6.5.10 Velocity Pressure. Velocity pressure, q_z, evaluated at height z shall be calculated by the following equation:

$$q_z = 0.00256\, K_z K_{zt} K_d V^2 I \ (\text{lb/ft}^2)$$

[In SI: $q_z = 0.613\, K_z K_{zt} K_d V^2 I \ (\text{N/m}^2)$; V in m/s]

(Eq. 6-15)

where K_d is the wind directionality factor defined in Section 6.5.4.4, K_z is the velocity pressure exposure coefficient defined in Section 6.5.6.6 and K_{zt} is the topographic factor defined in Section 6.5.7.2, and q_h is the velocity pressure calculated using Eq. 6-15 at mean roof height h.

The numerical coefficient 0.00256 (0.613 in SI) shall be used except where sufficient climatic data are available to justify the selection of a different value of this factor for a design application.

6.5.11 Pressure and Force Coefficients.

6.5.11.1 Internal Pressure Coefficient. Internal pressure coefficients, GC_{pi}, shall be determined from

Figure 6-5 based on building enclosure classifications determined from Section 6.5.9.

6.5.11.1.1 Reduction Factor for Large Volume Buildings, R_i. For a partially enclosed building containing a single, unpartitioned large volume, the internal pressure coefficient, GC_{pi}, shall be multiplied by the following reduction factor, R_i:

$$R_i = 1.0 \text{ or}$$

$$R_i = 0.5 \left(1 + \frac{1}{\sqrt{1 + \dfrac{V_i}{22,800 A_{og}}}} \right) \le 1.0$$

(Eq. 6-16)

where

A_{og} = total area of openings in the building envelope (walls and roof, in ft^2)

V_i = unpartitioned internal volume, in ft^3

6.5.11.2 External Pressure Coefficients.

6.5.11.2.1 Main Wind Force-Resisting Systems. External pressure coefficients for main wind force-resisting systems C_p are given in Figures 6-6, 6-7, and 6-8. Combined gust effect factor and external pressure coefficients, GC_{pf}, are given in Figure 6-10 for low-rise buildings. The pressure coefficient values and gust effect factor in Figure 6-10 shall not be separated.

6.5.11.2.2 Components and Cladding. Combined gust effect factor and external pressure coefficients for components and cladding GC_p are given in Figures 6-11 through 6-17. The pressure coefficient values and gust effect factor shall not be separated.

6.5.11.3 Force Coefficients. Force coefficients C_f are given in Figures 6-18 through 6-22.

6.5.11.4 Roof Overhangs.

6.5.11.4.1 Main Wind Force-Resisting System. Roof overhangs shall be designed for a positive pressure on the bottom surface of windward roof overhangs corresponding to $C_p = 0.8$ in combination with the pressures determined from using Figures 6-6 and 6-10.

6.5.11.4.2 Components and Cladding. For all buildings, roof overhangs shall be designed for pressures determined from pressure coefficients given in Figure 6-11B, C, and D.

6.5.11.5 Parapets.

6.5.11.5.1 Main Wind Force-Resisting System. The pressure coefficients for the effect of parapets on the MWFRS loads are given in Section 6.5.12.2.4

6.5.11.5.2 Components and Cladding. The pressure coefficients for the design of parapet component and cladding elements are taken from the wall and roof pressure coefficients as specified in Section 6.5.12.4.4.

6.5.12 Design Wind Loads on Enclosed and Partially Enclosed Buildings.

6.5.12.1 General.

6.5.12.1.1 Sign Convention. Positive pressure acts toward the surface and negative pressure acts away from the surface.

6.5.12.1.2 Critical Load Condition. Values of external and internal pressures shall be combined algebraically to determine the most critical load.

6.5.12.1.3 Tributary Areas Greater Than 700 ft^2 (65 m^2). Component and cladding elements with tributary areas greater than 700 ft^2 (65 m^2) shall be permitted to be designed using the provisions for main wind force resisting systems.

6.5.12.2 Main Wind Force-Resisting Systems.

6.5.12.2.1 Rigid Buildings of All Height. Design wind pressures for the main wind force-resisting system of buildings of all heights shall be determined by the following equation:

$$p = qGC_p - q_i(GC_{pi}) \text{ (lb/ft}^2\text{)(N/m}^2\text{)} \qquad \textbf{(Eq. 6-17)}$$

where

> $q = q_z$ for windward walls evaluated at height z above the ground;
> $q = q_h$ for leeward walls, side walls, and roofs, evaluated at height h;
> $q_i = q_h$ for windward walls, side walls, leeward walls, and roofs of enclosed buildings and for negative internal pressure evaluation in partially enclosed buildings;
> $q_i = q_z$ for positive internal pressure evaluation in partially enclosed buildings where height z is defined as the level of the highest opening in the building

that could affect the positive internal pressure. For buildings sited in wind-borne debris regions, glazing that is not impact resistant or protected with an impact-resistant covering, shall be treated as an opening in accordance with Section 6.5.9.3. For positive internal pressure evaluation, q_i may conservatively be evaluated at height h ($q_i = q_h$);

> G = gust effect factor from Section 6.5.8;
> C_p = external pressure coefficient from Figure 6-6 or 6-8;
> (GC_{pi}) = internal pressure coefficient from Figure 6-5

q and q_i shall be evaluated using exposure defined in Section 6.5.6.3. Pressure shall be applied simultaneously on windward and leeward walls and on roof surfaces as defined in Figures 6-6 and 6-8.

6.5.12.2.2 Low-Rise Building. Alternatively, design wind pressures for the main wind force-resisting system of low-rise buildings shall be determined by the following equation:

$$p = q_h[(GC_{pf}) - (GC_{pi})] \text{ (lb/ft}^2\text{)(N/m}^2\text{)}$$
$$\textbf{(Eq. 6-18)}$$

where

> q_h = velocity pressure evaluated at mean roof height h using exposure defined in Section 6.5.6.3;
> (GC_{pf}) = external pressure coefficient from Figure 6-10; and
> (GC_{pi}) = internal pressure coefficient from Figure 6-5.

6.5.12.2.3 Flexible Buildings. Design wind pressures for the main wind force-resisting system of flexible buildings shall be determined from the following equation:

$$p = qG_fC_p - q_i(GC_{pi}) \text{ (lb/ft}^2\text{)(N/m}^2\text{)}$$
$$\textbf{(Eq. 6-19)}$$

where q, q_i, C_p and (GC_{pi}) are as defined in Section 6.5.12.2.1 and, G_f = gust effect factor defined in Section 6.5.8.2

6.5.12.2.4 Parapets. The design wind pressure for the effect of parapets on main wind force-resisting systems of rigid, low-rise or flexible buildings with flat, gable, or hip roofs shall be determined by the following equation:

$$p_p = q_pGC_{pn} \text{ (lb/sf)} \qquad \textbf{(Eq. 6-20)}$$

where

p_p = combined net pressure on the parapet due to the combination of the net pressures from the front and back parapet surfaces. Plus (and minus) signs signify net pressure acting toward (and away from) the front (exterior) side of the parapet.

q_p = velocity pressure evaluated at the top of the parapet.

GC_{pn} = combined net pressure coefficient
= +1.8 for windward parapet
= −1.1 for leeward parapet

6.5.12.3 Design Wind Load Cases. The main wind force-resisting system of buildings of all heights, whose wind loads have been determined under the provisions of Sections 6.5.12.2.1 and 6.5.12.2.3, shall be designed for the wind load cases as defined in Figure 6-9. The eccentricity e for rigid structures shall be measured from the geometric center of the building face and shall be considered for each principal axis (e_X, e_Y). The eccentricity e for flexible structures shall be determined from the following equation and shall be considered for each principal axis (e_X, e_Y):

$$e = \frac{e_Q + 1.7 I_{\bar{Z}}\sqrt{(g_Q Q e_Q)^2 + (g_R R e_R)^2}}{1.7 I_{\bar{Z}}\sqrt{(g_Q Q)^2 + (g_R R)^2}} \quad \textbf{(Eq. 6-21)}$$

where

e_Q = eccentricity e as determined for rigid structures in Figure 6-9

e_R = distance between the elastic shear center and center of mass of each floor

$I_{\bar{Z}}$, g_Q, Q, g_R, R shall be as defined in Section 6.5.8

The sign of the eccentricity e shall be plus or minus, whichever causes the more severe load effect.

Exception: One-story buildings with h less than or equal to 30 ft, buildings two stories or less framed with light-framed construction and buildings two stories or less designed with flexible diaphragms need only be designed for Load Case 1 and Load Case 3 in Figure 6-9.

6.5.12.4 Components and Cladding.

6.5.12.4.1 Low-Rise Buildings and Buildings with $h \leq 60$ ft (18.3 m). Design wind pressures on component and cladding elements of low-rise buildings and buildings with $h \leq 60$ ft (18.3 m) shall be determined from the following equation:

$$p = q_h[(GC_p) - (GC_{pi})] \ (\text{lb/ft}^2)(\text{N/m}^2) \quad \textbf{(Eq. 6-22)}$$

where

q_h = velocity pressure evaluated at mean roof height h using exposure defined in Section 6.5.6.3

(GC_p) = external pressure coefficients given in Figures 6-11 through 6-16; and

(GC_{pi}) = internal pressure coefficient given in Figure 6-5.

6.5.12.4.2 Buildings with $h > 60$ ft (18.3 m). Design wind pressures on components and cladding for all buildings with $h > 60$ ft (18.3 m) shall be determined from the following equation:

$$p = q(GC_p) - q_i(GC_{pi}) \ (\text{lb/ft}^2)(\text{N/m}^2) \quad \textbf{(Eq. 6-23)}$$

where

$q = q_z$ for windward walls calculated at height z above the ground;

$q = q_h$ for leeward walls, side walls, and roofs, evaluated at height h;

$q_i = q_h$ for windward walls, side walls, leeward walls, and roofs of enclosed buildings and for negative internal pressure evaluation in partially enclosed buildings; and

$q_i = q_z$ for positive internal pressure evaluation in partially enclosed buildings where height z is defined as the level of the highest opening in the building that could affect the positive internal pressure. For buildings sited in wind-borne debris regions, glazing that is not impact resistant or protected with an impact-resistant covering, shall be treated as an opening in accordance with Section 6.5.9.3. For positive internal pressure evaluation, q_i may conservatively be evaluated at height h ($q_i = q_h$);

(GC_p) = external pressure coefficient from Figure 6-17; and

(GC_{pi}) = internal pressure coefficient given in Figure 6-5

q and q_i shall be evaluated using exposure defined in Section 6.5.6.3.

6.5.12.4.3 Alternative Design Wind Pressures for Components and Cladding in Buildings with 60 ft (18.3 m) < h < 90 ft (27.4 m). Alternative to the requirements of Section 6.5.12.4.2, the design of components and cladding for buildings with a mean

roof height greater than 60 ft (18.3 m) and less than 90 ft (27.4 m) values from Figure 6-11 through 6-17 shall be used only if the height to width ratio is one or less (except as permitted by Note 6 of Figure 6-17) and Eq. 6-22 is used.

6.5.12.4.4 Parapets. The design wind pressure on the components and cladding elements of parapets shall be designed by the following equation:

$$p = q_p(GC_p - GC_{pi}) \qquad \textbf{(Eq. 6-24)}$$

where

q_p = velocity pressure evaluated at the top of the parapet;

GC_p = external pressure coefficient from Figures 6-11 through 6-17; and

GC_{pi} = internal pressure coefficient from Figure 6-5, based on the porosity of the parapet envelope.

Two load cases shall be considered. Load Case A shall consist of applying the applicable positive wall pressure from Figure 6-11A or 6-17 to the front surface of the parapet while applying the applicable negative edge or corner zone roof pressure from Figure 6-11B through 6-17 to the back surface. Load Case B shall consist of applying the applicable positive wall pressure from Figure 6-11A or 6-17 to the back of the parapet surface, and applying the applicable negative wall pressure from Figure 6-11A or 6-17 to the front surface. Edge and corner zones shall be arranged as shown in Figures 6-11 through 6-17. GC_p shall be determined for appropriate roof angle and effective wind area from Figures 6-11 through 6-17. If internal pressure is present, both load cases should be evaluated under positive and negative internal pressure.

6.5.13 Design Wind Loads on Open Buildings and Other Structures. The design wind force for open buildings and other structures shall be determined by the following formula:

$$F = q_z GC_f A_f \ (lb)(N) \qquad \textbf{(Eq. 6-25)}$$

where

q_z = velocity pressure evaluated at height z of the centroid of area A_f using exposure defined in Section 6.5.6.3;

G = gust effect factor from Section 6.5.8;

C_f = net force coefficients from Figure 6-18 through 6-22; and

A_f = projected area normal to the wind except where C_f is specified for the actual surface area, ft^2 (m^2).

SECTION 6.6
METHOD 3 — WIND-TUNNEL PROCEDURE

6.6.1 Scope. Wind-tunnel tests shall be used where required by Section 6.5.2. Wind-tunnel testing shall be permitted in lieu of Methods 1 and 2 for any building or structure.

6.6.2 Test Conditions. Wind-tunnel tests, or similar tests employing fluids other than air, used for the determination of design wind loads for any building or other structure, shall be conducted in accordance with this section. Tests for the determination of mean and fluctuating forces and pressures shall meet all of the following conditions:

1. the natural atmospheric boundary layer has been modeled to account for the variation of wind speed with height;

2. the relevant macro (integral) length and micro length scales of the longitudinal component of atmospheric turbulence are modeled to approximately the same scale as that used to model the building or structure;

3. the modeled building or other structure and surrounding structures and topography are geometrically similar to their full-scale counterparts, except that, for low-rise buildings meeting the requirements of Section 6.5.1, tests shall be permitted for the modeled building in a single exposure site as defined in Section 6.5.6.3;

4. the projected area of the modeled building or other structure and surroundings is less than 8% of the test section cross-sectional area unless correction is made for blockage;

5. the longitudinal pressure gradient in the wind-tunnel test section is accounted for;

6. Reynolds number effects on pressures and forces are minimized; and

7. response characteristics of the wind-tunnel instrumentation are consistent with the required measurements.

6.6.3 Dynamic Response. Tests for the purpose of determining the dynamic response of a building or other structure shall be in accordance with Section 6.6.2. The structural model and associated analysis shall account for mass distribution, stiffness, and damping.

6.6.4 Limitations.

6.6.4.1 Limitations on Wind Speeds. Variation of basic wind speeds with direction shall not be permitted unless the analysis for wind speeds conforms to the requirements of Section 6.5.4.2.

SECTION 6.7
REFERENCES

Ref. 6-1 Standard Test Method for Performance of Exterior Windows, Curtain Walls, Doors and Storm Shutters Impacted by Missile(s) and Exposed to Cyclic Pressure Differentials, ASTM E1886-97, ASTM Inc., West Conshohocken, PA, 1997.

Ref. 6-2 Specification Standard for Performance of Exterior Windows, Glazed Curtain Walls, Doors and Storm Shutters Impacted by Windborne Debris in Hurricanes, ASTM E 1996-99, ASTM Inc., West Conshohocken, PA, 1999.

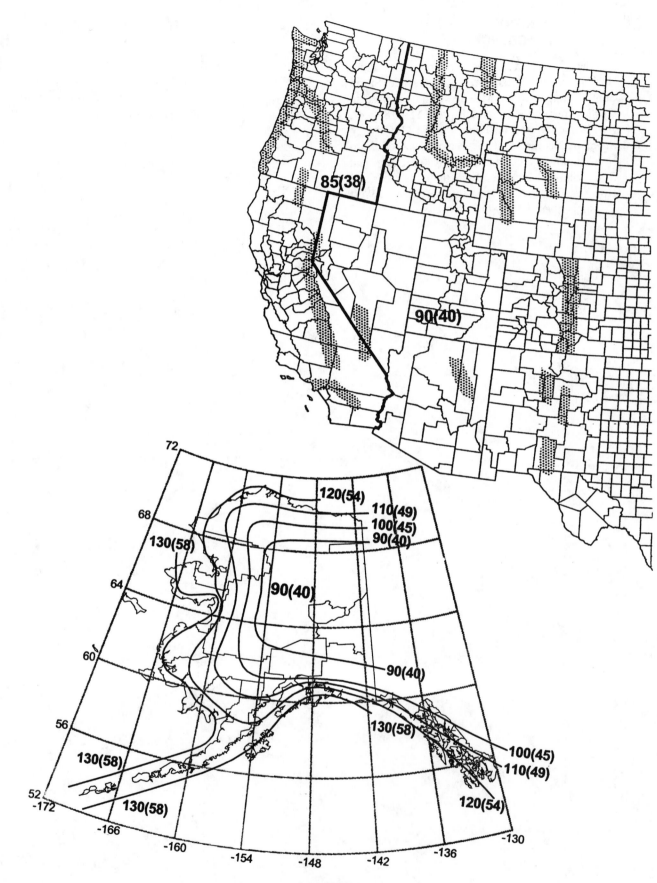

85(38)

90(40)

72

120(54)
110(49)
100(45)
90(40)

68

130(58)

64

90(40)

60

90(40)

56

130(58)

130(58)

130(58)

100(45)
110(49)

120(54)

52
-172
-166
-160
-154
-148
-142
-136
-130

FIGURE 6-1
BASIC WIND SPEED

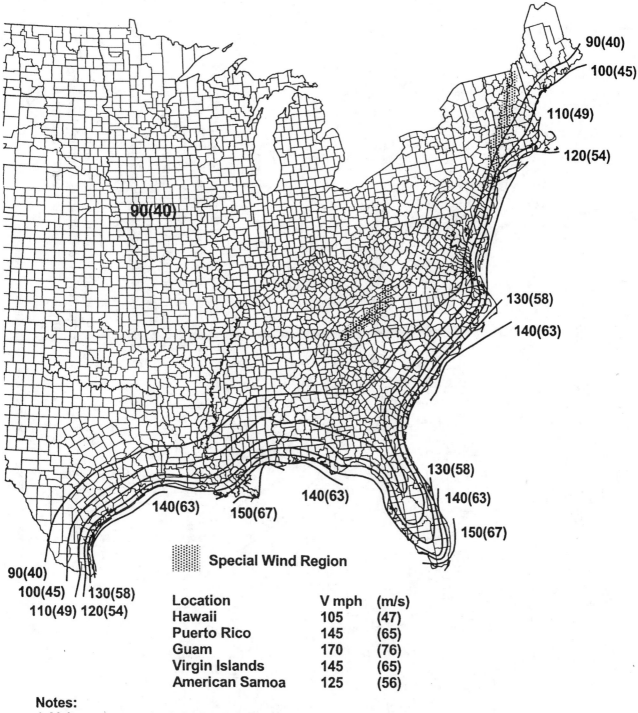

90(40)
100(45)
110(49)
120(54)
130(58)
140(63)

90(40)

130(58)
140(63)

140(63)
150(67)

140(63)
150(67)

90(40)
100(45)   130(58)
110(49) 120(54)

::::: **Special Wind Region**

Location	V mph	(m/s)
Hawaii	105	(47)
Puerto Rico	145	(65)
Guam	170	(76)
Virgin Islands	145	(65)
American Samoa	125	(56)

Notes:

1. Values are nominal design 3-second gust wind speeds in miles per hour (m/s) at 33 ft (10 m) above ground for Exposure C category.
2. Linear interpolation between wind contours is permitted.
3. Islands and coastal areas outside the last contour shall use the last wind speed contour of the coastal area.
4. Mountainous terrain, gorges, ocean promontories, and special wind regions shall be examined for unusual wind conditions.

FIGURE 6-1 — continued
BASIC WIND SPEED

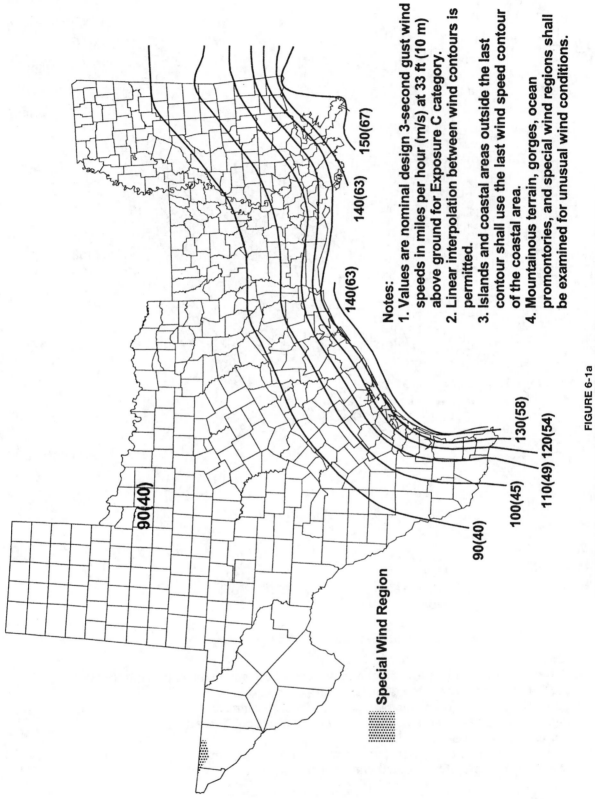

90(40)

Notes:
1. Values are nominal design 3-second gust wind speeds in miles per hour (m/s) at 33 ft (10 m) above ground for Exposure C category.
2. Linear interpolation between wind contours is permitted.
3. Islands and coastal areas outside the last contour shall use the last wind speed contour of the coastal area.
4. Mountainous terrain, gorges, ocean promontories, and special wind regions shall be examined for unusual wind conditions.

150(67)

140(63)

140(63)

130(58)

120(54)

110(49)

100(45)

90(40)

 Special Wind Region

FIGURE 6-1a
BASIC WIND SPEED — WESTERN GULF OF MEXICO HURRICANE COASTLINE

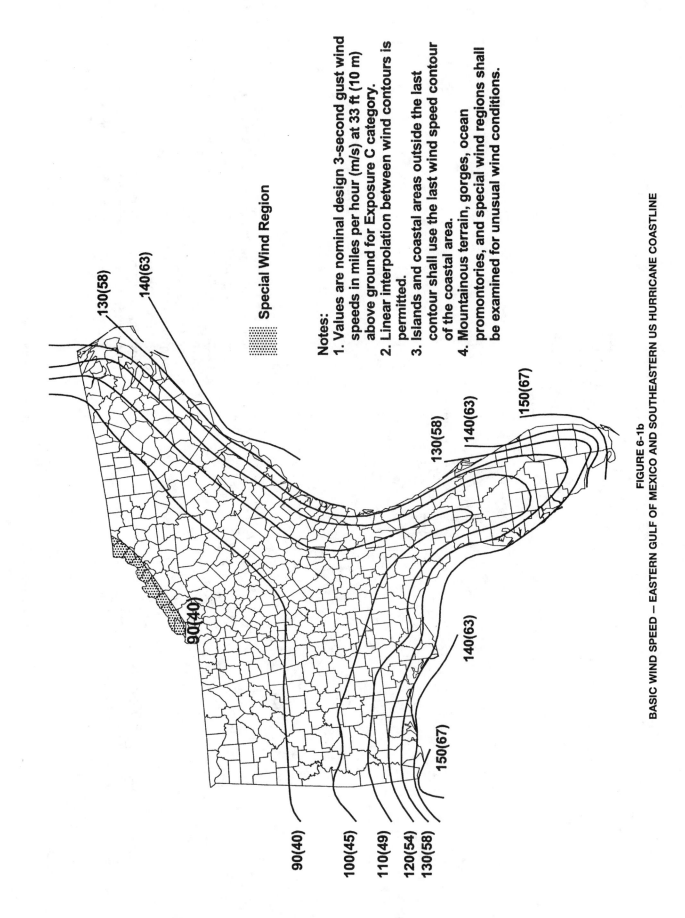

Notes:

1. Values are nominal design 3-second gust wind speeds in miles per hour (m/s) at 33 ft (10 m) above ground for Exposure C category.
2. Linear interpolation between wind contours is permitted.
3. Islands and coastal areas outside the last contour shall use the last wind speed contour of the coastal area.
4. Mountainous terrain, gorges, ocean promontories, and special wind regions shall be examined for unusual wind conditions.

Special Wind Region

130(58)
140(63)
90(40)
130(58)
140(63)
150(67)
140(63)
150(67)

90(40)
100(45)
110(49)
120(54)
130(58)

FIGURE 6-1b

BASIC WIND SPEED – EASTERN GULF OF MEXICO AND SOUTHEASTERN US HURRICANE COASTLINE

Minimum Design Loads for Buildings and Other Structures

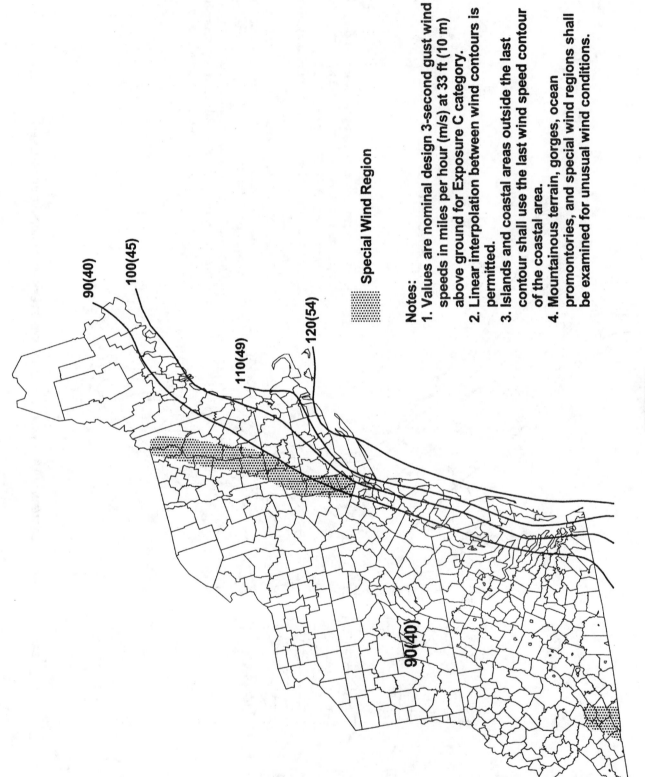

90(40)

100(45)

110(49)

120(54)

90(40)

Special Wind Region

Notes:
1. Values are nominal design 3-second gust wind speeds in miles per hour (m/s) at 33 ft (10 m) above ground for Exposure C category.
2. Linear interpolation between wind contours is permitted.
3. Islands and coastal areas outside the last contour shall use the last wind speed contour of the coastal area.
4. Mountainous terrain, gorges, ocean promontories, and special wind regions shall be examined for unusual wind conditions.

FIGURE 6-1c

BASIC WIND SPEED – MID AND NORTHERN ATLANTIC HURRICANE COASTLINE

Main Wind Force Resisting System – Method 1		h ≤ 60 ft.
Figure 6-2	**Design Wind Pressures**	**Walls & Roofs**
Enclosed Buildings		

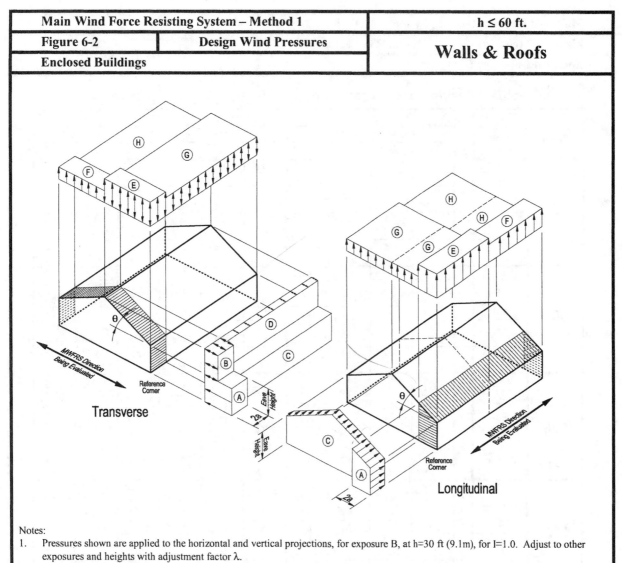

Notes:
1. Pressures shown are applied to the horizontal and vertical projections, for exposure B, at h=30 ft (9.1m), for I=1.0. Adjust to other exposures and heights with adjustment factor λ.
2. The load patterns shown shall be applied to each corner of the building in turn as the reference corner. (See Figure 6-10)
3. For the design of the longitudinal MWFRS use $\theta = 0°$, and locate the zone E/F, G/H boundary at the mid-length of the building.
4. Load cases 1 and 2 must be checked for $25° < \theta \leq 45°$. Load case 2 at 25° is provided only for interpolation between 25° to 30°.
5. Plus and minus signs signify pressures acting toward and away from the projected surfaces, respectively.
6. For roof slopes other than those shown, linear interpolation is permitted.
7. The total horizontal load shall not be less than that determined by assuming $p_S = 0$ in zones B & D.
8. The zone pressures represent the following:
   Horizontal pressure zones – Sum of the windward and leeward net (sum of internal and external) pressures on vertical projection of:
A - End zone of wall	C - Interior zone of wall
B - End zone of roof	D - Interior zone of roof

   Vertical pressure zones – Net (sum of internal and external) pressures on horizontal projection of:
E - End zone of windward roof	G - Interior zone of windward roof
F - End zone of leeward roof	H - Interior zone of leeward roof
9. Where zone E or G falls on a roof overhang on the windward side of the building, use E_{OH} and G_{OH} for the pressure on the horizontal projection of the overhang. Overhangs on the leeward and side edges shall have the basic zone pressure applied.
10. Notation:
    a: 10 percent of least horizontal dimension or 0.4h, whichever is smaller, but not less than either 4% of least horizontal dimension or 3 ft (0.9 m).
    h: Mean roof height, in feet (meters), except that eave height shall be used for roof angles <10°.
    θ: Angle of plane of roof from horizontal, in degrees.

Main Wind Force Resisting System – Method 1	h ≤ 60 ft.
Figure 6-2 (cont'd) Design Wind Pressures	**Walls & Roofs**
Enclosed Buildings	

Simplified Design Wind Pressure , p_{S30} (psf) *(Exposure B at h = 30 ft. with I = 1.0)*

Basic Wind Speed (mph)	Roof Angle (degrees)	Load Case	Zones									
			Horizontal Pressures				Vertical Pressures				Overhangs	
			A	B	C	D	E	F	G	H	EOH	GOH
85	0 to 5°	1	11.5	-5.9	7.6	-3.5	-13.8	-7.8	-9.6	-6.1	-19.3	-15.1
	10°	1	12.9	-5.4	8.6	-3.1	-13.8	-8.4	-9.6	-6.5	-19.3	-15.1
	15°	1	14.4	-4.8	9.6	-2.7	-13.8	-9.0	-9.6	-6.9	-19.3	-15.1
	20°	1	15.9	-4.2	10.6	-2.3	-13.8	-9.6	-9.6	-7.3	-19.3	-15.1
	25°	1	14.4	2.3	10.4	2.4	-6.4	-8.7	-4.6	-7.0	-11.9	-10.1
		2	------	------	------	------	-2.4	-4.7	-0.7	-3.0	------	------
	30 to 45	1	12.9	8.8	10.2	7.0	1.0	-7.8	0.3	-6.7	-4.5	-5.2
		2	12.9	8.8	10.2	7.0	5.0	-3.9	4.3	-2.8	-4.5	-5.2
90	0 to 5°	1	12.8	-6.7	8.5	-4.0	-15.4	-8.8	-10.7	-6.8	-21.6	-16.9
	10°	1	14.5	-6.0	9.6	-3.5	-15.4	-9.4	-10.7	-7.2	-21.6	-16.9
	15°	1	16.1	-5.4	10.7	-3.0	-15.4	-10.1	-10.7	-7.7	-21.6	-16.9
	20°	1	17.8	-4.7	11.9	-2.6	-15.4	-10.7	-10.7	-8.1	-21.6	-16.9
	25°	1	16.1	2.6	11.7	2.7	-7.2	-9.8	-5.2	-7.8	-13.3	-11.4
		2	------	------	------	------	-2.7	-5.3	-0.7	-3.4	------	------
	30 to 45	1	14.4	9.9	11.5	7.9	1.1	-8.8	0.4	-7.5	-5.1	-5.8
		2	14.4	9.9	11.5	7.9	5.6	-4.3	4.8	-3.1	-5.1	-5.8
100	0 to 5°	1	15.9	-8.2	10.5	-4.9	-19.1	-10.8	-13.3	-8.4	-26.7	-20.9
	10°	1	17.9	-7.4	11.9	-4.3	-19.1	-11.6	-13.3	-8.9	-26.7	-20.9
	15°	1	19.9	-6.6	13.3	-3.8	-19.1	-12.4	-13.3	-9.5	-26.7	-20.9
	20°	1	22.0	-5.8	14.6	-3.2	-19.1	-13.3	-13.3	-10.1	-26.7	-20.9
	25°	1	19.9	3.2	14.4	3.3	-8.8	-12.0	-6.4	-9.7	-16.5	-14.0
		2	------	------	------	------	-3.4	-6.6	-0.9	-4.2	------	------
	30 to 45	1	17.8	12.2	14.2	9.8	1.4	-10.8	0.5	-9.3	-6.3	-7.2
		2	17.8	12.2	14.2	9.8	6.9	-5.3	5.9	-3.8	-6.3	-7.2
110	0 to 5°	1	19.2	-10.0	12.7	-5.9	-23.1	-13.1	-16.0	-10.1	-32.3	-25.3
	10°	1	21.6	-9.0	14.4	-5.2	-23.1	-14.1	-16.0	-10.8	-32.3	-25.3
	15°	1	24.1	-8.0	16.0	-4.6	-23.1	-15.1	-16.0	-11.5	-32.3	-25.3
	20°	1	26.6	-7.0	17.7	-3.9	-23.1	-16.0	-16.0	-12.2	-32.3	-25.3
	25°	1	24.1	3.9	17.4	4.0	-10.7	-14.6	-7.7	-11.7	-19.9	-17.0
		2	------	------	------	------	-4.1	-7.9	-1.1	-5.1	------	------
	30 to 45	1	21.6	14.8	17.2	11.8	1.7	-13.1	0.6	-11.3	-7.6	-8.7
		2	21.6	14.8	17.2	11.8	8.3	-6.5	7.2	-4.6	-7.6	-8.7
120	0 to 5°	1	22.8	-11.9	15.1	-7.0	-27.4	-15.6	-19.1	-12.1	-38.4	-30.1
	10°	1	25.8	-10.7	17.1	-6.2	-27.4	-16.8	-19.1	-12.9	-38.4	-30.1
	15°	1	28.7	-9.5	19.1	-5.4	-27.4	-17.9	-19.1	-13.7	-38.4	-30.1
	20°	1	31.6	-8.3	21.1	-4.6	-27.4	-19.1	-19.1	-14.5	-38.4	-30.1
	25°	1	28.6	4.6	20.7	4.7	-12.7	-17.3	-9.2	-13.9	-23.7	-20.2
		2	------	------	------	------	-4.8	-9.4	-1.3	-6.0	------	------
	30 to 45	1	25.7	17.6	20.4	14.0	2.0	-15.6	0.7	-13.4	-9.0	-10.3
		2	25.7	17.6	20.4	14.0	9.9	-7.7	8.6	-5.5	-9.0	-10.3
130	0 to 5°	1	26.8	-13.9	17.8	-8.2	-32.2	-18.3	-22.4	-14.2	-45.1	-35.3
	10°	1	30.2	-12.5	20.1	-7.3	-32.2	-19.7	-22.4	-15.1	-45.1	-35.3
	15°	1	33.7	-11.2	22.4	-6.4	-32.2	-21.0	-22.4	-16.1	-45.1	-35.3
	20°	1	37.1	-9.8	24.7	-5.4	-32.2	-22.4	-22.4	-17.0	-45.1	-35.3
	25°	1	33.6	5.4	24.3	5.5	-14.9	-20.4	-10.8	-16.4	-27.8	-23.7
		2	------	------	------	------	-5.7	-11.1	-1.5	-7.1	------	------
	30 to 45	1	30.1	20.6	24.0	16.5	2.3	-18.3	0.8	-15.7	-10.6	-12.1
		2	30.1	20.6	24.0	16.5	11.6	-9.0	10.0	-6.4	-10.6	-12.1

Unit Conversions – 1.0 ft = 0.3048 m; 1.0 psf = 0.0479 kN/m^2

Figure 6-2 (cont'd)	Design Wind Pressures	**Walls & Roofs**
Enclosed Buildings		

Simplified Design Wind Pressure , p_{S30} (psf) *(Exposure B at h = 30 ft. with I = 1.0)*

Basic Wind Speed (mph)	Roof Angle (degrees)	Load Case	Zones									
			Horizontal Pressures				Vertical Pressures				Overhangs	
			A	B	C	D	E	F	G	H	E_{OH}	G_{OH}
140	0 to 5°	1	31.1	-16.1	20.6	-9.6	-37.3	-21.2	-26.0	-16.4	-52.3	-40.9
	10°	1	35.1	-14.5	23.3	-8.5	-37.3	-22.8	-26.0	-17.5	-52.3	-40.9
	15°	1	39.0	-12.9	26.0	-7.4	-37.3	-24.4	-26.0	-18.6	-52.3	-40.9
	20°	1	43.0	-11.4	28.7	-6.3	-37.3	-26.0	-26.0	-19.7	-52.3	-40.9
	25°	1	39.0	6.3	28.2	6.4	-17.3	-23.6	-12.5	-19.0	-32.3	-27.5
		2	-------	-------	-------	-------	-6.6	-12.8	-1.8	-8.2	-------	-------
	30 to 45	1	35.0	23.9	27.8	19.1	2.7	-21.2	0.9	-18.2	-12.3	-14.0
		2	35.0	23.9	27.8	19.1	13.4	-10.5	11.7	-7.5	-12.3	-14.0
150	0 to 5°	1	35.7	-18.5	23.7	-11.0	-42.9	-24.4	-29.8	-18.9	-60.0	-47.0
	10°	1	40.2	-16.7	26.8	-9.7	-42.9	-26.2	-29.8	-20.1	-60.0	-47.0
	15°	1	44.8	-14.9	29.8	-8.5	-42.9	-28.0	-29.8	-21.4	-60.0	-47.0
	20°	1	49.4	-13.0	32.9	-7.2	-42.9	-29.8	-29.8	-22.6	-60.0	-47.0
	25°	1	44.8	7.2	32.4	7.4	-19.9	-27.1	-14.4	-21.8	-37.0	-31.6
		2	-------	-------	-------	-------	-7.5	-14.7	-2.1	-9.4	-------	-------
	30 to 45	1	40.1	27.4	31.9	22.0	3.1	-24.4	1.0	-20.9	-14.1	-16.1
		2	40.1	27.4	31.9	22.0	15.4	-12.0	13.4	-8.6	-14.1	-16.1
170	0 to 5°	1	45.8	-23.8	30.4	-14.1	-55.1	-31.3	-38.3	-24.2	-77.1	-60.4
	10°	1	51.7	-21.4	34.4	-12.5	-55.1	-33.6	-38.3	-25.8	-77.1	-60.4
	15°	1	57.6	-19.1	38.3	-10.9	-55.1	-36.0	-38.3	-27.5	-77.1	-60.4
	20°	1	63.4	-16.7	42.3	-9.3	-55.1	-38.3	-38.3	-29.1	-77.1	-60.4
	25°	1	57.5	9.3	41.6	9.5	-25.6	-34.8	-18.5	-28.0	-47.6	-40.5
		2	-------	-------	-------	-------	-9.7	-18.9	-2.6	-12.1	-------	-------
	30 to 45	1	51.5	35.2	41.0	28.2	4.0	-31.3	1.3	-26.9	-18.1	-20.7
		2	51.5	35.2	41.0	28.2	19.8	-15.4	17.2	-11.0	-18.1	-20.7

Adjustment Factor
for Building Height and Exposure, λ

Mean roof height (ft)	Exposure		
	B	C	D
15	1.00	1.21	1.47
20	1.00	1.29	1.55
25	1.00	1.35	1.61
30	1.00	1.40	1.66
35	1.05	1.45	1.70
40	1.09	1.49	1.74
45	1.12	1.53	1.78
50	1.16	1.56	1.81
55	1.19	1.59	1.84
60	1.22	1.62	1.87

Unit Conversions – 1.0 ft = 0.3048 m; 1.0 psf = 0.0479 kN/m^2

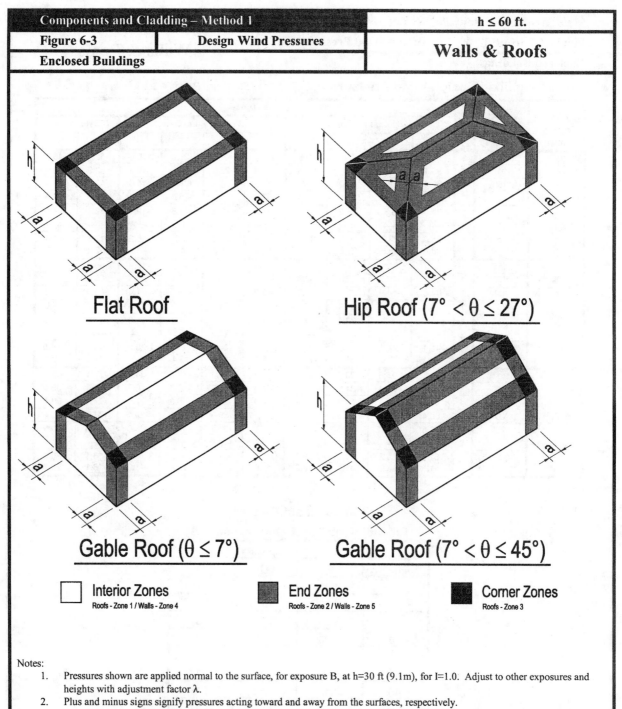

Flat Roof

Hip Roof (7° < θ ≤ 27°)

Gable Roof (θ ≤ 7°)

Gable Roof (7° < θ ≤ 45°)

☐ **Interior Zones**
Roofs - Zone 1 / Walls - Zone 4

▨ **End Zones**
Roofs - Zone 2 / Walls - Zone 5

■ **Corner Zones**
Roofs - Zone 3

Notes:

1. Pressures shown are applied normal to the surface, for exposure B, at h=30 ft (9.1m), for I=1.0. Adjust to other exposures and heights with adjustment factor λ.

2. Plus and minus signs signify pressures acting toward and away from the surfaces, respectively.

3. For hip roofs with θ ≤ 25°, Zone 3 shall be treated as Zone 2.

4. For effective wind areas between those given, value may be interpolated, otherwise use the value associated with the lower effective wind area.

5. Notation:

   a: 10 percent of least horizontal dimension or 0.4h, whichever is smaller, but not less than either 4% of least horizontal dimension or 3 ft (0.9 m).

   h: Mean roof height, in feet (meters), except that eave height shall be used for roof angles <10°.

   $θ$: Angle of plane of roof from horizontal, in degrees.

ASCE 7-02

| Figure 6-3 (cont'd) | Design Wind Pressures | **Walls & Roofs** |
| Enclosed Buildings | | |

Net Design Wind Pressure, p_{net30} (psf) *(Exposure B at h = 30 ft. with I = 1.0)*

	Zone	Effective wind area (sf)	Basic Wind Speed V (mph)															
			85		90		100		110		120		130		140		150		170	
Roof 0 to 7 degrees	1	10	5.3	-13.0	5.9	-14.6	7.3	-18.0	8.9	-21.8	10.5	-25.9	12.4	-30.4	14.3	-35.3	16.5	-40.5	21.1	-52.0
	1	20	5.0	-12.7	5.6	-14.2	6.9	-17.5	8.3	-21.2	9.9	-25.2	11.6	-29.6	13.4	-34.4	15.4	-39.4	19.8	-50.7
	1	50	4.5	-12.2	5.1	-13.7	6.3	-16.9	7.6	-20.5	9.0	-24.4	10.6	-28.6	12.3	-33.2	14.1	-38.1	18.1	-48.9
	1	100	4.2	-11.9	4.7	-13.3	5.8	-16.5	7.0	-19.9	8.3	-23.7	9.8	-27.8	11.4	-32.3	13.0	-37.0	16.7	-47.6
	2	10	5.3	-21.8	5.9	-24.4	7.3	-30.2	8.9	-36.5	10.5	-43.5	12.4	-51.0	14.3	-59.2	16.5	-67.9	21.1	-87.2
	2	20	5.0	-19.5	5.6	-21.8	6.9	-27.0	8.3	-32.6	9.9	-38.8	11.6	-45.6	13.4	-52.9	15.4	-60.7	19.8	-78.0
	2	50	4.5	-16.4	5.1	-18.4	6.3	-22.7	7.6	-27.5	9.0	-32.7	10.6	-38.4	12.3	-44.5	14.1	-51.1	18.1	-65.7
	2	100	4.2	-14.1	4.7	-15.8	5.8	-19.5	7.0	-23.6	8.3	-28.1	9.8	-33.0	11.4	-38.2	13.0	-43.9	16.7	-56.4
	3	10	5.3	-32.8	5.9	-36.8	7.3	-45.4	8.9	-55.0	10.5	-65.4	12.4	-76.8	14.3	-89.0	16.5	-102.2	21.1	-131.3
	3	20	5.0	-27.2	5.6	-30.5	6.9	-37.6	8.3	-45.5	9.9	-54.2	11.6	-63.6	13.4	-73.8	15.4	-84.7	19.8	-108.7
	3	50	4.5	-19.7	5.1	-22.1	6.3	-27.3	7.6	-33.1	9.0	-39.3	10.6	-46.2	12.3	-53.5	14.1	-61.5	18.1	-78.9
	3	100	4.2	-14.1	4.7	-15.8	5.8	-19.5	7.0	-23.6	8.3	-28.1	9.8	-33.0	11.4	-38.2	13.0	-43.9	16.7	-56.4
Roof > 7 to 27 degrees	1	10	7.5	-11.9	8.4	-13.3	10.4	-16.5	12.5	-19.9	14.9	-23.7	17.5	-27.8	20.3	-32.3	23.3	-37.0	30.0	-47.6
	1	20	6.8	-11.6	7.7	-13.0	9.4	-16.0	11.4	-19.4	13.6	-23.0	16.0	-27.0	18.5	-31.4	21.3	-36.0	27.3	-46.3
	1	50	6.0	-11.1	6.7	-12.5	8.2	-15.4	10.0	-18.6	11.9	-22.2	13.9	-26.0	16.1	-30.2	18.5	-34.6	23.8	-44.5
	1	100	5.3	-10.8	5.9	-12.1	7.3	-14.9	8.9	-18.1	10.5	-21.5	12.4	-25.2	14.3	-29.3	16.5	-33.6	21.1	-43.2
	2	10	7.5	-20.7	8.4	-23.2	10.4	-28.7	12.5	-34.7	14.9	-41.3	17.5	-48.4	20.3	-56.2	23.3	-64.5	30.0	-82.8
	2	20	6.8	-19.0	7.7	-21.4	9.4	-26.4	11.4	-31.9	13.6	-38.0	16.0	-44.6	18.5	-51.7	21.3	-59.3	27.3	-76.2
	2	50	6.0	-16.9	6.7	-18.9	8.2	-23.3	10.0	-28.2	11.9	-33.6	13.9	-39.4	16.1	-45.7	18.5	-52.5	23.8	-67.4
	2	100	5.3	-15.2	5.9	-17.0	7.3	-21.0	8.9	-25.5	10.5	-30.3	12.4	-35.6	14.3	-41.2	16.5	-47.3	21.1	-60.8
	3	10	7.5	-30.6	8.4	-34.3	10.4	-42.4	12.5	-51.3	14.9	-61.0	17.5	-71.6	20.3	-83.1	23.3	-95.4	30.0	-122.5
	3	20	6.8	-28.6	7.7	-32.1	9.4	-39.6	11.4	-47.9	13.6	-57.1	16.0	-67.0	18.5	-77.7	21.3	-89.2	27.3	-114.5
	3	50	6.0	-26.0	6.7	-29.1	8.2	-36.0	10.0	-43.5	11.9	-51.8	13.9	-60.8	16.1	-70.5	18.5	-81.0	23.8	-104.0
	3	100	5.3	-24.0	5.9	-26.9	7.3	-33.2	8.9	-40.2	10.5	-47.9	12.4	-56.2	14.3	-65.1	16.5	-74.8	21.1	-96.0
Roof > 27 to 45 degrees	1	10	11.9	-13.0	13.3	-14.6	16.5	-18.0	19.9	-21.8	23.7	-25.9	27.8	-30.4	32.3	-35.3	37.0	-40.5	47.6	-52.0
	1	20	11.6	-12.3	13.0	-13.8	16.0	-17.1	19.4	-20.7	23.0	-24.6	27.0	-28.9	31.4	-33.5	36.0	-38.4	46.3	-49.3
	1	50	11.1	-11.5	12.5	-12.8	15.4	-15.9	18.6	-19.2	22.2	-22.8	26.0	-26.8	30.2	-31.1	34.6	-35.7	44.5	-45.8
	1	100	10.8	-10.8	12.1	-12.1	14.9	-14.9	18.1	-18.1	21.5	-21.5	25.2	-25.2	29.3	-29.3	33.6	-33.6	43.2	-43.2
	2	10	11.9	-15.2	13.3	-17.0	16.5	-21.0	19.9	-25.5	23.7	-30.3	27.8	-35.6	32.3	-41.2	37.0	-47.3	47.6	-60.8
	2	20	11.6	-14.5	13.0	-16.3	16.0	-20.1	19.4	-24.3	23.0	-29.0	27.0	-34.0	31.4	-39.4	36.0	-45.3	46.3	-58.1
	2	50	11.1	-13.7	12.5	-15.3	15.4	-18.9	18.6	-22.9	22.2	-27.2	26.0	-32.0	30.2	-37.1	34.6	-42.5	44.5	-54.6
	2	100	10.8	-13.0	12.1	-14.6	14.9	-18.0	18.1	-21.8	21.5	-25.9	25.2	-30.4	29.3	-35.3	33.6	-40.5	43.2	-52.0
	3	10	11.9	-15.2	13.3	-17.0	16.5	-21.0	19.9	-25.5	23.7	-30.3	27.8	-35.6	32.3	-41.2	37.0	-47.3	47.6	-60.8
	3	20	11.6	-14.5	13.0	-16.3	16.0	-20.1	19.4	-24.3	23.0	-29.0	27.0	-34.0	31.4	-39.4	36.0	-45.3	46.3	-58.1
	3	50	11.1	-13.7	12.5	-15.3	15.4	-18.9	18.6	-22.9	22.2	-27.2	26.0	-32.0	30.2	-37.1	34.6	-42.5	44.5	-54.6
	3	100	10.8	-13.0	12.1	-14.6	14.9	-18.0	18.1	-21.8	21.5	-25.9	25.2	-30.4	29.3	-35.3	33.6	-40.5	43.2	-52.0
Wall	4	10	13.0	-14.1	14.6	-15.8	18.0	-19.5	21.8	-23.6	25.9	-28.1	30.4	-33.0	35.3	-38.2	40.5	-43.9	52.0	-56.4
	4	20	12.4	-13.5	13.9	-15.1	17.2	-18.7	20.8	-22.6	24.7	-26.9	29.0	-31.6	33.7	-36.7	38.7	-42.1	49.6	-54.1
	4	50	11.6	-12.7	13.0	-14.3	16.1	-17.6	19.5	-21.3	23.2	-25.4	27.2	-29.8	31.6	-34.6	36.2	-39.7	46.6	-51.0
	4	100	11.1	-12.2	12.4	-13.6	15.3	-16.8	18.5	-20.4	22.0	-24.2	25.9	-28.4	30.0	-33.0	34.4	-37.8	44.2	-48.6
	4	500	9.7	-10.8	10.9	-12.1	13.4	-14.9	16.2	-18.1	19.3	-21.5	22.7	-25.2	26.3	-29.3	30.2	-33.6	38.8	-43.2
	5	10	13.0	-17.4	14.6	-19.5	18.0	-24.1	21.8	-29.1	25.9	-34.7	30.4	-40.7	35.3	-47.2	40.5	-54.2	52.0	-69.6
	5	20	12.4	-16.2	13.9	-18.2	17.2	-22.5	20.8	-27.2	24.7	-32.4	29.0	-38.0	33.7	-44.0	38.7	-50.5	49.6	-64.9
	5	50	11.6	-14.7	13.0	-16.5	16.1	-20.3	19.5	-24.6	23.2	-29.3	27.2	-34.3	31.6	-39.8	36.2	-45.7	46.6	-58.7
	5	100	11.1	-13.5	12.4	-15.1	15.3	-18.7	18.5	-22.6	22.0	-26.9	25.9	-31.6	30.0	-36.7	34.4	-42.1	44.2	-54.1
	5	500	9.7	-10.8	10.9	-12.1	13.4	-14.9	16.2	-18.1	19.3	-21.5	22.7	-25.2	26.3	-29.3	30.2	-33.6	38.8	-43.2

Unit Conversions – 1.0 ft = 0.3048 m; 1.0 sf = 0.0929 m², 1.0 psf = 0.0479 kN/m²

Components and Cladding – Method 1		h ≤ 60 ft.
Figure 6-3 (cont'd)	**Design Wind Pressures**	**Walls & Roofs**
Enclosed Buildings		

Roof Overhang Net Design Wind Pressure , p_{net30} (psf)

(Exposure B at h = 30 ft. with I = 1.0)

	Zone	Effective Wind Area (sf)	Basic Wind Speed V (mph)							
			90	100	110	120	130	140	150	170
Roof 0 to 7 degrees	2	10	-21.0	-25.9	-31.4	-37.3	-43.8	-50.8	-58.3	-74.9
	2	20	-20.6	-25.5	-30.8	-36.7	-43.0	-49.9	-57.3	-73.6
	2	50	-20.1	-24.9	-30.1	-35.8	-42.0	-48.7	-55.9	-71.8
	2	100	-19.8	-24.4	-29.5	-35.1	-41.2	-47.8	-54.9	-70.5
	3	10	-34.6	-42.7	-51.6	-61.5	-72.1	-83.7	-96.0	-123.4
	3	20	-27.1	-33.5	-40.5	-48.3	-56.6	-65.7	-75.4	-96.8
	3	50	-17.3	-21.4	-25.9	-30.8	-36.1	-41.9	-48.1	-61.8
	3	100	-10.0	-12.2	-14.8	-17.6	-20.6	-23.9	-27.4	-35.2
Roof > 7 to 27 degrees	2	10	-27.2	-33.5	-40.6	-48.3	-56.7	-65.7	-75.5	-96.9
	2	20	-27.2	-33.5	-40.6	-48.3	-56.7	-65.7	-75.5	-96.9
	2	50	-27.2	-33.5	-40.6	-48.3	-56.7	-65.7	-75.5	-96.9
	2	100	-27.2	-33.5	-40.6	-48.3	-56.7	-65.7	-75.5	-96.9
	3	10	-45.7	-56.4	-68.3	-81.2	-95.3	-110.6	-126.9	-163.0
	3	20	-41.2	-50.9	-61.6	-73.3	-86.0	-99.8	-114.5	-147.1
	3	50	-35.3	-43.6	-52.8	-62.8	-73.7	-85.5	-98.1	-126.1
	3	100	-30.9	-38.1	-46.1	-54.9	-64.4	-74.7	-85.8	-110.1
Roof > 27 to 45 degrees	2	10	-24.7	-30.5	-36.9	-43.9	-51.5	-59.8	-68.6	-88.1
	2	20	-24.0	-29.6	-35.8	-42.6	-50.0	-58.0	-66.5	-85.5
	2	50	-23.0	-28.4	-34.3	-40.8	-47.9	-55.6	-63.8	-82.0
	2	100	-22.2	-27.4	-33.2	-39.5	-46.4	-53.8	-61.7	-79.3
	3	10	-24.7	-30.5	-36.9	-43.9	-51.5	-59.8	-68.6	-88.1
	3	20	-24.0	-29.6	-35.8	-42.6	-50.0	-58.0	-66.5	-85.5
	3	50	-23.0	-28.4	-34.3	-40.8	-47.9	-55.6	-63.8	-82.0
	3	100	-22.2	-27.4	-33.2	-39.5	-46.4	-53.8	-61.7	-79.3

Adjustment Factor
for Building Height and Exposure, λ

Mean roof height (ft)	Exposure		
	B	C	D
15	1.00	1.21	1.47
20	1.00	1.29	1.55
25	1.00	1.35	1.61
30	1.00	1.40	1.66
35	1.05	1.45	1.70
40	1.09	1.49	1.74
45	1.12	1.53	1.78
50	1.16	1.56	1.81
55	1.19	1.59	1.84
60	1.22	1.62	1.87

Unit Conversions – 1.0 ft = 0.3048 m; 1.0 sf = 0.0929 m^2; 1.0 psf = 0.0479 kN/m^2

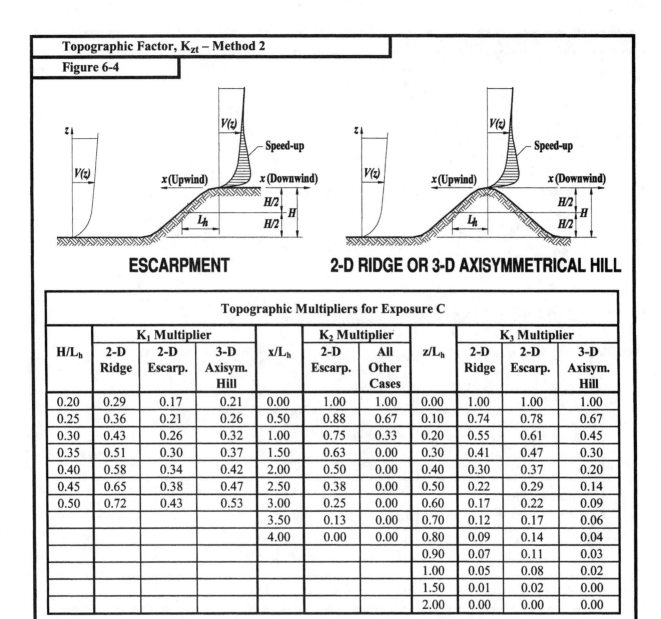

ESCARPMENT　　　**2-D RIDGE OR 3-D AXISYMMETRICAL HILL**

| | \multicolumn{3}{c}{**Topographic Multipliers for Exposure C**} | | | | | | | |

Topographic Multipliers for Exposure C

H/L_h	K_1 Multiplier			x/L_h	K_2 Multiplier		z/L_h	K_3 Multiplier		
	2-D Ridge	2-D Escarp.	3-D Axisym. Hill		2-D Escarp.	All Other Cases		2-D Ridge	2-D Escarp.	3-D Axisym. Hill
0.20	0.29	0.17	0.21	0.00	1.00	1.00	0.00	1.00	1.00	1.00
0.25	0.36	0.21	0.26	0.50	0.88	0.67	0.10	0.74	0.78	0.67
0.30	0.43	0.26	0.32	1.00	0.75	0.33	0.20	0.55	0.61	0.45
0.35	0.51	0.30	0.37	1.50	0.63	0.00	0.30	0.41	0.47	0.30
0.40	0.58	0.34	0.42	2.00	0.50	0.00	0.40	0.30	0.37	0.20
0.45	0.65	0.38	0.47	2.50	0.38	0.00	0.50	0.22	0.29	0.14
0.50	0.72	0.43	0.53	3.00	0.25	0.00	0.60	0.17	0.22	0.09
				3.50	0.13	0.00	0.70	0.12	0.17	0.06
				4.00	0.00	0.00	0.80	0.09	0.14	0.04
							0.90	0.07	0.11	0.03
							1.00	0.05	0.08	0.02
							1.50	0.01	0.02	0.00
							2.00	0.00	0.00	0.00

Notes:

1. For values of H/L_h, x/L_h and z/L_h other than those shown, linear interpolation is permitted.
2. For $H/L_h > 0.5$, assume $H/L_h = 0.5$ for evaluating K_1 and substitute $2H$ for L_h for evaluating K_2 and K_3.
3. Multipliers are based on the assumption that wind approaches the hill or escarpment along the direction of maximum slope.
4. Notation:

   H: Height of hill or escarpment relative to the upwind terrain, in feet (meters).

   L_h: Distance upwind of crest to where the difference in ground elevation is half the height of hill or escarpment, in feet (meters).

   K_1: Factor to account for shape of topographic feature and maximum speed-up effect.

   K_2: Factor to account for reduction in speed-up with distance upwind or downwind of crest.

   K_3: Factor to account for reduction in speed-up with height above local terrain.

   x: Distance (upwind or downwind) from the crest to the building site, in feet (meters).

   z: Height above local ground level, in feet (meters).

   μ: Horizontal attenuation factor.

   γ: Height attenuation factor.

Equations:

$$K_{zt} = (1 + K_1 K_2 K_3)^2$$

K_1 determined from table below

$$K_2 = \left(1 - \frac{|x|}{\mu L_h}\right)$$

$$K_3 = e^{-\gamma z/L_h}$$

Parameters for Speed-Up Over Hills and Escarpments						
Hill Shape	**$K_1/(H/L_h)$**			γ	μ	
	Exposure				**Upwind of Crest**	**Downwind of Crest**
	B	**C**	**D**			
2-dimensional ridges (or valleys with negative H in $K_1/(H/L_h)$	1.30	1.45	1.55	3	1.5	1.5
2-dimensional escarpments	0.75	0.85	0.95	2.5	1.5	4
3-dimensional axisym. hill	0.95	1.05	1.15	4	1.5	1.5

Main Wind Force Res. Sys. / Comp and Clad. – Method 2		All Heights
Figure 6-5	Internal Pressure Coefficient, GC$_{pi}$	**Walls & Roofs**
Enclosed, Partially Enclosed, and Open Buildings		

Enclosure Classification	GC$_{pi}$
Open Buildings	0.00
Partially Enclosed Buildings	+0.55 / -0.55
Enclosed Buildings	+0.18 / -0.18

Notes:

1. Plus and minus signs signify pressures acting toward and away from the internal surfaces, respectively.

2. Values of GC$_{pi}$ shall be used with q_z or q_h as specified in 6.5.12.

3. Two cases shall be considered to determine the critical load requirements for the appropriate condition:

   (i) a positive value of GC$_{pi}$ applied to all internal surfaces
   (ii) a negative value of GC$_{pi}$ applied to all internal surfaces

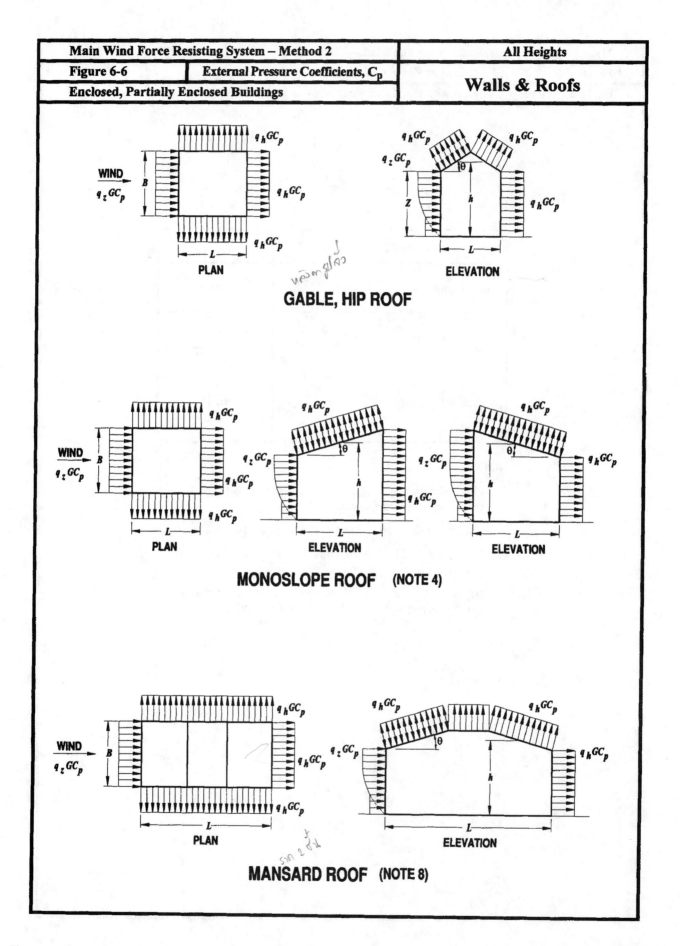

GABLE, HIP ROOF

MONOSLOPE ROOF (NOTE 4)

MANSARD ROOF (NOTE 8)

Main Wind Force Resisting System – Method 2	All Heights
Figure 6-6 (con't) **External Pressure Coefficients, C_p** **Enclosed, Partially Enclosed Buildings**	**Walls & Roofs**

Wall Pressure Coefficients, C_p

Surface	L/B	C_p	Use With
Windward Wall	All values	0.8	q_z
Leeward Wall	0-1	-0.5	q_h
	2	-0.3	
	≥4	-0.2	
Side Wall	All values	-0.7	q_h

Roof Pressure Coefficients, C_p, for use with q_h

Wind Direction		Windward								Leeward		
		Angle, θ (degrees)								Angle, θ (degrees)		
	h/L	10	15	20	25	30	35	45	≥60#	10	15	≥20
Normal to ridge for θ ≥ 10°	≤0.25	-0.7 / -0.18	-0.5 / 0.0*	-0.3 / 0.2	-0.2 / 0.3	-0.2 / 0.3	0.0* / 0.4	0.4	0.01 θ	-0.3	-0.5	-0.6
	0.5	-0.9 / -0.18	-0.7 / -0.18	-0.4 / 0.0*	-0.3 / 0.2	-0.2 / 0.2	-0.2 / 0.3	0.0* / 0.4	0.01 θ	-0.5	-0.5	-0.6
	≥1.0	-1.3** / -0.18	-1.0 / -0.18	-0.7 / -0.18	-0.5 / 0.0*	-0.3 / 0.2	-0.2 / 0.2	0.0* / 0.3	0.01 θ	-0.7	-0.6	-0.6

Normal to ridge for θ < 10 and Parallel to ridge for all θ	Horiz distance from windward edge		C_p	*Value is provided for interpolation purposes.
≤ 0.5	0 to h/2		-0.9, -0.18	
	h/2 to h		-0.9, -0.18	**Value can be reduced linearly with area over which it is applicable as follows
	h to 2 h		-0.5, -0.18	
	> 2h		-0.3, -0.18	

≥ 1.0	0 to h/2	-1.3**, -0.18	Area (sq ft)	Reduction Factor
			≤ 100 (9.3 sq m)	1.0
	> h/2	-0.7, -0.18	200 (23.2 sq m)	0.9
			≥ 1000 (92.9 sq m)	0.8

Notes:

1. Plus and minus signs signify pressures acting toward and away from the surfaces, respectively.
2. Linear interpolation is permitted for values of L/B, h/L and θ other than shown. Interpolation shall only be carried out between values of the same sign. Where no value of the same sign is given, assume 0.0 for interpolation purposes.
3. Where two values of C_p are listed, this indicates that the windward roof slope is subjected to either positive or negative pressures and the roof structure shall be designed for both conditions. Interpolation for intermediate ratios of h/L in this case shall only be carried out between C_p values of like sign.
4. For monoslope roofs, entire roof surface is either a windward or leeward surface.
5. For flexible buildings use appropriate G_f as determined by Section 6.5.8.
6. Refer to Figure 6-7 for domes and Figure 6-8 for arched roofs.
7. Notation:
   B: Horizontal dimension of building, in feet (meter), measured normal to wind direction.
   L: Horizontal dimension of building, in feet (meter), measured parallel to wind direction.
   h: Mean roof height in feet (meters), except that eave height shall be used for θ ≤ 10 degrees.
   z: Height above ground, in feet (meters).
   G: Gust effect factor.
   q_z, q_h: Velocity pressure, in pounds per square foot (N/m^2), evaluated at respective height.
   θ: Angle of plane of roof from horizontal, in degrees.
8. For mansard roofs, the top horizontal surface and leeward inclined surface shall be treated as leeward surfaces from the table.
9. Except for MWFRS's at the roof consisting of moment resisting frames, the total horizontal shear shall not be less than that determined by neglecting wind forces on roof surfaces.

#For roof slopes greater than 80°, use C_p = 0.8

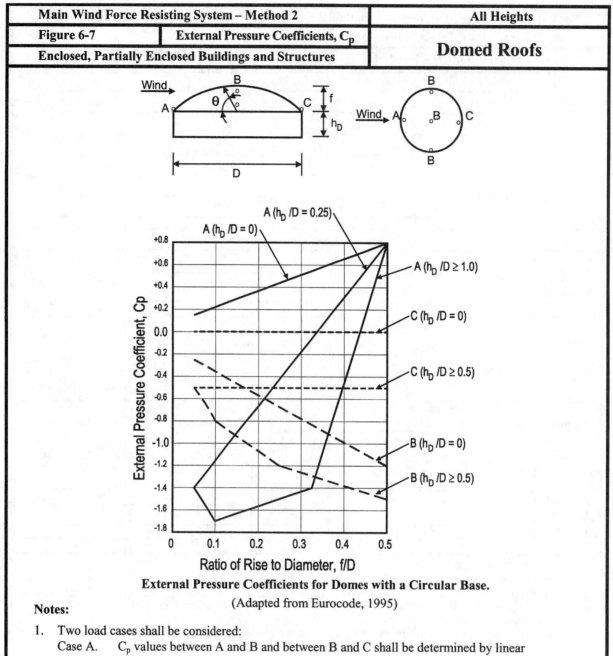

External Pressure Coefficients for Domes with a Circular Base.

(Adapted from Eurocode, 1995)

Notes:

1. Two load cases shall be considered:

   Case A. C_p values between A and B and between B and C shall be determined by linear interpolation along arcs on the dome parallel to the wind direction;

   Case B. C_p shall be the constant value of A for $\theta \leq 25$ degrees, and shall be determined by linear interpolation from 25 degrees to B and from B to C.

2. Values denote C_p to be used with $q_{(h_D+f)}$ where $h_D + f$ is the height at the top of the dome.

3. Plus and minus signs signify pressures acting toward and away from the surfaces, respectively.

4. C_p is constant on the dome surface for arcs of circles perpendicular to the wind direction; for example, the arc passing through B-B-B and all arcs parallel to B-B-B.

5. For values of h_D/D between those listed on the graph curves, linear interpolation shall be permitted.

6. $\theta = 0$ degrees on dome springline, $\theta = 90$ degrees at dome center top point. f is measured from springline to top.

7. The total horizontal shear shall not be less than that determined by neglecting wind forces on roof surfaces.

8. For f/D values less than 0.05, use Figure 6-6.

Main Wind Force Res. Sys. / Comp and Clad. – Method 2	All Heights
Figure 6-8 External Pressure Coefficients, C_p	**Arched Roofs**
Enclosed, Partially Enclosed Buildings and Structures	

Conditions	Rise-to-span ratio, r	C_p		
		Windward quarter	Center half	Leeward quarter
Roof on elevated structure	$0 < r < 0.2$	-0.9	$-0.7 - r$	-0.5
	$0.2 \leq r < 0.3*$	$1.5r - 0.3$	$-0.7 - r$	-0.5
	$0.3 \leq r \leq 0.6$	$2.75r - 0.7$	$-0.7 - r$	-0.5
Roof springing from ground level	$0 < r \leq 0.6$	$1.4r$	$-0.7 - r$	-0.5

*When the rise-to-span ratio is $0.2 \leq r \leq 0.3$, alternate coefficients given by $6r - 2.1$ shall also be used for the windward quarter.

Notes:

1. Values listed are for the determination of average loads on main wind force resisting systems.

2. Plus and minus signs signify pressures acting toward and away from the surfaces, respectively.

3. For wind directed parallel to the axis of the arch, use pressure coefficients from Fig. 6-6 with wind directed parallel to ridge.

4. For components and cladding: (1) At roof perimeter, use the external pressure coefficients in Fig. 6-11 with θ based on spring-line slope and (2) for remaining roof areas, use external pressure coefficients of this table multiplied by 0.87.

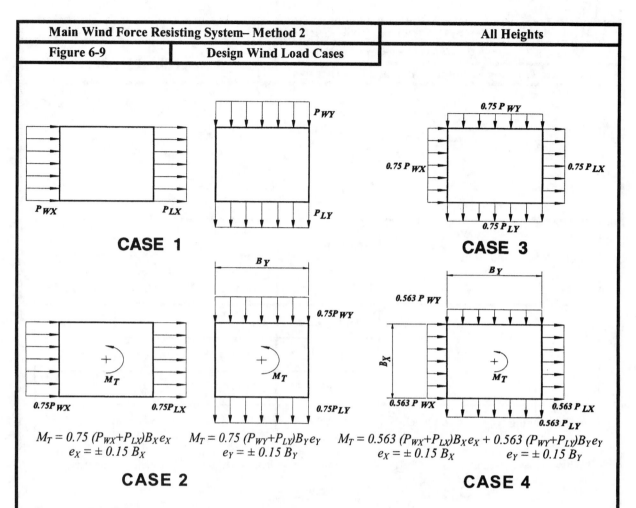

CASE 1

CASE 3

$M_T = 0.75 (P_{WX}+P_{LX})B_X e_X$
$e_X = \pm 0.15 B_X$

$M_T = 0.75 (P_{WY}+P_{LY})B_Y e_Y$
$e_Y = \pm 0.15 B_Y$

$M_T = 0.563 (P_{WX}+P_{LX})B_X e_X + 0.563 (P_{WY}+P_{LY})B_Y e_Y$
$e_X = \pm 0.15 B_X \qquad e_Y = \pm 0.15 B_Y$

CASE 2

CASE 4

Case 1. Full design wind pressure acting on the projected area perpendicular to each principal axis of the structure, considered separately along each principal axis.

Case 2. Three quarters of the design wind pressure acting on the projected area perpendicular to each principal axis of the structure in conjunction with a torsional moment as shown, considered separately for each principal axis.

Case 3. Wind loading as defined in Case 1, but considered to act simultaneously at 75% of the specified value.

Case 4. Wind loading as defined in Case 2, but considered to act simultaneously at 75% of the specified value.

Notes:

1. Design wind pressures for windward and leeward faces shall be determined in accordance with the provisions of 6.5.12.2.1 and 6.5.12.2.3 as applicable for building of all heights.
2. Diagrams show plan views of building.
3. Notation:

    P_{WX}, P_{WY}: Windward face design pressure acting in the x, y principal axis, respectively.

    P_{LX}, P_{LY}: Leeward face design pressure acting in the x, y principal axis, respectively.

    $e (e_X, e_Y)$: Eccentricity for the x, y principal axis of the structure, respectively.

    M_T: Torsional moment per unit height acting about a vertical axis of the building.

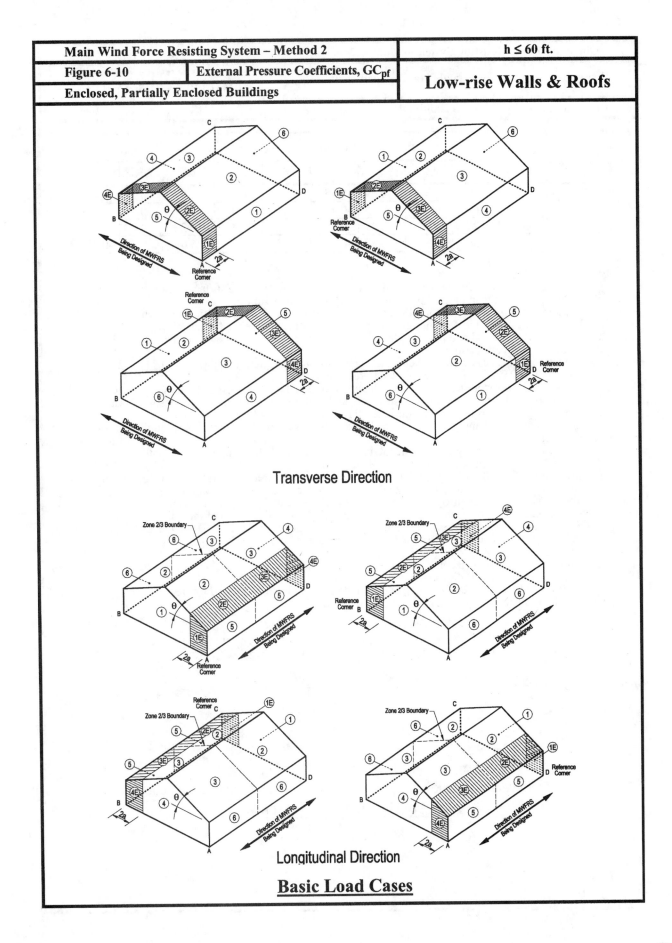

Transverse Direction

Longitudinal Direction

Basic Load Cases

| Main Wind Force Resisting System – Method 2 | | | | | | | | h ≤ 60 ft. | | |

Figure 6-10 (cont'd)	External Pressure Coefficients, GC_pf	Low-rise Walls & Roofs
Enclosed, Partially Enclosed Buildings		

Roof Angle θ (degrees)	Building Surface									
	1	2	3	4	5	6	1E	2E	3E	4E
0-5	0.40	-0.69	-0.37	-0.29	-0.45	-0.45	0.61	-1.07	-0.53	-0.43
20	0.53	-0.69	-0.48	-0.43	-0.45	-0.45	0.80	-1.07	-0.69	-0.64
30-45	0.56	0.21	-0.43	-0.37	-0.45	-0.45	0.69	0.27	-0.53	-0.48
90	0.56	0.56	-0.37	-0.37	-0.45	-0.45	0.69	0.69	-0.48	-0.48

Notes:
1. Plus and minus signs signify pressures acting toward and away from the surfaces, respectively.
2. For values of θ other than those shown, linear interpolation is permitted.
3. The building must be designed for all wind directions using the 8 loading patterns shown. The load patterns are applied to each building corner in turn as the Reference Corner.
4. Combinations of external and internal pressures (see Figure 6-5) shall be evaluated as required to obtain the most severe loadings.
5. For the torsional load cases shown below, the pressures in zones designated with a "T" (1T, 2T, 3T, 4T) shall be 25% of the full design wind pressures (zones 1, 3, 3, 4).
   Exception: One story buildings with h less than or equal to 30 ft (9.1m), buildings two stories or less framed with light frame construction, and buildings two stories or less designed with flexible diaphragms need not be designed for the torsional load cases.
   Torsional loading shall apply to all eight basic load patterns using the figures below applied at each reference corner.
6. Except for moment-resisting frames, the total horizontal shear shall not be less than that determined by neglecting wind forces on roof surfaces.
7. For the design of the MWFRS providing lateral resistance in a direction parallel to a ridge line or for flat roofs, use θ = 0° and locate the zone 2/3 boundary at the mid-length of the building.
8. The roof pressure coefficient GC_{pf}, when negative in Zone 2, shall be applied in Zone 2 for a distance from the edge of roof equal to 0.5 times the horizontal dimension of the building parallel to the direction of the MWFRS being designed or 2.5h, whichever is less; the remainder of Zone 2 extending to the ridge line shall use the pressure coefficient GC_{pf} for Zone 3.
9. Notation:
   a: 10 percent of least horizontal dimension or 0.4h, whichever is smaller, but not less than either 4% of least horizontal dimension or 3 ft (0.9 m).
   h: Mean roof height, in feet (meters), except that eave height shall be used for θ ≤ 10°.
   θ: Angle of plane of roof from horizontal, in degrees.

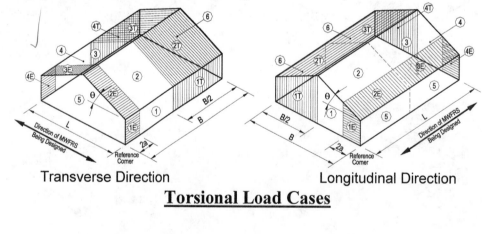

Transverse Direction   Longitudinal Direction

Torsional Load Cases

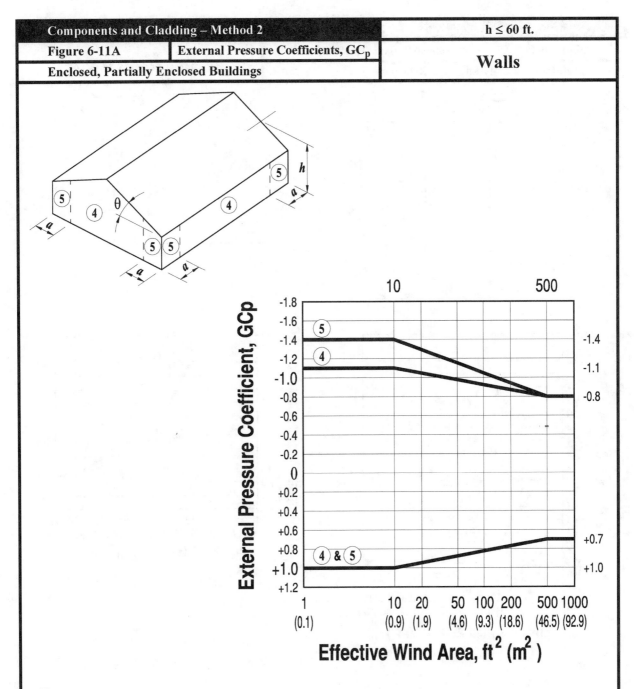

Notes:

1. Vertical scale denotes GC_p to be used with q_h.
2. Horizontal scale denotes effective wind area, in square feet (square meters).
3. Plus and minus signs signify pressures acting toward and away from the surfaces, respectively
4. Each component shall be designed for maximum positive and negative pressures.
5. Values of GC_p for walls shall be reduced by 10% when $\theta \le 10°$.
6. Notation:

   a: 10 percent of least horizontal dimension or 0.4h, whichever is smaller, but not less than either 4% of least horizontal dimension or 3 ft (0.9 m).

   h: Mean roof height, in feet (meters), except that eave height shall be used for $\theta \le 10°$.

   θ: Angle of plane of roof from horizontal, in degrees.

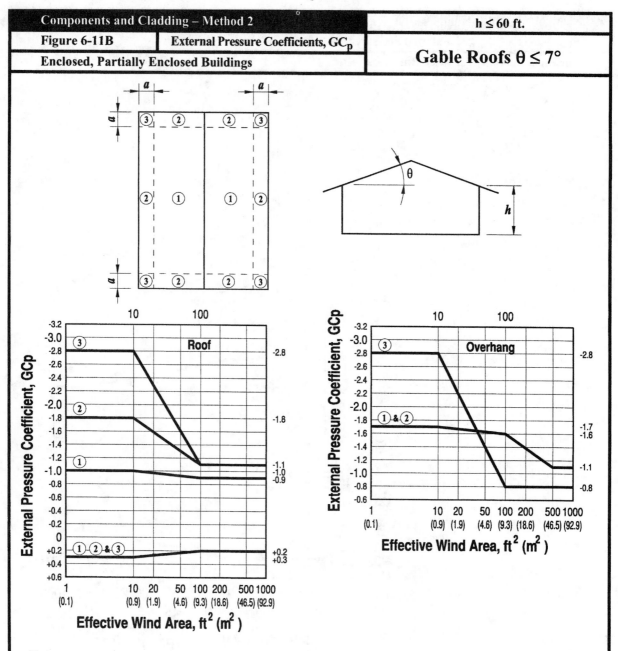

Notes:

1. Vertical scale denotes GC_p to be used with q_h.
2. Horizontal scale denotes effective wind area, in square feet (square meters).
3. Plus and minus signs signify pressures acting toward and away from the surfaces, respectively.
4. Each component shall be designed for maximum positive and negative pressures.
5. If a parapet equal to or higher than 3 ft (0.9m) is provided around the perimeter of the roof with θ ≤ 7°, Zone 3 shall be treated as Zone 2.
6. Values of GC_p for roof overhangs include pressure contributions from both upper and lower surfaces.
7. Notation:
   - a: 10 percent of least horizontal dimension or 0.4h, whichever is smaller, but not less than either 4% of least horizontal dimension or 3 ft (0.9 m).
   - h: Eave height shall be used for θ ≤ 10°.
   - θ: Angle of plane of roof from horizontal, in degrees.

Components and Cladding – Method 2		h ≤ 60 ft.
Figure 6-11C	**External Pressure Coefficients, GC$_p$**	**Gable/Hip Roofs 7°< θ ≤ 27°**
Enclosed, Partially Enclosed Buildings		

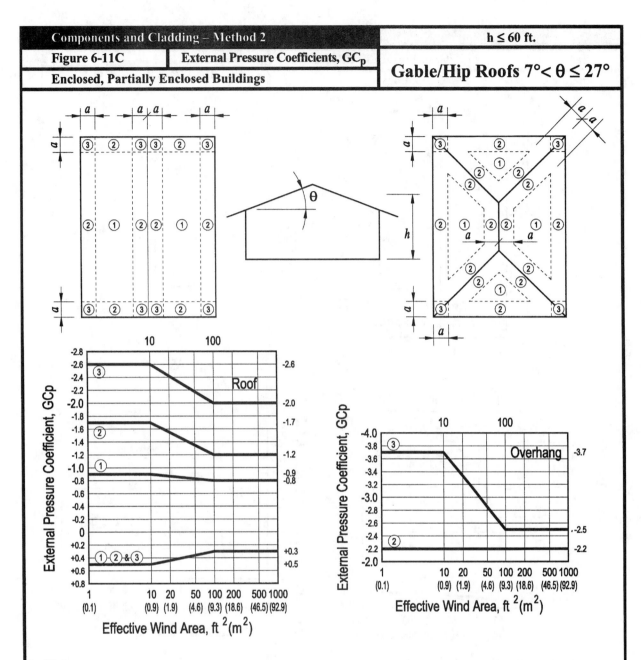

Notes:

1. Vertical scale denotes GC_p to be used with q_h.
2. Horizontal scale denotes effective wind area, in square feet (square meters).
3. Plus and minus signs signify pressures acting toward and away from the surfaces, respectively.
4. Each component shall be designed for maximum positive and negative pressures.
5. Values of GC_p for roof overhangs include pressure contributions from both upper and lower surfaces.
6. For hip roofs with 7° < θ ≤ 27°, edge/ridge strips and pressure coefficients for ridges of gabled roofs shall apply on each hip.
7. For hip roofs with θ ≤ 25°, Zone 3 shall be treated as Zone 2.
8. Notation:
   - a: 10 percent of least horizontal dimension or 0.4h, whichever is smaller, but not less than either 4% of least horizontal dimension or 3 ft (0.9 m).
   - h: Mean roof height, in feet (meters), except that eave height shall be used for θ ≤ 10°.
   - θ: Angle of plane of roof from horizontal, in degrees.

| Figure 6-11D | External Pressure Coefficients GC$_p$ | **Gable Roofs 27°< θ ≤ 45°** |
| Enclosed, Partially Enclosed Buildings | | |

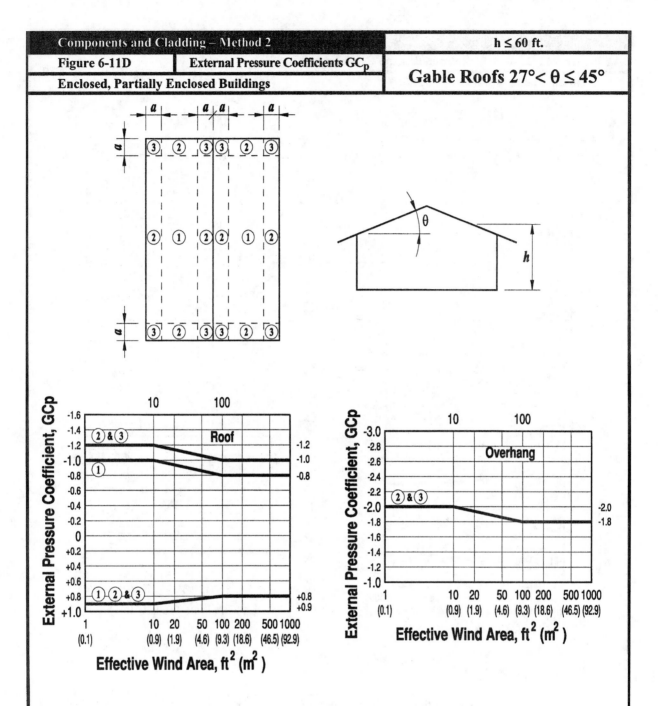

Notes:

1. Vertical scale denotes GC_p to be used with q_h.
2. Horizontal scale denotes effective wind area, in square feet (square meters).
3. Plus and minus signs signify pressures acting toward and away from the surfaces, respectively.
4. Each component shall be designed for maximum positive and negative pressures.
5. Values of GC_p for roof overhangs include pressure contributions from both upper and lower surfaces.
6. Notation:
   - *a*: 10 percent of least horizontal dimension or 0.4h, whichever is smaller, but not less than either 4% of least horizontal dimension or 3 ft (0.9 m).
   - *h*: Mean roof height, in feet (meters).
   - θ: Angle of plane of roof from horizontal, in degrees.

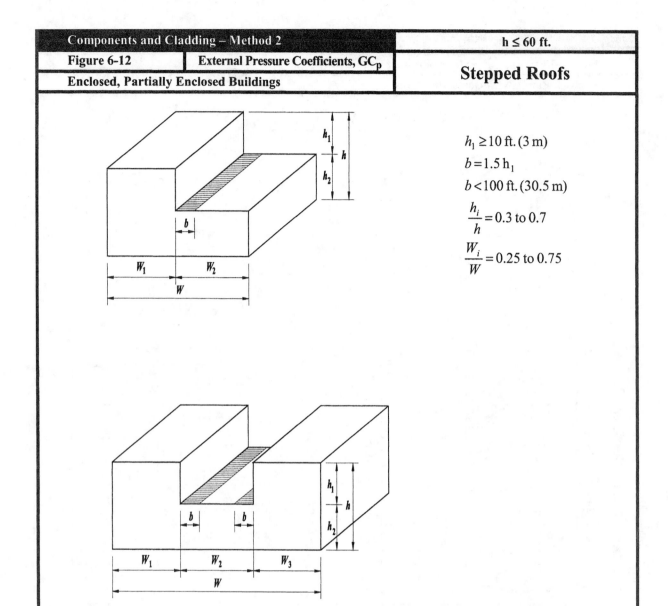

$h_1 \geq 10$ ft. (3 m)

$b = 1.5\,h_1$

$b < 100$ ft. (30.5 m)

$\dfrac{h_i}{h} = 0.3$ to 0.7

$\dfrac{W_i}{W} = 0.25$ to 0.75

Notes:

1. On the lower level of flat, stepped roofs shown in Fig. 6-12, the zone designations and pressure coefficients shown in Fig. 6-11B shall apply, except that at the roof-upper wall intersection(s), Zone 3 shall be treated as Zone 2 and Zone 2 shall be treated as Zone 1. Positive values of GC_p equal to those for walls in Fig. 6-11A shall apply on the cross-hatched areas shown in Fig. 6-12.

2. Notation:
   b: $1.5h_1$ in Fig. 6-12, but not greater than 100 ft (30.5 m).
   h: Mean roof height, in feet (meters).
   h_i: h_1 or h_2 in Fig. 6-12; $h = h_1 + h_2$; $h_1 \geq 10$ ft (3.1 m); $h_i/h = 0.3$ to 0.7.
   W: Building width in Fig. 6-12.
   W_i: W_1 or W_2 or W_3 in Fig. 6-12. $W = W_1 + W_2$ or $W_1 + W_2 + W_3$; $W_i/W = 0.25$ to 0.75.
   θ: Angle of plane of roof from horizontal, in degrees.

Figure 6-13	External Pressure Coefficients, GC_p	**Multispan Gable Roofs**
Enclosed, Partially Enclosed Buildings		

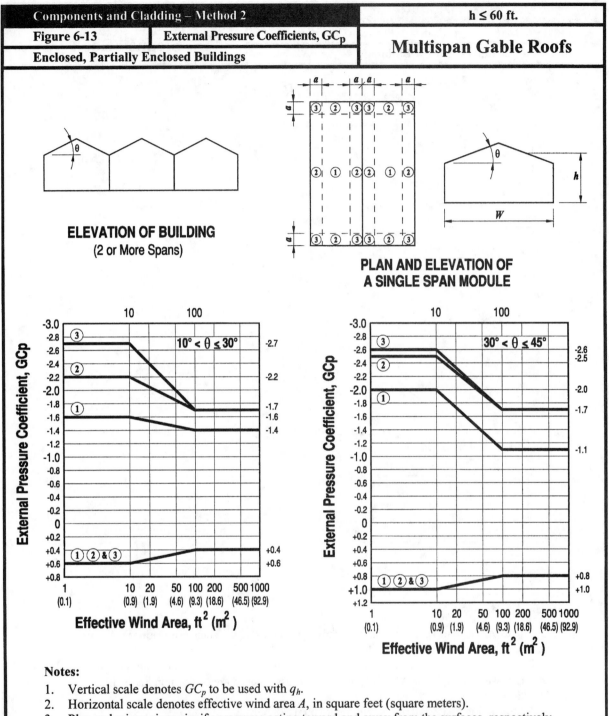

ELEVATION OF BUILDING
(2 or More Spans)

PLAN AND ELEVATION OF A SINGLE SPAN MODULE

Notes:

1. Vertical scale denotes GC_p to be used with q_h.
2. Horizontal scale denotes effective wind area A, in square feet (square meters).
3. Plus and minus signs signify pressures acting toward and away from the surfaces, respectively.
4. Each component shall be designed for maximum positive and negative pressures.
5. For $\theta \le 10°$, values of GC_p from Fig. 6-11 shall be used.
6. Notation:

   a: 10 percent of least horizontal dimension of a single-span module or $0.4h$, whichever is smaller, but not less than either 4 percent of least horizontal dimension of a single-span module or 3 ft (0.9 m).

   h: Mean roof height, in feet (meters), except that eave height shall be used for $\theta \le 10°$.

   W: Building module width, in feet (meters).

   θ: Angle of plane of roof from horizontal, in degrees.

ASCE 7-02

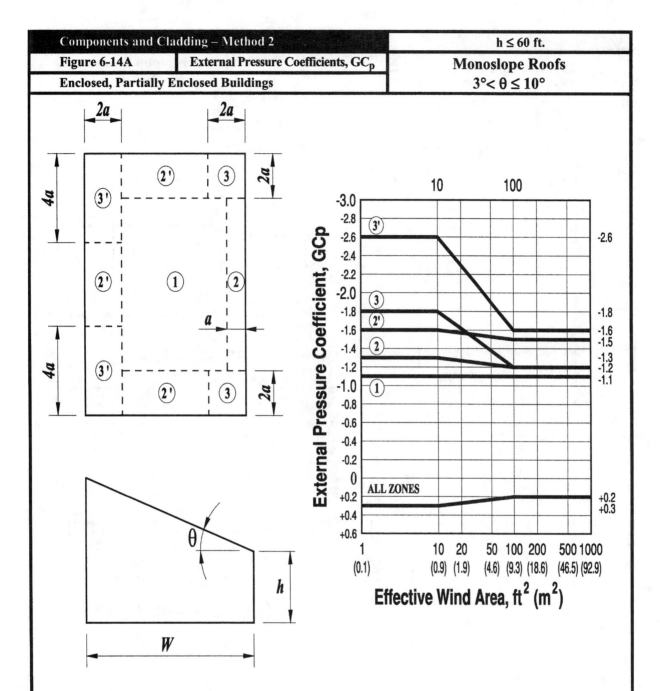

Notes:

1. Vertical scale denotes GC_p to be used with q_h.
2. Horizontal scale denotes effective wind area A, in square feet (square meters).
3. Plus and minus signs signify pressures acting toward and away from the surfaces, respectively.
4. Each component shall be designed for maximum positive and negative pressures.
5. For $θ ≤ 3°$, values of GC_p from Fig. 6-11B shall be used.
6. Notation:
   - a: 10 percent of least horizontal dimension or $0.4h$, whichever is smaller, but not less than either 4 percent of least horizontal dimension or 3 ft (0.9 m).
   - h: Eave height shall be used for $θ ≤ 10°$.
   - W: Building width, in feet (meters).
   - $θ$: Angle of plane of roof from horizontal, in degrees.

Figure 6-14B	External Pressure Coefficients, GC_p	Monoslope Roofs
Enclosed, Partially Enclosed Buildings		$10° < \theta \le 30°$

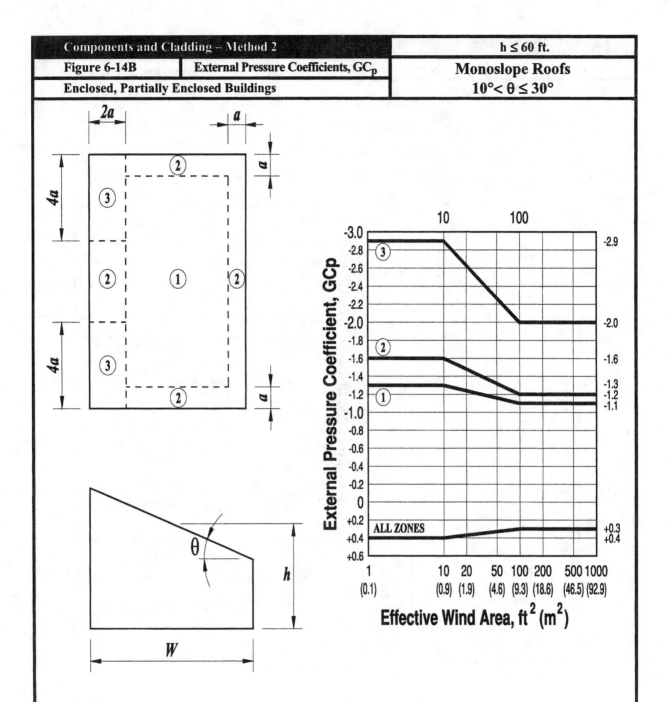

Notes:

1. Vertical scale denotes GC_p to be used with q_h.
2. Horizontal scale denotes effective wind area A, in square feet (square meters).
3. Plus and minus signs signify pressures acting toward and away from the surfaces, respectively.
4. Each component shall be designed for maximum positive and negative pressures.
5. Notation:

   a: 10 percent of least horizontal dimension or $0.4h$, whichever is smaller, but not less than either 4 percent of least horizontal dimension or 3 ft (0.9 m).

   h: Mean roof height, in feet (meters).

   W: Building width, in feet (meters).

   θ: Angle of plane of roof from horizontal, in degrees.

ASCE 7-02

| **Figure 6-15** | **External Pressure Coefficients, GC_p** | **Sawtooth Roofs** |
| **Enclosed, Partially Enclosed Buildings** | | |

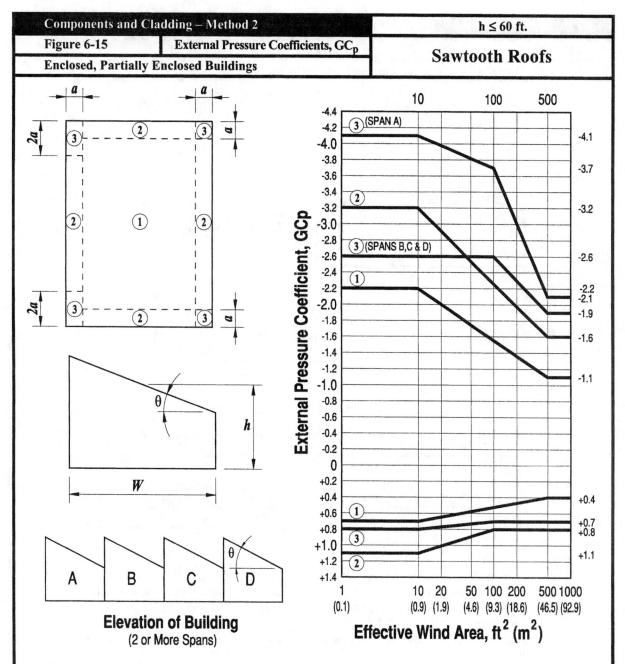

Elevation of Building
(2 or More Spans)

Effective Wind Area, ft² (m²)

Notes:

1. Vertical scale denotes GC_p to be used with q_h.
2. Horizontal scale denotes effective wind area A, in square feet (square meters).
3. Plus and minus signs signify pressures acting toward and away from the surfaces, respectively.
4. Each component shall be designed for maximum positive and negative pressures.
5. For $\theta \leq 10°$, values of GC_p from Fig. 6-11 shall be used.
6. Notation:
   - a: 10 percent of least horizontal dimension or $0.4h$, whichever is smaller, but not less than either 4 percent of least horizontal dimension or 3 ft (0.9 m).
   - h: Mean roof height, in feet (meters), except that eave height shall be used for $\theta \leq 10°$.
   - W: Building width, in feet (meters).
   - θ: Angle of plane of roof from horizontal, in degrees.

External Pressure Coefficients for Domes with a Circular Base			
	Negative Pressures	Positive Pressures	Positive Pressures
θ, degrees	0 – 90	0 – 60	61 – 90
GC$_p$	-0.9	+0.9	+0.5

Notes:

1. Values denote GC$_p$ to be used with $q_{(h_D+f)}$ where $h_D + f$ is the height at the top of the dome.
2. Plus and minus signs signify pressures acting toward and away from the surfaces, respectively.
3. Each component shall be designed for the maximum positive and negative pressures.
4. Values apply to $0 \leq h_D/D \leq 0.5$, $0.2 \leq f/D \leq 0.5$.
5. $\theta = 0$ degrees on dome springline, $\theta = 90$ degrees at dome center top point. f is measured from springline to top.

Figure 6-17 **External Pressure Coefficients, GCp**

Enclosed, Partially Enclosed Buildings

Walls & Roofs

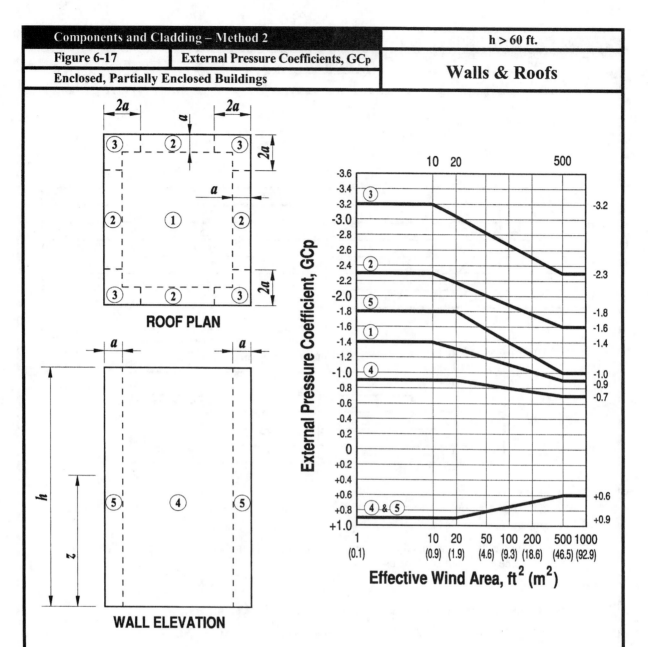

ROOF PLAN

WALL ELEVATION

Notes:

1. Vertical scale denotes GC_p to be used with appropriate q_z or q_h.
2. Horizontal scale denotes effective wind area A, in square feet (square meters).
3. Plus and minus signs signify pressures acting toward and away from the surfaces, respectively.
4. Use q_z with positive values of GC_p and q_h with negative values of GC_p.
5. Each component shall be designed for maximum positive and negative pressures.
6. Coefficients are for roofs with angle $\theta \leq 10°$. For other roof angles and geometry, use GC_p values from Fig. 6-11 and attendant q_h based on exposure defined in 6.5.6.
7. If a parapet equal to or higher than 3 ft (0.9m) is provided around the perimeter of the roof with $\theta \leq 10°$, Zone 3 shall be treated as Zone 2.
8. Notation:
   a: 10 percent of least horizontal dimension, but not less than 3 ft (0.9 m).
   h: Mean roof height, in feet (meters), except that eave height shall be used for $\theta \leq 10°$.
   z: height above ground, in feet (meters).
   θ: Angle of plane of roof from horizontal, in degrees.

Main Wind Force Resisting System – Method 2		All Heights
Figure 6-18	**Force Coefficients, C$_f$**	**Monoslope Roofs**
Open Buildings		

Roof angle θ degrees	L/B						
	5	**3**	**2**	**1**	**1/2**	**1/3**	**1/5**
10	0.2	0.25	0.3	0.45	0.55	0.7	0.75
15	0.35	0.45	0.5	0.7	0.85	0.9	0.85
20	0.5	0.6	0.75	0.9	1.0	0.95	0.9
25	0.7	0.8	0.95	1.15	1.1	1.05	0.95
30	0.9	1.0	1.2	1.3	1.2	1.1	1.0

Roof angle θ degrees	Center of Pressure X/L		
	L/B		
	2 to 5	**1**	**1/5 to 1/2**
10 to 20	0.35	0.3	0.3
25	0.35	0.35	0.4
30	0.35	0.4	0.45

Notes:

1. Wind forces act normal to the surface. Two cases shall be considered: (1) wind forces directed inward; and (2) wind forces directed outward.

2. The roof angle shall be assumed to vary ± 10° from the actual angle and the angle resulting in the greatest force coefficient shall be used.

3. Notation:

    B: dimension of roof measured normal to wind direction, in feet (meters);

    L: Dimension of roof measured parallel to wind direction, in feet (meters);

    X: Distance to center of pressure from windward edge of roof, in feet (meters); and

    θ: Angle of plane of roof from horizontal, in degrees.

Cross-Section	Type of Surface	h/D		
		1	7	25
Square (wind normal to face)	All	1.3	1.4	2.0
Square (wind along diagonal)	All	1.0	1.1	1.5
Hexagonal or octagonal	All	1.0	1.2	1.4
Round $(D\sqrt{q_z} > 2.5)$ $(D\sqrt{q_z} > 5.3, D \text{ in m}, q_z \text{ in N/m}^2)$	Moderately smooth	0.5	0.6	0.7
	Rough (D'/D = 0.02)	0.7	0.8	0.9
	Very rough (D'/D = 0.08)	0.8	1.0	1.2
Round $(D\sqrt{q_z} \leq 2.5)$ $(D\sqrt{q_z} \leq 5.3, D \text{ in m}, q_z \text{ in N/m}^2)$	All	0.7	0.8	1.2

Notes:

1. The design wind force shall be calculated based on the area of the structure projected on a plane normal to the wind direction. The force shall be assumed to act parallel to the wind direction.

2. Linear interpolation is permitted for *h/D* values other than shown.

3. Notation:

    D: diameter of circular cross-section and least horizontal dimension of square, hexagonal or octagonal cross-sections at elevation under consideration, in feet (meters);

    D': depth of protruding elements such as ribs and spoilers, in feet (meters); and

    h: height of structure, in feet (meters); and

    q_z: velocity pressure evaluated at height z above ground, in pounds per square foot (N/m^2).

At Ground Level		Above Ground Level	
v	C_f	M/N	C_f
≤3	1.2	≤6	1.2
5	1.3	10	1.3
8	1.4	16	1.4
10	1.5	20	1.5
20	1.75	40	1.75
30	1.85	60	1.85
≥40	2.0	≥80	2.0

Notes:

1. The term "signs" in notes below applies also to "freestanding walls".

2. Signs with openings comprising less than 30% of the gross area shall be considered as solid signs.

3. Signs for which the distance from the ground to the bottom edge is less than 0.25 times the vertical dimension shall be considered to be at ground level.

4. To allow for both normal and oblique wind directions, two cases shall be considered:

   a. resultant force acts normal to the face of the sign on a vertical line passing through the geometric center, and

   b. resultant force acts normal to the face of the sign at a distance from a vertical line passing through the geometric center equal to 0.2 times the average width of the sign.

5. Notation:

   v: ratio of height to width;

   M: larger dimension of sign, in feet (meters); and

   N: smaller dimension of sign, in feet (meters).

$\in$	Flat-Sided Members	Rounded Members	
		$D\sqrt{q_z} \leq 2.5$ $(D\sqrt{q_z} \leq 5.3)$	$D\sqrt{q_z} > 2.5$ $(D\sqrt{q_z} > 5.3)$
< 0.1	2.0	1.2	0.8
0.1 to 0.29	1.8	1.3	0.9
0.3 to 0.7	1.6	1.5	1.1

Notes:

1. Signs with openings comprising 30% or more of the gross area are classified as open signs.

2. The calculation of the design wind forces shall be based on the area of all exposed members and elements projected on a plane normal to the wind direction. Forces shall be assumed to act parallel to the wind direction.

3. The area A_f consistent with these force coefficients is the solid area projected normal to the wind direction.

4. Notation:

   $\in$: ratio of solid area to gross area;

   D: diameter of a typical round member, in feet (meters);

   q_z: velocity pressure evaluated at height z above ground in pounds per square foot (N/m²).

Tower Cross Section	C_f
Square	$4.0 \in^2 - 5.9 \in + 4.0$
Triangle	$3.4 \in^2 - 4.7 \in + 3.4$

Notes:

1. For all wind directions considered, the area A_f consistent with the specified force coefficients shall be the solid area of a tower face projected on the plane of that face for the tower segment under consideration.

2. The specified force coefficients are for towers with structural angles or similar flat-sided members.

3. For towers containing rounded members, it is acceptable to multiply the specified force coefficients by the following factor when determining wind forces on such members:

   $0.51 \in^2 + 0.57$, but not > 1.0

4. Wind forces shall be applied in the directions resulting in maximum member forces and reactions. For towers with square cross-sections, wind forces shall be multiplied by the following factor when the wind is directed along a tower diagonal:

   $1 + 0.75 \in$, but not > 1.2

5. Wind forces on tower appurtenances such as ladders, conduits, lights, elevators, etc., shall be calculated using appropriate force coefficients for these elements.

6. Loads due to ice accretion as described in Section 11 shall be accounted for.

7. Notation:

   $\in$: ratio of solid area to gross area of one tower face for the segment under consideration.

Category	Non-Hurricane Prone Regions and Hurricane Prone Regions with V = 85-100 mph and Alaska	Hurricane Prone Regions with V > 100 mph
I	0.87	0.77
II	1.00	1.00
III	1.15	1.15
IV	1.15	1.15

Note:

1. The building and structure classification categories are listed in Table 1-1.

Terrain Exposure Constants

Table 6-2

Exposure	α	z_g (ft)	$\hat{a}$	$\hat{b}$	$\bar{\alpha}$	$\bar{b}$	c	ℓ (ft)	$\bar{\in}$	z_{min} (ft)*
B	7.0	1200	1/7	0.84	1/4.0	0.45	0.30	320	1/3.0	30
C	9.5	900	1/9.5	1.00	1/6.5	0.65	0.20	500	1/5.0	15
D	11.5	700	1/11.5	1.07	1/9.0	0.80	0.15	650	1/8.0	7

*z_{min} = minimum height used to ensure that the equivalent height $\bar{z}$ is greater of $0.6h$ or z_{min}. For buildings with $h \leq z_{min}$, $\bar{z}$ shall be taken as z_{min}.

Height above ground level, z		Exposure (Note 1)			
		B		C	D
ft	(m)	Case 1	Case 2	Cases 1 & 2	Cases 1 & 2
0-15	(0-4.6)	0.70	0.57	0.85	1.03
20	(6.1)	0.70	0.62	0.90	1.08
25	(7.6)	0.70	0.66	0.94	1.12
30	(9.1)	0.70	0.70	0.98	1.16
40	(12.2)	0.76	0.76	1.04	1.22
50	(15.2)	0.81	0.81	1.09	1.27
60	(18)	0.85	0.85	1.13	1.31
70	(21.3)	0.89	0.89	1.17	1.34
80	(24.4)	0.93	0.93	1.21	1.38
90	(27.4)	0.96	0.96	1.24	1.40
100	(30.5)	0.99	0.99	1.26	1.43
120	(36.6)	1.04	1.04	1.31	1.48
140	(42.7)	1.09	1.09	1.36	1.52
160	(48.8)	1.13	1.13	1.39	1.55
180	(54.9)	1.17	1.17	1.43	1.58
200	(61.0)	1.20	1.20	1.46	1.61
250	(76.2)	1.28	1.28	1.53	1.68
300	(91.4)	1.35	1.35	1.59	1.73
350	(106.7)	1.41	1.41	1.64	1.78
400	(121.9)	1.47	1.47	1.69	1.82
450	(137.2)	1.52	1.52	1.73	1.86
500	(152.4)	1.56	1.56	1.77	1.89

Notes:

1. **Case 1:** a. All components and cladding.
   b. Main wind force resisting system in low-rise buildings designed using Figure 6-10.

   Case 2: a. All main wind force resisting systems in buildings except those in low-rise buildings designed using Figure 6-10.
   b. All main wind force resisting systems in other structures.

2. The velocity pressure exposure coefficient K_z may be determined from the following formula:

   For 15 ft. $\leq z \leq z_g$ For $z < 15$ ft.

   $K_z = 2.01 (z/z_g)^{2/\alpha}$ $K_z = 2.01 (15/z_g)^{2/\alpha}$

   Note: z shall not be taken less than 30 feet for Case 1 in exposure B.

3. α and z_g are tabulated in Table 6-2.

4. Linear interpolation for intermediate values of height z is acceptable.

5. Exposure categories are defined in 6.5.6.

Structure Type	Directionality Factor K_d*

Buildings
    Main Wind Force Resisting System
    Components and Cladding
	
	0.85
	0.85

Arched Roofs	0.85

Chimneys, Tanks, and Similar Structures
    Square
    Hexagonal
    Round
	
	0.90
	0.95
	0.95

Solid Signs	0.85
Open Signs and Lattice Framework	0.85

Trussed Towers
    Triangular, square, rectangular
    All other cross sections
	
	0.85
	0.95

*Directionality Factor K_d has been calibrated with combinations of loads specified in Section 2. This factor shall only be applied when used in conjunction with load combinations specified in 2.3 and 2.4.

SECTION 7.0
SNOW LOADS

SECTION 7.1
SYMBOLS AND NOTATIONS

β = gable roof drift parameter as determined from Eq. 7-3

C_e = exposure factor as determined from Table 7-2

C_s = slope factor as determined from Figure 7-2

C_t = thermal factor as determined from Table 7-3

h_b = height of balanced snow load determined by dividing p_f or p_s by γ, in ft (m)

h_c = clear height from top of balanced snow load to (1) closest point on adjacent upper roof, (2) top of parapet, or (3) top of a projection on the roof, in ft (m)

h_d = height of snow drift, in ft (m)

h_e = elevation difference between the ridge line and the eaves

h_o = height of obstruction above the surface of the roof, in ft (m)

I = importance factor as determined from Table 7-4

l_u = length of the roof upwind of the drift, in ft (m)

p_d = maximum intensity of drift surcharge load, in pounds per square ft (kn/m^2)

p_f = snow load on flat roofs ("flat" = roof slope $\leq 5°$), in lbs/ft^2 (kn/m^2)

p_g = ground snow load as determined from Figure 7-1 and Table 7-1; or a site-specific analysis, in lbs/ft^2 (kn/m^2)

p_s = sloped roof snow load, in pounds per square ft (kn/m^2)

s = separation distance between buildings, in ft (m)

θ = roof slope on the leeward side, in degrees

w = width of snow drift, in ft (m)

W = horizontal distance from eave to ridge, in ft (m)

γ = snow density, in pounds per cubic ft (kn/m^3) as determined from Eq. 7-4

SECTION 7.2
GROUND SNOW LOADS, p_g

Ground snow loads, p_g, to be used in the determination of design snow loads for roofs shall be as set forth in Figure 7-1 for the contiguous United States and Table 7-1 for Alaska. Site-specific case studies shall be made to determine ground snow loads in areas designated CS in Figure 7-1. Ground snow loads for sites at elevations above the limits indicated in Figure 7-1 and for all sites within the CS areas shall be approved by the authority having

jurisdiction. Ground snow load determination for such sites shall be based on an extreme-value statistical-analysis of data available in the vicinity of the site using a value with a 2% annual probability of being exceeded (50-year mean recurrence interval).

Snow loads are zero for Hawaii, except in mountainous regions as determined by the authority having jurisdiction.

SECTION 7.3
FLAT ROOF SNOW LOADS p_f

The snow load, p_f, on a roof with a slope equal to or less than 5° (1 in./ft = 4.76°) shall be calculated in lbs/ft^2 (kn/m^2) using the following formula:

$$p_f = 0.7 C_e C_t I p_g \qquad \textbf{(Eq. 7-1)}$$

but not less than the following minimum values for low-slope roofs as defined in Section 7.3.4:

where p_g is 20 lb/ft^2 (0.96 kN/m^2) or less,

$\qquad p_f = (I)p_g$ *(Importance factor times p_g)*

where p_g exceeds 20 lb/ft^2 (0.96 kN/m^2),

$\qquad p_f = 20(I)$ *(Importance factor times 20 lb/ft^2)*

7.3.1 Exposure Factor, C_e. The value for C_e shall be determined from Table 7-2.

7.3.2 Thermal Factor, C_t. The value for C_t shall be determined from Table 7-3.

7.3.3 Importance Factor, I. The value for I shall be determined from Table 7-4.

7.3.4 Minimum Values of p_f for Low-Slope Roofs. Minimum values of p_f shall apply to monoslope roofs with slopes less than 15 degrees, hip, and gable roofs with slopes less than or equal to $(70/W) + 0.5$ with W in ft (in SI: $21.3/W + 0.5$, with W in m), and curved roofs where the vertical angle from the eaves to the crown is less than 10 degrees.

SECTION 7.4
SLOPED ROOF SNOW LOADS, p_s

Snow loads acting on a sloping surface shall be assumed to act on the horizontal projection of that surface. The sloped

roof snow load, p_s, shall be obtained by multiplying the flat roof snow load, p_f, by the roof slope factor, C_s:

$$p_s = C_s p_f \qquad \text{(Eq. 7-2)}$$

Values of C_s for warm roofs, cold roofs, curved roofs, and multiple roofs are determined from Sections 7.4.1 through 7.4.4. The thermal factor, C_t, from Table 7-3 determines if a roof is "cold" or "warm." "Slippery surface" values shall be used only where the roof's surface is unobstructed and sufficient space is available below the eaves to accept all the sliding snow. A roof shall be considered unobstructed if no objects exist on it that prevent snow on it from sliding. Slippery surfaces shall include metal, slate, glass, and bituminous, rubber, and plastic membranes with a smooth surface. Membranes with an imbedded aggregate or mineral granule surface shall not be considered smooth. Asphalt shingles, wood shingles, and shakes shall not be considered slippery.

7.4.1 Warm Roof Slope Factor, C_s.
For warm roofs ($C_t \le 1.0$ as determined from Table 7-3) with an unobstructed slippery surface that will allow snow to slide off the eaves, the roof slope factor C_s shall be determined using the dashed line in Figure 7-2a, provided that for nonventilated warm roofs, their thermal resistance (R-value) equals or exceeds 30 ft^2·hr·°F/Btu (5.3 K·m^2/W) and for warm ventilated roofs, their R-value equals or exceeds 20 ft^2·hr·°F/Btu (3.5 K·m^2/W). Exterior air shall be able to circulate freely under a ventilated roof from its eaves to its ridge. For warm roofs that do not meet the aforementioned conditions, the solid line in Figure 7-2a shall be used to determine the roof slope factor C_s.

7.4.2 Cold Roof Slope Factor, C_s.
Cold roofs are those with a $C_t > 1.0$ as determined from Table 7-3. For cold roofs with $C_t = 1.1$ and an unobstructed slippery surface that will allow snow to slide off the eaves, the roof slope factor C_s shall be determined using the dashed line in Figure 7-2b. For all other cold roofs with $C_t = 1.1$, the solid line in Figure 7-2b shall be used to determine the roof slope factor C_s. For cold roofs with $C_t = 1.2$ and an unobstructed slippery surface that will allow snow to slide off the eaves, the roof slope factor C_s shall be determined using the dashed line on Figure 7-2c. For all other cold roofs with $C_t = 1.2$, the solid line in Figure 7-2c shall be used to determine the roof slope factor C_s.

7.4.3 Roof Slope Factor for Curved Roofs.
Portions of curved roofs having a slope exceeding 70 degrees shall be considered free of snow load, (i.e., $C_s = 0$). Balanced loads shall be determined from the balanced load diagrams in

Figure 7-3 with C_s determined from the appropriate curve in Figure 7-2.

7.4.4 Roof Slope Factor for Multiple Folded Plate, Sawtooth, and Barrel Vault Roofs.
Multiple folded plate, sawtooth, or barrel vault roofs shall have a $C_s = 1.0$, with no reduction in snow load because of slope (i.e., $p_s = p_f$).

7.4.5 Ice Dams and Icicles Along Eaves.
Two types of warm roofs that drain water over their eaves shall be capable of sustaining a uniformly distributed load of $2p_f$ on all overhanging portions there: those that are unventilated and have an R-value less than 30 ft^2·hr·°F/Btu (5.3 k·m^2/W) and those that are ventilated and have an R-value less than 20 ft^2·hr·°F/Btu (3.5 k·m^2/W). No other loads except dead loads shall be present on the roof when this uniformly distributed load is applied.

SECTION 7.5
PARTIAL LOADING

The effect of having selected spans loaded with the balanced snow load and remaining spans loaded with half the balanced snow load shall be investigated as follows:

7.5.1 Continuous Beam Systems.
Continuous beam systems shall be investigated for the effects of the three loadings shown in Figure 7-4:

Case 1: Full balanced snow load on either exterior span and half the balanced snow load on all other spans.

Case 2: Half the balanced snow load on either exterior span and full balanced snow load on all other spans.

Case 3: All possible combinations of full balanced snow load on any two adjacent spans and half the balanced snow load on all other spans. For this case there will be $(n-1)$ possible combinations where n equals the number of spans in the continuous beam system.

If a cantilever is present in any of the above cases, it shall be considered to be a span.

Partial load provisions need not be applied to structural members that span perpendicular to the ridge line in gable roofs with slopes greater than $70/W + 0.5$ with W in ft (in SI: $21.3/W + 0.5$, with W in m).

7.5.2 Other Structural Systems.
Areas sustaining only half the balanced snow load shall be chosen so as to produce the greatest effects on members being analyzed.

SECTION 7.6
UNBALANCED ROOF SNOW LOADS

Balanced and unbalanced loads shall be analyzed separately. Winds from all directions shall be accounted for when establishing unbalanced loads.

7.6.1 Unbalanced Snow Loads for Hip and Gable Roofs.

For hip and gable roofs with a slope exceeding $70°$ or with a slope less than $70/W + 0.5$ with W in ft (in SI: $21.3/W + 0.5$, with W in m), unbalanced snow loads are not required to be applied. For roofs with an eave to ridge distance, W, of 20 ft (6.1 m) or less, the structure shall be designed to resist an unbalanced uniform snow load on the leeward side equal to $1.5p_s/C_e$. For roofs with $W > 20$ ft (6.1 m), the structure shall be designed to resist an unbalanced uniform snow load on the leeward side equal to $1.2(1 + \beta/2)p_s/C_e$ with β given by Eq. 7-3.

$$\beta = \begin{cases} 1.0 & p_g \leq 20 \text{ lb/ft}^2 \\ 1.5 - 0.025p_g & 20 < p_g < 40 \text{ lb/ft}^2 \\ 0.5 & p_g \geq 40 \text{ lb/ft}^2 \end{cases}$$

(Eq. 7-3)

In SI:

$$\beta = \begin{cases} 1.0 & p_g \leq 0.97 \text{ kN/m}^2 \\ 1.5 - 0.52p_g & 0.97 < p_g < 1.93 \text{ kN/m}^2 \\ 0.5 & p_g \geq 1.93 \text{ kN/m}^2 \end{cases}$$

For the unbalanced situation with $W > 20$ ft (6.1 m), the windward side shall have a uniform load equal to $0.3p_s$. Balanced and unbalanced loading diagrams are presented in Figure 7-5.

7.6.2 Unbalanced Snow Loads for Curved Roofs.

Portions of curved roofs having a slope exceeding 70 degrees shall be considered free of snow load. If the slope of a straight line from the eaves (or the 70-degree point, if present) to the crown is less than 10 degrees or greater than 60 degrees, unbalanced snow loads shall not be taken into account.

Unbalanced loads shall be determined according to the loading diagrams in Figure 7-3. In all cases the windward side shall be considered free of snow. If the ground or another roof abuts a Case II or Case III (see Figure 7-3) curved roof at or within 3 ft (0.91 m) of its eaves, the snow load shall not be decreased between the 30-degree point and the eaves but shall remain constant at the 30-degree point value. This distribution is shown as a dashed line in Figure 7-3.

7.6.3 Unbalanced Snow Loads for Multiple Folded Plate, Sawtooth, and Barrel Vault Roofs.

Unbalanced loads shall be applied to folded plate, sawtooth, and barrel vaulted multiple roofs with a slope exceeding 3/8 in./ft (1.79 degrees). According to 7.4.4, $C_s = 1.0$ for such roofs, and the balanced snow load equals p_f. The unbalanced snow load shall increase from one-half the balanced load at the ridge or crown (i.e., $0.5p_f$) to two times the balanced load given in 7.4.4 divided by C_e at the valley (i.e., $2\ p_f/C_e$). Balanced and unbalanced loading diagrams for a sawtooth roof are presented in Figure 7-6. However, the snow surface above the valley shall not be at an elevation higher than the snow above the ridge. Snow depths shall be determined by dividing the snow load by the density of that snow from Eq. 7-4, which is in Section 7.7.1.

7.6.4 Unbalanced Snow Loads for Dome Roofs.

Unbalanced snow loads shall be applied to domes and similar rounded structures. Snow loads, determined in the same manner as for curved roofs in Section 7.6.2, shall be applied to the downwind 90-degree sector in plan view. At both edges of this sector, the load shall decrease linearly to zero over sectors of 22.5 degrees each. There shall be no snow load on the remaining 225-degree upwind sector.

SECTION 7.7
DRIFTS ON LOWER ROOFS
(AERODYNAMIC SHADE)

Roofs shall be designed to sustain localized loads from snow drifts that form in the wind shadow of (1) higher portions of the same structure and, (2) adjacent structures and terrain features.

7.7.1 Lower Roof of a Structure.

Snow that forms drifts comes from a higher roof or, with the wind from the opposite direction, from the roof on which the drift is located. These two kinds of drifts ("leeward" and "windward," respectively) are shown in Figure 7-7. The geometry of the surcharge load due to snow drifting shall be approximated by a triangle as shown in Figure 7-8. Drift loads shall be superimposed on the balanced snow load. If h_c/h_b is less than 0.2, drift loads are not required to be applied.

For leeward drifts, the drift height h_d shall be determined directly from Figure 7-9 using the length of the upper roof. For windward drifts, the drift height shall be determined by substituting the length of the lower roof for l_u in Figure 7-9 and using three-quarters of h_d as determined from Figure 7-9 as the drift height. The larger of these two heights shall be used in design. If this height is equal to or less than h_c, the drift width, w, shall equal $4h_d$ and the drift height shall equal h_d. If this height exceeds h_c, the drift width, w, shall equal $4h_d^2/h_c$ and the drift height shall equal h_c. However, the drift width, w, shall not be greater than $8h_c$. If the drift width, w, exceeds the width of the lower roof, the drift shall be truncated at the far edge of the roof, not reduced to zero there. The maximum intensity of the drift surcharge load, p_d, equals $h_d\gamma$ where snow density, γ, is defined in Eq. 7-4:

$$\gamma = 0.13p_g + 14 \text{ but not more than 30 pcf} \quad \textbf{(Eq. 7-4)}$$

(in SI: $\gamma = 0.426p_g + 2.2$ but not more than 4.7 kN/m^3)

This density shall also be used to determine h_b by dividing p_f (or p_s) by γ (in SI: also multiply by 102 to get the depth in m).

7.7.2 Adjacent Structures and Terrain Features. The requirements in Section 7.7.1 shall also be used to determine drift loads caused by a higher structure or terrain feature within 20 ft (6.1 m) of a roof. The separation distance, s, between the roof and adjacent structure or terrain feature shall reduce applied drift loads on the lower roof by the factor $(20 - s)/20$ where s is in ft $[(6.1 - s)/6.1$ where s is in m].

SECTION 7.8
ROOF PROJECTIONS

The method in Section 7.7.1 shall be used to calculate drift loads on all sides of roof projections and at parapet walls. The height of such drifts shall be taken as three-quarters the drift height from Figure 7-9 (i.e., 0.75 h_d) with l_u equal to the length of the roof upwind of the projection or parapet wall. If the side of a roof projection is less than 15 ft (4.6 m) long, a drift load is not required to be applied to that side.

SECTION 7.9
SLIDING SNOW

The load caused by snow sliding off a sloped roof onto a lower roof shall be determined for slippery upper roofs with slopes greater than $\frac{1}{4}$ on 12, and for other (i.e., non-slippery) upper roofs with slopes greater than 2 on 12. The total sliding load per unit length of eave shall be $0.4p_f W$, where W is the horizontal distance from the eave to ridge for the sloped upper roof. The sliding load shall be distributed uniformly on the lower roof over a distance of 15 ft from the upper roof eave. If the width of the lower roof is less than 15 ft, the sliding load shall be reduced proportionally.

The sliding snow load shall not be further reduced unless a portion of the snow on the upper roof is blocked from sliding onto the lower roof by snow already on the lower roof or is expected to slide clear of the lower roof.

Sliding loads shall be superimposed on the balanced snow load.

SECTION 7.10
RAIN-ON-SNOW SURCHARGE LOAD

For locations where p_g is 20 lb/ft² (0.96 kN/m²) or less but not zero, all roofs with a slope less than 1/2 in./ft (2.38°), shall have a 5 lb/ft² (0.24 kN/m²) rain-on-snow surcharge load applied to establish the design snow loads. Where the minimum flat roof design snow load from 7.3.4 exceeds p_f as determined by Eq. 7-1, the rain-on-snow surcharge load shall be reduced by the difference between these two values with a maximum reduction of 5 lb/ft² (0.24 kN/m²).

SECTION 7.11
PONDING INSTABILITY

Roofs shall be designed to preclude ponding instability. For roofs with a slope less than 1/4 in./ft (1.19°), roof deflections caused by full snow loads shall be investigated when determining the likelihood of ponding instability from rain-on-snow or from snow meltwater (see Section 8.4).

SECTION 7.12
EXISTING ROOFS

Existing roofs shall be evaluated for increased snow loads caused by additions or alterations. Owners or agents for owners of an existing lower roof shall be advised of the potential for increased snow loads where a higher roof is constructed within 20 ft (6.1 m). See footnote to Table 7-2 and Section 7.7.2.

This page intentionally left blank.

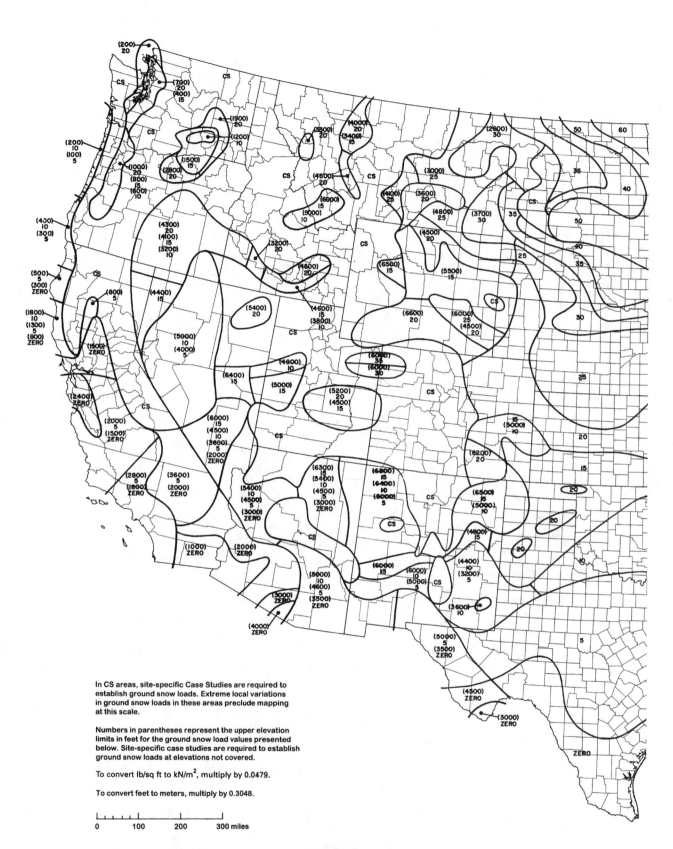

In CS areas, site-specific Case Studies are required to establish ground snow loads. Extreme local variations in ground snow loads in these areas preclude mapping at this scale.

Numbers in parentheses represent the upper elevation limits in feet for the ground snow load values presented below. Site-specific case studies are required to establish ground snow loads at elevations not covered.

To convert lb/sq ft to kN/m^2, multiply by 0.0479.

To convert feet to meters, multiply by 0.3048.

```
|   |   |   |   |
0   100  200  300 miles
```

FIGURE 7-1
GROUND SNOW LOADS, p_g FOR THE UNITED STATES (lB/SQ FT)

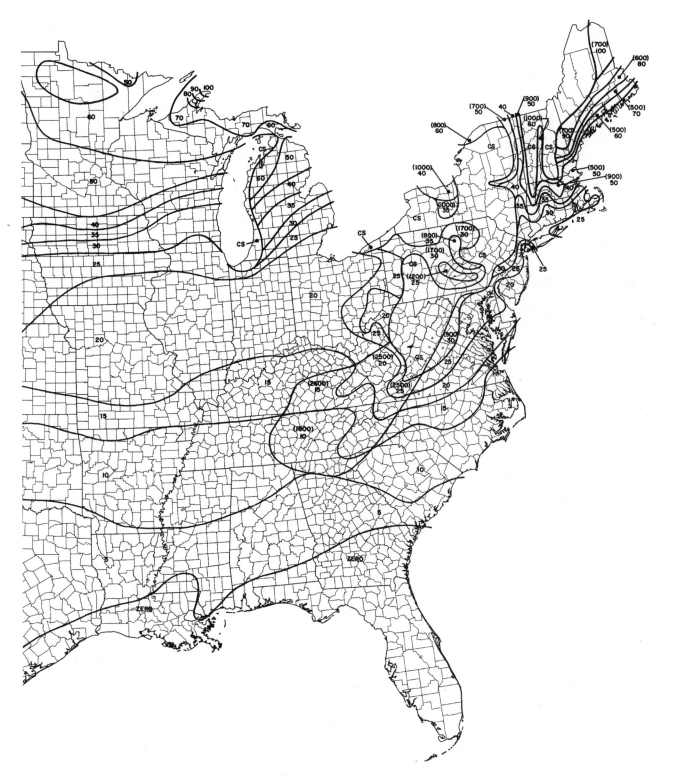

FIGURE 7-1 — continued
GROUND SNOW LOADS, p_g FOR THE UNITED STATES (IB/SQ FT)

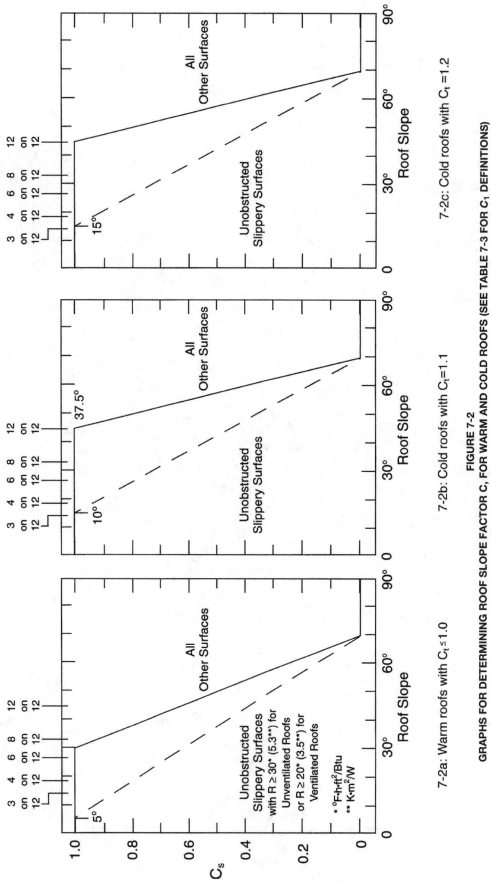

7-2a: Warm roofs with $C_t \leq 1.0$

7-2b: Cold roofs with $C_t = 1.1$

7-2c: Cold roofs with $C_t = 1.2$

FIGURE 7-2
GRAPHS FOR DETERMINING ROOF SLOPE FACTOR C_s FOR WARM AND COLD ROOFS (SEE TABLE 7-3 FOR C_t DEFINITIONS)

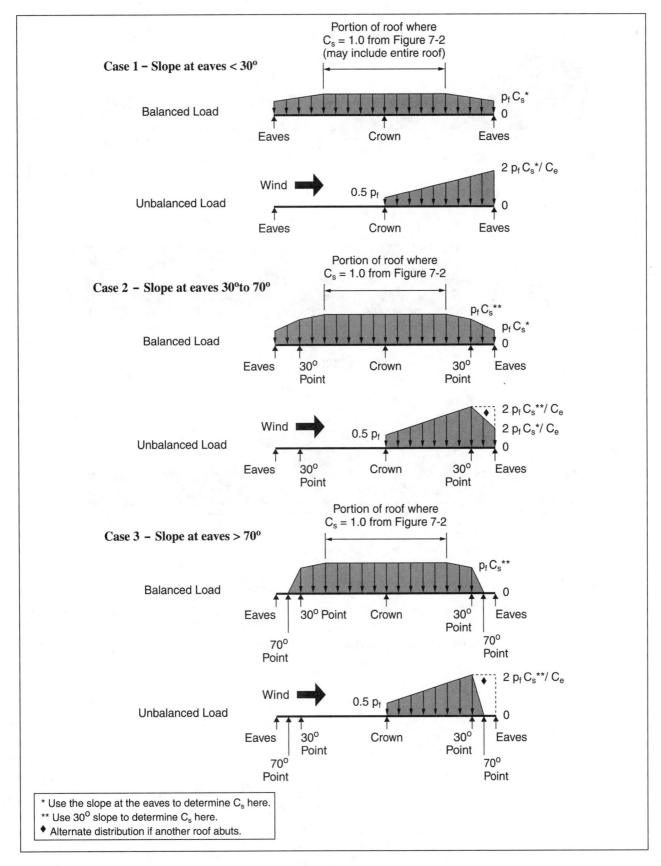

FIGURE 7-3
BALANCED AND UNBALANCED LOADS FOR CURVED ROOFS

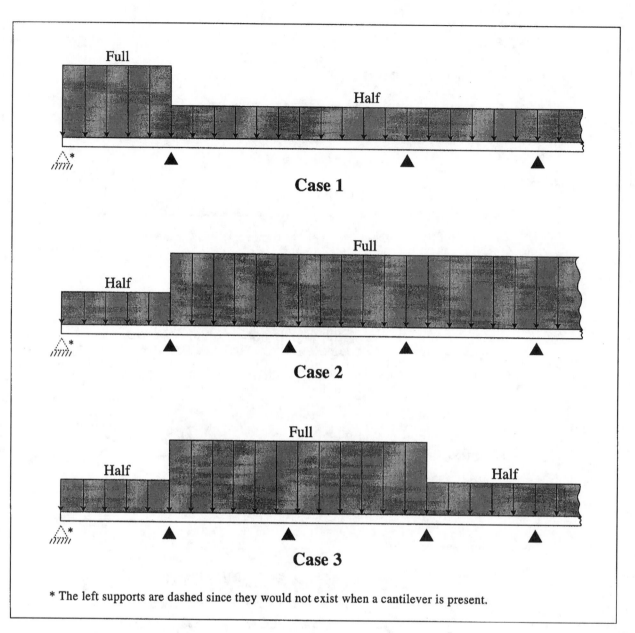

Full

Half

Case 1

Half

Full

Case 2

Full

Half

Half

Case 3

* The left supports are dashed since they would not exist when a cantilever is present.

FIGURE 7-4
PARTIAL LOADING DIAGRAMS FOR CONTINUOUS BEAMS

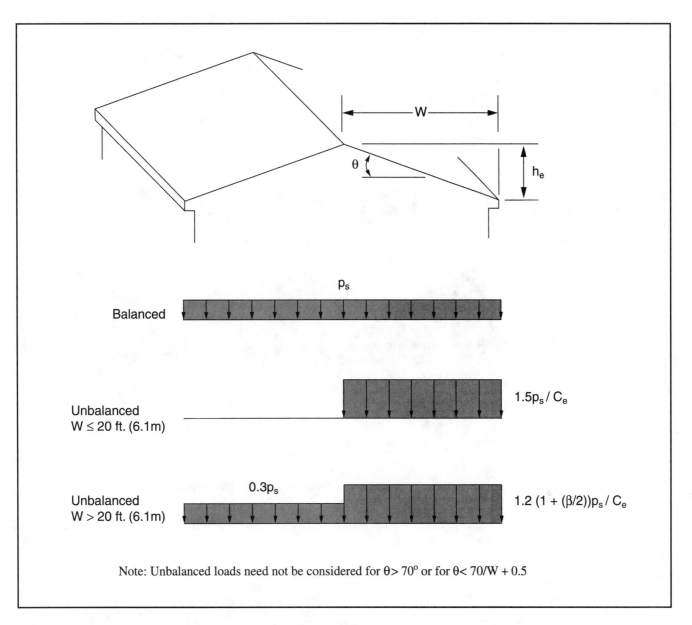

Note: Unbalanced loads need not be considered for $\theta > 70°$ or for $\theta < 70/W + 0.5$

FIGURE 7-5
BALANCED AND UNBALANCED SNOW LOADS FOR HIP AND GABLE ROOFS

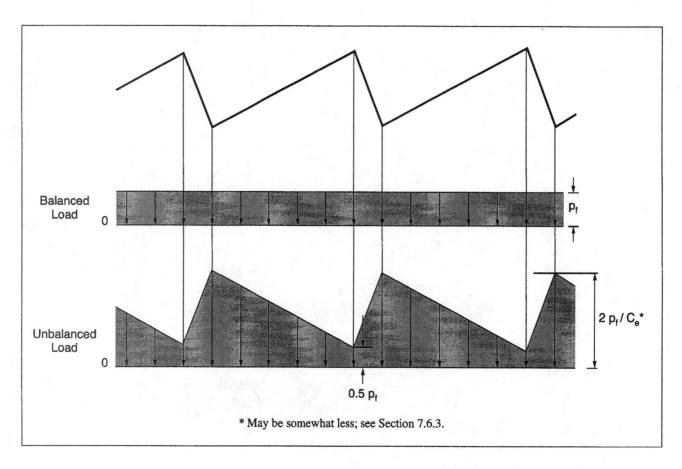

FIGURE 7-6
BALANCED AND UNBALANCED SNOW LOADS FOR A SAWTOOTH ROOF

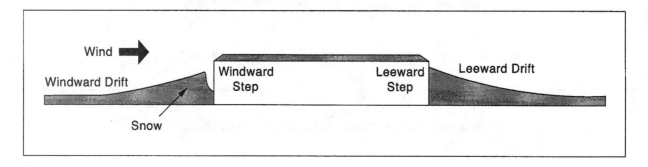

FIGURE 7-7
DRIFTS FORMED AT WINDWARD AND LEEWARD STEPS

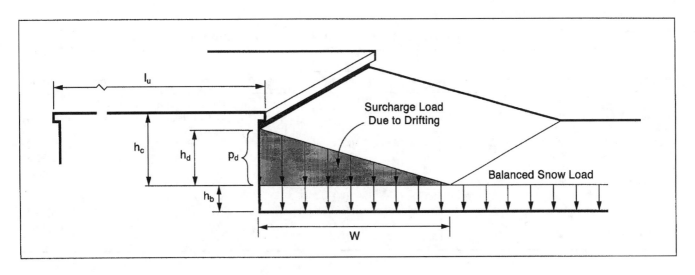

FIGURE 7-8
CONFIGURATION OF SNOW DRIFTS ON LOWER ROOFS

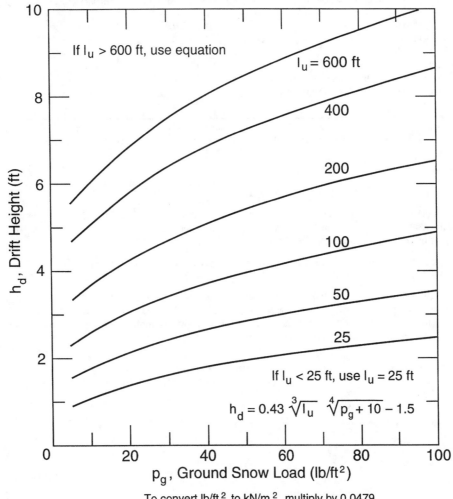

If l_u > 600 ft, use equation

l_u = 600 ft

400

200

100

50

25

If l_u < 25 ft, use l_u = 25 ft

$$h_d = 0.43 \sqrt[3]{l_u} \; \sqrt[4]{p_g + 10} - 1.5$$

To convert lb/ft^2 to kN/m^2, multiply by 0.0479.
To convert feet to meters, multiply by 0.3048.

FIGURE 7-9
GRAPH AND EQUATION FOR DETERMINING DRIFT HEIGHT, h_d

TABLE 7-1
GROUND SNOW LOADS, p_g, FOR ALASKAN LOCATIONS

Location	p_g lb/ft^2	(kN/m^2)	Location	p_g lb/ft^2	(kN/m^2)	Location	p_g lb/ft^2	(kN/m^2)
Adak	30	(1.4)	Galena	60	(2.9)	Petersburg	150	(7.2)
Anchorage	50	(2.4)	Gulkana	70	(3.4)	St Paul Islands	40	(1.9)
Angoon	70	(3.4)	Homer	40	(1.9)	Seward	50	(2.4)
Barrow	25	(1.2)	Juneau	60	(2.9)	Shemya	25	(1.2)
Barter Island	35	(1.7)	Kenai	70	(3.4)	Sitka	50	(2.4)
Bethel	40	(1.9)	Kodiak	30	(1.4)	Talkeetna	120	(5.8)
Big Delta	50	(2.4)	Kotzebue	60	(2.9)	Unalakleet	50	(2.4)
Cold Bay	25	(1.2)	McGrath	70	(3.4)	Valdez	160	(7.7)
Cordova	100	(4.8)	Nenana	80	(3.8)	Whittier	300	(14.4)
Fairbanks	60	(2.9)	Nome	70	(3.4)	Wrangell	60	(2.9)
Fort Yukon	60	(2.9)	Palmer	50	(2.4)	Yakutat	150	(7.2)

TABLE 7-2
EXPOSURE FACTOR, C_e

Terrain Category	Fully Exposed	Exposure of Roof* Partially Exposed	Sheltered
A (see Section 6.5.6)	N/A	1.1	1.3
B (see Section 6.5.6)	0.9	1.0	1.2
C (see Section 6.5.6)	0.9	1.0	1.1
D (see Section 6.5.6)	0.8	0.9	1.0
Above the treeline in windswept mountainous areas.	0.7	0.8	N/A
In Alaska, in areas where trees do not exist within a 2-mile (3 km) radius of the site.	0.7	0.8	N/A

The terrain category and roof exposure condition chosen shall be representative of the anticipated conditions during the life of the structure. An exposure factor shall be determined for each roof of a structure.

*Definitions

PARTIALLY EXPOSED. All roofs except as indicated below.

FULLY EXPOSED. Roofs exposed on all sides with no shelter** afforded by terrain, higher structures, or trees. Roofs that contain several large pieces of mechanical equipment, parapets that extend above the height of the balanced snow load (h_b), or other obstructions are not in this category.

SHELTERED. Roofs located tight in among conifers that qualify as obstructions.

**Obstructions within a distance of $10h_o$ provide "shelter," where h_o is the height of the obstruction above the roof level. If the only obstructions are a few deciduous trees that are leafless in winter, the "fully exposed" category shall be used except for terrain Category "A." Note that these are heights above the roof. Heights used to establish the terrain category in Section 6.5.3 are heights above the ground.

TABLE 7-3
THERMAL FACTOR, C_t

Thermal Condition*	C_t
All structures except as indicated below	1.0
Structures kept just above freezing and others with cold, ventilated roofs in which the thermal resistance (R-value) between the ventilated space and the heated space exceeds 25 F°·hr·sq ft/Btu (4.4 K·m²/W)	1.1
Unheated structures and structures intentionally kept below freezing	1.2
Continuously heated greenhouses** with a roof having a thermal resistance (R-value) less than 2.0 F°·hr·ft²/Btu(0.4 K·m²/W)	0.85

*These conditions shall be representative of the anticipated conditions during winters for the life of the structure.

**Greenhouses with a constantly maintained interior temperature of 50°F (10°C) or more at any point 3 ft above the floor level during winters and having either a maintenance attendant on duty at all times or a temperature alarm system to provide warning in the event of a heating failure.

TABLE 7-4
IMPORTANCE FACTOR,
I, (SNOW LOADS)

Category*	I
I	0.8
II	1.0
III	1.1
IV	1.2

*See Section 1.5 and Table 1-1.

SECTION 8.0
RAIN LOADS

SECTION 8.1
SYMBOLS AND NOTATION

R = rain load on the undeflected roof, in pounds per square ft (kilonewtons/m^2). When the phrase "undeflected roof" is used, deflections from loads (including dead loads) shall not be considered when determining the amount of rain on the roof.

d_s = depth of water on the undeflected roof up to the inlet of the secondary drainage system when the primary drainage system is blocked (i.e., the static head), in in. (mm).

d_h = additional depth of water on the undeflected roof above the inlet of the secondary drainage system at its design flow (i.e., the hydraulic head), in in. (mm).

SECTION 8.2
ROOF DRAINAGE

Roof drainage systems shall be designed in accordance with the provisions of the code having jurisdiction. The flow capacity of secondary (overflow) drains or scuppers shall not be less than that of the primary drains or scuppers.

SECTION 8.3
DESIGN RAIN LOADS

Each portion of a roof shall be designed to sustain the load of all rainwater that will accumulate on it if the primary drainage system for that portion is blocked plus the uniform load caused by water that rises above the inlet of the secondary drainage system at its design flow.

$$R = 5.2(d_s + d_h) \qquad \textbf{(Eq. 8-1)}$$

In SI: $R = 0.0098(d_s + d_h)$

If the secondary drainage systems contain drain lines, such lines and their point of discharge shall be separate from the primary drain lines.

SECTION 8.4
PONDING INSTABILITY

"Ponding" refers to the retention of water due solely to the deflection of relatively flat roofs. Roofs with a slope less than 1/4 in./ft (1.19 degrees) shall be investigated by structural analysis to ensure that they possess adequate stiffness to preclude progressive deflection (i.e., instability) as rain falls on them or meltwater is created from snow on them. The larger of snow load or rain load shall be used in this analysis. The primary drainage system within an area subjected to ponding shall be considered to be blocked in this analysis.

SECTION 8.5
CONTROLLED DRAINAGE

Roofs equipped with hardware to control the rate of drainage shall be equipped with a secondary drainage system at a higher elevation that limits accumulation of water on the roof above that elevation. Such roofs shall be designed to sustain the load of all rainwater that will accumulate on them to the elevation of the secondary drainage system, plus the uniform load caused by water that rises above the inlet of the secondary drainage system at its design flow (determined from Section 8.3).

Such roofs shall also be checked for ponding instability (determined from Section 8.4).

SECTION 9.0
EARTHQUAKE LOADS

[Note to user: This Section is based on the 2000 NEHRP Recommended Provisions for the Development of Seismic Regulations for New Buildings.]

SECTION 9.1
GENERAL PROVISIONS

9.1.1 Purpose. Section 9 presents criteria for the design and construction of buildings and similar structures subject to earthquake ground motions. The specified earthquake loads are based on post-elastic energy dissipation in the structure, and because of this fact, the provisions for design, detailing, and construction shall be satisfied even for structures and members for which load combinations that do not contain the earthquake effect indicate larger demands than combinations including earthquake.

9.1.2 Scope and Application.

9.1.2.1 Scope. Every building, and portion thereof, shall be designed and constructed to resist the effects of earthquake motions as prescribed by these provisions. Certain nonbuilding structures, as described in Section 9.14, are within the scope and shall be designed and constructed as required for buildings. Additions to existing structures also shall be designed and constructed to resist the effects of earthquake motions as prescribed by these provisions. Existing structures and alterations to existing structures need only comply with these provisions when required by Sections 9.1.2.2 and 9.1.2.3.

Exceptions:

1. Structures located where the mapped spectral response acceleration at 1-sec period, S_1, is less than or equal to 0.04 g and the mapped short period spectral response acceleration, S_S, is less than or equal to 0.15 g shall only be required to comply with Section 9.5.2.6.1.

2. Detached one- and two-family dwellings that are located where the mapped, short period, spectral response acceleration, S_S, is less than 0.4 g or where the Seismic Design Category determined in accordance with Section 9.4.2 is A, B, or C are exempt from the requirements of these provisions.

3. Detached one- and two-family wood frame dwellings not included in Exception 2 with

not more than 2 stories and satisfying the limitations of International Code Council (ICC) [Ref. 9.12-10] are only required to be constructed in accordance with Ref. 9.12-10.

4. Agricultural storage structures that are intended only for incidental human occupancy are exempt from the requirements of these provisions.

Special structures including, but not limited to, vehicular bridges, transmission towers, piers and wharves, hydraulic structures, and nuclear reactors require special consideration of their response characteristics and environment that is beyond the scope of these provisions.

9.1.2.2 Additions to Existing Structures. Additions shall be made to existing structures only as follows:

9.1.2.2.1 An addition that is structurally independent from an existing structure shall be designed and constructed in accordance with the seismic requirements for new structures.

9.1.2.2.2 An addition that is not structurally independent from an existing structure shall be designed and constructed such that the entire structure conforms to the seismic force-resistance requirements for new structures unless the following three conditions are complied with:

1. The addition shall comply with the requirements for new structures.

2. The addition shall not increase the seismic forces in any structural element of the existing structure by more than 5% unless the capacity of the element subject to the increased forces is still in compliance with these provisions.

3. The addition shall not decrease the seismic resistance of any structural element of the existing structure unless the reduced resistance is equal to or greater than that required for new structures.

9.1.2.3 Change of Use. When a change of use results in a structure being reclassified to a higher Seismic Use Group, the structure shall conform to the seismic requirements for new construction.

Exceptions:

1. When a change of use results in a structure being reclassified from Seismic Use Group I to Seismic Use Group II and the structure is located in a seismic map area where $S_{DS} < 0.33$, compliance with these provisions is not required.

2. Specific seismic detailing provisions of Appendix A required for a new structure are not required to be met when it can be shown that the level of performance and seismic safety is equivalent to that of a new structure. Such analysis shall consider the regularity, overstrength, redundancy, and ductility of the structure within the context of the existing and retrofit (if any) detailing provided.

9.1.2.4 Application of Provisions. Buildings and structures within the scope of these provisions shall be designed and constructed as required by this Section. When required by the authority having jurisdiction, design documents shall be submitted to determine compliance with these provisions.

9.1.2.4.1 New Buildings. New buildings and structures shall be designed and constructed in accordance with the quality assurance requirements of Section 9.4.3. The analysis and design of structural systems and components, including foundations, frames, walls, floors, and roofs shall be in accordance with the applicable requirements of Sections 9.5 and 9.7. Materials used in construction and components made of these materials shall be designed and constructed to meet the requirements of Sections 9.8 through 9.12. Architectural, electrical, and mechanical systems and components including tenant improvements shall be designed in accordance with Section 9.6.

9.1.2.5 Alternate Materials and Methods of Construction. Alternate materials and methods of construction to those prescribed in these provisions shall not be used unless approved by the authority having jurisdiction. Substantiating evidence shall be submitted demonstrating that the proposed alternate, for the purpose intended, will be at least equal in strength, durability, and seismic resistance.

9.1.3 Seismic Use Groups. All structures shall be assigned to Seismic Use Group: I, II, or III as specified in Table 9.1.3 corresponding to its Occupancy Category determined from Table 1-1:

9.1.3.1 High Hazard Exposure Structures. All buildings and structures assigned to Seismic Use Group III shall meet the following requirements:

TABLE 9.1.3
SEISMIC USE GROUP

		Seismic Use Group		
		I	II	III
Occupancy Category (Table 1-1)	I	X		
	II	X		
	III		X	
	IV			X

9.1.3.1.1 Seismic Use Group III Structure Protected Access. Where operational access to a Seismic Use Group III structure is required through an adjacent structure, the adjacent structure shall conform to the requirements for Seismic Use Group III structures. Where operational access is less than 10 ft from the interior lot line or another structure on the same lot, protection from potential falling debris from adjacent structures shall be provided by the owner of the Seismic Use Group III structure.

9.1.3.1.2 Seismic Use Group III Function. Designated seismic systems in Seismic Use Group III structures shall be provided with the capacity to function, in so far as practical, during and after an earthquake. Site-specific conditions as specified in Section 9.6.3.8 that could result in the interruption of utility services shall be considered when providing the capacity to continue to function.

9.1.3.2 This Section is intentionally left blank.

9.1.3.3 This Section is intentionally left blank.

9.1.3.4 Multiple Use. Structures having multiple uses shall be assigned the classification of the use having the highest Seismic Use Group except in structures having two or more portions which are structurally separated in accordance with Section 9.5.2.8, each portion shall be separately classified. Where a structurally separated portion of a structure provides access to, egress from, or shares life safety components with another portion having a higher Seismic Use Group, both portions shall be assigned the higher Seismic Use Group.

9.1.4 Occupancy Importance Factor. An occupancy importance factor, I, shall be assigned to each structure in accordance with Table 9.1.4.

9.2.1 Definitions. The definitions presented in this Section provide the meaning of the terms used in these provisions. Definitions of terms that have a specific meaning relative to

TABLE 9.1.4
OCCUPANCY IMPORTANCE FACTORS

Seismic Use Group	I
I	1.0
II	1.25
III	1.5

the use of wood, steel, concrete, or masonry are presented in the section devoted to the material (Sections A.9.8 through A.9.12, respectively).

ACTIVE FAULT. A fault determined to be active by the authority having jurisdiction from properly substantiated geotechnical data (e.g., most recent mapping of active faults by the U.S. Geological Survey).

ADDITION. An increase in building area, aggregate floor area, height, or number of stories of a structure.

ADJUSTED RESISTANCE (D'). The reference resistance adjusted to include the effects of all applicable adjustment factors resulting from end-use and other modifying factors. Time effect factor (λ) adjustments are not included.

ALTERATION. Any construction or renovation to an existing structure other than an addition.

APPENDAGE. An architectural component such as a canopy, marquee, ornamental balcony, or statuary.

APPROVAL. The written acceptance by the authority having jurisdiction of documentation that establishes the qualification of a material, system, component, procedure, or person to fulfill the requirements of these provisions for the intended use.

ARCHITECTURAL COMPONENT SUPPORT. Those structural members or assemblies of members, including braces, frames, struts, and attachments that transmit all loads and forces between architectural systems, components, or elements and the structure.

ATTACHMENTS. Means by which components and their supports are secured or connected to the seismic force-resisting system of the structure. Such attachments include anchor bolts, welded connections, and mechanical fasteners.

BASE. The level at which the horizontal seismic ground motions are considered to be imparted to the structure.

BASE SHEAR. Total design lateral force or shear at the base.

BASEMENT. A basement is any story below the lowest story above grade.

BOUNDARY ELEMENTS. Diaphragm and shear wall boundary members to which the diaphragm transfers forces. Boundary members include chords and drag struts at diaphragm and shear wall perimeters, interior openings, discontinuities, and re-entrant corners.

BOUNDARY MEMBERS. Portions along wall and diaphragm edges strengthened by longitudinal and transverse reinforcement. Boundary members include chords and drag struts at diaphragm and shear wall perimeters, interior openings, discontinuities, and re-entrant corners.

BUILDING. Any structure whose use could include shelter of human occupants.

CANTILEVERED COLUMN SYSTEM. A seismic force-resisting system in which lateral forces are resisted entirely by columns acting as cantilevers from the foundation.

COMPONENT. A part or element of an architectural, electrical, mechanical, or structural system.

> **Component, equipment.** A mechanical or electrical component or element that is part of a mechanical and/or electrical system within or without a building system.

> **Component, flexible.** Component, including its attachments, having a fundamental period greater than 0.06 sec.

> **Component, rigid.** Component, including its attachments, having a fundamental period less than or equal to 0.06 sec.

CONCRETE, PLAIN. Concrete that is either unreinforced or contains less reinforcement than the minimum amount specified in Ref. 9.9-1 for reinforced concrete.

CONCRETE, REINFORCED. Concrete reinforced with no less than the minimum amount required by Ref. 9.9-1, prestressed or nonprestressed, and designed on the assumption that the two materials act together in resisting forces.

CONFINED REGION. That portion of a reinforced concrete or reinforced masonry component in which the concrete or masonry is confined by closely spaced special transverse reinforcement restraining the concrete or masonry in directions perpendicular to the applied stress.

CONSTRUCTION DOCUMENTS. The written, graphic, electronic, and pictorial documents describing the design, locations, and physical characteristics of the project required to verify compliance with this Standard.

CONTAINER. A large-scale independent component used as a receptacle or vessel to accommodate plants, refuse, or similar uses, not including liquids.

COUPLING BEAM. A beam that is used to connect adjacent concrete wall elements to make them act together as a unit to resist lateral loads.

DEFORMABILITY. The ratio of the ultimate deformation to the limit deformation.

>   **High deformability element.** An element whose deformability is not less than 3.5 when subjected to four fully reversed cycles at the limit deformation.

>   **Limited deformability element.** An element that is neither a low deformability nor a high deformability element.

>   **Low deformability element.** An element whose deformability is 1.5 or less.

DEFORMATION.

>   **Limit deformation.** Two times the initial deformation that occurs at a load equal to 40% of the maximum strength.

>   **Ultimate deformation.** The deformation at which failure occurs and which shall be deemed to occur if the sustainable load reduces to 80% or less of the maximum strength.

DESIGN EARTHQUAKE GROUND MOTION. The earthquake effects that buildings and structures are specifically proportioned to resist as defined in Section 9.4.1.

DESIGN EARTHQUAKE. The earthquake effects that are two-thirds of the corresponding maximum considered earthquake.

DESIGNATED SEISMIC SYSTEMS. The seismic force-resisting system and those architectural, electrical, and mechanical systems or their components that require design in accordance with Section 9.6.1 and for which the component importance factor, I_p, is >1.0.

DIAPHRAGM. Roof, floor, or other membrane or bracing system acting to transfer the lateral forces to the vertical resisting elements. Diaphragms are classified as either flexible or rigid according to the requirements of Section 9.5.2.3.1.

DIAPHRAGM, BLOCKED. A diaphragm in which all sheathing edges not occurring on a framing member are supported on and fastened to blocking.

DIAPHRAGM BOUNDARY. A location where shear is transferred into or out of the diaphragm element. Transfer is either to a boundary element or to another force-resisting element.

DIAPHRAGM CHORD. A diaphragm boundary element perpendicular to the applied load that is assumed to take axial stresses due to the diaphragm moment in a manner analogous to the flanges of a beam. Also applies to shear walls.

DISPLACEMENT.

>   **Design displacement.** The design earthquake lateral displacement, excluding additional displacement due to actual and accidental torsion, required for design of the isolation system.

>   **Total design displacement.** The design earthquake lateral displacement, including additional displacement due to actual and accidental torsion, required for design of the isolation system or an element thereof.

>   **Total maximum displacement.** The maximum considered earthquake lateral displacement, including additional displacement due to actual and accidental torsion, required for verification of the stability of the isolation system or elements thereof, design of structure separations, and vertical load testing of isolator unit prototypes.

DISPLACEMENT RESTRAINT SYSTEM. A collection of structural elements that limits lateral displacement of seismically isolated structures due to the maximum considered earthquake.

DRAG STRUT (COLLECTOR, TIE, DIAPHRAGM STRUT). A diaphragm or shear wall boundary element parallel to the applied load that collects and transfers diaphragm shear forces to the vertical-force-resisting elements or distributes forces within the diaphragm or shear wall. A drag strut often is an extension of a boundary element that transfers forces into the diaphragm or shear wall. See Sections 9.5.2.6.3.1 and 9.5.2.6.4.1.

EFFECTIVE DAMPING. The value of equivalent viscous damping corresponding to energy dissipated during cyclic response of the isolation system.

EFFECTIVE STIFFNESS. The value of the lateral force in the isolation system, or an element thereof, divided by the corresponding lateral displacement.

ENCLOSURE. An interior space surrounded by walls.

EQUIPMENT SUPPORT. Those structural members or assemblies of members or manufactured elements, including braces, frames, legs, lugs, snuggers, hangers, or saddles, that transmit gravity loads and operating loads between the equipment and the structure.

ESSENTIAL FACILITY. A structure required for post-earthquake recovery.

FACTORED RESISTANCE ($\lambda \phi D$). Reference resistance multiplied by the time effect and resistance factors. This value must be adjusted for other factors such as size effects, moisture conditions, and other end-use factors.

FLEXIBLE EQUIPMENT CONNECTIONS. Those connections between equipment components that permit rotational and/or translational movement without degradation of performance. Examples include universal joints, bellows expansion joints, and flexible metal hose.

FRAME.

Braced frame. An essentially vertical truss, or its equivalent, of the concentric or eccentric type that is provided in a bearing wall, building frame, or dual system to resist seismic forces.

Concentrically braced frame (CBF). A braced frame in which the members are subjected primarily to axial forces.

Eccentrically braced frame (EBF). A diagonally braced frame in which at least one end of each brace frames into a beam a short distance from a beam-column joint or from another diagonal brace.

Ordinary concentrically braced frame (OCBF). A steel concentrically braced frame in which members and connections are designed in accordance with the provisions of Ref. 9.8-3 without modification.

Special concentrically braced frame (SCBF). A steel or composite steel and concrete concentrically braced frame in which members and connections are designed for ductile behavior. Special concentrically braced frames shall conform to Section A.9.8.1.3.1.

Moment frame.

Intermediate moment frame (IMF). A moment frame in which members and joints are capable of resisting forces by flexure as well as along the axis of the members. Intermediate moment frames of reinforced concrete shall conform to Ref. 9.9-1. Intermediate moment frames of structural steel construction shall conform to Section 9.8-3 or 10. Intermediate moment frames of composite construction shall conform to Ref. 9.10-3, Part II, Section 6.4b, 7, 8, and 10.

Ordinary moment frame (OMF). A moment frame in which members and joints are capable of resisting forces by flexure as well as along the axis of the members. Ordinary moment frames shall conform to Ref. 9.9-1, exclusive of Chapter 21, and Ref. 9.8-3, Section 12 or A.9.9.3.1.

Special moment frame (SMF). A moment frame in which members and joints are capable of resisting forces by flexure as well as along the axis of the members. Special moment frames shall conform to Ref. 9.8-3 or Ref. 9.9-1.

FRAME SYSTEM.

Building frame system. A structural system with an essentially complete space frame providing support for vertical loads. Seismic force resistance is provided by shear walls or braced frames.

Dual frame system. A structural system with an essentially complete space frame providing support for vertical loads. Seismic force resistance is provided by moment-resisting frames and shear walls or braced frames as prescribed in Section 9.5.2.2.1.

Space frame system. A structural system composed of interconnected members, other than bearing walls, that is capable of supporting vertical loads and, when designed for such an application, is capable of providing resistance to seismic forces.

GLAZED CURTAIN WALL. A nonbearing wall that extends beyond the edges of building floor slabs and includes a glazing material installed in the curtain wall framing.

GLAZED STOREFRONT. A nonbearing wall that is installed between floor slabs, typically including entrances, and includes a glazing material installed in the storefront framing.

GRADE PLANE. A reference plane representing the average of finished ground level adjoining the structure at all exterior walls. Where the finished ground level slopes away from the exterior walls, the reference plane shall be established by the lowest points within the area between the buildings and the lot line or, where the lot line is more than 6 ft (1829 mm) from the structure, between the structure and a point 6 ft (1829 mm) from the structure.

HAZARDOUS CONTENTS. A material that is highly toxic or potentially explosive and in sufficient quantity to pose a significant life safety threat to the general public if an uncontrolled release were to occur.

HIGH TEMPERATURE ENERGY SOURCE. A fluid, gas, or vapor whose temperature exceeds 220 °F.

INSPECTION, SPECIAL. The observation of the work by the special inspector to determine compliance with the approved construction documents and these standards.

Continuous special inspection. The full-time observation of the work by an approved special inspector who is present in the area where work is being performed.

Periodic special inspection. The part-time or intermittent observation of the work by an approved special inspector who is present in the area where work has been or is being performed.

INSPECTOR, SPECIAL (WHO SHALL BE IDENTIFIED AS THE OWNER'S INSPECTOR). A person

approved by the authority having jurisdiction to perform special inspection. The authority having jurisdiction shall have the option to approve the quality assurance personnel of a fabricator as a special inspector.

INVERTED PENDULUM-TYPE STRUCTURES. Structures that have a large portion of their mass concentrated near the top and, thus, have essentially one degree of freedom in horizontal translation. The structures are usually T-shaped with a single column supporting the beams or framing at the top.

ISOLATION INTERFACE. The boundary between the upper portion of the structure, which is isolated, and the lower portion of the structure, which moves rigidly with the ground.

ISOLATION SYSTEM. The collection of structural elements that includes all individual isolator units, all structural elements that transfer force between elements of the isolation system, and all connections to other structural elements. The isolation system also includes the wind-restraint system, energy-dissipation devices, and/or the displacement restraint system if such systems and devices are used to meet the design requirements of Section 9.13.

ISOLATOR UNIT. A horizontally flexible and vertically stiff structural element of the isolation system that permits large lateral deformations under design seismic load. An isolator unit may be used either as part of or in addition to the weight-supporting system of the structure.

JOINT. The geometric volume common to intersecting members.

LIGHT-FRAME CONSTRUCTION. A method of construction where the structural assemblies (e.g., walls, floors, ceilings, and roofs) are primarily formed by a system of repetitive wood or cold-formed steel framing members or subassemblies of these members (e.g., trusses).

LOAD.

   Dead load. The gravity load due to the weight of all permanent structural and nonstructural components of a building such as walls, floors, roofs, and the operating weight of fixed service equipment.

   Gravity load (W). The total dead load and applicable portions of other loads as defined in Section 9.5.3.

   Live load. The load superimposed by the use and occupancy of the building not including the wind load, earthquake load, or dead load, see Section 9.5.3.

MAXIMUM CONSIDERED EARTHQUAKE GROUND MOTION. The most severe earthquake effects considered by these standards as defined in Section 9.4.1.

NONBUILDING STRUCTURE. A structure, other than a building, constructed of a type included in Section 9.14 and within the limits of Section 9.14.1.1.

OCCUPANCY IMPORTANCE FACTOR. A factor assigned to each structure according to its Seismic Use Group as prescribed in Section 9.1.4.

OWNER. Any person, agent, firm, or corporation having a legal or equitable interest in the property.

PARTITION. A nonstructural interior wall that spans horizontally or vertically from support to support. The supports may be the basic building frame, subsidiary structural members, or other portions of the partition system.

***P*-DELTA EFFECT.** The secondary effect on shears and moments of structural members due to the action of the vertical loads induced by displacement of the structure resulting from various loading conditions.

QUALITY ASSURANCE PLAN. A detailed written procedure that establishes the systems and components subject to special inspection and testing. The type and frequency of testing and the extent and duration of special inspection are given in the quality assurance plan.

REFERENCE RESISTANCE (D). The resistance (force or moment as appropriate) of a member or connection computed at the reference end-use conditions.

REGISTERED DESIGN PROFESSIONAL. An architect or engineer, registered or licensed to practice professional architecture or engineering, as defined by the statutory requirements of the professional registrations laws of the state in which the project is to be constructed.

ROOFING UNIT. A unit of roofing tile or similar material weighing more than 1 pound.

SEISMIC DESIGN CATEGORY. A classification assigned to a structure based on its Seismic Use Group and the severity of the design earthquake ground motion at the site as defined in Section 9.4.2.

SEISMIC FORCE-RESISTING SYSTEM. That part of the structural system that has been considered in the design to provide the required resistance to the seismic forces prescribed herein.

SEISMIC FORCES. The assumed forces prescribed herein, related to the response of the structure to earthquake motions, to be used in the design of the structure and its components.

SEISMIC RESPONSE COEFFICIENT. Coefficient C_s as determined from Section 9.5.5.2.1.

SEISMIC USE GROUP. A classification assigned to a structure based on its use as defined in Section 9.1.3.

SHALLOW ANCHOR. Anchors with embedment length-to-diameter ratios of less than 8.

SHEAR PANEL. A floor, roof, or wall component sheathed to act as a shear wall or diaphragm.

SITE CLASS. A classification assigned to a site based on the types of soils present and their engineering properties as defined in Section 9.4.1.2.

SITE COEFFICIENTS. The values of F_a and F_v as indicated in Tables 9.4.1.2.4a and 9.4.1.2.4b, respectively.

SPECIAL TRANSVERSE REINFORCEMENT. Reinforcement composed of spirals, closed stirrups, or hoops and supplementary crossties provided to restrain the concrete and qualify the portion of the component, where used, as a confined region.

STORAGE RACKS. Include industrial pallet racks, moveable shelf racks, and stacker racks made of cold-formed or hot-rolled structural members. Does not include other types of racks such as drive-in and drive-through racks, cantilever racks, portable racks, or racks made of materials other than steel.

STORY. The portion of a structure between the top of two successive, finished floor surfaces and, for the topmost story, from the top of the floor finish to the top of the roof structural element.

STORY ABOVE GRADE. Any story having its finished floor surface entirely above grade, except that a story shall be considered as a story above grade where the finished floor surface of the story immediately above is more than 6 ft (1829 mm) above the grade plane, more than 6 ft (1829 mm) above the finished ground level for more than 40% of the total structure perimeter, or more than 12 ft (3658 mm) above the finished ground level at any point.

STORY DRIFT. The difference of horizontal deflections at the top and bottom of the story as determined in Section 9.5.5.7.1.

STORY DRIFT RATIO. The story drift, as determined in Section 9.5.5.7.1, divided by the story height.

STORY SHEAR. The summation of design lateral seismic forces at levels above the story under consideration.

STRENGTH.

 Design strength. Nominal strength multiplied by a strength reduction factor, ϕ.

 Nominal strength. Strength of a member or cross-section calculated in accordance with the requirements and assumptions of the strength design methods of this Standard (or the referenced standards) before application of any strength reduction factors.

Required strength. Strength of a member, cross-section, or connection required to resist factored loads or related internal moments and forces in such combinations as stipulated by this Standard.

STRUCTURE. That which is built or constructed and limited to buildings and nonbuilding structures as defined herein.

STRUCTURAL OBSERVATIONS. The visual observations performed by the registered design professional in responsible charge (or another registered design professional) to determine that the seismic force-resisting system is constructed in general conformance with the construction documents.

STRUCTURAL-USE PANEL. A wood-based panel product that meets the requirements of Ref. 9.12-3 or Ref. 9.12-5 and is bonded with a waterproof adhesive. Included under this designation are plywood, oriented strand board, and composite panels.

SUBDIAPHRAGM. A portion of a diaphragm used to transfer wall anchorage forces to diaphragm crossties.

TESTING AGENCY. A company or corporation that provides testing and/or inspection services. The person in charge of the special inspector(s) and the testing services shall be a registered design professional.

TIE-DOWN (HOLD-DOWN). A device used to resist uplift of the boundary elements of shear walls. These devices are intended to resist load without significant slip between the device and the shear wall boundary element or be shown with cyclic testing to not reduce the wall capacity or ductility.

TIME EFFECT FACTOR (λ). A factor applied to the adjusted resistance to account for effects of duration of load.

TORSIONAL FORCE DISTRIBUTION. The distribution of horizontal shear through a rigid diaphragm when the center of mass of the structure at the level under consideration does not coincide with the center of rigidity (sometimes referred to as diaphragm rotation).

TOUGHNESS. The ability of a material to absorb energy without losing significant strength.

UTILITY OR SERVICE INTERFACE. The connection of the structure's mechanical and electrical distribution systems to the utility or service company's distribution system.

VENEERS. Facings or ornamentation of brick, concrete, stone, tile, or similar materials attached to a backing.

WALL. A component that has a slope of 60 degrees or greater with the horizontal plane used to enclose or divide space.

Bearing wall. Any wall meeting either of the following classifications:

1. Any metal or wood stud wall that supports more than 100 pounds per linear ft (1459 N/m) of vertical load in addition to its own weight.

2. Any concrete or masonry wall that supports more than 200 pounds per linear ft (2919 N/m) of vertical load in addition to its own weight.

Cripple wall. Short stud wall, less than 8 ft (2400 mm) in height, between the foundation and the lowest framed floors with studs not less than 14 inches long — also known as a knee wall. Cripple walls can occur in both engineered structures and conventional construction.

Light-framed wall. A wall with wood or steel studs.

Light-framed wood shear wall. A wall constructed with wood studs and sheathed with material rated for shear resistance.

Nonbearing wall. Any wall that is not a bearing wall.

Nonstructural wall. All walls other than bearing walls or shear walls.

Shear wall (vertical diaphragm). A wall, bearing or nonbearing, designed to resist lateral seismic forces acting in the plane of the wall (sometimes referred to as a vertical diaphragm).

WALL SYSTEM, BEARING. A structural system with bearing walls providing support for all or major portions of the vertical loads. Shear walls or braced frames provide seismic force resistance.

WIND-RESTRAINT SYSTEM. The collection of structural elements that provides restraint of the seismic-isolated structure for wind loads. The wind-restraint system may be either an integral part of isolator units or a separate device.

9.2.2 Symbols. The unit dimensions used with the items covered by the symbols shall be consistent throughout except where specifically noted. The symbols and definitions presented in this Section apply to these provisions as indicated.

A, B, C, D, E, F = the Seismic Performance Categories as defined in Tables 9.4.2.1a and 9.4.2.1b

A, B, C, D, E, F = the Site Classes as defined in Section 9.4.1.2.1

A_{ch} = cross-sectional area (in.^{2} or mm^2) of a component measured to the outside of the special lateral reinforcement

A_0 = the area of the load-carrying foundation (ft^2 or m^2)

A_{sh} = total cross-sectional area of hoop reinforcement (in.^{2} or mm^2), including supplementary crossties, having a spacing of s_h and crossing a section with a core dimension of h_c

A_{vd} = required area of leg (in.^{2} or mm^2) of diagonal reinforcement

A_x = the torsional amplification factor, Section 9.5.5.5.2

a_d = the incremental factor related to P-delta effects in Section 9.5.5.7.2

a_p = the amplification factor related to the response of a system or component as affected by the type of seismic attachment, determined in Section 9.6.1.3

B_D = numerical coefficient as set forth in Table 9.13.3.3.1 for effective damping equal to β_D

B_M = numerical coefficient as set forth in Table 9.13.3.3.1 for effective damping equal to β_M

b = the shortest plan dimension of the structure, in ft (mm) measured perpendicular to d

C_d = the deflection amplification factor as given in Table 9.5.2.2

C_s = the seismic response coefficient determined in Sections 9.5.5.2.1 and 9.5.9.3.1

C_{sm} = the modal seismic response coefficient determined in Section 9.5.6.5 (dimensionless)

C_T = the building period coefficient in Section 9.5.3.3

C_{vx} = the vertical distribution factor as determined in Section 9.5.5.4

c = distance from the neutral axis of a flexural member to the fiber of maximum compressive strain (in. or mm)

D = the effect of dead load

D_D = design displacement, in in. (mm), at the center of rigidity of the isolation system in the direction under consideration as prescribed by Eq. 9.13.3.3.1

D_D' = design displacement, in in. (mm), at the center of rigidity of the isolation system in the direction under consideration, as prescribed by Eq. 9.13.4.2-1

D_M = maximum displacement, in in. (mm), at the center of rigidity of the isolation system in the direction under consideration, as prescribed by Eq. 9.13.3.3.3

D_M' = maximum displacement, in in. (mm), at the center of rigidity of the isolation system in the direction under consideration, as prescribed by Eq. 9.13.4.2-2

D_p = relative seismic displacement that the component must be designed to accommodate as defined in Section 9.6.1.4

D_s = the total depth of the stratum in Eq. 9.5.9.2.1.2-4 (ft or m)

D_{TD} = total design displacement, in in. (mm), of an element of the isolation system including both translational displacement at the center of rigidity and the component of torsional displacement in the direction under consideration as prescribed by Eq. 9.13.3.3.5-1

D_{TM} = total maximum displacement, in in. (mm), of an element of the isolation system including both translational displacement at the center of rigidity and the component of torsional displacement in the direction under consideration as prescribed by Eq. 9.13.3.3.5-2

d = overall depth of member (in. or mm) in Section 9.5

d = the longest plan dimension of the structure, in ft (mm), in Section 9.13

d_p = the longest plan dimension of the structure, in ft (mm)

E = the effect of horizontal and vertical earthquake-induced forces, in Sections 9.5.2.7 and 9.13

E_{loop} = energy dissipated in kip-in. (kN-mm), in an isolator unit during a full cycle of reversible load over a test displacement range from Δ^+ to Δ^-, as measured by the area enclosed by the loop of the force-deflection curve

e = the actual eccentricity, in ft (mm), measured in plan between the center of mass of the structure above the isolation interface and the center of rigidity of the isolation system, plus accidental eccentricity, in ft (mm), taken as 5% of the maximum building dimension perpendicular to the direction of force under consideration

F_a = acceleration-based site coefficient (at 0.3-sec period)

F^- = maximum negative force in an isolator unit during a single cycle of prototype testing at a displacement amplitude of Δ^-

F^+ = positive force in kips (kN) in an isolator unit during a single cycle of prototype testing at a displacement amplitude of Δ^+

F_i, F_n, F_x = the portion of the seismic base shear, V, induced at Level i, n, or x, respectively, as determined in Section 9.5.5.4

F_p = the seismic force acting on a component of a structure as determined in Section 9.6.1.3

F_v = velocity-based site coefficient (at 1.0-sec period)

F_x = total force distributed over the height of the structure above the isolation interface as prescribed by Eq. 9.13.3.5

F_{xm} = the portion of the seismic base shear, V_m, induced at Level x as determined in Section 9.5.5.6

f_c' = specified compressive strength of concrete used in design

f_s' = ultimate tensile strength (psi or MPa) of the bolt, stud, or insert leg wires. For A307 bolts or A108 studs, it is permitted to be assumed to be 60,000 psi (415 MPa).

f_y = specified yield strength of reinforcement (psi or MPa)

f_{yh} = specified yield stress of the special lateral reinforcement (psi or kPa)

$G = \gamma v_s^2/g$ = the average shear modulus for the soils beneath the foundation at large strain levels (psf or Pa)

$G_0 = \gamma v_{s0}^2/g$ = the average shear modulus for the soils beneath the foundation at small strain levels (psf or Pa)

g = the acceleration due to gravity

H = thickness of soil

h = the height of a shear wall measured as the maximum clear height from top of foundation to bottom of diaphragm framing above, or the maximum clear height from top of diaphragm to bottom of diaphragm framing above

$\bar{h}$ = the effective height of the building as determined in Section 9.5.9.2 or 9.5.9.3 (ft or m)

h_c = the core dimension of a component measured to the outside of the special lateral reinforcement (in. or mm)

h_i, h_n, h_x = the height above the base Level i, n, or x, respectively

h_{sx} = the story height below Level $x = (h_x - h_{x-1})$

I = the occupancy importance factor in Section 9.1.4

I_0 = the static moment of inertia of the load-carrying foundation, see Section 9.5.9.2.1 (in.^{4} or mm^4)

I_p = the component importance factor as prescribed in Section 9.6.1.5

i = the building level referred to by the subscript i; $i = 1$ designates the first level above the base

K_p = the stiffness of the component or attachment, Section 9.6.3.3

K_y = the lateral stiffness of the foundation as defined in Section 9.5.9.2.1.1 (lb/in. or N/m)

K_θ = the rocking stiffness of the foundation as defined in Section 9.5.9.2.1.1 (ft-lb/degree or N-m/rad)

KL/r = the lateral slenderness of a compression member measured in terms of its effective buckling length, KL, and the least radius of gyration of the member cross-section, r

K_{Dmax} = maximum effective stiffness, in kips/in. (kN/mm), of the isolation system at the design displacement in the horizontal direction under consideration as prescribed by Eq. 9.13.9.5.1-1

K_{Dmin} = minimum effective stiffness, in kips/in. (kN/mm), of the isolation system at the design displacement in the horizontal direction under consideration as prescribed by Eq. 9.13.9.5.1-2

K_{max} = maximum effective stiffness, in kips/in. (kN/mm), of the isolation system at the maximum displacement in the horizontal direction under consideration as prescribed by Eq. 9.13.9.5.1-3

K_{min} = minimum effective stiffness, in kips/in. (kN/mm), of the isolation system at the maximum displacement in the horizontal direction under consideration, as prescribed by Eq. 9.13.9.5.1-4

k = the distribution exponent given in Section 9.5.5.4

k = the stiffness of the building as determined in Section 9.5.9.2.1.1 (lb/ft or N/m)

k_{eff} = effective stiffness of an isolator unit, as prescribed by Eq. 9.13.9.3-1

L = the overall length of the building (ft or m) at the base in the direction being analyzed

L = length of bracing member (in. or mm) in Section A.9.8

L = the effect of live load in Section 9.13

L_0 = the overall length of the side of the foundation in the direction being analyzed, Section 9.5.9.2.1.2 (ft or m)

l = the dimension of a diaphragm perpendicular to the direction of application of force. For open-front structures, l is the length from the edge of the diaphragm at the open-front to the vertical resisting elements parallel to the direction of the applied force. For a cantilevered diaphragm, l is the length of the cantilever.

M_f = the foundation overturning design moment as defined in Section 9.5.5.6 (ft-kip or kN-m)

M_0, M_{01} = the overturning moment at the foundation-soil interface as determined in Sections 9.5.9.2.3 and 9.5.9.3.2 (ft-lb or N-m)

M_t = the torsional moment resulting from the location of the building masses, Section 9.5.5.5.2

M_{ta} = the accidental torsional moment as determined in Section 9.5.5.5.2

M_x = the building overturning design moment at Level x as defined in Section 9.5.5.6

m = a subscript denoting the mode of vibration under consideration; i.e., $m = 1$ for the fundamental mode

N = number of stories, Section 9.5.5.3.2

N = standard penetration resistance, ASTM D1536-84.

$\overline{N}$ = average field standard penetration resistance for the top 100 ft (30 m), see Section 9.4.1.2.1

N_{ch} = average standard penetration resistance for cohesionless soil layers for the top 100 ft (30 m), see Section 9.4.1.2.1

n = designates the level that is uppermost in the main portion of the building

P_D = required axial strength on a column resulting from application of dead load, D, in Section 9.5 (kip or kN)

P_E = required axial strength on a column resulting from application of the amplified earthquake load, E', in Section 9.5 (kip or kN)

P_L = required axial strength on a column resulting from application of live load, L, in Section 9.5 (kip or kN)

P_n = nominal axial load strength (lb or N) in A.9.8

P_n = the algebraic sum of the shear wall and the minimum gravity loads on the joint surface acting simultaneously with the shear (lb or N)

P_u^* = required axial strength on a brace (kip or kN) in Section A.9.8

P_x = the total unfactored vertical design load at, and above, Level x for use in Section 9.5.5.7.2

PI = plasticity index, ASTM D4318−93.

Q_E = the effect of horizontal seismic (earthquake-induced) forces, Section 9.5.2.7

Q_v = the load equivalent to the effect of the horizontal and vertical shear strength of the vertical segment, Section A.9.8

R = the response modification coefficient as given in Table 9.5.2.2

R_p = the component response modification factor as defined in Section 9.6.1.3

r = the characteristic length of the foundation as defined in Section 9.5.9.2.1

r_a = the characteristic foundation length as defined by Eq. 9.5.9.2.1.1-5 (ft or m)

r_m = the characteristic foundation length as defined by Eq. 9.5.9.2.1.1-6 (ft or m)

r_x = the ratio of the design story shear resisted by the most heavily loaded single element in the story, in direction x, to the total story shear

S_S = the mapped maximum considered earthquake, 5% damped, spectral response acceleration at short periods as defined in Section 9.4.1.2

S_1 = the mapped maximum considered earthquake, 5% damped, spectral response acceleration at a period of 1 sec as defined in Section 9.4.1.2

S_{DS} = the design, 5% damped, spectral response acceleration at short periods as defined in Section 9.4.1.2

S_{D1} = the design, 5% damped, spectral response acceleration at a period of 1 sec as defined in Section 9.4.1.2

S_{MS} = the maximum considered earthquake, 5% damped, spectral response acceleration at short periods adjusted for site class effects as defined in Section 9.4.1.2

S_{M1} = the maximum considered earthquake, 5% damped, spectral response acceleration at a period of 1 sec adjusted for site class effects as defined in Section 9.4.1.2

$\overline{S}_u$ = average undrained shear strength in top 100 ft (30 m); see Section 9.4.1.2, ASTM D2166-91 or ASTM D2850-87.

s_h = spacing of special lateral reinforcement (in. or mm)

T = the fundamental period of the building as determined in Section 9.5.5.2.1

$\tilde{T}, \tilde{T}_1$ = the effective fundamental period (sec) of the building as determined in Sections 9.5.9.2.1.1 and 9.5.9.3.1

T_a = the approximate fundamental period of the building as determined in Section 9.5.5.3

T_D = effective period, in seconds (sec), of the seismically isolated structure at the design displacement in the direction under consideration as prescribed by Eq. 9.13.3.3.2

T_m = the modal period of vibration of the mth mode of the building as determined in Section 9.5.6.5

T_M = effective period, in seconds (sec), of the seismically isolated structure at the maximum displacement in the direction under consideration as prescribed by Eq. 9.13.3.3.4

T_p = the fundamental period of the component and its attachment, Section 9.6.3.3

$T_0 = 0.2S_{D1}/S_{DS}$

$T_S = S_{D1}/S_{DS}$

T_M = effective period, in seconds (sec), of the seismically isolated structure at the maximum displacement in the direction under consideration as prescribed by Eq. 9.13.3.3.4

T_4 = net tension in steel cable due to dead load, prestress, live load, and seismic load, Section A.9.8.8

V = the total design lateral force or shear at the base, Section 9.5.5.2

V_b = the total lateral seismic design force or shear on elements of the isolation system or elements below the isolation system as prescribed by Eq. 9.13.3.4.1

V_s = the total lateral seismic design force or shear on elements above the isolation system as prescribed by Eq. 9.13.3.4.2

V_t = the design value of the seismic base shear as determined in Section 9.5.6.8

V_u = required shear strength (lb or N) due to factored loads in Section 9.6

V_x = the seismic design shear in story x as determined in Section 9.5.5.5 or 9.5.6.8

$\overline{V}_1$ = the portion of the seismic base shear, V, contributed by the fundamental mode, Section 9.5.9.3 (kip or kN)

ΔV = the reduction in V as determined in Section 9.5.9.2 (kip or kN)

ΔV_1 = the reduction in V_1 as determined in Section 9.5.9.3 (kip or kN)

v_s = the average shear wave velocity for the soils beneath the foundation at large strain levels, Section 9.5.9.2 (ft/s or m/s)

$\bar{v}_s$ = average shear wave velocity in top 100 ft (30 m), see Section 9.4.1.2

v_{so} = the average shear wave velocity for the soils beneath the foundation at small strain levels, Section 9.5.9.2 (ft/s or m/s)

W = the total gravity load of the building as defined in Section 9.5.5.2. For calculation of seismic-isolated building period, W is the total seismic dead load weight of the building as defined in Sections 9.5.9.2 and 9.5.9.3 (kip or kN).

$\overline{W}$ = the effective gravity load of the building as defined in Sections 9.5.9.2 and 9.5.9.3 (kip or kN)

W_c = the gravity load of a component of the building

W_D = the energy dissipated per cycle at the story displacement for the design earthquake, Section 9.13.3.2

W_m = the effective modal gravity load determined in accordance with Eq. 9.5.6.5-2.

W_p = component operating weight (lb or N)

w = the width of a diaphragm or shear wall in the direction of application of force. For sheathed diaphragms, the width shall be defined as the dimension between the outside faces of the tension and compression chords.

w = moisture content (in percent), ASTM D2216-92 [3]

w_i, w_n, w_x = the portion of W that is located at or assigned to Level i, n, or x, respectively

x = the level under consideration

$x = 1$ designates the first level above the base

y = elevations difference between points of attachment in Section 9.6

y = the distance, in ft (mm), between the center of rigidity of the isolation system rigidity and the element of interest measured perpendicular to the direction of seismic loading under consideration, Section 9.13

z = the level under consideration; $x = 1$ designates the first level above the base

α = the relative weight density of the structure and the soil as determined in Section 9.5.9.2.1.1

α = angle between diagonal reinforcement and longitudinal axis of the member (degree or rad)

β = ratio of shear demand to shear capacity for the story between Level x and $x - 1$

$\tilde{\beta}$ = the fraction of critical damping for the coupled structure-foundation system, determined in Section 9.5.9.2.1

β_D = effective damping of the isolation system at the design displacement as prescribed by Eq. 9.13.9.5.2-1

β_M = effective damping of the isolation system at the maximum displacement as prescribed by Eq. 9.13.5.2-2

β_0 = the foundation damping factor as specified in Section 9.5.9.2.1

β_{eff} = effective damping of the isolation system as prescribed by Eq. 9.13.9.3-2

γ = the average unit weight of soil (lb/ft^3 or kg/m^3)

Δ = the design story drift as determined in Section 9.5.5.7.1

Δ_a = the allowable story drift as specified in Section 9.5.2.8

Δ_m = the design modal story drift determined in Section 9.5.6.6

Δ^+ = maximum positive displacement of an isolator unit during each cycle of prototype testing

Δ^- = maximum negative displacement of an isolator unit during each cycle of prototype testing

δ_{max} = the maximum displacement at Level x, considering torsion, Section 9.5.5.5.2

δ_{avg} = the average of the displacements at the extreme points of the structure at Level x, Section 9.5.5.5.2

δ_x = the deflection of Level x at the center of the mass at and above Level x, Eq. 9.5.5.7.1

δ_{xe} = the deflection of Level x at the center of the mass at and above Level x determined by an elastic analysis, Section 9.5.5.7.1

δ_{xem} = the modal deflection of Level x at the center of the mass at and above Level x determined by an elastic analysis, Section 9.5.6.6

δ_{xm} = the modal deflection of Level x at the center of the mass at and above Level x as determined by Eqs. 9.5.4.6-3 and 9.13.3.2-1

$\tilde{\delta}_x, \tilde{\delta}_{x1}$ = the deflection of Level x at the center of the mass at and above Level x, Eqs. 9.5.9.2.3-1 and 9.5.9.3.2-1 (in. or mm)

θ = the stability coefficient for P-delta effects as determined in Section 9.5.5.7.2

τ = the overturning moment reduction factor, Eq. 9.5.3.6

ρ = a reliability coefficient based on the extent of structural redundance present in a building as defined in Section 9.5.2.7

ρ_s = spiral reinforcement ratio for precast, prestressed piles in Sections A.9.7.4.4.5 and A.9.7.5.4.4

ρ_x = a reliability coefficient based on the extent of structural redundancy present in the seismic force-resisting system of a building in the x direction

λ = time effect factor

ϕ = the capacity reduction factor

ϕ = the strength reduction factor or resistance factor

ϕ_{im} = the displacement amplitude at the ith level of the building for the fixed-base condition when vibrating in its mth mode, Section 9.5.6.5

Ω_0 = overstrength factor as defined in Table 9.5.2.2

ΣE_D = total energy dissipated, in kip-ins. (kN-mm), in the isolation system during a full cycle of response at the design displacement, D_D

ΣE_M = total energy dissipated, in kip-ins. (kN-mm), in the isolation system during a full cycle of response at the maximum displacement, D_M

$\Sigma |F_D^+|_{max}$ = sum, for all isolator units, of the maximum absolute value of force, in kips (kN), at a positive displacement equal to D_D

$\Sigma |F_D^+|_{min}$ = sum, for all isolator units, of the minimum absolute value of force, in kips (kN), at a positive displacement equal to D_D

$\Sigma |F_D^-|_{max}$ = sum, for all isolator units, of the maximum absolute value of force, in kips (kN), at a negative displacement equal to D_D

$\Sigma |F_D^-|_{min}$ = sum, for all isolator units, of the minimum absolute value of force, in kips (kN), at a negative displacement equal to D_D

$\Sigma |F_M^+|_{max}$ = sum, for all isolator units, of the maximum absolute value of force, in kips (kN), at a positive displacement equal to D_M

$\Sigma |F_M^+|_{min}$ = sum, for all isolator units, of the minimum absolute value of force, in kips (kN), at a positive displacement equal to D_M

$\Sigma |F_M^-|_{max}$ = sum, for all isolator units, of the maximum absolute value of force, in kips (kN), at a negative displacement equal to D_M

$\Sigma |F_M^-|_{min}$ = sum, for all isolator units, of the minimum absolute value of force, in kips (kN), at a negative displacement equal to D_M

SECTION 9.3

This Section is intentionally left blank.

9.4.1 Procedures for Determining Maximum Considered Earthquake and Design Earthquake Ground Motion Accelerations and Response Spectra. Ground motion accelerations, represented by response spectra and coefficients derived from these spectra, shall be determined in accordance with the general procedure of Section 9.4.1.2 or the site-specific procedure of Section 9.4.1.3. The general procedure in which spectral response acceleration parameters for the maximum considered earthquake ground motions are derived using Figure 9.4.1.1a through 9.4.1.1j, modified by site coefficients to include local site effects and scaled to design values, are permitted to be used for any structure except as specifically indicated in these provisions. The site-specific procedure also is permitted to be used for any structure and shall be used where specifically required by these provisions.

9.4.1.1 Maximum Considered Earthquake Ground Motions. The maximum considered earthquake ground motions shall be as represented by the mapped spectral response acceleration at short periods, S_S, and at 1-sec, S_1, obtained from Figure 9.4.1.1a through 9.4.1.1j and adjusted for Site Class effects using the site coefficients of Section 9.4.1.2.4. When a site-specific procedure is used, maximum considered earthquake ground motion shall be determined in accordance with Section 9.4.1.3.

9.4.1.2 General Procedure for Determining Maximum Considered Earthquake and Design Spectral Response Accelerations. The mapped maximum considered earthquake spectral response acceleration at short periods (S_S) and at 1-sec (S_1) shall be determined from Figure 9.4.1.1a through 9.4.1.1j.

For buildings and structures included in the scope of this Standard as specified in Section 9.1.2.1, the Site Class shall be determined in accordance with Section 9.4.1.2.1. The maximum considered earthquake spectral response accelerations adjusted for Site Class effects, S_{MS} and S_{M1}, shall be determined in accordance with Section 9.4.1.2.4 and the design spectral response accelerations, S_{DS} and S_{D1}, shall be determined in accordance with Section 9.4.1.2.5. The general response spectrum, when required by these provisions, shall be determined in accordance with Section 9.4.1.2.6.

9.4.1.2.1 Site Class Definitions. The site shall be classified as one of the following classes:

A = Hard rock with measured shear wave velocity, $\overline{v}_s > 5000$ ft/s (1500 m/s)

B = Rock with 2500 ft/s $< \overline{v}_s \leq 5000$ ft/s (760 m/s $< \overline{v}_s \leq 1500$ m/s)

TABLE 9.4.1.2
SITE CLASSIFICATION

Site Class	$\bar{v}_s$	$\bar{N}$ or $\bar{N}_{ch}$	$\bar{s}_u$
A Hard rock	>5000 ft/s (>1500 m/s)	not applicable	not applicable
B Rock	2500 to 5000 ft/s (760 to 1500 m/s)	not applicable	not applicable
C Very dense soil and soft rock	1200 to 2500 ft/s (370 to 760 m/s)	> 50	> 2000 psf (> 100 kPa)
D Stiff soil	600 to 1200 ft/s (180 to 370 m/s)	15 to 50	1000 to 2000 psf (50 to 100 kPa)
E Soil	<600 ft/s (<180 m/s)	<15	<1000 psf (<50 kPa)
	Any profile with more than 10 ft of soil having the following characteristics: — Plasticity index PI > 20, — Moisture content $w \geq 40\%$, and — Undrained shear strength $\bar{s}_u < 500$ psf		
F Soils requiring site-specific evaluation	1. Soils vulnerable to potential failure or collapse 2. Peats and/or highly organic clays 3. Very high plasticity clays 4. Very thick soft/medium clays		

C = Very dense soil and soft rock with 1200 ft/s $\leq$ $\bar{v}_s \leq$ 2500 ft/s (370 m/s $\leq \bar{v}_s \leq$ 760 m/s) or $\bar{N}$ or $\bar{N}_{ch} >$ 50 or $\bar{s}_u \geq$ 2000 psf (100 kPa)

D = Stiff soil with 600 ft/s $\leq \bar{v}_s \leq$ 1200 ft/s (180 m/s $\leq \bar{v}_s \leq$ 370 m/s) or with 15 $\leq \bar{N}$ or $\bar{N}_{ch} \leq$ 50 or 1000 psf $\leq \bar{s}_u \leq$ 2000 psf (50 kPa $\leq \bar{s}_u \leq$ 100 kPa)

E = A soil profile with $\bar{v}_s <$ 600 ft/s (180 m/s) or any profile with more than 10 ft (3 m) of soft clay. Soft clay is defined as soil with $PI >$ 20, $w \geq$ 40%, and $s_u <$ 500 psf (25 kPa)

F = Soils requiring site-specific evaluations:

1. Soils vulnerable to potential failure or collapse under seismic loading such as liquefiable soils, quick and highly sensitive clays, collapsible weakly cemented soils.

   Exception: For structures having fundamental periods of vibration equal to or less than 0.5-sec, site-specific evaluations are not required to determine spectral accelerations for liquefiable soils. Rather, the Site Class may be determined in accordance with Section 9.4.1.2.2 and the corresponding values of F_a and F_v determined from Tables 9.4.1.2.4a and 9.4.1.2.4b.

2. Peats and/or highly organic clays ($H >$ 10 ft [3 m] of peat and/or highly organic clay where H = thickness of soil).

3. Very high plasticity clays ($H >$ 25 ft [7.6 m] with $PI >$ 75).

4. Very thick soft/medium stiff clays ($H >$ 120 ft [37 m]).

   Exception: When the soil properties are not known in sufficient detail to determine the Site Class, Class D shall be used. Site Class E shall be used when the authority having jurisdiction determines that Site Class E is present at the site or in the event that Site E is established by geotechnical data.

The following standards are referenced in the provisions for determining the seismic coefficients:

[1] ASTM. "Test Method for Penetration Test and Split-Barrel Sampling of Soils." *ASTM D1586-84*, 1984.

[2] ASTM. "Test Method for Liquid Limit, Plastic Limit, and Plasticity Index of Soils." *ASTM D4318-93*, 1993.

[3] ASTM. "Test Method for Laboratory Determination of Water (Moisture) Content of Soil and Rock." *ASTM D2216-92*, 1992.

[4] ASTM. "Test Method for Unconfined Compressive Strength of Cohesive Soil." *ASTM D2166-91*, 1991.

[5] ASTM. "Test Method for Unconsolidated, Undrained Compressive Strength of Cohesive Soils in Triaxial Compression." *ASTM D2850-87*, 1987.

This page intentionally left blank.

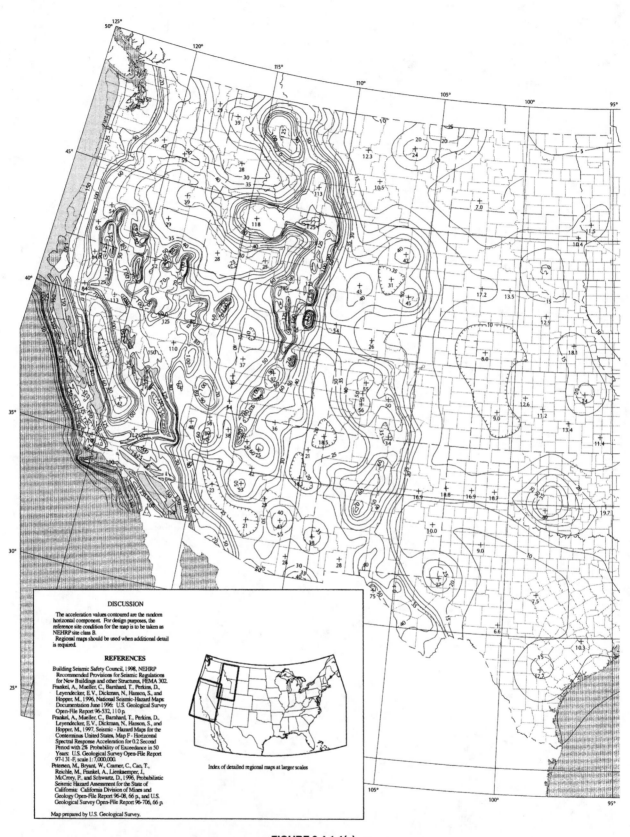

Index of detailed regional maps at larger scales

FIGURE 9.4.1.1(a)
MAXIMUM CONSIDERED EARTHQUAKE GROUND MOTION FOR
CONTERMINOUS UNITED STATES. OF 0.2 s SPECTRAL RESPONSE
ACCELERATION (5% OF CRITICAL DAMPING), SITE CLASS B

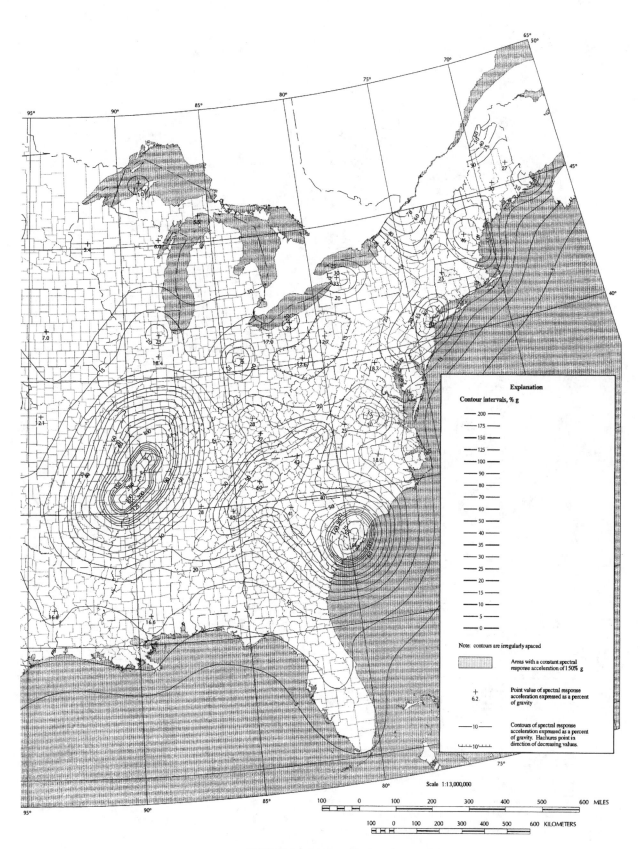

FIGURE 9.4.1.1(a) — continued
MAXIMUM CONSIDERED EARTHQUAKE GROUND MOTION FOR
CONTERMINOUS UNITED STATES. OF 0.2 s SPECTRAL RESPONSE
ACCELERATION (5% OF CRITICAL DAMPING), SITE CLASS B

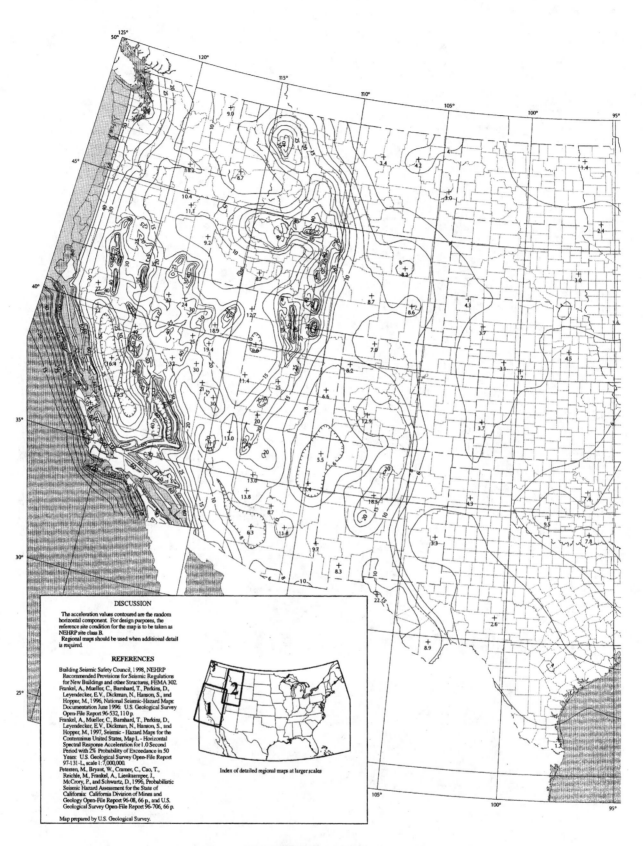

FIGURE 9.4.1.1(b)
MAXIMUM CONSIDERED EARTHQUAKE GROUND MOTION FOR
CONTERMINOUS UNITED STATES OF 1.0 s SPECTRAL RESPONSE
ACCELERATION (5% OF CRITICAL DAMPING), SITE CLASS B

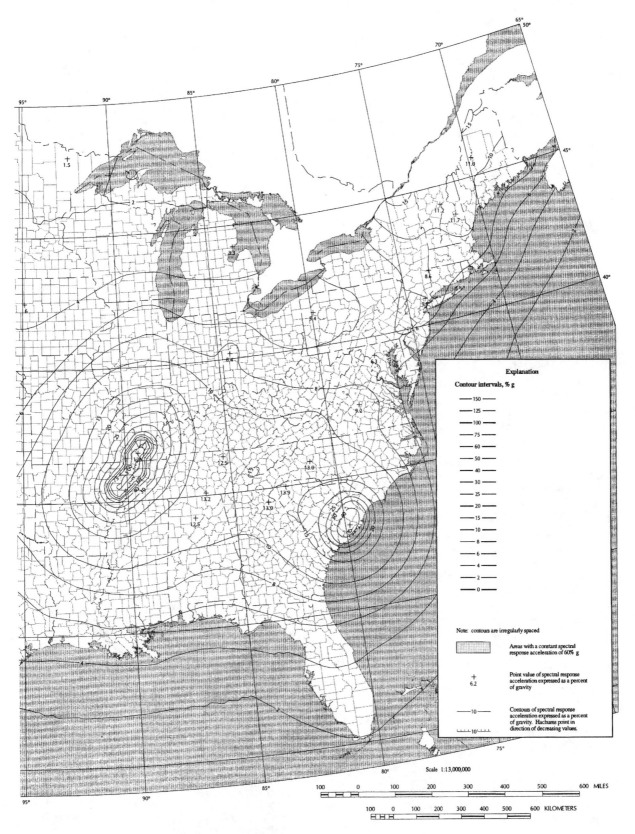

FIGURE 9.4.1.1(b) — continued
MAXIMUM CONSIDERED EARTHQUAKE GROUND MOTION FOR
CONTERMINOUS UNITED STATES OF 1.0s SPECTRAL RESPONSE
ACCELERATION (5% OF CRITICAL DAMPING), SITE CLASS B

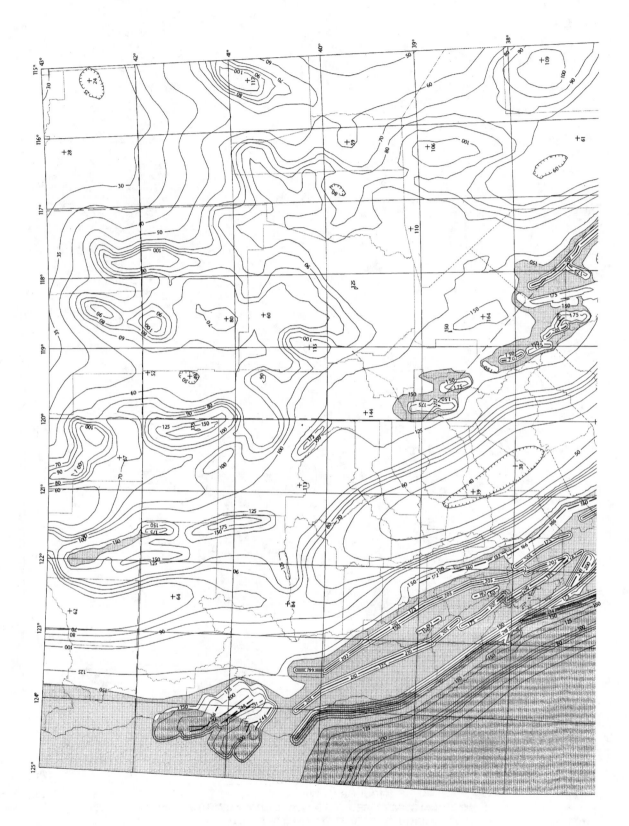

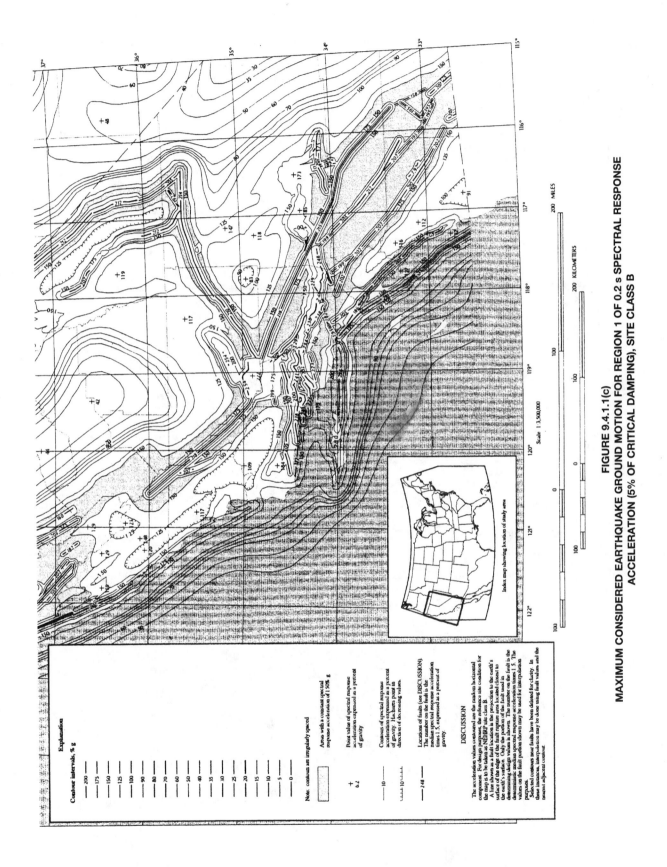

FIGURE 9.4.1.1(c)

MAXIMUM CONSIDERED EARTHQUAKE GROUND MOTION FOR REGION 1 OF 0.2 s SPECTRAL RESPONSE ACCELERATION (5% OF CRITICAL DAMPING), SITE CLASS B

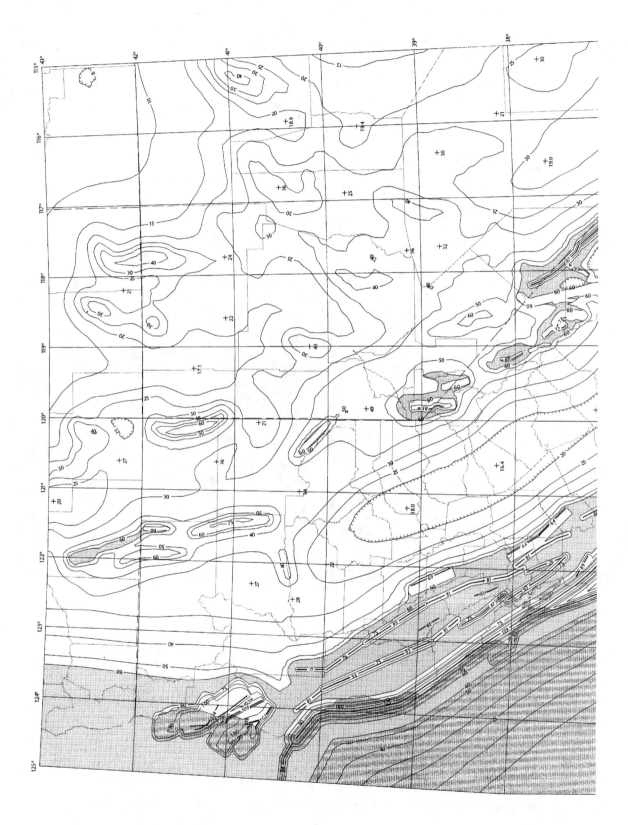

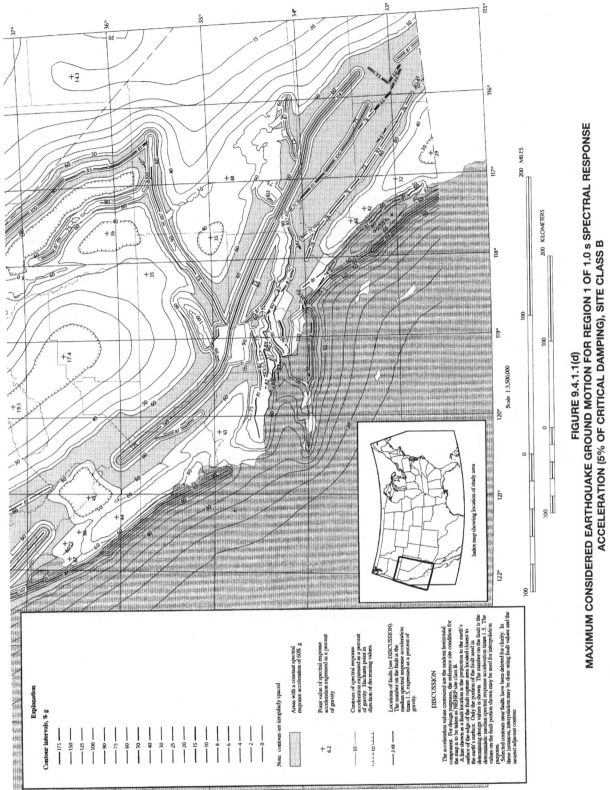

FIGURE 9.4.1.1(d)

MAXIMUM CONSIDERED EARTHQUAKE GROUND MOTION FOR REGION 1 OF 1.0 s SPECTRAL RESPONSE ACCELERATION (5% OF CRITICAL DAMPING), SITE CLASS B

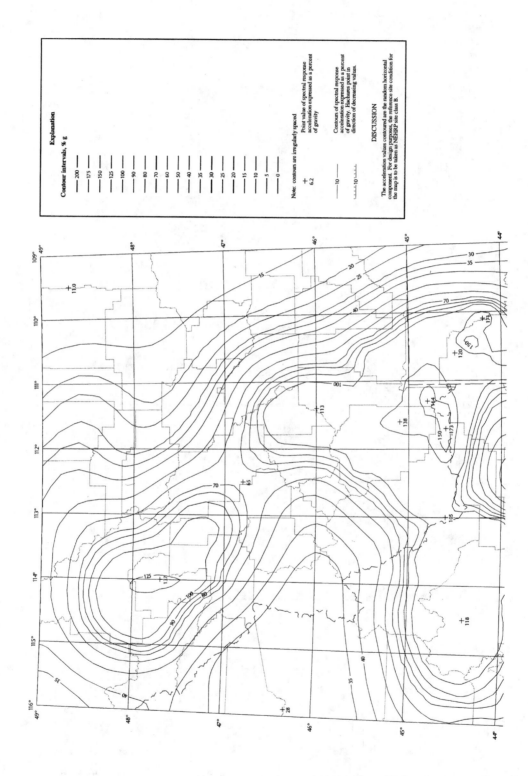

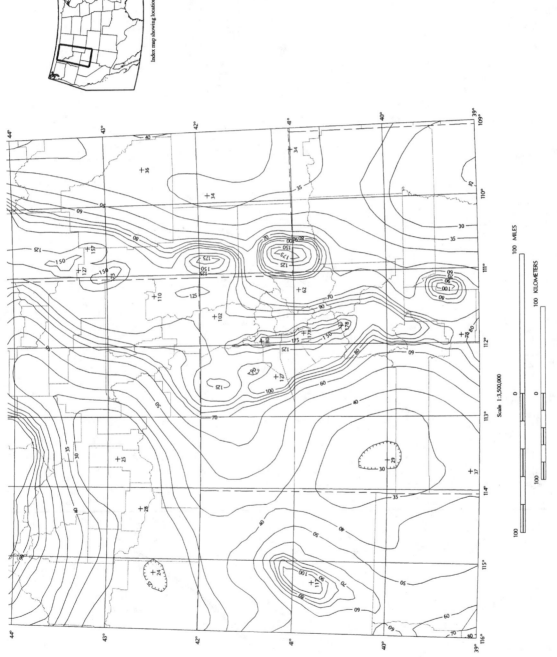

FIGURE 9.4.1.1(e)
MAXIMUM CONSIDERED EARTHQUAKE GROUND MOTION FOR REGION 2 OF 0.2 s SPECTRAL RESPONSE
ACCELERATION (5% OF CRITICAL DAMPING), SITE CLASS B

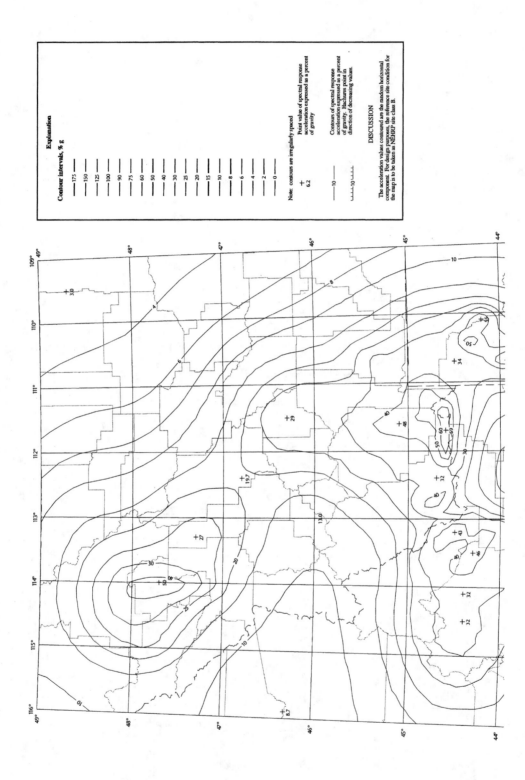

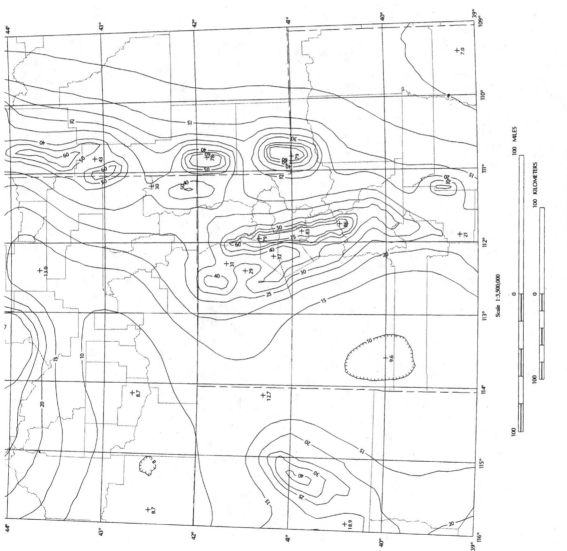

FIGURE 9.4.1.1(f)

MAXIMUM CONSIDERED EARTHQUAKE GROUND MOTION FOR REGION 2 OF 1.0 s SPECTRAL RESPONSE

ACCELERATION (5% OF CRITICAL DAMPING), SITE CLASS B

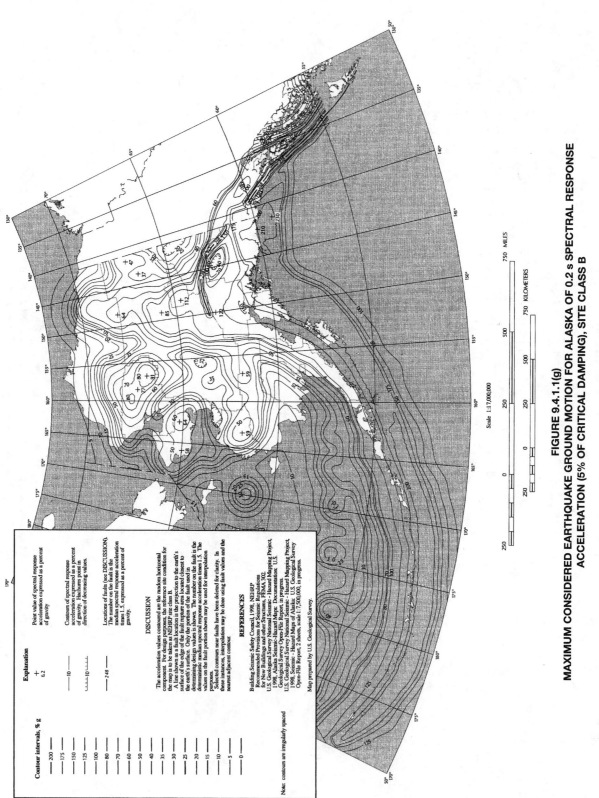

Scale 1:17,000,000

FIGURE 9.4.1.1(g)
MAXIMUM CONSIDERED EARTHQUAKE GROUND MOTION FOR ALASKA OF 0.2 s SPECTRAL RESPONSE
ACCELERATION (5% OF CRITICAL DAMPING), SITE CLASS B

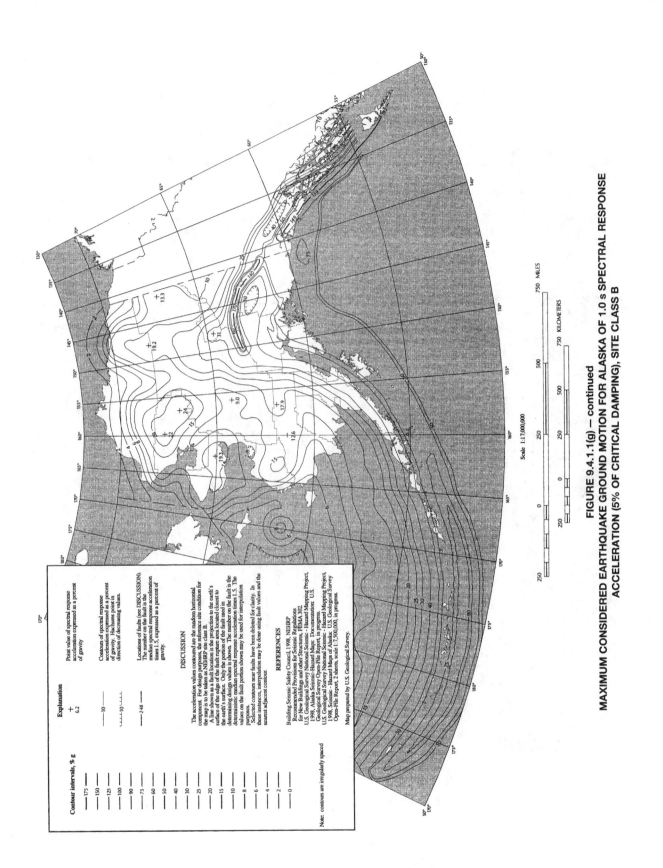

Explanation

+ 6.2   Point value of spectral response acceleration expressed as a percent of gravity

——10——   Contours of spectral response acceleration expressed as a percent of gravity. Hachures point in direction of decreasing values.

········10········

——248——   Locations of faults (see DISCUSSION). The number on the fault is the median spectral response acceleration times 1.5, expressed as a percent of gravity.

Contour intervals, % g

——175——
——150——
——125——
——100——
——90——
——75——
——60——
——50——
——40——
——30——
——25——
——20——
——15——
——10——
——8——
——6——
——4——
——2——
——0——

Note: contours are irregularly spaced

DISCUSSION

The acceleration values contoured are the random horizontal component. For design purposes, the reference site condition for the map is to be taken as NEHRP site class B.

A line shown as a fault location is the projection to the earth's surface of the edge of the fault rupture area located closest to the earth's surface. Only the portion of the fault used in determining design values is shown. The number on the fault is the deterministic median spectral response acceleration times 1.5. The values on the fault portion shown may be used for interpolation purposes.

Selected contours near faults have been deleted for clarity. In these instances, interpolation may be done using fault values and the nearest adjacent contour.

REFERENCES

Building Seismic Safety Council, 1998, NEHRP Recommended Provisions for Seismic Regulations for New Buildings and other Structures, FEMA 302.
U.S. Geological Survey National Seismic - Hazard Mapping Project, 1998, Alaska Seismic-Hazard Maps: Documentation: U.S. Geological Survey Open-File Report, in progress.
U.S. Geological Survey National Seismic - Hazard Mapping Project, 1998, Seismic - Hazard Maps of Alaska: U.S. Geological Survey Open-File Report, 2 sheets, scale 1:7,500,000, in progress.

Map prepared by U.S. Geological Survey.

Scale 1:17,000,000

FIGURE 9.4.1.1(g) — continued
MAXIMUM CONSIDERED EARTHQUAKE GROUND MOTION FOR ALASKA OF 1.0 s SPECTRAL RESPONSE
ACCELERATION (5% OF CRITICAL DAMPING), SITE CLASS B

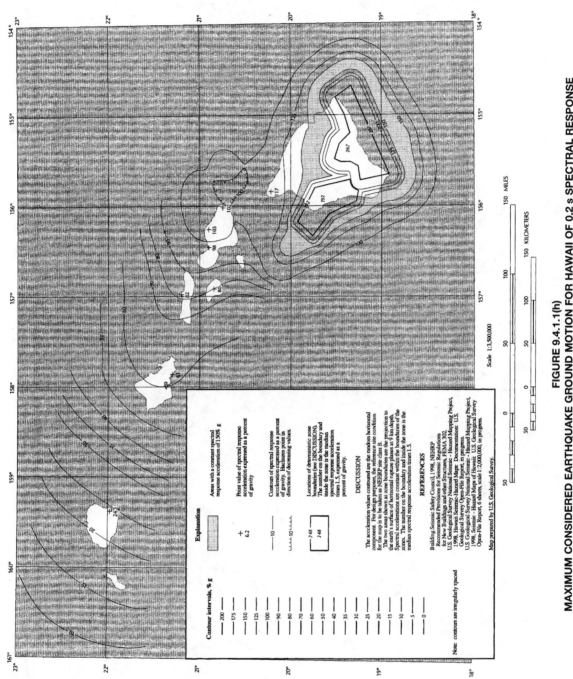

FIGURE 9.4.1.1(h)

MAXIMUM CONSIDERED EARTHQUAKE GROUND MOTION FOR HAWAII OF 0.2 s SPECTRAL RESPONSE
ACCELERATION (5% OF CRITICAL DAMPING), SITE CLASS B

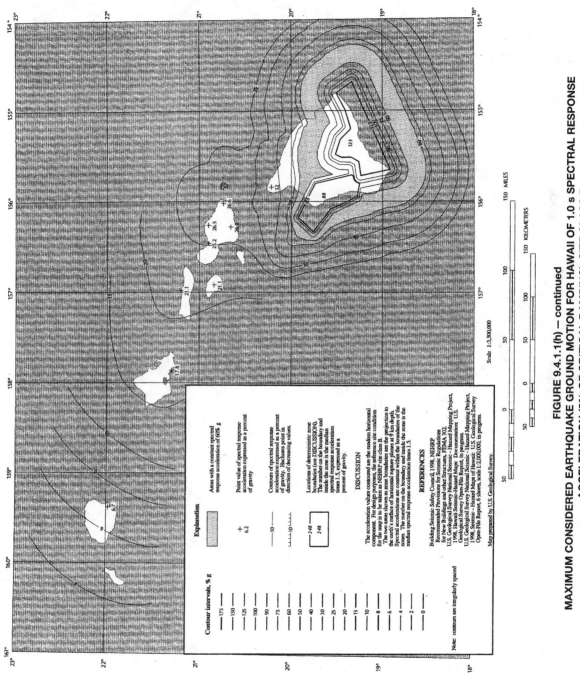

FIGURE 9.4.1.1(h) — continued

MAXIMUM CONSIDERED EARTHQUAKE GROUND MOTION FOR HAWAII OF 1.0 s SPECTRAL RESPONSE
ACCELERATION (5% OF CRITICAL DAMPING), SITE CLASS B

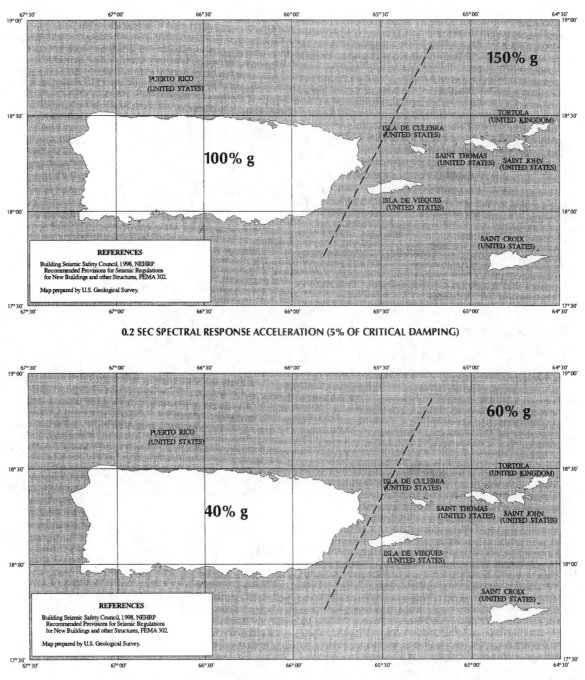

0.2 SEC SPECTRAL RESPONSE ACCELERATION (5% OF CRITICAL DAMPING)

1.0 SEC SPECTRAL RESPONSE ACCELERATION (5% OF CRITICAL DAMPING)

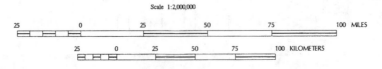

Scale 1:2,000,000

FIGURE 9.4.1.1(i)
MAXIMUM CONSIDERED EARTHQUAKE GROUND MOTION FOR PUERTO RICO, CULEBRA, VIEQUES, ST. THOMAS, ST. JOHN, AND ST. CROIX OF 0.2 AND 1.0 s SPECTRAL RESPONSE ACCELERATION (5% OF CRITICAL DAMPING), SITE CLASS B

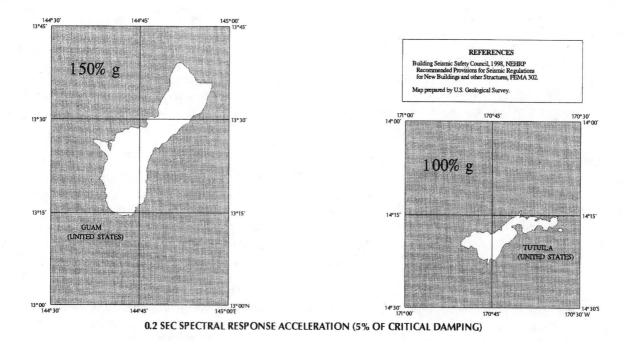

0.2 SEC SPECTRAL RESPONSE ACCELERATION (5% OF CRITICAL DAMPING)

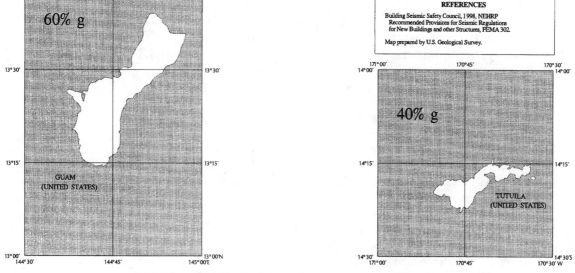

1.0 SEC SPECTRAL RESPONSE ACCELERATION (5% OF CRITICAL DAMPING)

Scale 1:1,000,000

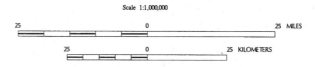

FIGURE 9.4.1.1(j)
MAXIMUM CONSIDERED EARTHQUAKE GROUND MOTION FOR GUAM
AND TUTUILLA OF 0.2 AND 1.0 S SPECTRAL RESPONSE ACCELERATION
(5% OF CRITICAL DAMPING), SITE CLASS B

9.4.1.2.2 Steps for Classifying a Site. The Site Class of a site shall be determined using the following steps:

Step 1: Check for the four categories of Site Class F requiring site-specific evaluation. If the site corresponds to any of these categories, classify the site as Site Class F and conduct a site-specific evaluation.

Step 2: Check for the existence of a total thickness of soft clay > 10 ft (3 m) where a soft clay layer is defined by $s_u < 500$ psf (25 kPa), $w \geq 40\%$, and $PI > 20$. If this criterion is satisfied, classify the site as Site Class E.

Step 3: Categorize the site using one of the following three methods with $\overline{v}_s$, $\overline{N}$, and $\overline{s}_u$ computed in all cases as specified by the definitions in Section 9.4.1.2.3.

    a. The $\overline{v}_s$ method: Determine $\overline{v}_s$ for the top 100 ft (30 m) of soil. Compare the value of $\overline{v}_s$ with those given in Section 9.4.1.2 and Table 9.4.1.2 and assign the corresponding Site Class.

    $\overline{v}_s$ for rock, Site Class B, shall be measured on-site or estimated by a geotechnical engineer or engineering geologist/seismologist for competent rock with moderate fracturing and weathering. $\overline{v}_s$ for softer and more highly fractured and weathered rock shall be measured on-site or shall be classified as Site Class C. The classification of hard rock, Site Class A, shall be supported by on-site measurements of $\overline{v}_s$ or on profiles of the same rock type in the same formation with an equal or greater degree of weathering and fracturing. Where hard rock conditions are known to be continuous to a depth of at least 100 ft (30 m), surficial measurements of v_s are not prohibited from being extrapolated to assess $\overline{v}_s$.

    The rock categories, Site Classes A and B, shall not be assigned to a site if there is more than 10 ft (3 m) of soil between the rock surface and the bottom of the spread footing or mat foundation.

    b. The $\overline{N}$ method: Determine $\overline{N}$ for the top 100 ft (30 m) of soil. Compare the value of $\overline{N}$ with those given in Section 9.4.1.2 and Table 9.4.1.2 and assign the corresponding Site Class.

    c. The $\overline{s}_u$ method: For cohesive soil layers, determine $\overline{s}_u$ for the top 100 ft (30 m) of soil. For cohesionless soil layers, determine $\overline{N}_{ch}$ for the top 100 ft (30 m) of soil.

Cohesionless soil is defined by a $PI < 20$ where cohesive soil is defined by a $PI > 20$. Compare the values of $\overline{s}_u$ and $\overline{N}_{ch}$ with those given in Section 9.4.1.2 and Table 9.4.1.2 and assign the corresponding Site Class. When the $\overline{N}_{ch}$ and $\overline{s}_u$ criteria differ, assign the category with the softer soil (Site Class E soil is softer than D).

9.4.1.2.3 Definitions of Site Class Parameters. The definitions presented below apply to the upper 100 ft (30 m) of the site profile. Profiles containing distinctly different soil layers shall be subdivided into those layers designated by a number that ranges from 1 to n at the bottom where there are a total of n distinct layers in the upper 100 ft (30 m). Where some of the n layers are cohesive and others are not, k is the number of cohesive layers and m is the number of cohesionless layers. The symbol i refers to any one of the layers between 1 and n.

v_{si} is the shear wave velocity in ft/s (m/s).

d_i is the thickness of any layer between 0 and 100 ft (30 m).

$\overline{v}_s$ is

$$\overline{v}_s = \frac{\sum_{i=1}^{n} d_i}{\sum_{i=1}^{n} \frac{d_i}{v_{si}}} \qquad \text{(Eq. 9.4.1.2-1)}$$

whereby $\sum_{i=1}^{n} d_i$ is equal to 100 ft (30 m)

N_i is the standard penetration resistance, ASTM D1586-84 not to exceed 100 blows/ft as directly measured in the field without corrections.

$\overline{N}$ is:

$$\overline{N} = \frac{\sum_{i=1}^{n} d_i}{\sum_{i=1}^{n} \frac{d_i}{N_i}} \qquad \text{(Eq. 9.4.1.2-2)}$$

$\overline{N}_{ch}$ is:

$$\overline{N}_{ch} = \frac{d_s}{\sum_{i=1}^{m} \frac{d_i}{N_i}} \qquad \text{(Eq. 9.4.1.2-3)}$$

whereby $\sum_{i=1}^{m} d_i = d_s$. (Use only d_i and N_i for cohesionless soils.)

d_s is the total thickness of cohesionless soil layers in the top 100 ft (30 m)

s_{ui} is the undrained shear strength in psf (kPa), not to exceed 5000 psf (240 kPa), ASTM D2166-91 or D2850-87.

$\bar{s}_u$ is

$$\bar{s}_u = \frac{d_c}{\sum\limits_{i=1}^{k} \dfrac{d_i}{s_{ui}}} \qquad \textbf{(Eq. 9.4.1.2-4)}$$

whereby $\sum\limits_{i=1}^{k} d_i = d_c$.

d_c is the total thickness $(100 - d_s)$ of cohesive soil layers in the top 100 ft (30 m)

PI is the plasticity index, ASTM D4318-93

w is the moisture content in percent, ASTM D2216-92

9.4.1.2.4 Site Coefficients and Adjusted Maximum Considered Earthquake Spectral Response Acceleration Parameters.
The maximum considered earthquake spectral response acceleration for short periods (S_{MS}) and at 1-sec (S_{M1}), adjusted for site class effects, shall be determined by Eqs. 9.4.1.2.4-1 and 9.4.1.2.4-2, respectively.

$$S_{MS} = F_a S_s \qquad \textbf{(Eq. 9.4.1.2.4-1)}$$

$$S_{M1} = F_v S_1 \qquad \textbf{(Eq. 9.4.1.2.4-2)}$$

where

S_1 = the mapped maximum considered earthquake spectral response acceleration at a period of 1-sec as determined in accordance with Section 9.4.1

S_S = the mapped maximum considered earthquake spectral response acceleration at short periods as determined in accordance with Section 9.4.1

where site coefficients F_a and F_v are defined in Tables 9.4.1.2.4a and b, respectively.

9.4.1.2.5 Design Spectral Response Acceleration Parameters.
Design earthquake spectral response acceleration at short periods, S_{DS}, and at 1-sec period, S_{D1}, shall be determined from Eqs. 9.4.1.2.5-1 and 9.4.1.2.5-2, respectively.

$$S_{DS} = \frac{2}{3} S_{MS} \qquad \textbf{(Eq. 9.4.1.2.5-1)}$$

$$S_{D1} = \frac{2}{3} S_{M1} \qquad \textbf{(Eq. 9.4.1.2.5-2)}$$

9.4.1.2.6 General Procedure Response Spectrum.
Where a design response spectrum is required by these provisions and site-specific procedures are not used, the design response spectrum curve shall be developed as indicated in Figure 9.4.1.2.6 and as follows:

1. For periods less than or equal to T_0, the design spectral response acceleration, S_a, shall be taken as given by Eq. 9.4.1.2.6-1:

$$S_a = S_{DS} \left(0.4 + 0.6 \frac{T}{T_0} \right)$$

$$\textbf{(Eq. 9.4.1.2.6-1)}$$

TABLE 9.4.1.2.4a
VALUES OF F_a AS A FUNCTION OF SITE CLASS AND MAPPED SHORT PERIOD MAXIMUM CONSIDERED EARTHQUAKE SPECTRAL ACCELERATION

Site Class	Mapped Maximum Considered Earthquake Spectral Response Acceleration at Short Periods				
	$S_S \leq 0.25$	$S_S = 0.5$	$S_S = 0.75$	$S_S = 1.0$	$S_S \geq 1.25$
A	0.8	0.8	0.8	0.8	0.8
B	1.0	1.0	1.0	1.0	1.0
C	1.2	1.2	1.1	1.0	1.0
D	1.6	1.4	1.2	1.1	1.0
E	2.5	1.7	1.2	0.9	0.9
F	a	a	a	a	a

Note: Use straight-line interpolation for intermediate values of S_S.
a Site-specific geotechnical investigation and dynamic site response analyses shall be performed except that for structures with periods of vibration equal to or less than 0.5-seconds, values of F_a for liquefiable soils may be assumed equal to the values for the site class determined without regard to liquefaction in Step 3 of Section 9.4.1.2.2.

TABLE 9.4.1.2.4b
**VALUES OF F_v AS A FUNCTION OF SITE CLASS AND MAPPED
1-SECOND PERIOD MAXIMUM CONSIDERED EARTHQUAKE
SPECTRAL ACCELERATION**

Site Class	Mapped Maximum Considered Earthquake Spectral Response Acceleration at 1-Second Periods				
	$S_1 \leq 0.1$	$S_1 = 0.2$	$S_1 = 0.3$	$S_1 = 0.4$	$S_1 \geq 0.5$
A	0.8	0.8	0.8	0.8	0.8
B	1.0	1.0	1.0	1.0	1.0
C	1.7	1.6	1.5	1.4	1.3
D	2.4	2.0	1.8	1.6	1.5
E	3.5	3.2	2.8	2.4	2.4
F	a	a	a	a	a

Note: Use straight-line interpolation for intermediate values of S_1.

[a] Site-specific geotechnical investigation and dynamic site response analyses shall be performed except that for structures with periods of vibration equal to or less than 0.5-seconds, values of F_v for liquefiable soils may be assumed equal to the values for the site class determined without regard to liquefaction in Step 3 of Section 9.4.1.2.2.

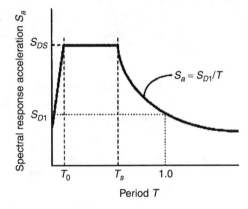

**FIGURE 9.4.1.2.6
DESIGN RESPONSE SPECTRUM**

2. For periods greater than or equal to T_0 and less than or equal to T_S, the design spectral response acceleration, S_a, shall be taken as equal to S_{DS}.

3. For periods greater than T_S, the design spectral response acceleration, S_a, shall be taken as given by Eq. 9.4.1.2.6-2:

$$S_a = \frac{S_{D1}}{T} \qquad \text{(Eq. 9.4.1.2.6-2)}$$

where

S_{DS} = the design spectral response acceleration at short periods

S_{D1} = the design spectral response acceleration at 1-sec period, in units of g-sec

T = the fundamental period of the structure (sec)

$T_0 = 0.2 S_{D1}/S_{DS}$ and

$T_S = S_{D1}/S_{DS}$.

9.4.1.3 Site-Specific Procedure for Determining Ground Motion Accelerations. A site-specific study shall account for the regional seismicity and geology, the expected recurrence rates and maximum magnitudes of events on known faults and source zones, the location of the site with respect to these near source effects, if any, and the characteristics of subsurface site conditions.

9.4.1.3.1 Probabilistic Maximum Considered Earthquake. When site-specific procedures are utilized, the maximum considered earthquake ground motion shall be taken as that motion represented by a 5% damped acceleration response spectrum having a 2% probability of exceedance within a 50-year period. The maximum considered earthquake spectral response acceleration at any period S_{aM} shall be taken from that spectrum.

Exception: Where the spectral response ordinates or a 5% damped spectrum having a 2% probability of exceedance within a 50-year period at periods of 1.0- or 0.2-sec exceed the corresponding ordinate of the deterministic limit of Section 9.4.1.3.2, the maximum considered earthquake ground motion shall be taken as the lesser of the probabilistic maximum considered earthquake ground motion or the deterministic maximum considered earthquake ground motion of Section 9.4.1.3.3 but shall not

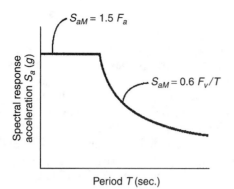

FIGURE 9.4.1.3.2
DETERMINISTIC LIMIT ON MAXIMUM CONSIDERED
EARTHQUAKE RESPONSE SPECTRUM

be taken less than the deterministic limit ground motion of Section 9.4.1.3.2.

9.4.1.3.2 Deterministic Limit on Maximum Considered Earthquake Ground Motion. The deterministic limit on maximum considered earthquake ground motion shall be taken as the response spectrum determined in accordance with Figure 9.4.1.3.2, where F_a and F_v are determined in accordance with Section 9.4.1.2.4, with the value of S_S taken as 1.5 g and the value of S_1 taken as 0.6 g.

9.4.1.3.3 Deterministic Maximum Considered Earthquake Ground Motion. The deterministic maximum considered earthquake ground motion response spectrum shall be calculated as 150% of the median spectral response accelerations (S_{aM}) at all periods resulting from a characteristic earthquake on any known active fault within the region.

9.4.1.3.4 Site-Specific Design Ground Motion. Where site-specific procedures are used to determine the maximum considered earthquake ground motion response spectrum, the design spectral response acceleration at any period shall be determined from Eq. 9.4.1.3.4:

$$S_a = \frac{2}{3} S_{aM} \qquad \textbf{(Eq. 9.4.1.3.4)}$$

and shall be greater than or equal to 80% of the S_a determined by the general response spectrum in Section 9.4.1.2.6.

9.4.1.3.5 Design Acceleration Parameters. Where the site-specific procedure is used to determine the design ground motion in accordance with Section 9.4.1.3.4, the parameter S_{DS} shall be taken as the spectral acceleration, S_a, obtained from the site-specific spectra at a period of 0.2 sec, except that

it shall not be taken as less than 90% of the peak spectral acceleration, S_a, at any period. The parameter S_{D1} shall be taken as the greater of the spectral acceleration, S_a, at a period of 1 sec or two times the spectral acceleration, S_a, at a period of 2 sec. The parameters S_{MS} and S_{M1} shall be taken as 1.5 times S_{DS} and S_{D1}, respectively.

9.4.2 Seismic Design Category. Structures shall be assigned a Seismic Design Category in accordance with Section 9.4.2.1.

9.4.2.1 Determination of Seismic Design Category. All structures shall be assigned to a Seismic Design Category based on their Seismic Use Group and the design spectral response acceleration coefficients, S_{DS} and S_{D1}, determined in accordance with Section 9.4.1.2.5. Each building and structure shall be assigned to the most severe Seismic Design Category in accordance with Table 9.4.2.1a or 9.4.2.1b, irrespective of the fundamental period of vibration of the structure, T.

9.4.2.2 Site Limitation for Seismic Design Categories E and F. A structure assigned to Category E or F shall not be sited where there is a known potential for an active fault to cause rupture of the ground surface at the structure.

> **Exception:** Detached one- and two-family dwellings of light-frame construction.

9.4.3 Quality Assurance. The performance required of structures in Seismic Design Categories C, D, E, or F requires that special attention be paid to quality assurance during construction. Refer to A.9.3 for supplementary provisions.

TABLE 9.4.2.1a
SEISMIC DESIGN CATEGORY BASED ON SHORT
PERIOD RESPONSE ACCELERATIONS

Value of S_{DS}	Seismic Use Group		
	I	II	III
$S_{DS} < 0.167g$	A	A	A
$0.167g \leq S_{DS} < 0.33g$	B	B	C
$0.33g \leq S_{DS} < 0.50g$	C	C	D
$0.50g \leq S_{DS}$	D[a]	D[a]	D[a]

[a] Seismic Use Group I and II structures located on sites with mapped maximum considered earthquake spectral response acceleration at 1-second period, S_1, equal to or greater than 0.75g shall be assigned to Seismic Design Category E and Seismic Use Group III structures located on such sites shall be assigned to Seismic Design Category F.

Value of S_{D1}	Seismic Use Group		
	I	II	III
$S_{D1} < 0.067g$	A	A	A
$0.067g \leq S_{D1} < 0.133g$	B	B	C
$0.133g \leq S_{D1} < 0.20g$	C	C	D
$0.20g \leq S_{D1}$	D[a]	D[a]	D[a]

[a] Seismic Use Group I and II structures located on sites with mapped maximum considered earthquake spectral response acceleration at 1-second period, S_1, equal to or greater than 0.75g shall be assigned to Seismic Design Category E and Seismic Use Group III structures located on such sites shall be assigned to Seismic Design Category F.

SECTION 9.5
STRUCTURAL DESIGN CRITERIA, ANALYSIS, AND PROCEDURES

9.5.1 This Section Has Been Intentionally Left Blank.

9.5.2 Structural Design Requirements.

9.5.2.1 Design Basis. The seismic analysis and design procedures to be used in the design of structures and their components shall be as prescribed in this Section. The structure shall include complete lateral and vertical force-resisting systems capable of providing adequate strength, stiffness, and energy dissipation capacity to withstand the design ground motions within the prescribed limits of deformation and strength demand. The design ground motions shall be assumed to occur along any horizontal direction of a structure. The adequacy of the structural systems shall be demonstrated through the construction of a mathematical model and evaluation of this model for the effects of design ground motions. The design seismic forces, and their distribution over the height of the structure, shall be established in accordance with one of the applicable procedures indicated in Section 9.5.2.5 and the corresponding internal forces and deformations in the members of the structure shall be determined. An approved alternative procedure shall not be used to establish the seismic forces and their distribution unless the corresponding internal forces and deformations in the members are determined using a model consistent with the procedure adopted.

Individual members shall be provided with adequate strength to resist the shears, axial forces, and moments determined in accordance with these provisions, and connections shall develop the strength of the connected members or the forces indicated above. The deformation of the structure shall not exceed the prescribed

limits when the structure is subjected to the design seismic forces.

A continuous load path, or paths, with adequate strength and stiffness shall be provided to transfer all forces from the point of application to the final point of resistance. The foundation shall be designed to resist the forces developed and accommodate the movements imparted to the structure by the design ground motions. The dynamic nature of the forces, the expected ground motion, and the design basis for strength and energy dissipation capacity of the structure shall be included in the determination of the foundation design criteria.

Allowable Stress Design is permitted to be used to evaluate sliding, overturning, and soil bearing at the soil-structure interface regardless of the design approach used in the design of the structure.

9.5.2.2 Basic Seismic Force-Resisting Systems. The basic lateral and vertical seismic force-resisting system shall conform to one of the types indicated in Table 9.5.2.2. Each type is subdivided by the types of vertical element used to resist lateral seismic forces. The structural system used shall be in accordance with the Seismic Design Category and height limitations indicated in Table 9.5.2.2. The appropriate response modification, coefficient, R, system overstrength factor, Ω_0, and the deflection amplification factor (C_d) indicated in Table 9.5.2.2 shall be used in determining the base shear, element design forces, and design story drift. Special framing requirements are indicated in Section 9.5.2.6 and Sections 9.8, 9.9, 9.10, 9.11, and 9.12 for structures assigned to the various Seismic Design Categories.

Seismic force-resisting systems that are not contained in Table 9.5.2.2 shall be permitted if analytical and test data are submitted that establish the dynamic characteristics and demonstrate the lateral force resistance and energy dissipation capacity to be equivalent to the structural systems listed in Table 9.5.2.2 for equivalent response modification coefficient, R, system overstrength coefficient, Ω_0, and deflection amplification factor, C_d, values.

9.5.2.2.1 Dual System. For a dual system, the moment frame shall be capable of resisting at least 25% of the design seismic forces. The total seismic-force resistance is to be provided by the combination of the moment frame and the shear walls or braced frames in proportion to their rigidities.

9.5.2.2.2 Combinations of Framing Systems. Different seismic force-resisting systems are permitted along the two orthogonal axes of the structure. Combinations of seismic force-resisting systems shall comply with the requirements of this Section.

Basic Seismic Force-Resisting System	Response Modification Coefficient, R^a	System Over-strength Factor, $\Omega_0{}^g$	Deflection Amplification Factor, $C_d{}^b$	Structural System Limitations and Building Height (ft) Limitations[c] Seismic Design Category				
				A&B	C	D^d	E^e	F^e
Bearing Wall Systems								
Ordinary steel concentrically braced frames	4	2	$3\frac{1}{2}$	NL	NL	35^k	35^k	NP^k
Special reinforced concrete shear walls	5	$2\frac{1}{2}$	5	NL	NL	160	160	100
Ordinary reinforced concrete shear walls	4	$2\frac{1}{2}$	4	NL	NL	NP	NP	NP
Detailed plain concrete shear walls	$2\frac{1}{2}$	$2\frac{1}{2}$	2	NL	NP	NP	NP	NP
Ordinary plain concrete shear walls	$1\frac{1}{2}$	$2\frac{1}{2}$	$1\frac{1}{2}$	NL	NP	NP	NP	NP
Special reinforced masonry shear walls	5	$2\frac{1}{2}$	$3\frac{1}{2}$	NL	NL	160	160	100
Intermediate reinforced masonry shear walls	$3\frac{1}{2}$	$2\frac{1}{2}$	$2\frac{1}{4}$	NL	NL	NP	NP	NP
Ordinary reinforced masonry shear walls	2	$2\frac{1}{2}$	$1\frac{3}{4}$	NL	160	NP	NP	NP
Detailed plain masonry shear walls	2	$2\frac{1}{2}$	$1\frac{3}{4}$	NL	NP	NP	NP	NP
Ordinary plain masonry shear walls	$1\frac{1}{2}$	$2\frac{1}{2}$	$1\frac{1}{4}$	NL	NP	NP	NP	NP
Light-framed walls sheathed with wood structural panels rated for shear resistance or steel sheets	6	3	4	NL	NL	65	65	65
Light-framed walls with shear panels of all other materials	2	$2\frac{1}{2}$	2	NL	NL	35	NP	NP
Light-framed wall systems using flat strap bracing	4	2	$3\frac{1}{2}$	NL	NL	65	65	65
Building Frame Systems								
Steel eccentrically braced frames, moment resisting, connections at columns away from links	8	2	4	NL	NL	160	160	100
Steel eccentrically braced frames, non-moment resisting, connections at columns away from links	7	2	4	NL	NL	160	160	100
Special steel concentrically braced frames	6	2	5	NL	NL	160	160	100
Ordinary steel concentrically braced frames	5	2	$4\frac{1}{2}$	NL	NL	35^k	35^k	NP^k
Special reinforced concrete shear walls	6	$2\frac{1}{2}$	5	NL	NL	160	160	100
Ordinary reinforced concrete shear walls	5	$2\frac{1}{2}$	$4\frac{1}{2}$	NL	NL	NP	NP	NP
Detailed plain concrete shear walls	3	$2\frac{1}{2}$	$2\frac{1}{2}$	NL	NP	NP	NP	NP
Ordinary plain concrete shear walls	2	$2\frac{1}{2}$	2	NL	NP	NP	NP	NP
Composite eccentrically braced frames	8	2	4	NL	NL	160	160	100
Composite concentrically braced frames	5	2	$4\frac{1}{2}$	NL	NL	160	160	100
Ordinary composite braced frames	3	2	3	NL	NL	NP	NP	NP
Composite steel plate shear walls	$6\frac{1}{2}$	$2\frac{1}{2}$	$5\frac{1}{2}$	NL	NL	160	160	100
Special composite reinforced concrete shear walls with steel elements	6	$2\frac{1}{2}$	5	NL	NL	160	160	100
Ordinary composite reinforced concrete shear walls with steel elements	5	$2\frac{1}{2}$	$4\frac{1}{4}$	NL	NL	NP	NP	NP
Special reinforced masonry shear walls	$5\frac{1}{2}$	$2\frac{1}{2}$	4	NL	NL	160	160	100
Intermediate reinforced masonry shear walls	4	$2\frac{1}{2}$	4	NL	NL	NP	NP	NP
Ordinary reinforced masonry shear walls	$2\frac{1}{2}$	$2\frac{1}{2}$	$2\frac{1}{4}$	NL	160	NP	NP	NP

Basic Seismic Force-Resisting System	Response Modification Coefficient, R^a	System Over-strength Factor, $W_0{}^g$	Deflection Amplification Factor, $C_d{}^b$	Structural System Limitations and Building Height (ft) Limitations[c]				
				Seismic Design Category				
				A&B	C	D^d	E^e	F^e
Detailed plain masonry shear walls	$2\frac{1}{2}$	$2\frac{1}{2}$	$2\frac{1}{4}$	NL	160	NP	NP	NP
Ordinary plain masonry shear walls	$1\frac{1}{2}$	$2\frac{1}{2}$	$1\frac{1}{4}$	NL	NP	NP	NP 65	NP
Light-framed walls sheathed with wood structural panels rated for shear resistance or steel sheets	$6\frac{1}{2}$	$2\frac{1}{2}$	$4\frac{1}{4}$	NL	NL	65	65	65
Light-framed walls with shear panels of all other materials	$2\frac{1}{2}$	$2\frac{1}{2}$	$2\frac{1}{2}$	NL	NL	35	NP	NP
Moment Resisting Frame Systems								
Special steel moment frames	8	3	$5\frac{1}{2}$	NL	NL	NL	NL	NL
Special steel truss moment frames	7	3	$5\frac{1}{2}$	NL	NL	160	100	NP
Intermediate steel moment frames	4.5	3	4	NL	NL	35^h	$NP^{h,\,i}$	$NP^{h,\,i}$
Ordinary steel moment frames	3.5	3	3	NL	NL	$NP^{h,\,i}$	$NP^{h,\,i}$	$NP^{h,\,i}$
Special reinforced concrete moment frames	8	3	$5\frac{1}{2}$	NL	NL	NL	NL	NL
Intermediate reinforced concrete moment frames	5	3	$4\frac{1}{2}$	NL	NL	NP	NP	NP
Ordinary reinforced concrete moment frames	3	3	$2\frac{1}{2}$	NL	NP	NP	NP	NP
Special composite moment frames	8	3	$5\frac{1}{2}$	NL	NL	NL	NL	NL
Intermediate composite moment frames	5	3	$4\frac{1}{2}$	NL	NL	NP	NP	NP
Composite partially restrained moment frames	6	3	$5\frac{1}{2}$	160	160	100	NP	NP
Ordinary composite moment frames	3	3	$2\frac{1}{2}$	NL	NP	NP	NP	NP
Special masonry moment frames	$5\frac{1}{2}$	3	5	NL	NL	160	160	100
Dual Systems with Special Moment Frames Capable of Resisting at Least 25% of Prescribed Seismic Forces								
Steel eccentrically braced frames, moment resisting connections, at columns away from links	8	$2\frac{1}{2}$	4	NL	NL	NL	NL	NL
Steel eccentrically braced frames, non-moment resisting connections, at columns away from links	7	$2\frac{1}{2}$	4	NL	NL	NL	NL	NL
Special steel concentrically braced frames	8	$2\frac{1}{2}$	$6\frac{1}{2}$	NL	NL	NL	NL	NL
Special reinforced concrete shear walls	8	$2\frac{1}{2}$	$6\frac{1}{2}$	NL	NL	NL	NL	NL
Ordinary reinforced concrete shear walls	7	$2\frac{1}{2}$	6	NL	NL	NP	NP	NP
Composite eccentrically braced frames	8	$2\frac{1}{2}$	4	NL	NL	NL	NL	NL
Composite concentrically braced frames	6	$2\frac{1}{2}$	5	NL	NL	NL	NL	NL
Composite steel plate shear walls	8	$2\frac{1}{2}$	$6\frac{1}{2}$	NL	NL	NL	NL	NL
Special composite reinforced concrete shear walls with steel elements	8	$2\frac{1}{2}$	$6\frac{1}{2}$	NL	NL	NL	NL	NL

Basic Seismic Force-Resisting System	Response Modification Coefficient, R^a	System Over-strength Factor, $W_0{}^g$	Deflection Amplification Factor, $C_d{}^b$	Structural System Limitations and Building Height (ft) Limitations[c] Seismic Design Category				
				A&B	C	D[d]	E[e]	F[e]
Ordinary composite reinforced concrete shear walls with steel elements	7	$2\frac{1}{2}$	6	NL	NL	NP	NP	NP
Special reinforced masonry shear walls	7	3	$6\frac{1}{2}$	NL	NL	NL	NL	NL
Intermediate reinforced masonry shear walls	6	$2\frac{1}{2}$	5	NL	NL	NL	NL	NL
Ordinary steel concentrically braced frames	6	$2\frac{1}{2}$	5	NL	NL	NL	NL	NL
Dual Systems with Intermediate Moment Frames Capable of Resisting at Least 25% of Prescribed Seismic Forces								
Special steel concentrically braced frames[f]	$4\frac{1}{2}$	$2\frac{1}{2}$	$4\frac{1}{2}$	NL	NL	35	NP	NP[h,i]
Special reinforced concrete shear walls	6	$2\frac{1}{2}$	5	NL	NL	160	100	100
Ordinary reinforced masonry shear walls	3	3	$2\frac{1}{2}$	NL	160	NP	NP	NP
Intermediate reinforced masonry shear walls	5	3	$4\frac{1}{2}$	NL	NL	NP	NP	NP
Composite concentrically braced frames	5	$2\frac{1}{2}$	$4\frac{1}{2}$	NL	NL	160	100	NP
Ordinary composite braced frames	4	$2\frac{1}{2}$	3	NL	NL	NP	NP	NP
Ordinary composite reinforced concrete shear walls with steel elements	5	3	$4\frac{1}{2}$	NL	NL	NP	NP	NP
Ordinary steel concentrically braced frames	5	$2\frac{1}{2}$	$4\frac{1}{2}$	NL	NL	160	100	NP
Ordinary reinforced concrete shear walls	$5\frac{1}{2}$	$2\frac{1}{2}$	$4\frac{1}{2}$	NL	NL	NP	NP	NP
Inverted Pendulum Systems and Cantilevered Column Systems								
Special steel moment frames	$2\frac{1}{2}$	2	$2\frac{1}{2}$	NL	NL	NL	NL	NL
Ordinary steel moment frames	$1\frac{1}{4}$	2	$2\frac{1}{2}$	NL	NL	NP	NP	NP
Special reinforced concrete moment frames	$2\frac{1}{2}$	2	$1\frac{1}{4}$	NL	NL	NL	NL	NL
Structural Steel Systems Not Specifically Detailed for Seismic Resistance	3	3	3	NL	NL	NP	NP	NP

[a] Response modification coefficient, R, for use throughout the standard. Note R reduces forces to a strength level, not an allowable stress level.

[b] Deflection amplification factor, C_d, for use in Sections 9.5.3.7.1 and 9.5.3.7.2

[c] NL = Not Limited and NP = Not Permitted. For metric units use 30 m for 100 ft and use 50 m for 160 ft. Heights are measured from the base of the structure as defined in Section 9.2.1.

[d] See Section 9.5.2.2.4.1 for a description of building systems limited to buildings with a height of 240 ft (75 m) or less.

[e] See Sections 9.5.2.2.4 and 9.5.2.2.4.5 for building systems limited to buildings with a height of 160 ft (50 m) or less.

[f] Ordinary moment frame is permitted to be used in lieu of intermediate moment frame in Seismic Design Categories B and C.

[g] The tabulated value of the overstrength factor, W_0, may be reduced by subtracting $\frac{1}{2}$ for structures with flexible diaphragms but shall not be taken as less than 2.0 for any structure.

[h] Steel ordinary moment frames and intermediate moment frames are permitted in single-story buildings up to a height of 60 ft, when the moment joints of field connections are constructed of bolted end plates and the dead load of the roof does not exceed 15 psf.

[i] Steel ordinary moment frames are permitted in buildings up to a height of 35 ft where the dead load of the walls, floors, and roofs does not exceed 15 psf.

[k] Steel ordinary concentrically braced frames are permitted in single-story buildings up to a height of 60 ft when the dead load of the roof does not exceed 15 psf and in penthouse structures.

9.5.2.2.2.1 *R* and Ω_0 Factors. The response modification, coefficient, *R*, in the direction under consideration at any story shall not exceed the lowest response modification coefficient, *R*, for the seismic force-resisting system in the same direction considered above that story excluding penthouses. For other than dual systems, where a combination of different structural systems is utilized to resist lateral forces in the same direction, the value of *R* used in that direction shall not be greater than the least value of any of the systems utilized in the same direction. If a system other than a dual system with a response modification coefficient, *R*, with a value of less than 5 is used as part of the seismic force-resisting system in any direction of the structure, the lowest such value shall be used for the entire structure. The system overstrength factor, Ω_0, in the direction under consideration at any story shall not be less than the largest value of this factor for the seismic force-resisting system in the same direction considered above that story.

Exceptions:

1. The limit does not apply to supported structural systems with a weight equal to or less than 10% of the weight of the structure.
2. Detached one- and two-family dwellings of light-frame construction.

9.5.2.2.2.2 Combination Framing Detailing Requirements. The detailing requirements of Section 9.5.2.6 required by the higher response modification coefficient, *R*, shall be used for structural components common to systems having different response modification coefficients.

9.5.2.2.3 Seismic Design Categories B and C. The structural framing system for structures assigned to Seismic Design Categories B and C shall comply with the structure height and structural limitations in Table 9.5.2.2.

9.5.2.2.4 Seismic Design Categories D and E. The structural framing system for a structure assigned to Seismic Design Categories D and E shall comply with Section 9.5.2.2.3 and the additional provisions of this Section.

9.5.2.2.4.1 Increased Building Height Limit. The height limits in Table 9.5.2.2 are permitted to be increased to 240 ft (75 m) in buildings that have steel braced frames or concrete cast-in-place shear walls and that meet the requirements of

this Section. In such buildings the braced frames or cast-in-place special reinforced concrete shear walls in any one plane shall resist no more than 60% of the total seismic forces in each direction, neglecting torsional effects. The seismic force in any braced frame or shear wall in any one plane resulting from torsional effects shall not exceed 20% of the total seismic force in that braced frame or shear wall.

9.5.2.2.4.2 Interaction Effects. Moment resisting frames that are enclosed or adjoined by more rigid elements not considered to be part of the seismic force-resisting system shall be designed so that the action or failure of those elements will not impair the vertical load and seismic force-resisting capability of the frame. The design shall provide for the effect of these rigid elements on the structural system at structure deformations corresponding to the design story drift (Δ) as determined in Section 9.5.3.7. In addition, the effects of these elements shall be considered when determining whether a structure has one or more of the irregularities defined in Section 9.5.2.3.

9.5.2.2.4.3 Deformational Compatibility. Every structural component not included in the seismic force-resisting system in the direction under consideration shall be designed to be adequate for the vertical load-carrying capacity and the induced moments and shears resulting from the design story drift (Δ) as determined in accordance with Section 9.5.3.7; see also Section 9.5.2.8.

Exception: Reinforced concrete frame members not designed as part of the seismic force-resisting system shall comply with Section 21.9 of Ref. 9.9-1.

When determining the moments and shears induced in components that are not included in the seismic force-resisting system in the direction under consideration, the stiffening effects of adjoining rigid structural and nonstructural elements shall be considered and a rational value of member and restraint stiffness shall be used.

9.5.2.2.4.4 Special Moment Frames. A special moment frame that is used but not required by Table 9.5.2.2 shall not be discontinued and supported by a more rigid system with a lower response modification coefficient (*R*) unless the requirements of Sections 9.5.2.6.2.4 and 9.5.2.6.4.2 are met. Where a special moment frame is required by Table 9.5.2.2, the frame shall be continuous to the foundation.

9.5.2.2.5 Seismic Design Category F. The framing systems of structures assigned to Seismic Design Category F shall conform to the requirements of Section 9.5.2.2.4 for Seismic Design Categories D and E and to the additional requirements and limitations of this Section. The increased height limit of Section 9.5.2.2.4.1 for braced frame or shear wall systems shall be reduced from 240 ft (75 m) to 160 ft (50 m).

9.5.2.3 Structure Configuration. Structures shall be classified as regular or irregular based on the criteria in this Section. Such classification shall be based on the plan and vertical configuration.

9.5.2.3.1 Diaphragm Flexibility. A diaphragm shall be considered flexible for the purposes of distribution of story shear and torsional moment when the computed maximum in-plane deflection of the diaphragm itself under lateral load is more than two times the average drift of adjoining vertical elements of the lateral force-resisting system of the associated story under equivalent tributary lateral load. The loadings used for this calculation shall be those prescribed by Section 9.5.5.

9.5.2.3.2 Plan Irregularity. Structures having one or more of the irregularity types listed in Table 9.5.2.3.2 shall be designated as having plan structural irregularity. Such structures assigned to the Seismic Design Categories listed in Table 9.5.2.3.2 shall comply with the requirements in the sections referenced in that table.

9.5.2.3.3 Vertical Irregularity. Structures having one or more of the irregularity types listed in Table 9.5.2.3.3 shall be designated as having vertical irregularity. Such structures assigned to the Seismic

TABLE 9.5.2.3.2
PLAN STRUCTURAL IRREGULARITIES

	Irregularity Type and Description	Reference Section	Seismic Design Category Application
1a.	Torsional Irregularity Torsional irregularity is defined to exist where the maximum story drift, computed including accidental torsion, at one end of the structure transverse to an axis is more than 1.2 times the average of the story drifts at the two ends of the structure. Torsional irregularity requirements in the reference sections apply only to structures in which the diaphragms are rigid or semirigid.	9.5.2.6.4.2 9.5.5.5.2	D, E, and F C, D, E, and F
1b.	Extreme Torsional Irregularity Extreme Torsional Irregularity is defined to exist where the maximum story drift, computed including accidental torsion, at one end of the structure transverse to an axis is more than 1.4 times the average of the story drifts at the two ends of the structure. Extreme torsional irregularity requirements in the reference sections apply only to structures in which the diaphragms are rigid or semirigid.	9.5.2.6.4.2 9.5.5.5.2 9.5.2.6.5.1	D C and D E and F
2.	Re-entrant Corners Plan configurations of a structure and its lateral force-resisting system contain re-entrant corners, where both projections of the structure beyond a re-entrant corner are greater than 15% of the plan dimension of the structure in the given direction.	9.5.2.6.4.2	D, E, and F
3.	Diaphragm Discontinuity Diaphragms with abrupt discontinuities or variations in stiffness, including those having cutout or open areas greater than 50% of the gross enclosed diaphragm area, or changes in effective diaphragm stiffness of more than 50% from one-story to the next.	9.5.2.6.4.2	D, E, and F
4.	Out-of-Plane Offsets Discontinuities in a lateral force-resistance path, such as out-of-plane offsets of the vertical elements.	9.5.2.6.4.2 9.5.2.6.2.11	D, E, and F B, C, D, E, or F
5.	Nonparallel Systems The vertical lateral force-resisting elements are not parallel to or symmetric about the major orthogonal axes of the lateral force-resisting system.	9.5.2.6.3.1	C, D, E, and F

TABLE 9.5.2.3.3
VERTICAL STRUCTURAL IRREGULARITIES

Irregularity Type and Description	Reference Section	Seismic Design Category Application
1a. Stiffness Irregularity: Soft Story A soft story is one in which the lateral stiffness is less than 70% of that in the story above or less than 80% of the average stiffness of the three stories above.	9.5.2.5.1	D, E, and F
1b. Stiffness Irregularity: Extreme Soft Story An extreme soft story is one in which the lateral stiffness is less than 60% of that in the story above or less than 70% of the average stiffness of the three stories above.	9.5.2.5.1 9.5.2.6.5.1	D, E and F
2. Weight (Mass) Irregularity Mass irregularity shall be considered to exist where the effective mass of any story is more than 150% of the effective mass of an adjacent story. A roof that is lighter than the floor below need not be considered.	9.5.2.5.1	D, E, and F
3. Vertical Geometric Irregularity Vertical geometric irregularity shall be considered to exist where the horizontal dimension of the lateral force-resisting system in any story is more than 130% of that in an adjacent story.	9.5.2.5.1	D, E, and F
4. In-Plane Discontinuity in Vertical Lateral Force-Resisting Elements In-plane discontinuity in vertical lateral force-resisting elements shall be considered to exist where an in-plane offset of the lateral force-resisting elements is greater than the length of those elements or there exists a reduction in stiffness of the resisting element in the story below.	9.5.2.6.2.11 9.5.2.6.4.2	B, C, D, E, and F D, E, and F
5. Discontinuity in Lateral Strength: Weak Story A weak story is one in which the story lateral strength is less than 80% of that in the story above. The story strength is the total strength of all seismic-resisting elements sharing the story shear for the direction under consideration.	9.5.2.6.2.2 9.5.2.6.5.1	B, C, D, E, and F E and F

Design Categories listed in Table 9.5.2.3.3 shall comply with the requirements in the sections referenced in that table.

Exceptions:

1. Vertical structural irregularities of Types 1a, 1b, or 2 in Table 9.5.2.3.3 do not apply where no story drift ratio under design lateral seismic force is greater than 130% of the story drift ratio of the next story above. Torsional effects need not be considered in the calculation of story drifts. The story drift ratio relationship for the top 2 stories of the structure are not required to be evaluated.

2. Irregularities Types 1a, 1b, and 2 of Table 9.5.2.3.3 are not required to be considered for 1-story buildings in any Seismic Design Category or for 2-story buildings in Seismic Design Categories A, B, C, or D.

9.5.2.4 Redundancy. A reliability factor, ρ, shall be assigned to all structures in accordance with this Section, based on the extent of structural redundancy inherent in the lateral force-resisting system.

9.5.2.4.1 Seismic Design Categories A, B, and C. For structures in Seismic Design Categories A, B, and C, the value of ρ is 1.0.

9.5.2.4.2 Seismic Design Category D. For structures in Seismic Design Category D, ρ shall be taken as the largest of the values of ρ_x calculated at each story "x" of the structure in accordance with Eq. 9.5.2.4.2-1 as follows:

$$\rho_x = 2 - \frac{20}{r_{max_x}\sqrt{A_x}} \qquad \textbf{(Eq. 9.5.2.4.2-1)}$$

where

r_{max_x} = the ratio of the design story shear resisted by the single element carrying the most shear force in the story to the total story shear, for a given direction of loading. For

braced frames, the value of r_{max_x} is equal to the lateral force component in the most heavily loaded brace element divided by the story shear. For moment frames, r_{max_x} shall be taken as the maximum of the sum of the shears in any two adjacent columns in the plane of a moment frame divided by the story shear. For columns common to two bays with moment resisting connections on opposite sides at the level under consideration, 70% of the shear in that column may be used in the column shear summation. For shear walls, r_{max_x} shall be taken equal to shear in the most heavily loaded wall or wall pier multiplied by $10/l_w$ (the metric coefficient is $3.3/l_w$) where l_w is the wall or wall pier length in ft (m) divided by the story shear and where the ratio $10/l_w$ need not be taken greater than 1.0 for buildings of light-frame construction. For dual systems, r_{max_x} shall be taken as the maximum value as defined above considering all lateral-load–resisting elements in the story. The lateral loads shall be distributed to elements based on relative rigidities considering the interaction of the dual system. For dual systems, the value of ρ need not exceed 80% of the value calculated above.

A_x = the floor area in ft^2 of the diaphragm level immediately above the story

The value of ρ need not exceed 1.5, which may be used for any structure. The value of ρ shall not be taken as less than 1.0.

Exception: For structures with lateral-force-resisting systems in any direction consisting solely of special moment frames, the lateral-force-resisting system shall be configured such that the value of ρ calculated in accordance with this Section does not exceed 1.25.

The metric equivalent of Eq. 9.5.2.4.2 is:

$$\rho_x = 2 - \frac{6.1}{r_{\max_x}\sqrt{A_x}} \qquad \textbf{(Eq. 9.5.2.4.2)}$$

where A_x is in m^2.

9.5.2.4.3 Seismic Design Categories E and F.
For structures in Seismic Design Categories E and F, the value of r shall be calculated as indicated in Section 9.5.2.4.2, above.

Exception: For structures with lateral force-resisting systems in any direction consisting

solely of special moment frames, the lateral-force-resisting system shall be configured such that the value of ρ calculated in accordance with Section 9.5.2.4.2 does not exceed 1.1.

9.5.2.5 Analysis Procedures.
A structural analysis conforming to one of the types permitted in Section 9.5.2.5.1 shall be made for all structures. Application of loading shall be as indicated in Section 9.5.2.5.2.1, and as required by the selected analysis procedure. All members of the structure's seismic force-resisting system and their connections shall have adequate strength to resist the forces, Q_E, predicted by the analysis, in combination with other loads, as required by Section 9.5.2.7. Drifts predicted by the analysis shall be within the limits specified by Section 9.5.2.8. If a nonlinear analysis is performed, component deformation demands shall not exceed limiting values as indicated in Section 9.5.7.3.2.

Exception: For structures designed using the index force analysis procedure of Section 9.5.3 or the simplified analysis procedure of Section 9.5.4, drift need not be evaluated.

9.5.2.5.1 Analysis Procedures.
The structural analysis required by Section 9.5.2.5 shall consist of one of the types permitted in Table 9.5.2.5.1, based on the structure's Seismic Design Category, structural system, dynamic properties and regularity, or with the approval of the authority having jurisdiction, an alternative generally accepted procedure shall be permitted to be used.

9.5.2.5.2 Application of Loading.
The directions of application of seismic forces used in the design shall be those which will produce the most critical load effects. It shall be permitted to satisfy this requirement using the procedures of Section 9.5.2.5.2.1 for Seismic Design Category A and B, Section 9.5.2.5.2.2 for Seismic Design Category C, and Section 9.5.2.5.2.3 for Seismic Design Categories D, E, and F. All structural components and their connections shall be provided with strengths sufficient to resist the effects of the seismic forces prescribed herein. Loads shall be combined as prescribed in Section 9.5.2.7.

9.5.2.5.2.1 Seismic Design Categories A and B.
For structures assigned to Seismic Design Category A and B, the design seismic forces are permitted to be applied separately in each of two orthogonal directions and orthogonal interaction effects may be neglected.

9.5.2.5.2.2 Seismic Design Category C.
Loading applied to structures assigned to Seismic Design

TABLE 9.5.2.5.1
PERMITTED ANALYTICAL PROCEDURES

Seismic Design Category	Structural Characteristics	Index Force Analysis Section 9.5.3	Simplified Analysis Section 9.5.4	Equivalent Lateral Force Analysis Section 9.5.5	Modal Response Spectrum Analysis Section 9.5.6	Linear Response History Analysis Section 9.5.7	Nonlinear Response History Analysis Section 9.5.8
A	All structures	P	P	P	P	P	P
B, C	SUG-1 buildings of light-framed construction not exceeding three stories in height	NP	P	P	P	P	P
	Other SUG-1 buildings not exceeding two stories in height	NP	P	P	P	P	P
	All other structures	NP	NP	P	P	P	P
D, E, F	SUG-1 buildings of light-framed construction not exceeding three stories in height	NP	P	P	P	P	
	Other SUG-1 buildings not exceeding two stories in height	NP	P	P	P	P	P
	Regular structures with $T < 3.5\ T_s$ and all structures of light-frame construction	NP	NP	P	P	P	P
	Irregular structures with $T < 3.5\ T_s$ and having only plan irregularities type 2, 3, 4, or 5 of Table 9.5.2.3.2 or vertical irregularities type 4 or 5 of Table 9.5.2.3.3	NP	NP	P	P	P	P
	All other structures	NP	NP	NP	P	P	P

Notes: P—indicates permitted, NP—indicates not permitted

Category C shall, as a minimum, conform to the requirements of Section 9.5.2.5.2.1 for Seismic Design Categories A and B and the requirements of this Section. Structures that have plan structural irregularity Type 5 in Table 9.5.2.3.2 shall be analyzed for seismic forces using a three-dimensional representation and either of the following procedures:

a. The structure shall be analyzed using the equivalent lateral force analysis procedure of Section 9.5.5, the modal response spectrum analysis procedure of Section 9.5.6, or the linear response history analysis procedure of Section 9.5.7, as permitted under Section 9.5.2.5.1, with the loading applied independently in any two orthogonal directions and the most critical load effect due to direction of application of seismic forces on the structure may be assumed to be satisfied if components and their foundations are designed for the following combination of prescribed loads: 100% of the forces for one direction plus 30% of the forces for the perpendicular direction; the combination requiring the maximum component strength shall be used.

b. The structure shall be analyzed using the linear response history analysis procedure of Section 9.5.7 or the nonlinear response history analysis procedure of Section 9.5.8, as permitted by Section 9.5.2.5.1, with

orthogonal pairs of ground motion acceleration histories applied simultaneously.

9.5.2.5.2.3 Seismic Design Categories D, E, and F. Structures assigned to Seismic Design Categories D, E, and F shall, as a minimum, conform to the requirements of Section 9.5.2.5.2.2 In addition, any column or wall that forms part of two or more intersecting seismic force-resisting systems and is subjected to axial load due to seismic forces acting along either principal plan axis equaling or exceeding 20% of the axial design strength of the column or wall shall be designed for the most critical load effect due to application of seismic forces in any direction. Either of the procedures of Section 9.5.2.5.2a or b shall be permitted to be used to satisfy this requirement. Two-dimensional analyses shall be permitted for structures with flexible diaphragms.

9.5.2.6 Design and Detailing Requirements. The design and detailing of the components of the seismic force-resisting system shall comply with the requirements of this Section. Foundation design shall conform to the applicable requirements of Section 9.7. The materials and the systems composed of those materials shall conform to the requirements and limitations of Sections 9.8 through 9.12 for the applicable category.

9.5.2.6.1 Seismic Design Category A. The design and detailing of structures assigned to Category A shall comply with the requirements of this Section.

9.5.2.6.1.1 Load Path Connections. All parts of the structure between separation joints shall be interconnected to form a continuous path to the seismic force-resisting system, and the connections shall be capable of transmitting the seismic force (F_p) induced by the parts being connected. Any smaller portion of the structure shall be tied to the remainder of the structure with elements having a design strength capable of transmitting a seismic force of 0.133 times the short period design spectral response acceleration coefficient, S_{DS}, times the weight of the smaller portion or 5% of the portion's weight, whichever is greater. This connection force does not apply to the overall design of the lateral-force-resisting system. Connection design forces need not exceed the maximum forces that the structural system can deliver to the connection.

A positive connection for resisting a horizontal force acting parallel to the member shall be provided for each beam, girder, or truss to its support.

The connection shall have a minimum strength of 5% of the dead plus live load reaction. One means to provide the strength is to use connecting elements such as slabs.

9.5.2.6.1.2 Anchorage of Concrete or Masonry Walls. Concrete and masonry walls shall be anchored to the roof and all floors and members that provide lateral support for the wall or which are supported by the wall. The anchorage shall provide a direct connection between the walls and the roof or floor construction. The connections shall be capable of resisting the horizontal forces specified in Section 9.5.2.6.1.1 but not less than a minimum strength level, horizontal force of 280 lbs/ linear ft (4.09 kN/m) of wall substituted for E in the load combinations of Section 2.3.2 or 2.4.1.

9.5.2.6.2 Seismic Design Category B. Structures assigned to Seismic Design Category B shall conform to the requirements of Section 9.5.2.6.1 for Seismic Design Category A and the requirements of this Section.

9.5.2.6.2.1 *P*-Delta Effects. *P*-delta effects shall be included where required by Section 9.5.5.7.2.

9.5.2.6.2.2 Openings. Where openings occur in shear walls, diaphragms, or other plate-type elements, reinforcement at the edges of the openings shall be designed to transfer the stresses into the structure. The edge reinforcement shall extend into the body of the wall or diaphragm a distance sufficient to develop the force in the reinforcement. The extension must be sufficient in length to allow dissipation or transfer of the force without exceeding the shear and tension capacity of the diaphragm or the wall.

9.5.2.6.2.3 Direction of Seismic Load. The direction of application of seismic forces used in design shall be that which will produce the most critical load effect in each component. This requirement will be deemed satisfied if the design seismic forces are applied separately and independently in each of two orthogonal directions.

9.5.2.6.2.4 Discontinuities in Vertical System. Structures with a discontinuity in lateral capacity, vertical irregularity Type 5 as defined in Table 9.5.2.3.3, shall not be more than 2 stories or 30 ft (9 m) in height where the "weak" story has a calculated strength of less than 65% of the story above.

Exception: The limit does not apply where the "weak" story is capable of resisting a total seismic force equal to Ω_0 times the design force prescribed in Section 9.5.3.

9.5.2.6.2.5 Nonredundant Systems. The design of a structure shall consider the potentially adverse effect that the failure of a single member, connection, or component of the seismic force-resisting system will have on the stability of the structure; see Section 1.4.

9.5.2.6.2.6 Collector Elements. Collector elements shall be provided that are capable of transferring the seismic forces originating in other portions of the structure to the element providing the resistance to those forces.

9.5.2.6.2.7 Diaphragms. The deflection in the plane of the diaphragm, as determined by engineering analysis, shall not exceed the permissible deflection of the attached elements. Permissible deflection shall be that deflection which will permit the attached element to maintain its structural integrity under the individual loading and continue to support the prescribed loads.

Floor and roof diaphragms shall be designed to resist F_p where F_p is the larger of:

1. The portion of the design seismic force at the level of the diaphragm that depends on the diaphragm for transmission to the vertical elements of the seismic force-resisting system, or

2. $F_p = 0.2 S_{DS} I w_p + V_{px}$ **(Eq. 9.5.2.6.2.7)**

where

F_p = the seismic force induced by the parts

I = occupancy importance factor (Table 9.1.4)

S_{DS} = the short period site design spectral response acceleration coefficient, Section 9.4.1

w_p = the weight of the diaphragm and other elements of the structure attached to it

V_{px} = the portion of the seismic shear force at the level of the diaphragm, required to be transferred to the components of the vertical seismic force-resisting system because of the offsets or changes in stiffness of the vertical components above or below the diaphragm

Diaphragms shall be designed for both the shear and bending stresses resulting from these forces. Diaphragms shall have ties or struts to distribute the wall anchorage forces into the diaphragm. Diaphragm connections shall be positive, mechanical, or welded type connections.

At diaphragm discontinuities, such as openings and re-entrant corners, the design shall ensure that the dissipation or transfer of edge (chord) forces combined with other forces in the diaphragm is within shear and tension capacity of the diaphragm.

9.5.2.6.2.8 Anchorage of Concrete or Masonry Walls. Exterior and interior bearing walls and their anchorage shall be designed for a force normal to the surface equal to 40% of the short period design spectral response acceleration, S_{DS}, times the occupancy importance factor, I, multiplied by the weight of wall (W_c) associated with the anchor, with a minimum force of 10% of the weight of the wall. Interconnection of wall elements and connections to supporting framing systems shall have sufficient ductility, rotational capacity, or sufficient strength to resist shrinkage, thermal changes, and differential foundation settlement when combined with seismic forces. The connections shall also satisfy Section 9.5.2.6.1.2.

The anchorage of concrete or masonry walls to supporting construction shall provide a direct connection capable of resisting the greater of the force $0.4 S_{DS} I W_c$ as given above or $400 S_{DS} I$ lbs/linear ft ($5.84 S_{DS} I$ kN/m) of wall or the force specified in Section 9.5.2.6.1.2. Walls shall be designed to resist bending between anchors where the anchor spacing exceeds 4 ft (1219 mm).

9.5.2.6.2.9 Inverted Pendulum-Type Structures. Supporting columns or piers of inverted pendulum-type structures shall be designed for the bending moment calculated at the base determined using the procedures given in Section 9.5.3 and varying uniformly to a moment at the top equal to one-half the calculated bending moment at the base.

9.5.2.6.2.10 Anchorage of Nonstructural Systems. When required by Section 9.6, all portions or components of the structure shall be anchored for the seismic force, F_p, prescribed therein.

9.5.2.6.2.11 Elements Supporting Discontinuous Walls or Frames. Columns, beams, trusses, or slabs supporting discontinuous walls or frames of structures having plan irregularity Type 4 of Table 9.5.2.3.2 or vertical irregularity Type 4 of

Table 9.5.2.3.3 shall have the design strength to resist the maximum axial force that can develop in accordance with the special seismic loads of Section 9.5.2.7.1.

9.5.2.6.3 Seismic Design Category C. Structures assigned to Category C shall conform to the requirements of Section 9.5.2.6.2 for Category B and to the requirements of this Section.

9.5.2.6.3.1 Collector Elements. Collector elements shall be provided that are capable of transferring the seismic forces originating in other portions of the structure to the element providing the resistance to those forces. Collector elements, splices, and their connections to resisting elements shall resist the special seismic loads of Section 9.5.2.7.1.

Exception: In structures or portions thereof braced entirely by light-frame shear walls, collector elements, splices, and connections to resisting elements need only be designed to resist forces in accordance with Section 9.5.2.6.4.4.

The quantity $\Omega_0 E$ in Eq. 9.5.2.7.1-1 need not exceed the maximum force that can be transferred to the collector by the diaphragm and other elements of the lateral force-resisting system.

9.5.2.6.3.2 Anchorage of Concrete or Masonry Walls. Concrete or masonry walls shall be anchored to all floors, roofs, and members that provide out-of-plane lateral support for the wall or that are supported by the wall. The anchorage shall provide a positive direct connection between the wall and floor, roof, or supporting member capable of resisting horizontal forces specified in this Section for structures with flexible diaphragms or with Section 9.6.1.3 (using a_p of 1.0 and R_p of 2.5) for structures with diaphragms that are not flexible.

Anchorage of walls to flexible diaphragms shall have the strength to develop the out-of-plane force given by Eq. 9.5.2.6.3.2:

$$F_p = 0.8 S_{DS} I W_p \qquad \textbf{(Eq. 9.5.2.6.3.2)}$$

where

$F_p =$ the design force in the individual anchors
$S_{DS} =$ the design spectral response acceleration at short periods per Section 9.4.1.2.5
$I =$ the occupancy importance factor per Section 9.1.4

$W_p =$ the weight of the wall tributary to the anchor

Diaphragms shall be provided with continuous ties or struts between diaphragm chords to distribute these anchorage forces into the diaphragms. Added chords may be used to form subdiaphragms to transmit the anchorage forces to the main continuous cross ties. The maximum length-to-width ratio of the structural subdiaphragm shall be $2\frac{1}{2}$ to 1. Connections and anchorages capable of resisting the prescribed forces shall be provided between the diaphragm and the attached components. Connections shall extend into the diaphragm a sufficient distance to develop the force transferred into the diaphragm.

The strength design forces for steel elements of the wall anchorage system, other than anchor bolts and reinforcing steel, shall be 1.4 times the forces otherwise required by this Section.

In wood diaphragms, the continuous ties shall be in addition to the diaphragm sheathing. Anchorage shall not be accomplished by use of toe nails or nails subject to withdrawal nor shall wood ledgers of framing be used in cross-grain bending or cross-grain tension. The diaphragm sheathing shall not be considered effective as providing the ties or struts required by this Section.

In metal deck diaphragms, the metal deck shall not be used as the continuous ties required by this Section in the direction perpendicular to the deck span.

Diaphragm to wall anchorage using embedded straps shall be attached to, or hooked around, the reinforcing steel or otherwise terminated so as to effectively transfer forces to the reinforcing steel.

When elements of the wall anchorage system are loaded eccentrically or are not perpendicular to the wall, the system shall be designed to resist all components of the forces induced by the eccentricity.

When pilasters are present in the wall, the anchorage force at the pilasters shall be calculated considering the additional load transferred from the wall panels to the pilasters. However, the minimum anchorage force at a floor or roof shall not be reduced.

9.5.2.6.4 Seismic Design Category D. Structures assigned to Category D shall conform to the requirements of Section 9.5.2.6.3 for Category C and to the requirements of this Section.

9.5.2.6.4.1 Collector Elements. In addition to the requirements of Section 9.5.2.6.3.1, collector elements, splices, and their connections to resisting

elements shall resist the forces determined in accordance with Section 9.5.2.6.4.4.

9.5.2.6.4.2 Plan or Vertical Irregularities. When the ratio of the strength provided in any story to the strength required is less than two-thirds of that ratio for the story immediately above, the potentially adverse effect shall be analyzed and the strengths shall be adjusted to compensate for this effect.

For structures having a plan structural irregularity of Type 1, 2, 3, or 4 in Table 9.5.2.3.2 or a vertical structural irregularity of Type 4 in Table 9.5.2.3.3, the design forces determined from Section 9.5.5.2 shall be increased 25% for connections of diaphragms to vertical elements and to collectors and for connections of collectors to the vertical elements. Collectors and their connections also shall be designed for these increased forces unless they are designed for the special seismic loads of Section 9.5.2.7.1, in accordance with Section 9.5.2.6.3.1.

9.5.2.6.4.3 Vertical Seismic Forces. The vertical component of earthquake ground motion shall be considered in the design of horizontal cantilever and horizontal prestressed components. The load combinations used in evaluating such components shall include E as defined by Eqs. 9.5.2.7-1 and 9.5.2.7-2. Horizontal cantilever structural components shall be designed for a minimum net upward force of 0.2 times the dead load in addition to the applicable load combinations of Section 9.5.2.7.

9.5.2.6.4.4 Diaphragms. The deflection in the plane of the diaphragm shall not exceed the permissible deflection of the attached elements. Permissible deflection shall be that deflection that will permit the attached elements to maintain structural integrity under the individual loading and continue to support the prescribed loads. Floor and roof diaphragms shall be designed to resist design seismic forces determined in accordance with Eq. 9.5.2.6.4.4 as follows:

$$F_{px} = \frac{\sum_{i=x}^{n} F_i}{\sum_{i=x}^{n} w_i} w_{px} \qquad \textbf{(Eq. 9.5.2.6.4.4)}$$

where

F_{px} = the diaphragm design force
F_i = the design force applied to Level i
w_i = the weight tributary to Level i
w_{px} = the weight tributary to the diaphragm at Level x

The force determined from Eq. 9.5.2.6.4.4 need not exceed $0.4S_{DS}Iw_{px}$ but shall not be less than $0.2S_{DS}Iw_{px}$. When the diaphragm is required to transfer design seismic force from the vertical resisting elements above the diaphragm to other vertical resisting elements below the diaphragm due to offsets in the placement of the elements or to changes in relative lateral stiffness in the vertical elements, these forces shall be added to those determined from Eq. 9.5.2.6.4.4.

9.5.2.6.5 Seismic Design Categories E and F. Structures assigned to Seismic Design Categories E and F shall conform to the requirements of Section 9.5.2.6.4 for Seismic Design Category D and to the requirements of this Section.

9.5.2.6.5.1 Plan or Vertical Irregularities. Structures having plan irregularity Type 1b of Table 9.5.2.3.1 or vertical irregularities Type 1b or 5 of Table 9.5.2.3.3 shall not be permitted.

9.5.2.7 Combination of Load Effects. The effects on the structure and its components due to seismic forces shall be combined with the effects of other loads in accordance with the combinations of load effects given in Section 2. For use with those combinations, the earthquake-induced force effect shall include vertical and horizontal effects as given by Eq. 9.5.2.7-1 or 9.5.2.7-2, as applicable. The vertical seismic effect term $0.2S_{DS}D$ need not be included where S_{DS} is equal to or less than 0.125 in Eqs. 9.5.2.7-1, 9.5.2.7-2, 9.5.2.7.1-1, and 9.5.2.7.1-2. The vertical seismic effect term $0.2S_{DS}D$ need not be included in Eq. 9.5.2.7-2 when considering foundation overturning.

for load combination 5 in Section 2.3.2 or load combination 5 in Section 2.4.1:

$$E = \rho Q_E + 0.2S_{DS}D \qquad \textbf{(Eq. 9.5.2.7-1)}$$

for load combination 7 in Section 2.3.2 or load combination 8 in 2.4.1:

$$E = \rho Q_E - 0.2S_{DS}D \qquad \textbf{(Eq. 9.5.2.7-2)}$$

where

E = the effect of horizontal and vertical earthquake-induced forces
S_{DS} = the design spectral response acceleration at short periods obtained from Section 9.4.1.2.5
D = the effect of dead load, D
Q_E = the effect of horizontal seismic (earthquake-induced) forces
ρ = the reliability factor

9.5.2.7.1 Special Seismic Load. Where specifically indicated in this Standard, the special seismic load of Eq. 9.5.2.7.1-1 shall be used to compute E for use in load combination 5 in Section 2.3.2 or load combination 3 in 2.4.1 and the special seismic load of Eq. 9.5.2.7.1-2 shall be used to compute E in load combination 7 in Section 2.3.2 or load combination 5 in Section 2.4.1:

$$E = \Omega_o Q_E + 0.2 S_{DS} D \qquad \text{(Eq. 9.5.2.7.1-1)}$$

$$E = \Omega_o Q_E - 0.2 S_{DS} D \qquad \text{(Eq. 9.5.2.7.1-2)}$$

The value of the quantity $\Omega_o Q_E$ in Eqs. 9.5.2.7.1-1 and 9.5.2.7.1-2 need not be taken greater than the capacity of other elements of the structure to transfer force to the component under consideration.

Where allowable stress design methodologies are used with the special load of this Section applied in load combinations 3 or 5 of Section 2.4.1, allowable stresses are permitted to be determined using an allowable stress increase of 1.2. This increase shall not be combined with increases in allowable stresses or load combination reductions otherwise permitted by this standard or the material reference standard except that combination with the duration of load increases permitted in Ref. 9-12.1 is permitted.

9.5.2.8 Deflection, Drift Limits, and Building Separation. The design story drift (Δ) as determined in Section 9.5.5.7 or 9.5.6.6, shall not exceed the allowable story drift (Δ_a) as obtained from Table 9.5.2.8 for any story. For structures with significant torsional deflections, the maximum drift shall include torsional effects. All portions of the structure shall be designed and constructed to act as an integral unit in resisting seismic forces unless separated structurally by a distance sufficient to avoid damaging contact under total deflection (δ_x) as determined in Section 9.5.5.7.1

9.5.3 Index Force Analysis Procedure for Seismic Design of Buildings. See Section 9.5.2.5.1 for limitations on the use of this procedure. An index force analysis shall consist of the application of static lateral index forces to a linear mathematical model of the *structure*, independently in each of two orthogonal directions. The lateral index forces shall be as given by Eq. 9.5.3-1 and shall be applied simultaneously at each floor level. For purposes of analysis, the *structure* shall be considered to be fixed at the *base*:

$$F_x = 0.01 w_x \qquad \text{(Eq. 9.5.3-1)}$$

where

F_x = the design lateral force applied at story x

w_x = the portion of the total gravity load of the *structure*, W, located or assigned to Level x

W = the effective seismic weight of the structure, including the total dead load and other loads listed below:

1. In areas used for storage, a minimum of 25% of the floor live load (floor live load in public garages and open parking structures need not be included.)
2. Where an allowance for partition load is included in the floor load design, the actual partition weight or a minimum weight of 10 psf (0.48 kN/m²) of floor area, whichever is greater.
3. Total operating weight of permanent equipment.
4. 20% of flat roof snow load where flat roof snow load exceeds 30 psf (1.44 kN/m²).

TABLE 9.5.2.8
ALLOWABLE STORY DRIFT, Δ_a [a]

Structure	Seismic Use Group		
	I	II	III
Structures, other than masonry shear wall or masonry wall frame structures, four stories or less with interior walls, partitions, ceilings and exterior wall systems that have been designed to accommodate the story drifts.	$0.025 h_{sx}$ [b]	$0.020 h_{sx}$	$0.015 h_{sx}$
Masonry cantilever shear wall structures [c]	$0.010 h_{sx}$	$0.010 h_{sx}$	$0.010 h_{sx}$
Other masonry shear wall structures	$0.007 h_{sx}$	$0.007 h_{sx}$	$0.007 h_{sx}$
Masonry wall frame structures	$0.013 h_{sx}$	$0.013 h_{sx}$	$0.010 h_{sx}$
All other structures	$0.020 h_{sx}$	$0.015 h_{sx}$	$0.010 h_{sx}$

[a] h_{sx} is the story height below Level x.
[b] There shall be no drift limit for single-story structures with interior walls, partitions, ceilings, and exterior wall systems that have been designed to accommodate the story drifts. The structure separation requirement of Section 9.5.2.8 is not waived.
[c] Structures in which the basic structural system consists of masonry shear walls designed as vertical elements cantilevered from their base or foundation support which are so constructed that moment transfer between shear walls (coupling) is negligible.

9.5.4 Simplified Analysis Procedure for Seismic Design of Buildings. See Section 9.5.2.5.1 for limitations on the use of this procedure. For purposes of this analysis procedure, a building is considered to be fixed at the base.

9.5.4.1 Seismic Base Shear. The seismic base shear, V, in a given direction shall be determined in accordance with the following formula:

$$V = \frac{1.2 S_{DS}}{R} W \qquad \text{(Eq. 9.5.4.1)}$$

where

S_{DS} = the design elastic response acceleration at short period as determined in accordance with Section 9.4.1.2.5
R = the response modification factor from Table 9.5.2.2
W = the effective seismic weight of the structure as defined in Section 9.5.3

9.5.4.2 Vertical Distribution. The forces at each level shall be calculated using the following formula:

$$F_x = \frac{1.2 S_{DS}}{R} w_x \qquad \text{(Eq. 9.5.4.2)}$$

where

w_x = the portion of the effective seismic weight of the structure, W, at level x

9.5.4.3 Horizontal Distribution. Diaphragms constructed of wood structural panels or untopped steel decking are permitted to be considered as flexible.

9.5.4.4 Design Drift. For the purposes of Section 9.5.2.8, the design story drift, Δ, shall be taken as 1% of the story height unless a more exact analysis is provided.

9.5.5 Equivalent Lateral Force Procedure.

9.5.5.1 General. Section 9.5.5 provides required minimum standards for the equivalent lateral force procedure of seismic analysis of structures. An equivalent lateral force analysis shall consist of the application of equivalent static lateral forces to a linear mathematical model of the *structure*. The directions of application of lateral forces shall be as indicated in Section 9.5.2.5.2. The lateral forces applied in each direction shall be the total seismic base shear given by Section 9.5.5.2 and shall be distributed vertically in accordance with the provisions of Section 9.5.5.3. For purposes of analysis, the structure is considered to be fixed at the base. See Section 9.5.2.5 for limitations on the use of this procedure.

9.5.5.2 Seismic Base Shear. The seismic base shear (V) in a given direction shall be determined in accordance with the following equation:

$$V = C_s W \qquad \text{(Eq. 9.5.5.2-1)}$$

where

C_s = the seismic response coefficient determined in accordance with Section 9.5.5.2.1
W = the total dead load and applicable portions of other loads as indicated in Section 9.5.3

9.5.5.2.1 Calculation of Seismic Response Coefficient. When the fundamental period of the structure is computed, the seismic design coefficient (C_s) shall be determined in accordance with the following equation:

$$C_s = \frac{S_{DS}}{R/I} \qquad \text{(Eq. 9.5.5.2.1-1)}$$

where

S_{DS} = the design spectral response acceleration in the short period range as determined from Section 9.4.1.2.5
R = the response modification factor in Table 9.5.2.2
I = the occupancy importance factor determined in accordance with Section 9.1.4

A soil-structure interaction reduction shall not be used unless Section 9.5.9 or another generally accepted procedure approved by the authority having jurisdiction is used.

Alternatively, the seismic response coefficient, (C_s), need not be greater than the following equation:

$$C_s = \frac{S_{D1}}{T(R/I)} \qquad \text{(Eq. 9.5.5.2.1-2)}$$

but shall not be taken less than

$$C_s = 0.044 S_{DS} I \qquad \text{(Eq. 9.5.5.2.1-3)}$$

nor for buildings and structures in Seismic Design Categories E and F

$$C_s = \frac{0.5 S_1}{R/I} \qquad \text{(Eq. 9.5.5.2.1-4)}$$

where I and R are as defined above and

S_{D1} = the design spectral response acceleration at a period of 1.0 sec, in units of g-sec, as determined from Section 9.4.1.2.5
T = the fundamental period of the structure (sec) determined in Section 9.5.5.3

S_1 = the mapped maximum considered earthquake spectral response acceleration determined in accordance with Section 9.4.1

A soil-structure interaction reduction is permitted when determined using Section 9.5.9.

For regular structures 5 stories or less in height and having a period, T, of 0.5 sec or less, the seismic response coefficient, C_s shall be permitted to be calculated using values of 1.5 g and 0.6 g, respectively, for the mapped maximum considered earthquake spectral response accelerations S_S and S_1.

9.5.5.3 Period Determination.
The fundamental period of the structure (T) in the direction under consideration shall be established using the structural properties and deformational characteristics of the resisting elements in a properly substantiated analysis. The fundamental period (T) shall not exceed the product of the coefficient for upper limit on calculated period (C_u) from Table 9.5.5.3 and the approximate fundamental period (T_a) determined from Eq. 9.5.5.3-1. As an alternative to performing an analysis to determine the fundamental period (T), it shall be permitted to use the approximate building period, (T_a), calculated in accordance with Section 9.5.5.3.2, directly.

9.5.5.3.1 Upper Limit on Calculated Period.
The fundamental building period (T) determined in a properly substantiated analysis shall not exceed the product of the coefficient for upper limit on calculated period (C_u) from Table 9.5.5.3.1 and the approximate fundamental period (T_a) determined in accordance with Section 9.5.5.3.2.

9.5.5.3.2 Approximate Fundamental Period.
The approximate fundamental period (T_a), in seconds,

TABLE 9.5.5.3.1
COEFFICIENT FOR UPPER LIMIT ON CALCULATED PERIOD

Design Spectral Response Acceleration at 1 Second, S_{D1}	Coefficient C_u
≥ 0.4	1.4
0.3	1.4
0.2	1.5
0.15	1.6
0.1	1.7
≤ 0.05	1.7

shall be determined from the following equation:

$$T_a = C_t h_n{}^x \qquad \textbf{(Eq. 9.5.5.3.2-1)}$$

where h_n is the height in ft above the base to the highest level of the structure and the coefficients C_t and x are determined from Table 9.5.5.3.2.

Alternatively, it shall be permitted to determine the approximate fundamental period (T_a), in seconds, from the following equation for structures not exceeding 12 stories in height in which the lateral-force-resisting system consists entirely of concrete or steel moment resisting frames and the story height is at least 10 ft (3 m):

$$T_a = 0.1N \qquad \textbf{(Eq. 9.5.5.3.2-1a)}$$

where N = number of stories

The approximate fundamental period, T_a, in seconds for masonry or concrete shear wall structures shall be permitted to be determined from

TABLE 9.5.5.3.2
VALUES OF APPROXIMATE PERIOD PARAMETERS C_t AND x

Structure Type	C_t	x
Moment resisting frame systems of steel in which the frames resist 100% of the required seismic force and are not enclosed or adjoined by more rigid components that will prevent the frames from deflecting when subjected to seismic forces	0.028(0.068)[a]	0.8
Moment resisting frame systems of reinforced concrete in which the frames resist 100% of the required seismic force and are not enclosed or adjoined by more rigid components that will prevent the frame from deflecting when subjected to seismic forces	0.016(0.044)[a]	0.9
Eccentrically braced steel frames	0.03(0.07)[a]	0.75
All other structural systems	0.02(0.055)	0.75

[a] —metric equivalents are shown in parentheses

Eq. 9.5.5.3.2-2 as follows:

$$T_a = \frac{0.0019}{\sqrt{C_W}} h_n \qquad \textbf{(Eq. 9.5.5.3.2-2)}$$

where h_n is as defined above and C_w is calculated from Eq. 9.5.5.3.2-3 as follows:

$$C_W = \frac{100}{A_B} \sum_{i=1}^{n} \left(\frac{h_n}{h_i}\right)^2 \frac{A_i}{\left[1 + 0.83 \left(\frac{h_i}{D_i}\right)^2\right]}$$

$$\textbf{(Eq. 9.5.5.3.2-3)}$$

where

A_B = the base area of the structure ft^2
A_i = the area of shear wall "i" in ft^2
D_i = the length of shear wall "i" in ft
n = the number of shear walls in the building effective in resisting lateral forces in the direction under consideration

9.5.5.4 Vertical Distribution of Seismic Forces. The lateral seismic force (F_x) (kip or kN) induced at any level shall be determined from the following equations:

$$F_x = C_{vx} V \qquad \textbf{(Eq. 9.5.5.4-1)}$$

and

$$C_{vx} = \frac{w_x h_x^k}{\sum_{i=1}^{n} w_i h_i^k} \qquad \textbf{(Eq. 9.5.5.4-2)}$$

where

C_{vx} = vertical distribution factor
V = total design lateral force or shear at the base of the structure, (kip or kN)
w_i and w_x = the portion of the total gravity load of the structure (W) located or assigned to Level i or x
h_i and h_x = the height (ft or m) from the base to Level i or x
k = an exponent related to the structure period as follows:
for structures having a period of 0.5 sec or less, $k = 1$
for structures having a period of 2.5 sec or more, $k = 2$
for structures having a period between 0.5 and 2.5 seconds, k shall be 2 or shall be determined by linear interpolation between 1 and 2

9.5.5.5 Horizontal Shear Distribution and Torsion. The seismic design story shear in any story (V_x) (kip or

kN) shall be determined from the following equation:

$$V_x = \sum_{i=x}^{n} F_i \qquad \textbf{(Eq. 9.5.5.5)}$$

where F_i = the portion of the seismic base shear (V) (kip or kN) induced at Level i.

9.5.5.5.1 Direct Shear. The seismic design story shear (V_x) (kip or kN) shall be distributed to the various vertical elements of the seismic force-resisting system in the story under consideration based on the relative lateral stiffness of the vertical resisting elements and the diaphragm.

9.5.5.5.2 Torsion. Where diaphragms are not flexible, the design shall include the torsional moment (M_t) (kip or kN) resulting from the location of the structure masses plus the accidental torsional moments (M_{ta}) (kip or kN) caused by assumed displacement of the mass each way from its actual location by a distance equal to 5% of the dimension of the structure perpendicular to the direction of the applied forces. Where earthquake forces are applied concurrently in two orthogonal directions, the required 5% displacement of the center of mass need not be applied in both of the orthogonal directions at the same time, but shall be applied in the direction that produces the greater effect.

Structures of Seismic Design Categories C, D, E, and F, where Type 1 torsional irregularity exists as defined in Table 9.5.2.3.2, shall have the effects accounted for by multiplying M_{ta} at each level by a torsional amplification factor (A_x) determined from the following equation:

$$A_x = \left(\frac{\delta_{max}}{1.2\delta_{avg}}\right)^2 \qquad \textbf{(Eq. 9.5.5.5.2)}$$

where

δ_{max} = the maximum displacement at Level x (in. or mm)
δ_{avg} = the average of the displacements at the extreme points of the structure at Level x (in. or mm)

Exception: The torsional and accidental torsional moment need not be amplified for structures of light-frame construction.

The torsional amplification factor (A_x) is not required to exceed 3.0. The more severe loading for each element shall be considered for design.

9.5.5.6 Overturning. The structure shall be designed to resist overturning effects caused by the seismic

forces determined in Section 9.5.4.4. At any story, the increment of overturning moment in the story under consideration shall be distributed to the various vertical elements of the lateral force-resisting system in the same proportion as the distribution of the horizontal shears to those elements.

The overturning moments at Level x (M_x) (kip·ft or kN·m) shall be determined from the following equation:

$$M_x = \sum_{i=x}^{n} F_i(h_i - h_x) \qquad \textbf{(Eq. 9.5.5.6)}$$

where

F_i = the portion of the seismic base shear (V) induced at Level i

h_i and h_x = the height (in ft or m) from the base to Level i or x

The foundations of structures, except inverted pendulum-type structures, shall be permitted to be designed for 75% of the foundation overturning design moment (M_f) (kip·ft or kN·m) at the foundation–soil interface determined using the equation for the overturning moment at Level x (M_x) (kip·ft or kN·m).

9.5.5.7 Drift Determination and P-Delta Effects. Story drifts and, where required, member forces and moments due to P-delta effects shall be determined in accordance with this Section. Determination of story drifts shall be based on the application of the design seismic forces to a mathematical model of the physical structure. The model shall include the stiffness and strength of all elements that are significant to the distribution of forces and deformations in the structure and shall represent the spatial distribution of the mass and stiffness of the structure. In addition, the model shall comply with the following:

1. Stiffness properties of reinforced concrete and masonry elements shall consider the effects of cracked sections, and

2. For steel moment resisting frame systems, the contribution of panel zone deformations to overall story drift shall be included.

9.5.5.7.1 Story Drift Determination. The design story drift (Δ) shall be computed as the difference of the deflections at the top and bottom of the story under consideration. Where allowable stress design is used, Δ shall be computed using code-specified earthquake forces without reduction.

Exception: For structures of Seismic Design Categories C, D, E, and F having plan irregularity Types 1a or 1b of Table 9.5.2.3.2, the design story drift, D, shall be computed as the largest difference of the deflections along any of the edges of the structure at the top and bottom of the story under consideration.

The deflections of Level x at the center of the mass (δ_x) (in. or mm) shall be determined in accordance with the following equation:

$$\delta_x = \frac{C_d \delta_{xe}}{I} \qquad \textbf{(Eq. 9.5.5.7.1)}$$

where

C_d = the deflection amplification factor in Table 9.5.2.2

δ_{xe} = the deflections determined by an elastic analysis

I = the importance factor determined in accordance with Section 9.1.4

The elastic analysis of the seismic force-resisting system shall be made using the prescribed seismic design forces of Section 9.5.5.4. For the purpose of this Section, the value of the base shear, V, used in Eq. 9.5.5.2-1 need not be limited by the value obtained from Eq. 9.5.5.2.1-3.

For determining compliance with the story drift limitation of Section 9.5.2.8, the deflections at the center of mass of Level x (δ_x) (in. or mm) shall be calculated as required in this Section. For the purposes of this drift analysis only, the upper-bound limitation specified in Section 9.5.5.3 on the computed fundamental period, T, in seconds, of the building does not apply for computing forces and displacements.

Where applicable, the design story drift (Δ) (in. or mm) shall be increased by the incremental factor relating to the P-delta effects as determined in Section 9.5.5.7.2.

When calculating drift, the redundancy coefficient, ρ, is not used.

9.5.5.7.2 P-Delta Effects. P-delta effects on story shears and moments, the resulting member forces and moments, and the story drifts induced by these effects are not required to be considered when the stability coefficient (θ) as determined by the following equation is equal to or less than 0.10:

$$\theta = \frac{P_x \Delta}{V_x h_{sx} C_d} \qquad \textbf{(Eq. 9.5.5.7.2-1)}$$

where

P_x = the total vertical design load at and above Level x. (kip or kN); when computing P_x, no individual load factor need exceed 1.0

Δ = the design story drift as defined in Section 9.5.3.7.1 occurring simultaneously with V_x, (in. or mm)

V_x = the seismic shear force acting between Levels x and $x - 1$, (kip or kN)

h_{sx} = the story height below Level x, (in. or mm)

C_d = the deflection amplification factor in Table 9.5.2.2

The stability coefficient (θ) shall not exceed θ_{max} determined as follows:

$$\theta_{max} = \frac{0.5}{\beta C_d} \leq 0.25 \qquad \textbf{(Eq. 9.5.5.7.2-2)}$$

where β is the ratio of shear demand to shear capacity for the story between Level x and $x - 1$. This ratio may be conservatively taken as 1.0.

When the stability coefficient (θ) is greater than 0.10 but less than or equal to θ_{max}, the incremental factor related to P-delta effects (a_d) shall be determined by rational analysis. To obtain the story drift for including the P-delta effect, the design story drift determined in Section 9.5.5.7.1 shall be multiplied by $1.0/(1 - \theta)$.

When θ is greater than θ_{max}, the structure is potentially unstable and shall be redesigned.

When the P-delta effect is included in an automated analysis, Eq. 9.5.5.7.2-2 must still be satisfied, however, the value of θ computed from Eq. 9.5.5.7.2-1 using the results of the P-delta analysis may be divided by $(1 + \theta)$ before checking Eq. 9.5.5.7.2-2.

9.5.6 Modal Analysis Procedure.

9.5.6.1 General. Section 9.5.6 provides required standards for the modal analysis procedure of seismic analysis of structures. See Section 9.5.2.5 for requirements for use of this procedure. The symbols used in this method of analysis have the same meaning as those for similar terms used in Section 9.5.3, with the subscript m denoting quantities in the mth mode.

9.5.6.2 Modeling. A mathematical model of the structure shall be constructed that represents the spatial distribution of mass and stiffness throughout the structure.

For regular structures with independent orthogonal seismic force-resisting systems, independent two-dimensional models are permitted to be constructed to represent each system. For irregular structures or structures without independent orthogonal systems, a three-dimensional model incorporating a minimum of three dynamic degrees of freedom consisting of translation in two orthogonal plan directions and torsional rotation about the vertical axis shall be included at each level of the structure. Where the diaphragms are not rigid compared to the vertical elements of the lateral force-resisting system, the model should include representation of the diaphragm's flexibility and such additional dynamic degrees of freedom as are required to account for the participation of the diaphragm in the structure's dynamic response. In addition, the model shall comply with the following:

1. Stiffness properties of concrete and masonry elements shall consider the effects of cracked sections, and

2. For steel moment frame systems, the contribution of panel zone deformations to overall story drift shall be included.

9.5.6.3 Modes. An analysis shall be conducted to determine the natural modes of vibration for the structure, including the period of each mode, the modal shape vector Φ, the modal participation factor, and modal mass. The analysis shall include a sufficient number of modes to obtain a combined modal mass participation of at least 90% of the actual mass in each of two orthogonal directions.

9.5.6.4 Periods. The required periods, mode shapes, and participation factors of the structure in the direction under consideration shall be calculated by established methods of structural analysis for the fixed-base condition using the masses and elastic stiffnesses of the seismic force-resisting system.

9.5.6.5 Modal Base Shear. The portion of the base shear contributed by the mth mode (V_m) shall be determined from the following equations:

$$V_m = C_{sm} W_m \qquad \textbf{(Eq. 9.5.6.5-1)}$$

$$W_m = \frac{\left(\displaystyle\sum_{i=1}^{n} w_i \phi_{im}\right)^2}{\displaystyle\sum_{i=1}^{n} w_i \phi_{im}^2} \qquad \textbf{(Eq. 9.5.6.5-2)}$$

where

C_{sm} = the modal seismic design coefficient determined below

W_m = the effective modal gravity load

w_i = the portion of the total gravity load of the structure at Level i

ϕ_{im} = the displacement amplitude at the ith level of the structure when vibrating in its mth mode

The modal seismic design coefficient (C_{sm}) shall be determined in accordance with the following equation:

$$C_{sm} = \frac{S_{am}}{R/I} \qquad \textbf{(Eq. 9.5.6.5-3)}$$

where

S_{am} = the design spectral response acceleration at period T_m determined from either the general design response spectrum of Section 9.4.1.2.6 or a site-specific response spectrum per Section 9.4.1.3

R = the response modification factor determined from Table 9.5.2.2

I = the occupancy importance factor determined in accordance with Section 9.1.4

T_m = the modal period of vibration (in seconds) of the mth mode of the structure

Exception: When the general design response spectrum of Section 9.4.1.2.6 is used for structures where any modal period of vibration (T_m) exceeds 4.0 sec, the modal seismic design coefficient (C_{sm}) for that mode shall be determined by the following equation:

$$C_{sm} = \frac{4S_{D1}}{(R/I)T_m^2} \qquad \textbf{(Eq. 9.5.6.5-4)}$$

The reduction due to soil-structure interaction as determined in Section 9.5.5.3 is permitted to be used.

9.5.6.6 Modal Forces, Deflections, and Drifts. The modal force (F_{xm}) at each level shall be determined by the following equations:

$$F_{xm} = C_{vxm} V_m \qquad \textbf{(Eq. 9.5.6.6-1)}$$

and

$$C_{vxm} = \frac{w_x \phi_{xm}}{\sum\limits_{i=1}^{n} w_i \phi_{im}} \qquad \textbf{(Eq. 9.5.6.6-2)}$$

where

C_{vxm} = the vertical distribution factor in the mth mode

V_m = the total design lateral force or shear at the base in the mth mode

w_i and w_x = the portion of the total gravity load of the structure (W) located or assigned to Level i or x

ϕ_{xm} = the displacement amplitude at the xth level of the structure when vibrating in its mth mode

ϕ_{im} = the displacement amplitude at the ith level of the structure when vibrating in its mth mode

The modal deflection at each level (δ_{xm}) shall be determined by the following equations:

$$\delta_{xm} = \frac{C_d \delta_{xem}}{I} \qquad \textbf{(Eq. 9.5.6.6-3)}$$

and

$$\delta_{xem} = \left(\frac{g}{4\pi^2}\right)\left(\frac{T_m^2 F_{xm}}{w_x}\right) \qquad \textbf{(Eq. 9.5.6.6-4)}$$

where

C_d = the deflection amplification factor determined from Table 9.5.2.2

δ_{xem} = the deflection of Level x in the mth mode at the center of the mass at Level x determined by an elastic analysis

g = the acceleration due to gravity (ft^2/sec)

I = the occupancy importance factor determined in accordance with Section 9.1.4

T_m = the modal period of vibration, in seconds, of the mth mode of the structure

F_{xm} = the portion of the seismic base shear in the mth mode, induced at Level x, and

w_x = the portion of the total gravity load of the structure (W) located or assigned to Level x

The modal drift in a story (Δ_m) shall be computed as the difference of the deflections (δ_{xm}) at the top and bottom of the story under consideration.

9.5.6.7 Modal Story Shears and Moments. The story shears, story overturning moments, and the shear forces and overturning moments in vertical elements of the structural system at each level due to the seismic forces determined from the appropriate equation in Section 9.5.6.6 shall be computed for each mode by linear static methods.

9.5.6.8 Design Values. The design value for the modal base shear (V_t), each of the story shear, moment and drift quantities, and the deflection at each level shall be determined by combining their modal values as obtained from Sections 9.5.6.6 and 9.5.6.7. The combination shall be carried out by taking the square root of the sum of the squares of each of the modal values or where closely spaced periods in the translational and torsional modes result in significant cross-correlation of the modes, the complete quadratic combination (CQC) method, in accordance with ASCE-4, shall be used.

A base shear (V) shall be calculated using the equivalent lateral force procedure in Section 9.5.5. For the purpose of this calculation, a fundamental period of the structure (T), in seconds, shall not exceed the coefficient for upper limit on the calculated period (C_u) times the approximate fundamental period of the structure (T_a). Where the design value for the modal base shear (V_t) is less than 85% of the calculated base shear (V) using the equivalent lateral force procedure, the design story shears, moments, drifts, and floor deflections shall be multiplied by the following modification factor:

$$0.85\frac{V}{V_t} \qquad \textbf{(Eq. 9.5.6.8)}$$

where

V = the equivalent lateral force procedure base shear, calculated in accordance with this section and Section 9.5.5, and

V_t = the modal base shear, calculated in accordance with this section

9.5.6.9 Horizontal Shear Distribution. The distribution of horizontal shear shall be in accordance with the requirements of Section 9.5.5.5 except that amplification of torsion per Section 9.5.5.5.2 is not required for that portion of the torsion, A_x, included in the dynamic analysis model.

9.5.6.10 Foundation Overturning. The foundation overturning moment at the foundation-soil interface may be reduced by 10%.

9.5.6.11 P-Delta Effects. The P-delta effects shall be determined in accordance with Section 9.5.5.7. The story drifts and base shear used to determine the story shears shall be determined in accordance with Section 9.5.5.7.1.

9.5.7 Linear Response History Analysis Procedure. A linear response history analysis shall consist of an analysis of a linear mathematical model of the *structure* to determine its response, through methods of numerical integration, to suites of ground motion acceleration histories compatible with the design response spectrum for the site. The analysis shall be performed in accordance with the provisions of this Section. For purposes of analysis, the structure shall be permitted to be considered to be fixed at the *base*, or alternatively, it shall be permitted to use realistic assumptions with regard to the stiffness of foundations. See Section 9.5.2.1 for limitations on the use of this procedure.

9.5.7.1 Modeling. Mathematical models shall conform to the requirements of Section 9.5.6.1.

9.5.7.2 Ground Motion. A suite of not less than three appropriate ground motions shall be used in the analysis. Ground motion shall conform to the requirements of this Section.

9.5.7.2.1 Two-Dimensional Analysis. When two-dimensional analyses are performed, each ground motion shall consist of a horizontal acceleration history, selected from an actual recorded event. Appropriate acceleration histories shall be obtained from records of events having magnitudes, fault distance, and source mechanisms that are consistent with those that control the *maximum considered earthquake*. Where the required number of appropriate recorded

ground motion records are not available, appropriate simulated ground motion records shall be used to make up the total number required. The ground motions shall be scaled such that the average value of the 5% damped response spectra for the suite of motions is not less than the design response spectrum for the site, determined in accordance with Section 9.4.1.3 for periods ranging from $0.2T$ to $1.5T$ sec where T is the natural period of the *structure* in the fundamental mode for the direction of response being analyzed.

9.5.7.2.2 Three-Dimensional Analysis. When three-dimensional analyses are performed, ground motions shall consist of pairs of appropriate horizontal ground motion acceleration components that shall be selected and scaled from individual recorded events. Appropriate ground motions shall be selected from events having magnitudes, fault distance, and source mechanisms that are consistent with those that control the *maximum considered earthquake*. Where the required number of recorded ground motion pairs are not available, appropriate simulated ground motion pairs shall be used to make up the total number required. For each pair of horizontal ground motion components, the square root of the sum of the squares (SRSS) of the 5% damped response spectrum of the scaled horizontal components shall be constructed. Each pair of motions shall be scaled such that the average value of the SRSS spectra from all horizontal component pairs is not less than 1.3 times the 5% damped design response spectrum determined in accordance with Section 9.4.1.3 for periods ranging from $0.2T$ to $1.5T$ seconds, where T is the natural period of the fundamental mode of the *structure*.

9.5.7.3 Response Parameters. For each ground motion analyzed, the individual response parameters shall be scaled by the quantity I/R, where I is the occupancy importance factor determined in accordance with Section 1.4 and R is the Response Modification Coefficient selected in accordance with Section 9.5.2.2. The maximum value of the base shear, V_i, member forces, Q_{Ei}, and interstory drifts, d_i at each story, scaled as indicated above shall be determined. When the maximum scaled base shear predicted by the analysis, V_i, is less than given by Eq. 9.5.5.1.2-3, or in Seismic Design Categories E and F, Eq. 9.5.5.1.2-4, the scaled member forces, Q_{ei}, shall be additionally scaled by the factor:

$$\frac{V}{V_i} \qquad \textbf{(Eq. 9.5.7.3)}$$

where V is the minimum base shear determined in accordance with Eq. 9.5.5.2.1-3 or for structures in Seismic Design Category E or F, Eq. 9.5.5.2.1-4.

If at least seven ground motions are analyzed, the design member forces, Q_E, used in the load combinations of Section 9.5.2.7, and the design interstory drift, D, used in the evaluation of drift in accordance with Section 5.2.8 shall be permitted to be taken respectively as the average of the scaled Q_{Ei} and D_i values determined from the analyses and scaled as indicated above. If less than seven ground motions are analyzed, the design member forces, Q_E, and the design interstory drift, D, shall be taken as the maximum value of the scaled Q_{Ei} and D_i values determined from the analyses.

Where these provisions require the consideration of the special load combinations of Section 9.5.2.7, the value of $\Omega_0 Q_E$ need not be taken larger than the maximum of the unscaled value, Q_{Ei}, obtained from the suite of analyses.

9.5.8 Nonlinear Response History Analysis.
A nonlinear response history analysis shall consist of an analysis of a mathematical model of the *structure* that directly accounts for the nonlinear hysteretic behavior of the structure's components to determine its response through methods of numerical integration to suites of ground motion acceleration histories compatible with the design response spectrum for the site. The analysis shall be performed in accordance with this Section. See Section 9.5.2.1 for limitations on the use of this procedure.

9.5.8.1 Modeling.
A mathematical model of the structure shall be constructed that represents the spatial distribution of mass throughout the structure. The hysteretic behavior of elements shall be modeled consistent with suitable laboratory test data and shall account for all significant yielding, strength degradation, stiffness degradation, and hysteretic pinching indicated by such test data. Strength of elements shall be based on expected values considering material overstrength, strain hardening, and also hysteretic strength degradation. Linear properties, consistent with the provisions of Section 9.5.6.1, shall be permitted to be used for those elements demonstrated by the analysis to remain within their linear range of response. The structure shall be assumed to have a fixed base, or alternatively, it shall be permitted to use realistic assumptions with regard to the stiffness and load-carrying characteristics of the foundations consistent with site-specific soils data and rational principles of engineering mechanics.

For regular structures with independent orthogonal seismic force-resisting systems, independent two-dimensional models shall be permitted to be constructed to represent each system. For structures having plan irregularities Types 1a, 1b, 4 or 5 of Table 9.5.2.3.2, or structures without independent orthogonal systems, a three-dimensional model incorporating a minimum of three dynamic degrees of freedom consisting of translation in two orthogonal plan directions and torsional

rotation about the vertical axis at each level of the structure shall be used. Where the diaphragms are not rigid compared to the vertical elements of the lateral-force-resisting system, the model should include representation of the diaphragm's flexibility and such additional dynamic degrees of freedom as are required to account for the participation of the diaphragm in the structure's dynamic response.

9.5.8.2 Ground Motion and Other Loading.
Ground motion shall conform to the provisions of Section 9.5.7.2. The structure shall be analyzed for the effects of these ground motions simultaneously with the effects of dead load in combination with not less than 25% of the required live loads.

9.5.8.3 Response Parameters.
For each ground motion analyzed, individual response parameters consisting of the maximum value of the individual member forces, Q_{Ei}, member inelastic deformations, D_i and interstory drifts, D_i at each story shall be determined.

If at least seven ground motions are analyzed, the design values of member forces, Q_E, member inelastic deformations, D and interstory drift, D shall be permitted to be taken respectively as the average of the scaled Q_{Ei}, g_i, and D_i values determined from the analyses. If less than seven ground motions are analyzed, the design member forces, Q_E, design member inelastic deformations, g and the design interstory drift, D, shall be taken as the maximum value of the scaled Q_{Ei}, g_i, and D_i values determined from the analyses.

9.5.8.3.1 Member Strength.
The adequacy of members to resist the combination of load effects of Section 9.5.2.7 need not be evaluated.

> **Exception:** Where this Standard requires the consideration of the special seismic loads of Section 9.5.2.7.1. In such evaluations, the maximum value of Q_{Ei} obtained from the suite of analyses shall be taken in place of the quantity $\Omega_0 Q_E$.

9.5.8.3.2 Member Deformation.
The adequacy of individual members and their connections to withstand the estimated design deformation values, g_i, as predicted by the analyses shall be evaluated based on laboratory test data for similar components. The effects of gravity and other loads on member deformation capacity shall be considered in these evaluations. Member deformation shall not exceed two-thirds of a value that results in loss of ability to carry gravity loads or that results in deterioration of member strength to less than the 67% of the peak value.

9.5.8.3.3 Interstory Drift. The design interstory drift obtained from the analyses shall not exceed 125% of the drift limit specified in Section 9.5.2.8.

9.5.8.4 Design Review. A design review of the seismic force-resisting system and the structural analysis shall be performed by an independent team of *registered design professionals* in the appropriate disciplines and others experienced in seismic analysis methods and the theory and application of nonlinear seismic analysis and structural behavior under extreme cyclic loads. The design review shall include, but not be limited to, the following:

1. Review of any site-specific seismic criteria employed in the analysis including the development of site-specific spectra and ground motion time histories

2. Review of acceptance criteria used to demonstrate the adequacy of structural elements and systems to withstand the calculated force and deformation demands, together with that laboratory and other data used to substantiate these criteria.

3. Review of the preliminary design including the selection of structural system and the configuration of structural elements.

4. Review of the final design of the entire structural system and all supporting analyses.

9.5.9 Soil-Structure Interaction.

9.5.9.1 General. If the option to incorporate the effects of soil-structure interaction is exercised, the requirements of this Section shall be used in the determination of the design earthquake forces and the corresponding displacements of the structure. The use of these provisions will decrease the design values of the base shear, lateral forces, and overturning moments but may increase the computed values of the lateral displacements and the secondary forces associated with the P-delta effects.

The provisions for use with the equivalent lateral force procedure are given in Section 9.5.9.2. and those for use with the modal analysis procedure are given in Section 9.5.9.3.

9.5.9.2 Equivalent Lateral Force Procedure. The following requirements are supplementary to those presented in Section 9.5.5.

9.5.9.2.1 Base Shear. To account for the effects of soil-structure interaction, the base shear (V) determined from shall be reduced to:

$$\tilde{V} = V - \Delta V \qquad \textbf{(Eq. 9.5.9.2.1-1)}$$

The reduction (ΔV) shall be computed as follows and shall not exceed $0.3V$:

$$\Delta V = \left[C_s - \tilde{C}_s \left(\frac{0.05}{\tilde{B}} \right)^{0.4} \right] \overline{W} \leq 0.3V$$

$$\textbf{(Eq. 9.5.9.2.1-2)}$$

where

C_s = the seismic design coefficient computed from Eqs. 9.5.5.2.1-1 and 9.5.5.2.1-2 using the fundamental natural period of the fixed-base structure (T or T_a) as specified in Section 9.5.3.3

$\tilde{C}_s$ = the value of C_s computed from Eqs. 9.5.5.2.1-1 and 9.5.5.2.1-2 using the fundamental natural period of the flexibly supported structure (T) defined in Section 9.5.5.2.1.1

$\tilde{B}$ = the fraction of critical damping for the structure-foundation system determined in Section 9.5.5.2.1.2, and

$\overline{W}$ = the effective gravity load of the structure, which shall be taken as 0.7W, except that for structures where the gravity load is concentrated at a single level, it shall be taken equal to W

9.5.9.2.1.1 Effective Building Period. The effective period ($\tilde{T}$) shall be determined as follows:

$$\tilde{T} = T \sqrt{ 1 + \frac{\overline{k}}{K_y} \left(1 + \frac{K_y \overline{h}^2}{K_\theta} \right) }$$

$$\textbf{(Eq. 9.5.9.2.1.1-1)}$$

where

T = the fundamental period of the structure as determined in Section 9.5.5.3

k = the stiffness of the structure when fixed at the base, defined by the following:

$$\overline{k} = 4\pi^2 \left(\frac{\overline{W}}{g T^2} \right)$$

$$\textbf{(Eq. 9.5.9.2.1.1-2)}$$

$\overline{h}$ = the effective height of the structure which shall be taken as 0.7 times the total height (h_n) except that for structure where the gravity load is effectively concentrated at a single level, it shall be taken as the height to that level

K_y = the lateral stiffness of the foundation
defined as the horizontal force at the
level of the foundation necessary to
produce a unit deflection at that level, the
force and the deflection being measured
in the direction in which the structure is
analyzed

K_θ = the rocking stiffness of the foundation
defined as the moment necessary to
produce a unit average rotation of the
foundation, the moment and rotation
being measured in the direction in which
the structure is analyzed, and

g = the acceleration of gravity

The foundation stiffnesses (K_y and K_θ) shall
be computed by established principles of foundation
mechanics using soil properties that are
compatible with the soil strain levels associated
with the design earthquake motion. The average
shear modulus (G) for the soils beneath
the foundation at large strain levels and the
associated shear wave velocity (v_s) needed in
these computations shall be determined from
Table 9.5.9.2.1.1a where

v_{so} = the average shear wave velocity for the
soils beneath the foundation at small strain
levels ($10^{-3}\%$ or less)

$G_o = \gamma v_{so}^2/g$ = the average shear modulus for
the soils beneath the foundation at small
strain levels

γ = the average unit weight of the soils

Alternatively, for structures supported on mat
foundations that rest at or near the ground surface
or are embedded in such a way that the side wall
contact with the soil are not considered to remain
effective during the design ground motion, the
effective period of the structure is permitted to be
determined from:

$$\tilde{T} = T\sqrt{1 + \frac{25\alpha r_a \bar{h}}{v_s^2 T^2}\left(1 + \frac{1.12 r_a \bar{h}^2}{\alpha_\theta r_m^3}\right)}$$

(Eq. 9.5.9.2.1.1-3)

TABLE 9.5.9.2.1.1a
VALUES OF G/G_o AND v_s/v_{so}

	Spectral Response Acceleration, S_{D1}			
	≤ 0.10	≤ 0.15	≤ 0.20	≥ 0.30
Value of G/G_o	0.81	0.64	0.49	0.42
Value of v_s/v_{so}	0.9	0.8	0.7	0.65

TABLE 9.5.9.2.1.1b
VALUES OF α_θ

$r_m/v_s T$	α_θ
<0.05	1.0
0.15	0.85
0.35	0.7
0.5	0.6

where

α = the relative weight density of the structure
and the soil defined by

$$\alpha = \frac{\bar{W}}{\gamma A_o \bar{h}}$$ **(Eq. 9.5.9.2.1.1-4)**

r_a and r_m = characteristic foundation lengths
defined by

$$r_a = \sqrt{\frac{A_o}{\pi}}$$ **(Eq. 9.5.9.2.1.1-5)**

and

$$r_m = \sqrt[4]{\frac{4I_o}{\pi}}$$ **(Eq. 9.5.9.2.1.1-6)**

where

A_o = the area of the load-carrying foundation

I_o = the static moment of inertia of the
load-carrying foundation about a horizontal
centroidal axis normal to the direction in
which the structure is analyzed

α_θ = dynamic foundation stiffness modifier
for rocking as determined from
Table 9.5.9.2.1.1b

where

r_m = characteristic foundation length as determined
by Eq. 9.5.9.2.1.1-6

v_s = shear wave velocity

T = fundamental period as determined in
Section 9.5.5.3

9.5.9.2.1.2 Effective Damping. The effective
damping factor for the structure-foundation system
$\tilde{\beta}$ shall be computed as follows:

$$\tilde{\beta} = \beta_o + \frac{0.05}{(\tilde{T}/T)^3}$$ **(Eq. 9.5.9.2.1.2-1)**

where

β_o = the foundation damping factor as specified
in Figure 9.5.9.2.1.2

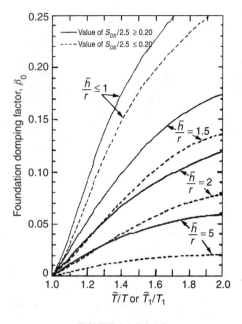

FIGURE 9.5.9.2.1.2
FOUNDATION DAMPING FACTOR

The values of β_o corresponding to $A_v = 0.15$ in Figure 9.5.9.2.1.2 shall be determined by averaging the results obtained from the solid lines and the dashed lines.

The quantity r in Figure 9.5.9.2.1.2 is a characteristic foundation length that shall be determined as follows:

For $\bar{h}/L_o \leq 0.5$,

$$r = r_a = \sqrt{\frac{A_o}{\pi}} \qquad \textbf{(Eq. 9.5.9.2.1.2-2)}$$

For $\bar{h}/L_o \geq 1$,

$$r = r_m = \sqrt[4]{\frac{4I_o}{\pi}} \qquad \textbf{(Eq. 9.5.9.2.1.2-3)}$$

where

L_o = the overall length of the side of the foundation in the direction being analyzed

A_o = the area of the load-carrying foundation, and

I_o = the static moment of inertia of the load-carrying foundation about a horizontal centroidal axis normal to the direction in which the structure is analyzed

For intermediate values of $\bar{h}/L_o$, the value of r shall be determined by linear interpolation.

Exception: For structures supported on point bearing piles and in all other cases where the foundation soil consists of a soft stratum of reasonably uniform properties underlain by a much

stiffer, rock-like deposit with an abrupt increase in stiffness, the factor β_o in Eq. 9.5.9.2.1.2-1 shall be replaced by β_o' if $4D_s/v_s T < 1$ where D_s is the total depth of the stratum. β_o' shall be determined as follows:

$$\beta_o' = \left(\frac{4D_s}{V_s \tilde{T}}\right)^2 \beta_o \qquad \textbf{(Eq. 9.5.9.2.1.2-4)}$$

The value of $\tilde{\beta}$ computed from Eq. 9.5.9.2.1.2-1, both with or without the adjustment represented by Eq. 9.5.9.2.1.2-4, shall in no case be taken as less than $\tilde{\beta} = 0.05$ or greater than $\tilde{\beta} = 0.20$.

9.5.9.2.2 Vertical Distribution of Seismic Forces. The distribution over the height of the structure of the reduced total seismic force ($\tilde{V}$) shall be considered to be the same as for the structure without interaction.

9.5.9.2.3 Other Effects. The modified story shears, overturning moments, and torsional effects about a vertical axis shall be determined as for structures without interaction using the reduced lateral forces.
The modified deflections ($\tilde{\delta}_x$) shall be determined as follows:

$$\tilde{\delta}_x = \frac{\tilde{V}}{V}\left[\frac{M_o h_x}{K_\theta} + \delta_x\right] \qquad \textbf{(Eq. 9.5.9.2.3-1)}$$

where

M_o = the overturning moment at the base determined in accordance with Section 9.5.3.6 using the unmodified seismic forces and not including the reduction permitted in the design of the foundation

h_x = the height above the base to the level under consideration

δ_x = the deflections of the fixed-base structure as determined in Section 9.5.3.7.1 using the unmodified seismic forces

The modified story drifts and P-delta effects shall be evaluated in accordance with the provisions of Section 9.5.3.7 using the modified story shears and deflections determined in this Section.

9.5.9.3 Modal Analysis Procedure. The following provisions are supplementary to those presented in Section 9.5.6.

9.5.9.3.1 Modal Base Shears. To account for the effects of soil-structure interaction, the base shear corresponding to the fundamental mode of vibration (V_1) shall be reduced to:

$$\tilde{V}_1 = V_1 - \Delta V_1 \qquad \textbf{(Eq. 9.5.9.3.1-1)}$$

The reduction (ΔV_1) shall be computed in accordance with Eq. 9.5.9.2.1-2 with $\overline{W}$ taken as equal to the gravity load $\overline{W}_1$ defined by Eq. 9.5.6.5-2, C_s computed from Eq. 9.5.6.5-3 using the fundamental period of the fixed-base structure (T_1), and C_s computed from Eq. 9.5.6.5-3 using the fundamental period of the elastically supported structure ($\tilde{T}_1$).

The period $\tilde{T}_1$ shall be determined from Eq. 9.5.9.2.1.1-1, or from Eq. 9.5.9.2.1.1-3 when applicable, taking $T = \tilde{T}_1$, evaluating $\overline{k}$ from Eq. 9.5.9.2.1.1-2 with $\overline{W} = \overline{W}_1$, and computing $\overline{h}$ as follows:

$$\overline{h} = \frac{\sum\limits_{i=1}^{n} w_i \varphi_{i1} h_i}{\sum\limits_{i=1}^{n} w_i \varphi_{i1}} \qquad \textbf{(Eq. 9.5.9.3.1-2)}$$

The above designated values of $\overline{W}$, $\overline{h}$, T, and $\tilde{T}$ also shall be used to evaluate the factor α from Eq. 9.5.9.2.1.1-4 and factor β_o from Figure 9.5.9.2.1.2. No reduction shall be made in the shear components contributed by the higher modes of vibration. The reduced base shear ($\overline{V}_1$) shall in no case be taken less than $0.7V_1$.

9.5.9.3.2 Other Modal Effects. The modified modal seismic forces, story shears, and overturning moments shall be determined as for structures without interaction using the modified base shear ($\overline{V}_1$) instead of V_1. The modified modal deflections ($\tilde{\delta}$) shall be determined as follows:

$$\tilde{\delta}_{xl} = \frac{\tilde{V}_1}{V_1}\left[\frac{M_{01}h_x}{K_\theta} + \delta_{x1}\right] \qquad \textbf{(Eq. 9.5.9.3.2-1)}$$

and

$$\tilde{\delta}_{xm} = \delta_{xm} \qquad \textbf{(Eq. 9.5.9.3.2-2)}$$

for $m = 2, 3, \ldots$
where

M_{o1} = the overturning base moment for the fundamental mode of the fixed-base structure, as determined in Section 9.5.6.7 using the unmodified modal base shear V_1

δ_{xm} = the modal deflections at Level x of the fixed-base structure as determined in Section 9.5.6.6 using the unmodified modal shears, V_m

The modified modal drift in a story ($\tilde{\Delta}_m$) shall be computed as the difference of the deflections ($\tilde{\delta}_{xm}$) at the top and bottom of the story under consideration.

9.5.9.3.3 Design Values. The design values of the modified shears, moments, deflections, and story

drifts shall be determined as for structures without interaction by taking the square root of the sum of the squares of the respective modal contributions. In the design of the foundation, it shall be permitted to reduce the overturning moment at the foundation-soil interface determined in this manner by 10% as for structures without interaction.

The effects of torsion about a vertical axis shall be evaluated in accordance with the provisions of Section 9.5.6.5 and the P-delta effects shall be evaluated in accordance with the provisions of Section 9.5.6.7.2 using the story shears and drifts determined in Section 9.5.9.3.2.

SECTION 9.6
ARCHITECTURAL, MECHANICAL, AND ELECTRICAL COMPONENTS AND SYSTEMS

9.6.1 General. Section 9.6 establishes minimum design criteria for architectural, mechanical, electrical, and non-structural systems, components, and elements permanently attached to structures including supporting structures and attachments (hereinafter referred to as "components"). The design criteria establish minimum equivalent static force levels and relative displacement demands for the design of components and their attachments to the structure, recognizing ground motion and structural amplification, component toughness and weight, and performance expectations. Seismic Design Categories for structures are defined in Section 9.4.2. For the purposes of this Section, components shall be considered to have the same Seismic Design Category as that of the structure that they occupy or to which they are attached unless otherwise noted.

This Section also establishes minimum seismic design force requirements for nonbuilding structures that are supported by other structures where the weight of the nonbuilding structure is less than 25% of the combined weight of the nonbuilding structure and the supporting structure. Seismic design requirements for nonbuilding structures that are supported by other structures where the weight of the nonbuilding structure is 25% or more of the combined weight of the nonbuilding structure and supporting structure are prescribed in Section 9.14. Seismic design requirements for nonbuilding structures that are supported at grade are prescribed in Section 9.14; however, the minimum seismic design forces for nonbuilding structures that are supported by another structure shall be determined in accordance with the requirements of Section 9.6.1.3 with R_p equal to the value of R specified in Section 9.14 and $a_p = 2.5$ for nonbuilding structures with flexible dynamic characteristics and $a_p = 1.0$ for nonbuilding structures with rigid dynamic characteristics. The distribution of lateral forces for the supported nonbuilding structure and all nonforce requirements specified in Section 9.14 shall apply to supported nonbuilding structures.

In addition, all components are assigned a component importance factor (I_p) in this chapter. The default value for I_p is 1.00 for typical components in normal service. Higher values for I_p are assigned for components, which contain hazardous substances, must have a higher level of assurance of function, or otherwise require additional attention because of their life safety characteristics. Component importance factors are prescribed in Section 9.6.1.5.

All architectural, mechanical, electrical, and other nonstructural components in structures shall be designed and constructed to resist the equivalent static forces and displacements determined in accordance with this Section. The design and evaluation of support structures and architectural components and equipment shall consider their flexibility as well as their strength.

Exception: The following components are exempt from the requirements of this Section:

1. All components in Seismic Design Category A.
2. Architectural components in Seismic Design Category B other than parapets supported by bearing walls or shear walls provided that the importance factor (I_p) is equal to 1.0.
3. Mechanical and electrical components in Seismic Design Category B.
4. Mechanical and electrical components in structures assigned to Seismic Design Category C provided that the importance factor (I_p) is equal to 1.0.
5. Mechanical and electrical components in Seismic Design Categories D, E, and F where $I_p = 1.0$ and flexible connections between the components and associated ductwork, piping, and conduit are provided and that are mounted at 4 ft (1.22 m) or less above a floor level and weigh 400 lb (1780 N) or less.
6. Mechanical and electrical components in Seismic Design Categories D, E, and F weighing 20 lb (95 N) or less where $I_p = 1.0$ and flexible connections between the components and associated ductwork, piping, and conduit are provided, or for distribution systems, weighing 5 lb/ft (7 N/m) or less.

The functional and physical interrelationship of components and their effect on each other shall be designed so that the failure of an essential or nonessential architectural, mechanical, or electrical component shall not cause the failure of a nearby essential architectural, mechanical, or electrical component.

9.6.1.1 Reference Standards.

9.6.1.1.1 Consensus Standards. The following references are consensus standards and are to be considered part of these provisions to the extent referred to in this chapter:

Reference 9.6-1	American Society of Mechanical Engineers (ASME), *ASME A17.1, Safety Code For Elevators and Escalators*, 1996.
Reference 9.6-2	American Society of Mechanical Engineers (ASME), Boiler And Pressure Vessel Code, including addendums through 1997.
Reference 9.6-3	American Society For Testing and Materials (ASTM), ASTM C635, *Standard Specification for the Manufacture, Performance, and Testing of Metal Suspension Systems For Acoustical Tile And Lay-in Panel Ceilings*, 1997.
Reference 9.6-4	American Society For Testing And Materials (ASTM), ASTM C636, *Standard Practice for Installation of Metal Ceiling Suspension Systems for Acoustical Tile And Lay-in Panels*, 1996.
Reference 9.6-5	American National Standards Institute/American Society of Mechanical Engineers, *ASME B31.1-98, Power Piping*.
Reference 9.6-6	American Society of Mechanical Engineers, *ASME B31.3-96, Process Piping*.
Reference 9.6-7	American Society of Mechanical Engineers, ASME B31.4-92, *Liquid Transportation Systems for Hydrocarbons, Liquid Petroleum Gas, Anhydrous Ammonia, and Alcohols*.
Reference 9.6-8	American Society of Mechanical Engineers, *ASME B31.5-92, Refrigeration Piping*.
Reference 9.6-9	American Society of Mechanical Engineers, *ASME B31.9-96, Building Services Piping*.
Reference 9.6-10	American Society of Mechanical Engineers, *ASME B31.11-89* (Reaffirmed 1998), *Slurry Transportation Piping Systems*.
Reference 9.6-11	American Society of Mechanical Engineers, *ASME B31.8-95, Gas Transmission and Distribution Piping Systems*.
Reference 9.6-12	Institute of Electrical and Electronic Engineers (IEEE), Standard 344, *Recommended Practice for Seismic Qualification of Class 1E Equipment for Nuclear Power Generating Stations*, 1987.
Reference 9.6-13	National Fire Protection Association (NFPA), NFPA-13, *Standard*

for the Installation of Sprinkler Systems, 1999.

9.6.1.1.2 Accepted Standards. The following references are standards developed within the industry and represent acceptable procedures for design and construction:

Reference 9.6-14 American Society of Heating, Ventilating, and Air Conditioning (ASHRAE), "Seismic Restraint Design," 1999.

Reference 9.6-15 Manufacturer's Standardization Society of the Valve and Fitting Industry (MSS). SP-58, "Pipehangers and Supports— Materials, Design, and Manufacture," 1988.

Reference 9.6-16 Ceilings and Interior Systems Construction Association (CISCA), "Recommendations for Direct-Hung Acoustical Tile and Lay-in Panel Ceilings," *Seismic Zones 0-2,* 1991.

Reference 9.6-17 Ceilings and Interior Systems Construction Association (CISCA), "Recommendations for Direct-Hung Acoustical Tile and Lay-in Panel Ceilings," *Seismic Zones 3-4,* 1991.

Reference 9.6-18 Sheet Metal and Air Conditioning Contractors National Association (SMACNA), *HVAC Duct Construction Standards, Metal and Flexible,* 1995.

Reference 9.6-19 Sheet Metal and Air Conditioning Contractors National Association (SMACNA), *Rectangular Industrial Duct Construction Standards,* 1980.

Reference 9.6-20 Sheet Metal and Air Conditioning Contractors National Association (SMACNA), *Seismic Restraint Manual Guidelines for Mechanical Systems,* 1991, including Appendix B, 1998.

Reference 9.6-21 American Architectural Manufacturers Association (AAMA), "Recommended Dynamic Test Method for Determining the Seismic Drift Causing Glass Fallout from a Wall System," *Publication No. AAMA 501.6-2001.*

9.6.1.2 Component Force Transfer. Components shall be attached such that the component forces are transferred to the structure. Component seismic attachments shall be bolted, welded, or otherwise positively fastened without consideration of frictional resistance produced by the effects of gravity. A continuous load path of sufficient strength and stiffness between the component and the supporting structure shall be provided. Local elements of the supporting structure shall be designed and constructed for the component forces where they control the design of the elements or their connections. The component forces shall be those determined in Section 9.6.1.3, except that modifications to F_p and R_p due to anchorage conditions need not be considered. The design documents shall include sufficient information relating to the attachments to verify compliance with the requirements of this chapter.

9.6.1.3 Seismic Forces. Seismic forces (F_p) shall be determined in accordance with Eq. 9.6.1.3-1:

$$F_p = \frac{0.4a_p S_{DS} W_p}{R_p/I_p}\left(1 + 2\frac{z}{h}\right) \qquad \textbf{(Eq. 9.6.1.3-1)}$$

F_p is not required to be taken as greater than

$$F_p = 1.6 S_{DS} I_p W_p \qquad \textbf{(Eq. 9.6.1.3-2)}$$

and F_p shall not be taken as less than

$$F_p = 0.3 S_{DS} I_p W_p \qquad \textbf{(Eq. 9.6.1.3-3)}$$

where

F_p = seismic design force centered at the component's center of gravity and distributed relative to component's mass distribution

S_{DS} = spectral acceleration, short period, as determined from Section 9.4.1.2.5

a_p = component amplification factor that varies from 1.00 to 2.50 (select appropriate value from Table 9.6.2.2 or 9.6.3.2)

I_p = component importance factor that varies from 1.00 to 1.50 (see Section 9.6.1.5)

W_p = component operating weight

R_p = component response modification factor that varies from 1.50 to 5.00 (select appropriate value from Tables 9.6.2.2 or 9.6.3.2)

z = height in structure of point of attachment of component with respect to the base. For items at or below the base, z shall be taken as 0. The value of z/h need not exceed 1.0

h = average roof height of structure with respect to the base

The force (F_p) shall be applied independently longitudinally, and laterally in combination with service loads

associated with the component. Combine horizontal and vertical load effects as indicated in Section 9.5.2.7 substituting F_p for the term Q_E. The reliability/redundancy factor, ρ, is permitted to be taken equal to 1.

When positive and negative wind loads exceed F_p for nonbearing exterior wall, these wind loads shall govern the design. Similarly, when the building code horizontal loads exceed F_p for interior partitions, these building code loads shall govern the design.

In lieu of the forces determined in accordance with Eq. 9.6.1.3-1, accelerations at any level may be determined by the modal analysis procedures of Section 9.5.6 with $R = 1.0$. Seismic forces shall be in accordance with Eq. 9.6.1.3-4:

$$F_p = \frac{a_i a_p W_p}{R_p/I_p} A_x \qquad \textbf{(Eq. 9.6.1.3-4)}$$

Where a_i is the acceleration at level i obtained from the modal analysis and where A_x is the torsional amplification factor determined by Eq. 9.5.5.5.2. Upper and lower limits of F_p determined by Eq. 9.6.1.3-2 and -3 shall apply.

9.6.1.4 Seismic Relative Displacements.
Seismic relative displacements (D_p) shall be determined in accordance with the following equations:

For two connection points on the same Structure A or the same structural system, one at a height h_x and the other at a height h_y, D_p shall be determined as

$$D_p = \delta_{xA} - \delta_{yA} \qquad \textbf{(Eq. 9.6.1.4-1)}$$

Alternatively, D_p shall be permitted to be determined using modal procedures described in Section 9.5.6.8, using the difference in story deflections calculated for each mode and then combined using appropriate modal combination procedures. D_p is not required to be taken as greater than

$$D_p = (h_x - h_y)\Delta_{aA}/h_{sx} \qquad \textbf{(Eq. 9.6.1.4-2)}$$

For two connection points on separate Structures A or B or separate structural systems, one at a height h_x and the other at a height h_y, D_p shall be determined as

$$D_p = |\delta_{xA}| + |\delta_{yB}| \qquad \textbf{(Eq. 9.6.1.4-3)}$$

D_p is not required to be taken as greater than

$$D_p = h_x\Delta_{aA}/h_{sx} + h_y\Delta_{aB}/h_{sx}$$
$$\textbf{(Eq. 9.6.1.4-4)}$$

where

D_p = relative seismic displacement that the component must be designed to accommodate

δ_{xA} = deflection at building Level x of Structure A, determined by an elastic analysis as defined in Section 9.5.5.7.1

δ_{yA} = deflection at building Level y of Structure A, determined by an elastic analysis as defined in Section 9.5.5.7.1

δ_{yB} = deflection at building Level y of Structure B, determined by an elastic analysis as defined in Section 9.5.5.7.1

h_x = height of Level x to which upper connection point is attached

h_y = height of Level y to which lower connection point is attached

Δ_{aA} = allowable story drift for Structure A as defined in Table 9.5.2.8

Δ_{aB} = allowable story drift for Structure B as defined in Table 9.5.2.8

h_{sx} = story height used in the definition of the allowable drift Δ_a in Table 9.5.2.8, note that Δ_a/h_{sx} = the drift index

The effects of seismic relative displacements shall be considered in combination with displacements caused by other loads as appropriate.

9.6.1.5 Component Importance Factor.
The component importance factor (I_p) shall be selected as follows:

$I_p = 1.5$ life safety component required to function after an earthquake (e.g., fire protection sprinkler system)

$I_p = 1.5$ component that contains hazardous content

$I_p = 1.5$ storage racks in structures open to the public (e.g., warehouse retails stores)

$I_p = 1.0$ all other components

In addition, for structures in Seismic Use Group III:

$I_p = 1.5$ all components needed for continued operation of the facility or whose failure could impair the continued operation of the facility

9.6.1.6 Component Anchorage.
Components shall be anchored in accordance with the following provisions.

9.6.1.6.1
The force in the connected part shall be determined based on the prescribed forces for the component specified in Section 9.6.1.3. Where component anchorage is provided by shallow expansion anchors, shallow chemical anchors, or shallow (low deformability) cast-in-place anchors, a value of $R_p = 1.5$ shall be used in Section 9.6.1.3 to determine the forces in the connected part.

9.6.1.6.2
Anchors embedded in concrete or masonry shall be proportioned to carry the least of the following:

a. The design strength of the connected part,

b. 1.3 times the force in the connected part due to the prescribed forces, or

c. The maximum force that can be transferred to the connected part by the component structural system.

9.6.1.6.3 Determination of forces in anchors shall take into account the expected conditions of installation including eccentricities and prying effects.

9.6.1.6.4 Determination of force distribution of multiple anchors at one location shall take into account the stiffness of the connected system and its ability to redistribute loads to other anchors in the group beyond yield.

9.6.1.6.5 Powder driven fasteners shall not be used for tension load applications in Seismic Design Categories D, E, and F unless approved for such loading.

9.6.1.6.6 The design strength of anchors in concrete shall be determined in accordance with the provisions of Section 9.9.

9.6.1.7 Construction Documents. Construction documents shall be prepared to comply with the requirements of this Standard, as indicated in Table 9.6.1.7.

9.6.2 Architectural Component Design.

9.6.2.1 General. Architectural systems, components, or elements (hereinafter referred to as "components") listed in Table 9.6.2.2 and their attachments shall meet the requirements of Sections 9.6.2.2 through 9.6.2.9.

9.6.2.2 Architectural Component Forces and Displacements. Architectural components shall meet the force requirements of Section 9.6.1.3 and Table 9.6.2.2.

Components supported by chains or otherwise suspended from the structural system above are not required to meet the lateral seismic force requirements and seismic relative displacement requirements of this Section provided that they cannot be damaged to become a hazard or cannot damage any other component when subject to seismic motion and they have ductile or articulating connections to the structure at the point of attachment. The gravity design load for these items shall be three times their operating load.

9.6.2.3 Architectural Component Deformation. Architectural components that could pose a life safety hazard shall be designed for the seismic relative displacement requirements of Section 9.6.1.4. Architectural components shall be designed for vertical deflection due to joint rotation of cantilever structural members.

TABLE 9.6.1.7
CONSTRUCTION DOCUMENTS

Component Description	Section Reference		Required Seismic Design Categories
	Quality Assurance	Design	
Exterior wall panels, including anchorage	A.9.3.3.9 No. 1	9.6.2.4	D, E, F
Suspended ceiling system, including anchorage	A.9.3.3.9 No. 2	9.6.2.6	D, E, F
Access floors, including anchorage	A.9.3.3.9 No. 2	9.6.2.7	D, E, F
Steel storage racks, including anchorage	A.9.3.3.9 No. 2	9.6.2.9	D, E, F
Glass in glazed curtain walls, glazed storefronts, and interior glazed partitions, including anchorage	A.9.3.3.9 No. 3	9.6.2.10	D, E, F
HVAC ductwork containing hazardous materials, including anchorage	A.9.3.3.10 No. 4	9.6.3.10	C, D, E
Piping systems and mechanical units containing flammable, combustible, or highly toxic materials	A.9.3.3.10 No. 3	9.6.3.11 9.6.3.12 9.6.3.13	C, D, E, F
Anchorage of electrical equipment for emergency or standby power systems	A.9.3.3.10 No. 1	9.6.3.14	C, D, E, F
Anchorage of all other electrical equipment	A.9.3.3.10 No. 2	9.6.3.14	E, F
Project-specific requirements for mechanical and electrical components and their anchorage	A.9.3.4.5	9.6.3	C, D, E, F

TABLE 9.6.2.2
ARCHITECTURAL COMPONENT COEFFICIENTS

Architectural Component or Element	a_p[a]	R_p[b]
Interior Nonstructural Walls and Partitions		
Plain (unreinforced) masonry walls	1	1.5
All other walls and partitions	1	2.5
Cantilever Elements (Unbraced or Braced to Structural Frame Below Its Center of Mass)		
Parapets and cantilever interior nonstructural walls	2.5	2.5
Chimneys and stacks when laterally braced or supported by the structural frame	2.5	2.5
Cantilever Elements (Braced to Structural Frame Above Its Center of Mass)		
Parapets	1.0	2.5
Chimneys and stacks	1.0	2.5
Exterior nonstructural walls	1.0[b]	2.5
Exterior Nonstructural Wall Elements and Connections		
Wall element	1	2.5
Body of wall panel connections	1	2.5
Fasteners of the connecting system	1.25	1
Veneer		
Limited deformability elements and attachments	1	2.5
Low deformability elements and attachments	1	2.5
Penthouses (Except when Framed by an Extension of the Building Frame)	2.5	3.5
Ceilings		
All	1	2.5
Cabinets		
Storage cabinets and laboratory equipment	1	2.5
Access Floors		
Special access floors (designed in accordance with Section 9.6.2.7.2)	1	2.5
All other	1	1.5
Appendages and Ornamentations	2.5	2.5
Signs and Billboards	2.5	2.5
Other Rigid Components		
High deformability elements and attachments	1	3.5
Limited deformability elements and attachments	1	2.5
Low deformability materials and attachments	1	1.5
Other Flexible Components	2.5	3.5
High deformability elements and attachments	2.5	2.5
Limited deformability elements and attachments	2.5	1.5
Low deformability materials and attachments		

[a] A lower value for a_p shall not be used unless justified by detailed dynamic analysis. The value for a_p shall not be less than 1.00. The value of $a_p = 1$ is for equipment generally regarded as rigid and rigidly attached. The value of $a_p = 2.5$ is for equipment generally regarded as flexible or flexibly attached. See Section 9.2.1 for definitions of rigid and flexible.

[b] Where flexible diaphragms provide lateral support for walls and partitions, the design forces for anchorage to the diaphragm shall be as specified in Section 9.5.2.6.

9.6.2.4 Exterior Nonstructural Wall Elements and Connections.

9.6.2.4.1 General. Exterior nonstructural wall panels or elements that are attached to or enclose the structure shall be designed to resist the forces in accordance with Eq. 9.6.1.3-1 or 9.6.1.3-2, and shall accommodate movements of the structure resulting from response to the design basis ground motion, D_p, or temperature changes. Such elements shall be supported by means of positive and direct structural supports or by mechanical connections and fasteners. The support system shall be designed in accordance with the following:

a. Connections and panel joints shall allow for the story drift caused by relative seismic displacements (D_p) determined in Section 9.6.1.4, or 1/2 in. (13 mm), whichever is greatest.

b. Connections to permit movement in the plane of the panel for story drift shall be sliding connections using slotted or oversize holes, connections that permit movement by bending of steel, or other connections that provide equivalent sliding or ductile capacity.

c. The connecting member itself shall have sufficient ductility and rotation capacity to preclude fracture of the concrete or brittle failures at or near welds.

d. All fasteners in the connecting system such as bolts, inserts, welds, and dowels and the body of the connectors shall be designed for the force (F_p) determined by Eq. 9.6.1.3-2 with values of R_p and a_p taken from Table 9.6.2.2 applied at the center of mass of the panel.

e. Anchorage using flat straps embedded in concrete or masonry shall be attached to or hooked around reinforcing steel or otherwise terminated so as to effectively transfer forces to the reinforcing steel or to assure that pullout of anchorage is not the initial failure mechanism.

9.6.2.4.2 Glass. Glass in glazed curtain walls and storefronts shall be designed and installed in accordance with Section 9.6.2.10.

9.6.2.5 Out-of-Plane Bending. Transverse or out-of-plane bending or deformation of a component or system that is subjected to forces as determined in Section 9.6.2.2 shall not exceed the deflection capability of the component or system.

9.6.2.6 Suspended Ceilings. Suspended ceilings shall be designed to meet the seismic force requirements of Section 9.6.2.6.1. In addition, suspended ceilings shall meet the requirements of either industry standard construction as modified in Section 9.6.2.6.2 or integral construction as specified in Section 9.6.2.6.3.

9.6.2.6.1 Seismic Forces. Suspended ceilings shall be designed to meet the force requirements of Section 9.6.1.3.

The weight of the ceiling, W_p, shall include the ceiling grid and panels; light fixtures if attached to, clipped to, or laterally supported by the ceiling grid; and other components which are laterally supported by the ceiling. W_p shall be taken as not less than 4 lbs/ft^2 (19 N/m^2).

The seismic force, F_p, shall be transmitted through the ceiling attachments to the building structural elements or the ceiling-structure boundary.

Design of anchorage and connections shall be in accordance with these provisions.

9.6.2.6.2 Industry Standard Construction. Unless designed in accordance with Section 9.6.2.6.3, suspended ceilings shall be designed and constructed in accordance with this Section.

9.6.2.6.2.1 Seismic Design Category C. Suspended ceilings in Seismic Design Category C shall be designed and installed in accordance with the CISCA recommendations for seismic Zones 0-2, (Ref. 9.6-16), except that seismic forces shall be determined in accordance with Sections 9.6.1.3 and 9.6.2.6.1.

Sprinkler heads and other penetrations in Seismic Design Category C shall have a minimum of 1/4 in. (6 mm) clearance on all sides.

9.6.2.6.2.2 Seismic Design Categories D, E, and F. Suspended ceilings in Seismic Design Categories D, E, and F shall be designed and installed in accordance with the CISCA recommendations for seismic Zones 3–4 (Ref. 9.6-17) and the additional requirements listed in this subsection.

a. A heavy duty T-bar grid system shall be used.

b. The width of the perimeter supporting closure angle shall be not less than 2.0 in. (50 mm). In each orthogonal horizontal direction, one end of the ceiling grid shall be attached to the closure angle. The other end in each horizontal direction shall have a 3/4 in. (19 mm) clearance from the wall and shall rest upon and be free to slide on a closure angle.

c. For ceiling areas exceeding 1000 ft^2 (92.9 m^2), horizontal restraint of the ceiling to the structural system shall be provided. The tributary areas of the horizontal restraints shall be approximately equal.

> **Exception:** Rigid braces are permitted to be used instead of diagonal splay wires. Braces and attachments to the structural system above shall be adequate to limit relative lateral deflections at point of attachment of ceiling grid to less than 1/4 in. (6 mm) for the loads prescribed in Section 9.6.1.3.

d. For ceiling areas exceeding 2500 ft^2 (232 m^2), a seismic separation joint or full height partition that breaks the ceiling up into areas not exceeding 2500 ft^2 shall be provided unless structural analyses are performed of the ceiling bracing system for the prescribed seismic forces which demonstrate ceiling system penetrations and closure angles provide sufficient clearance to accommodate the additional movement. Each area shall be provided with closure angles in accordance with Item b and horizontal restraints or bracing in accordance with Item c.

e. Except where rigid braces are used to limit lateral deflections, sprinkler heads and other penetrations shall have a 2 in. (50 mm) oversize ring, sleeve, or adapter through the ceiling tile to allow for free movement of at least 1 in. (25 mm) in all horizontal directions. Alternatively, a swing joint that can accommodate 1 in. (25 mm) of ceiling movement in all horizontal directions are permitted to be provided at the top of the sprinkler head extension.

f. Changes in ceiling plan elevation shall be provided with positive bracing.

g. Cable trays and electrical conduits shall be supported independently of the ceiling.

h. Suspended ceilings shall be subject to the special inspection requirements of Section A.9.3.3.9 of this Standard.

9.6.2.6.3 Integral Ceiling/Sprinkler Construction. As an alternative to providing large clearances around sprinkler system penetrations through ceiling systems, the sprinkler system and ceiling grid are permitted to be designed and tied together as an integral unit. Such a design shall consider the mass and flexibility of all elements involved, including: ceiling system, sprinkler system, light fixtures, and mechanical (HVAC) appurtenances. The design shall be performed by a registered design professional.

9.6.2.7 Access Floors.

9.6.2.7.1 General. Access floors shall be designed to meet the force provisions of Section 9.6.1.3 and the additional provisions of this Section. The weight of the access floor, W_p, shall include the weight of the floor system, 100% of the weight of all equipment fastened to the floor, and 25% of the weight of all equipment supported by, but not fastened to the floor. The seismic force, F_p, shall be transmitted

from the top surface of the access floor to the supporting structure.

Overturning effects of equipment fastened to the access floor panels also shall be considered. The ability of "slip on" heads for pedestals shall be evaluated for suitability to transfer overturning effects of equipment.

When checking individual pedestals for overturning effects, the maximum concurrent axial load shall not exceed the portion of W_p assigned to the pedestal under consideration.

9.6.2.7.2 Special Access Floors. Access floors shall be considered to be "special access floors" if they are designed to comply with the following considerations:

1. Connections transmitting seismic loads consist of mechanical fasteners, concrete anchors, welding, or bearing. Design load capacities comply with recognized design codes and/or certified test results.

2. Seismic loads are not transmitted by friction, produced solely by the effects of gravity, powder-actuated fasteners (shot pins), or adhesives.

3. The design analysis of the bracing system includes the destabilizing effects of individual members buckling in compression.

4. Bracing and pedestals are of structural or mechanical shape produced to ASTM specifications that specify minimum mechanical properties. Electrical tubing shall not be used.

5. Floor stringers that are designed to carry axial seismic loads and that are mechanically fastened to the supporting pedestals are used.

9.6.2.8 Partitions.

9.6.2.8.1 General. Partitions that are tied to the ceiling and all partitions greater than 6 ft (1.8 m) in height shall be laterally braced to the building structure. Such bracing shall be independent of any ceiling splay bracing. Bracing shall be spaced to limit horizontal deflection at the partition head to be compatible with ceiling deflection requirements as determined in Section 9.6.2.6 for suspended ceilings and Section 9.6.2.2 for other systems.

9.6.2.8.2 Glass. Glass in glazed partitions shall be designed and installed in accordance with Section 9.6.2.10.

9.6.2.9 Steel Storage Racks. Steel storage racks supported at the base of the structure shall be designed

to meet the force requirements of Section 9.14. Steel storage racks supported above the base of the structure shall be designed to meet the force requirements of Sections 9.6.1 and 9.6.2.

9.6.2.10 Glass in Glazed Curtain Walls, Glazed Storefronts, and Glazed Partitions.

9.6.2.10.1 General. Glass in glazed curtain walls, glazed storefronts, and glazed partitions shall meet the relative displacement requirement of Eq. 9.6.2.10.1-1:

$$\Delta_{fallout} \geq 1.25 D_p I \qquad \textbf{(Eq. 9.6.2.10.1-1)}$$

or 0.5 in. (13 mm), whichever is greater where

$\Delta_{fallout}$ = the relative seismic displacement (drift) causing glass fallout from the curtain wall, storefront wall, or partition (Section 9.6.2.10.2)

D_p = the relative seismic displacement that the component must be designed to accommodate (Eq. 9.6.1.4-1). D_p shall be applied over the height of the glass component under consideration

I = the occupancy importance factor (Table 9.1.4)

Exceptions:

1. Glass with sufficient clearances from its frame such that physical contact between the glass and frame will not occur at the design drift, as demonstrated by Eq. 9.6.2.10.1-2, shall be exempted from the provisions of Eq. 9.6.2.10.1-1:

$$D_{clear} \geq 1.25 D_p$$
$$\textbf{(Eq. 9.6.2.10.1-2)}$$

where

$$D_{clear} = 2c_1 \left(1 + \frac{h_p c_2}{b_p c_1} \right)$$

h_p = the height of the rectangular glass
b_p = the width of the rectangular glass
c_1 = the clearance (gap) between the vertical glass edges and the frame, and
c_2 = the clearance (gap) between the horizontal glass edges and the frame

2. Fully tempered monolithic glass in Seismic Use Groups I and II located no more than 10 ft (3 m) above a walking surface shall be exempted from the provisions of Eq. 9.6.2.10.1-1.

3. Annealed or heat-strengthened laminated glass in single thickness with interlayer no less than 0.030 in. (0.76 mm) that is captured mechanically in a wall system glazing pocket, and whose perimeter is secured to the frame by a wet glazed gunable curing elastomeric sealant perimeter bead of 1/2 in. (13 mm) minimum glass contact width, or other approved anchorage system shall be exempted from the provisions of Eq. 9.6.2.10.1-1.

9.6.2.10.2 Seismic Drift Limits for Glass Components. $\Delta_{fallout}$, the drift causing glass fallout from the curtain wall, storefront, or partition shall be determined in accordance with Ref. 9.6-21, or by engineering analysis.

9.6.3 Mechanical and Electrical Component Design.

9.6.3.1 General. Attachments and equipment supports for the mechanical and electrical systems, components, or elements (hereinafter referred to as "components") shall meet the requirements of Sections 9.6.3.2 through 9.6.3.16.

9.6.3.2 Mechanical and Electrical Component Forces and Displacements. Mechanical and electrical components shall meet the force and seismic relative displacement requirements of Sections 9.6.1.3, 9.6.1.4, and Table 9.6.3.2.

Components supported by chains or otherwise suspended from the structural system above are not required to meet the lateral seismic force requirements and seismic relative displacement requirements of this Section provided they are designed to prevent damage to themselves or causing damage to any other component when subject to seismic motion. Such supports shall have ductile or articulating connections to the structure at the point of attachment. The gravity design load for these items shall be three times their operating load.

9.6.3.3 Mechanical and Electrical Component Period. The fundamental period of the mechanical and electrical component (and its attachment to the building), T_p, shall be determined by the following equation provided that the component and attachment can be reasonably represented analytically by a simple spring and mass single degree of freedom system:

$$T_p = 2\pi \sqrt{\frac{W_p}{K_p g}} \qquad \textbf{(Eq. 9.6.3.3)}$$

TABLE 9.6.3.2
**MECHANICAL AND ELECTRICAL COMPONENTS SEISMIC
COEFFICIENTS**

Mechanical and Electrical Component or Element[b]	a_p[a]	R_p
General Mechanical Equipment		
Boilers and furnaces	1.0	2.5
Pressure vessels on skirts and free-standing	2.5	2.5
stacks	2.5	2.5
Cantilevered chimneys	2.5	2.5
Other	1.0	2.5
Manufacturing and Process Machinery		
General	1.0	2.5
Conveyors (non-personnel)	2.5	2.5
Piping Systems		
High deformability elements and attachments	1.0	3.5
Limited deformability elements and attachments	1.0	2.5
Low deformability elements and attachments	1.0	1.5
HVAC Systems		
Vibration isolated	2.5	2.5
Nonvibration isolated	1.0	2.5
Mounted in-line with ductwork	1.0	2.5
Other	1.0	2.5
Elevator Components	1.0	2.5
Escalator Components	1.0	2.5
Trussed Towers (free-standing or guyed)	2.5	2.5
General Electrical		
Distribution systems (bus ducts, conduit, cable tray)	2.5	5.0
Equipment	1.0	2.5
Lighting Fixtures	1.0	1.5

[a] A lower value for a_p shall not be used unless justified by detailed dynamic analyses. The value for a_p shall not be less than 1.00. The value of $a_p = 1$ is for equipment generally regarded as rigid or rigidly attached. The value of $a_p = 2.5$ is for equipment generally regarded as flexible or flexibly attached. See Section 9.2.2 for definitions of rigid and flexible.
[b] Components mounted on vibration isolation systems shall have a bumper restraint or snubber in each horizontal direction. The design force shall be taken as $2F_p$ if the maximum clearance (air gap) between the equipment support frame and restraint is greater than 1/4 in. If the maximum clearance is specified on the construction documents to be not greater than 1/4 in., the design force may be taken as F_p.

where

T_p = component fundamental period
W_p = component operating weight
g = gravitational acceleration
K_p = stiffness of resilient support system of the component and attachment, determined in terms of load per unit deflection at the center of gravity of the component

Note that consistent units must be used.

Otherwise, determine the fundamental period of the component in seconds (T_p) from experimental test data or by a properly substantiated analysis.

9.6.3.4 Mechanical and Electrical Component Attachments. The stiffness of mechanical and electrical component attachments shall be designed such that the load path for the component performs its intended function.

9.6.3.5 Component Supports. Mechanical and electrical component supports and the means by which they are attached to the component shall be designed for the forces determined in Section 9.6.1.3 and in conformance with Sections 9.8 through 9.12, as appropriate, for the materials comprising the means of attachment. Such supports include structural members, braces, frames, skirts, legs, saddles, pedestals, cables, guys, stays, snubbers, and tethers, as well as element forged or cast as a part of the mechanical or electrical component. If standard or proprietary supports are used, they shall be designed by either load rating (i.e., testing) or for the calculated seismic forces. In addition, the stiffness of the support, when appropriate, shall be designed such that the seismic load path for the component performs its intended function.

Component supports shall be designed to accommodate the seismic relative displacements between points of support determined in accordance with Section 9.6.1.4.

In addition, the means by which supports are attached to the component, except when integral (i.e., cast or forged), shall be designed to accommodate both the forces and displacements determined in accordance with Sections 9.6.1.3 and 9.6.1.4. If the value of $I_p = 1.5$ for the component, the local region of the support attachment point to the component shall be evaluated for the effect of the load transfer on the component wall.

9.6.3.6 Component Certification. Architectural, mechanical, and electrical components shall comply with the force requirements of Section 9.6. Components designated with an I_p greater than 1.0 in Seismic Design Category C, D, E, and F shall meet additional requirements of Section A.9.3.4.5 and in particular, mechanical and electrical equipment which must remain operable following the design earthquake shall demonstrate operability by shake table testing or experience data.

The manufacturer's certificate of compliance indicating compliance with this Section shall be submitted to the authority having jurisdiction when required by the contract documents or when required by the regulatory agency.

9.6.3.7 Utility and Service Lines at Structure Interfaces. At the interface of adjacent structures or portions of the same structure that may move independently, utility lines shall be provided with adequate flexibility to accommodate the anticipated differential movement between the portions that move independently. Differential displacement calculations shall be determined in accordance with Section 9.6.1.4.

9.6.3.8 Site-Specific Considerations. The possible interruption of utility service shall be considered in relation to designated seismic systems in Seismic Use Group III as defined in Section 9.1.3.4. Specific attention shall be given to the vulnerability of underground utilities and utility interfaces between the structure and the ground where Site Class E or F soil is present, and where the seismic coefficient S_{DS} at the underground utility or at the base of the structure is equal to or greater than 0.33.

9.6.3.9 Storage Tanks Mounted in Structures. Storage tanks, including their attachments and supports, shall be designed to meet the force requirements of Section 9.14.

9.6.3.10 HVAC Ductwork. Attachments and supports for HVAC ductwork systems shall be designed to meet the force and displacement provisions of Sections 9.6.1.3 and 9.6.1.4 and the additional provisions of this Section. In addition to their attachments and supports, ductwork systems designated as having an $I_p = 1.5$ themselves shall be designed to meet the force and displacement provisions of Sections 9.6.1.3 and 9.6.1.4 and the additional provisions of this Section. Where HVAC ductwork runs between structures that could displace relative to one another and for seismically isolated structures where the HVAC ductwork crosses the seismic isolation interface, the HVAC ductwork shall be designed to accommodate the seismic relative displacements specified in Section 9.6.1.4.

Seismic restraints are not required for HVAC ducts with $I_p = 1.0$ if either of the following conditions are met:

a. HVAC ducts are suspended from hangers 12 in. (305 mm) or less in length from the top of the duct to the supporting structure. The hangers shall be detailed to avoid significant bending of the hangers and their attachments,

or

b. HVAC ducts have a cross-sectional area of less than 6 ft^2 (0.557 m^2).

HVAC duct systems fabricated and installed in accordance with standards approved by the authority having jurisdiction shall be deemed to meet the lateral bracing requirements of this Section.

Equipment items installed in-line with the duct system (e.g., fans, heat exchangers, and humidifiers) weighing more than 75 lb (344 N) shall be supported and laterally braced independent of the duct system and shall meet the force requirements of Section 9.6.1.3. Appurtenances such as dampers, louvers, and diffusers shall be positively attached with mechanical fasteners. Unbraced piping attached to in-line equipment shall be provided with adequate flexibility to accommodate differential displacements.

9.6.3.11 Piping Systems. Attachments and supports for piping systems shall be designed to meet the force and displacement provisions of Sections 9.6.1.3 and 9.6.1.4 and the additional provisions of this Section. In addition to their attachments and supports, piping systems designated as having $I_p = 1.5$ themselves shall be designed to meet the force and displacement provisions of Sections 9.6.1.3 and 9.6.1.4 and the additional provisions of this Section. Where piping systems are attached to structures that could displace relative to one another and for seismically isolated structures where the piping system crosses the seismic isolation interface, the piping system shall be designed to accommodate the seismic relative displacements specified in Section 9.6.1.4.

Seismic effects that shall be considered in the design of a piping system include the dynamic effects of the piping system, its contents, and, when appropriate, its supports. The interaction between the piping system and the supporting structures, including other mechanical and electrical equipment shall also be considered.

9.6.3.11.1 Pressure Piping Systems. Pressure piping systems designed and constructed in accordance with ASME B31, *Code for Pressure Piping* (Ref. 9.6-3) shall be deemed to meet the force, displacement, and other provisions of this Section. In lieu of specific force and displacement provisions provided in the ASME B31, the force and displacement provisions of Sections 9.6.1.3 and 9.6.1.4 shall be used.

9.6.3.11.2 Fire Protection Sprinkler Systems. Fire protection sprinkler systems designed and constructed in accordance with NFPA 13, *Standard for the Installation of Sprinkler Systems* (Ref. 9.6-12) shall be deemed to meet the other requirements of this Section, except the force and displacement requirements of Sections 9.6.1.3 and 9.6.1.4 shall be satisfied.

9.6.3.11.3 Other Piping Systems. Piping designated as having an $I_p = 1.5$ but not designed and constructed in accordance with ASME B31 (Ref. 9.6-3) or NFPA 13 (Ref. 9.6-12) shall meet the following:

a. The design strength for seismic loads in combination with other service loads and appropriate environmental effects shall not exceed the following:

1. For piping systems constructed with ductile materials (e.g., steel, aluminum, or copper), 90% of the piping material yield strength.

2. For threaded connections with ductile materials, 70% of the piping material yield strength.

3. For piping constructed with nonductile materials (e.g., plastic, cast iron, or ceramics), 25% of the piping material minimum specified tensile strength.

4. For threaded connections in piping constructed with non-ductile materials, 20% of the piping material minimum specified tensile strength.

b. Provisions shall be made to mitigate seismic impact for piping components constructed of nonductile materials or in cases where material ductility is reduced (e.g., low temperature applications).

c. Piping shall be investigated to ensure that the piping has adequate flexibility between support attachment points to the structure, ground, other mechanical and electrical equipment, or other piping.

d. Piping shall be investigated to ensure that the interaction effects between it and other piping or constructions are acceptable.

9.6.3.11.4 Supports and Attachments for Other Piping. Attachments and supports for piping not designed and constructed in accordance with ASME B31 [Ref. 9.6-3] or NFPA 13 [Ref. 9.6-12] shall meet the following provisions:

a. Attachments and supports transferring seismic loads shall be constructed of materials suitable for the application and designed and constructed in accordance with a nationally recognized structural code such as, when constructed of steel, the AISC *Manual of Steel Construction* (Ref. 9.8-1 or 9.8-2) or MSS SP-58, *Pipe Hangers and Supports–Materials, Design, and Manufacture* (Ref. 9.6-15).

b. Attachments embedded in concrete shall be suitable for cyclic loads.

c. Rod hangers shall not be used as seismic supports unless the length of the hanger from the supporting structure is 12 in. (305 mm) or less. Rod hangers shall not be constructed in a manner that subjects the rod to bending moments.

d. Seismic supports are not required for:

1. Ductile piping in Seismic Design Category D, E, or F designated as having an $I_p = 1.5$ and a nominal pipe size of 1 in. (25 mm) or less when provisions are made to protect the piping from impact or to avoid the impact of larger piping or other mechanical equipment.

2. Ductile piping in Seismic Design Category C designated as having an $I_p = 1.5$ and a nominal pipe size of 2 in. (50 mm)

or less when provisions are made to protect the piping from impact or to avoid the impact of larger piping or other mechanical equipment.

3. Ductile piping in Seismic Design Category D, E, or F designated as having an $I_p = 1.0$ and a nominal pipe size of 3 in. (75 mm) or less.

4. Ductile piping in Seismic Design Category A, B, or C designated as having an $I_p = 1.0$ and a nominal pipe size of 6 in. (150 mm) or less.

e. Seismic supports shall be constructed so that support engagement is maintained.

9.6.3.12 Boilers and Pressure Vessels. Attachments and supports for boilers and pressure vessels shall be designed to meet the force and displacement provisions of Sections 9.6.1.3 and 9.6.1.4 and the additional provisions of this Section. In addition to their attachments and supports, boilers and pressure vessels designated as having an $I_p = 1.5$ themselves shall be designed to meet the force and displacement provisions of Sections 9.6.1.3 and 9.6.1.4.

The seismic design of a boiler or pressure vessel shall include analysis of the following: the dynamic effects of the boiler or pressure vessel, its contents, and its supports; sloshing of liquid contents; loads from attached components such as piping; and the interaction between the boiler or pressure vessel and its support.

9.6.3.12.1 ASME Boilers and Pressure Vessels. Boilers or pressure vessels designed in accordance with the ASME *Boiler and Pressure Vessel Code* (Ref. 9.6-4) shall be deemed to meet the force, displacement, and other requirements of this Section. In lieu of the specific force and displacement provisions provided in the ASME code, the force and displacement provisions of Sections 9.6.1.3 and 9.6.1.4 shall be used.

9.6.3.12.2 Other Boilers and Pressure Vessels. Boilers and pressure vessels designated as having an $I_p = 1.5$ but not constructed in accordance with the provisions of the ASME code (Ref. 9.6-4) shall meet the following provisions:

a. The design strength for seismic loads in combination with other service loads and appropriate environmental effects shall not exceed the following:

1. For boilers and pressure vessels constructed with ductile materials (e.g., steel, aluminum, or copper), 90% of the material minimum specified yield strength.

2. For threaded connections in boilers or pressure vessels or their supports constructed with ductile materials, 70% of the material minimum specified yield strength.

3. For boilers and pressure vessels constructed with nonductile materials (e.g., plastic, cast iron, or ceramics), 25% of the material minimum specified tensile strength.

4. For threaded connections in boilers or pressure vessels or their supports constructed with nonductile materials, 20% of the material minimum specified tensile strength.

b. Provisions shall be made to mitigate seismic impact for boiler and pressure vessel components constructed of nonductile materials or in cases where material ductility is reduced (e.g., low temperature applications).

c. Boilers and pressure vessels shall be investigated to ensure that the interaction effects between them and other constructions are acceptable.

9.6.3.12.3 Supports and Attachments for Other Boilers and Pressure Vessels. Attachments and supports for boilers and pressure vessels shall meet the following provisions:

a. Attachments and supports transferring seismic loads shall be constructed of materials suitable for the application and designed and constructed in accordance with nationally recognized structural code such as, when constructed of steel, the AISC *Manual of Steel Construction* (Ref. 9.8-1 or 9.8-2).

b. Attachments embedded in concrete shall be suitable for cyclic loads.

c. Seismic supports shall be constructed so that support engagement is maintained.

9.6.3.13 Mechanical Equipment, Attachments, and Supports. Attachments and supports for mechanical equipment not covered in Sections 9.6.3.8 through 9.6.3.12 or 9.6.3.16 shall be designed to meet the force and displacement provisions of Sections 9.6.1.3 and 9.6.1.4 and the additional provisions of this Section. In addition to their attachments and supports, such mechanical equipment designated as having an $I_p = 1.5$, itself, shall be designated to meet the force and displacement provisions of Sections 9.6.1.3 and 9.6.1.4 and the additional provisions of this Section.

The seismic design of mechanical equipment, attachments, and their supports shall include analysis of the following: the dynamic effects of the equipment, its contents, and, when appropriate, its supports. The interaction

between the equipment and the supporting structures, including other mechanical and electrical equipment, shall also be considered.

9.6.3.13.1 Mechanical Equipment. Mechanical equipment designated as having an $I_p = 1.5$ shall meet the following provisions.

a. The design strength for seismic loads in combination with other service loads and appropriate environmental effects shall not exceed the following:

1. For mechanical equipment constructed with ductile materials (e.g., steel, aluminum, or copper), 90% of the equipment material minimum specified yield strength.

2. For threaded connections in equipment constructed with ductile materials, 70% of the material minimum specified yield strength.

3. For mechanical equipment constructed with nonductile materials (e.g., plastic, cast iron, or ceramics), 25% of the equipment material minimum tensile strength.

4. For threaded connections in equipment constructed with nonductile materials, 20% of the material minimum specified yield strength.

b. Provisions shall be made to mitigate seismic impact for equipment components constructed of nonductile materials or in cases where material ductility is reduced (e.g., low temperature applications).

c. The possibility for loadings imposed on the equipment by attached utility or service lines due to differential motions of points of support from separate structures shall be evaluated.

9.6.3.13.2 Attachments and Supports for Mechanical Equipment. Attachments and supports for mechanical equipment shall meet the following provisions:

a. Attachments and supports transferring seismic loads shall be constructed of materials suitable for the application and designed and constructed in accordance with a nationally recognized structural code such as, when constructed of steel, AISC, *Manual of Steel Construction* (Ref. 9.8-1 or 9.8-2).

b. Friction clips shall not be used for anchorage attachment.

c. Expansion anchors shall not be used for mechanical equipment rated over 10 hp (7.45 kW).

Exception: Undercut expansion anchors.

d. Drilled and grouted-in-place anchors for tensile load applications shall use either expansive cement or expansive epoxy grout.

e. Supports shall be specifically evaluated if weak-axis bending of light-gauge support steel is relied on for the seismic load path.

f. Components mounted on vibration isolation systems shall have a bumper restraint or snubber in each horizontal direction. The design force shall be taken as $2F_p$. The intent is to prevent excessive movement and to avoid fracture of support springs and any nonductile components of the isolators.

g. Seismic supports shall be constructed so that support engagement is maintained.

9.6.3.14 Electrical Equipment, Attachments, and Supports. Attachments and supports for electrical equipment shall be designed to meet the force and displacement provisions of Sections 9.6.1.3 and 9.6.1.4 and the additional provisions of this Section. In addition to their attachments and supports, electrical equipment designated as having $I_p = 1.5$, itself, shall be designed to meet the force and displacement provisions of Sections 9.6.1.3 and 9.6.1.4 and the additional provisions of this Section.

The seismic design of other electrical equipment shall include analysis of the following: the dynamic effects of the equipment, its contents, and when appropriate, its supports. The interaction between the equipment and the supporting structures, including other mechanical and electrical equipment, shall also be considered. Where conduit, cable trays, or similar electrical distribution components are attached to structures that could displace relative to one another and for seismically isolated structures where the conduit or cable trays cross the seismic isolation interface, the conduit or cable trays shall be designed to accommodate the seismic relative displacement specified in Section 9.6.1.4.

9.6.3.14.1 Electrical Equipment. Electrical equipment designated as having an $I_p = 1.5$ shall meet the following provisions:

a. The design strength for seismic loads in combination with other service loads and appropriate environmental effects shall not exceed the following:

1. For electrical equipment constructed with ductile material (e.g., steel, aluminum, or copper), 90% of the equipment material minimum specified yield strength.

2. For threaded connections in equipment constructed with ductile materials, 70% of the material minimum specified yield strength.

3. For electrical equipment constructed with nonductile materials (e.g., plastic, cast iron, or ceramics), 25% of the equipment material minimum tensile strength.

4. For threaded connections in equipment constructed with nonductile materials, 20% of the material minimum specified yield strength.

b. Provisions shall be made to mitigate seismic impact for equipment components constructed of nonductile materials or in cases where material ductility is reduced (e.g., low-temperature applications).

c. The possibility for loadings imposed on the equipment by attached utility or service lines due to differential motion of points of support from separate structures shall be evaluated.

d. Batteries on racks shall have wraparound restraints to ensure that the batteries will not fall off the rack. Spacers shall be used between restraints and cells to prevent damage to cases. Racks shall be evaluated for sufficient lateral and longitudinal load capacity.

e. Internal coils of dry type transformers shall be positively attached to their supporting substructure within the transformer enclosure.

f. Slide out components in electrical control panels shall have a latching mechanism to hold contents in place.

g. Structural design of electrical cabinets shall be in conformance with standards of the industry that are acceptable to the authority having jurisdiction. Large cutouts in the lower shear panel shall be specifically evaluated if an evaluation is not provided by the manufacturer.

h. The attachment of additional items weighing more than 100 lb (440 N) shall be specifically evaluated if not provided by the manufacturer.

9.6.3.14.2 Attachments and Supports for Electrical Equipment. Attachments and supports for electrical equipment shall meet the following provisions:

a. Attachments and supports transferring seismic loads shall be constructed of materials suitable for the application and designed and constructed in accordance with a nationally recognized structural code such as, when constructed of steel, AISC, *Manual of Steel Construction* (Ref. 9.8-1 or 9.8-2).

b. Friction clips shall not be used for anchorage attachment.

c. Oversized washers shall be used at bolted connections through the base sheet metal if the base is not reinforced with stiffeners.

d. Supports shall be specifically evaluated if weak-axis bending of light-gauge support steel is relied on for the seismic load path.

e. The supports for linear electrical equipment such as cable trays, conduit, and bus ducts shall be designed to meet the force and displacement provisions of Sections 9.6.1.3 and 9.6.1.4 only if any of the following situations apply:

Supports are cantilevered up from the floor,

Supports include bracing to limit deflection,

Supports are constructed as rigid welded frames,

Attachments into concrete utilize nonexpanding insets, shot pins, or cast iron embedments, or

Attachments utilize spot welds, plug welds, or minimum size welds as defined by AISC (Ref. 9.8-1 or 9.8-2)

f. Components mounted on vibration isolation systems shall have a bumper restraint or snubber in each horizontal direction, and vertical restraints shall be provided where required to resist overturning. Isolator housings and restraints shall not be constructed of cast iron or other materials with limited ductility. (See additional design force requirements in Table 9.6.3.2.) A viscoelastic pad or similar material of appropriate thickness shall be used between the bumper and equipment item to limit the impact load.

9.6.3.15 Alternative Seismic Qualification Methods. As an alternative to the analysis methods implicit in the design methodology described above, equipment testing is an acceptable method to determine seismic capacity. Thus, adaptation of a nationally recognized standard for qualification by testing that is acceptable to the authority having jurisdiction is an acceptable alternative, so long as the equipment seismic capacity equals or exceeds the demand expressed in Sections 9.6.1.3 and 9.6.1.4.

9.6.3.16 Elevator Design Requirements.

9.6.3.16.1 Reference Document. Elevators shall meet the force and displacement provisions of Section 9.6.3.2 unless exempted by either Section 9.1.2.1 or Section 9.6.1. Elevators designed in accordance with the seismic provisions of the ASME *Safety Code for Elevators and Escalators* (Ref. 9.6-1) shall be deemed to meet the seismic force requirements of this Section, except as modified below.

9.6.3.16.2 Elevators and Hoistway Structural System. Elevators and hoistway structural systems shall be designed to meet the force and displacement provisions of Sections 9.6.1.3 and 9.6.1.4.

9.6.3.16.3 Elevator Machinery and Controller Supports and Attachments. Elevator machinery and controller supports and attachments shall be designed to meet with the force and displacement provisions of Sections 9.6.1.3 and 9.6.1.4.

9.6.3.16.4 Seismic Controls. Seismic switches shall be provided for all elevators addressed by Section 9.6.3.16.1 including those meeting the requirements of the ASME reference, provided they operate with a speed of 150 ft/min (46 m/min) or greater.

Seismic switches shall provide an electrical signal indicating that structural motions are of such a magnitude that the operation of elevators may be impaired. Upon activation of the seismic switch, elevator operations shall conform to provisions in the ASME *Safety Code for Elevators and Escalators* (Ref. 9.6-1) except as noted below. The seismic switch shall be located at or above the highest floor serviced by the elevators. The seismic switch shall have two horizontal perpendicular axes of sensitivity. Its trigger level shall be set to 30% of the acceleration of gravity.

In facilities where the loss of the use of an elevator is a life safety issue, the elevator shall only be used after the seismic switch has triggered provided that:

1. The elevator shall operate no faster than the service speed,
2. Before the elevator is occupied, it is operated from top to bottom and back to top to verify that it is operable, and
3. The individual putting the elevator back in service shall ride the elevator from top to bottom and back to top to verify acceptable performance.

9.6.3.16.5 Retainer Plates. Retainer plates are required at the top and bottom of the car and counterweight.

SECTION 9.7
FOUNDATION DESIGN REQUIREMENTS

9.7.1 General. This Section includes only those foundation requirements that are specifically related to seismic-resistant construction. It assumes compliance with other basic requirements, which include, but are not limited to, the extent of the foundation investigation, fills present or to be placed in the area of the structure, slope stability, subsurface drainage, settlement control, and pile, requirements. Except as specifically noted, the term "pile" as used in Sections 9.7.4.4 and 9.7.5.4 includes foundation piers, caissons, and piles, and the term "pile cap" includes the elements to which piles are connected, including grade beams and mats.

9.7.2 Seismic Design Category A. There are no special requirements for the foundations of structures assigned to Category A.

9.7.3 Seismic Design Category B. The determination of the site coefficient, Section 9.4.1.2.4, shall be documented and the resisting capacities of the foundations, subjected to the prescribed seismic forces of Sections 9.1 through 9.9, shall meet the following requirements.

9.7.3.1 Structural Components. The design strength of foundation components subjected to seismic forces alone or in combination with other prescribed loads and their detailing requirements shall conform to the requirements of Sections 9.8 through 9.12. The strength of foundation components shall not be less than that required for forces acting without seismic forces.

9.7.3.2 Soil Capacities. The capacity of the foundation soil in bearing, or the capacity of the soil interface between pile or pier and the soil, shall be sufficient to support the structure with all prescribed loads, without seismic forces, taking due account of the settlement that the structure can withstand. For the load combination, including earthquake as specified in Section 9.5.2.7, the soil capacities must be sufficient to resist loads at acceptable strains considering both the short duration of loading and the dynamic properties of the soil.

9.7.4 Seismic Design Category C. Foundations for structures assigned to Category C shall conform to all of the requirements for Categories A and B and to the additional requirements of this Section.

9.7.4.1 Investigation. When required by the authority having jurisdiction, a written geotechnical or geologic report shall be submitted. This report shall include, in addition to the requirements of Section 9.7.1 and the evaluations required in Section 9.7.3, the results of an investigation to evaluate the following potential earthquake hazards:

1. Slope instability
2. Liquefaction
3. Lateral spreading
4. Surface rupture

The investigation shall contain recommendations for appropriate foundation designs or other measures to mitigate the effects of the above hazards.

9.7.4.2 Pole-Type Structures. When construction employing posts or poles as columns embedded in earth or embedded in concrete footings in the earth is used to resist lateral loads, the depth of embedment required for posts or poles to resist seismic forces shall be determined by means of the design criteria established in the foundation investigation report.

9.7.4.3 Foundation Ties. Individual pile caps, drilled piers, or caissons shall be interconnected by ties. All ties shall have a design strength in tension or compression greater than a force equal to 10% of S_{DS} times the larger pile cap or column factored dead plus factored live load unless it is demonstrated that equivalent restraint will be provided by reinforced concrete beams within slabs on grade or reinforced concrete slabs on grade or confinement by competent rock, hard cohesive soils, very dense granular soils, or other approved means.

9.7.4.4 Special Pile Requirements. Concrete piles, concrete-filled steel pipe piles, drilled piers, or caissons require minimum bending, shear, tension, and elastic strain capacities. Refer to Section A.9.7.4.4 for supplementary provisions.

9.7.5 Foundation Requirements for Seismic Design Categories D, E, and F. Foundations for structures assigned to Seismic Design Categories D, E, and F shall conform to all of the requirements for Seismic Design Category C construction and to the additional requirements of this Section. Design and construction of concrete foundation components shall conform to the requirements of Ref. 9.9-1, Section 21.8, except as modified by the requirements of this Section.

Exception: Detached 1- and 2-family dwellings of light-frame construction not exceeding 2 stories in height above grade need only comply with the requirements for Sections 9.7.4 and 9.7.5.3.

9.7.5.1 Investigation. The owner shall submit to the authority having jurisdiction a written report that includes an evaluation of the items in Section 9.7.4.1 and the determination of lateral pressures on basement and retaining walls due to earthquake motions.

9.7.5.2 Foundation Ties. Individual spread footings founded on soil defined in Section 9.4.1.2.1 as Site Class E or F shall be interconnected by ties. Ties shall conform to Section 9.7.4.3.

9.7.5.3 Liquefaction Potential and Soil Strength Loss. The geotechnical report required by Section 9.7.5.1 shall assess potential consequences of liquefaction and soil strength loss, including estimation of differential settlement, lateral movement or reduction in foundation soil-bearing capacity, and shall discuss mitigation measures. Such measures shall be given consideration in the design of the structure and can include, but are not limited to, ground stabilization, selection of appropriate foundation type and depths, selection of appropriate structural systems to accommodate anticipated displacements, or any combination of these measures. The potential for liquefaction shall be evaluated for site peak ground accelerations, magnitudes, and source characteristics consistent with the design earthquake ground motions defined in Section 9.4.1.3.4. Where deemed appropriate by the authority having jurisdiction, a site-specific geotechnical report is not required when prior evaluations of nearby sites with similar soil conditions provide sufficient direction relative to the proposed construction.

The potential for liquefaction and soil strength loss shall be evaluated for site peak ground accelerations, magnitudes, and source characteristics consistent with the design earthquake ground motions. Peak ground acceleration may be determined based on a site-specific study taking into account soil amplification effects or, in the absence of such a study, peak ground accelerations shall be assumed equal to $S_S/2.5$.

9.7.5.4 Special Pile and Grade Beam Requirements. Piling shall be designed and constructed to withstand maximum imposed curvatures from earthquake ground motions and structure response. Curvatures shall include free-field soil strains (without the structure) modified for soil-pile-structure interaction coupled with pile deformations induced by lateral pile resistance to structure seismic forces. Concrete piles in Site Class E or F shall be designed and detailed in accordance with Sections 21.4.4.1, 21.4.4.2, and 21.4.4.3 of Ref. 9.9-1 within seven pile diameters of the pile cap and the interfaces of soft to medium stiff clay or liquefiable strata. Refer to Section A.9.7.5 for supplementary provisions in addition to those given in Section A.9.7.4.4. Batter piles and their connections shall be capable of resisting forces and moments from the special seismic load combinations of Section 9.5.2.7.1. For precast, prestressed concrete piles, detailing provisions as given in Section A.9.7.5.4.4 shall apply.

Section 21.8.3.3 of Ref. 9.9-1 need not apply when grade beams have the required strength to resist the forces from the special seismic loads of Section 9.5.2.7.1. Section 21.8.4.4(a) of Ref. 9.9-1 need not apply to concrete piles. Section 21.8.4.4(b) of ACI 318 need not apply to precast, prestressed concrete piles.

Design of anchorage of piles into the pile cap shall consider the combined effect of axial forces due to uplift and bending moments due to fixity to the pile cap. For piles required to resist uplift forces or provide rotational restraint, anchorage into the pile cap shall be capable of developing the following:

1. In the case of uplift, the lesser of the nominal tensile strength of the longitudinal reinforcement in a concrete pile, or the nominal tensile strength of a steel pile, or 1.3 times the pile pullout resistance, or the axial tension force resulting from the special seismic loads of Section 9.5.2.7.1.

2. In the case of rotational restraint, the lesser of the axial and shear forces and moments resulting from the special seismic loads of Section 9.5.2.7.1 or development of the full axial, bending, and shear nominal strength of the pile.

Splices of pile segments shall develop the nominal strength of the pile section, but the splice need not develop the nominal strength of the pile in tension, shear, and bending when it has been designed to resist axial and shear forces and moments from the special seismic loads of Section 9.5.2.7.1.

Pile moments, shears, and lateral deflections used for design shall be established considering the interaction of the shaft and soil. Where the ratio of the depth of embedment of the pile-to-the-pile diameter or width is less than or equal to 6, the pile may be assumed to be flexurally rigid with respect to the soil.

Pile group effects from soil on lateral pile nominal strength shall be included where pile center-to-center spacing in the direction of lateral force is less than eight pile diameters or widths. Pile group effects on vertical nominal strength shall be included where pile center-to-center spacing is less than three pile diameters or widths.

SECTION 9.8
STEEL

9.8.1 Reference Documents. The design, construction, and quality of steel components that resist seismic forces shall conform to the requirements of the references listed in this Section except that modifications are necessary to make the references compatible with the provisions of this document. Appendix A.9.8 provides supplementary provisions for this compatibility.

Reference 9.8-1   American Institute of Steel Construction (AISC), *Load and Resistance Factor Design Specification for Structural Steel Buildings (LRFD)*, 1999.

Reference 9.8-2   American Institute of Steel Construction, *Allowable Stress Design and Plastic Design Specification for Structural Steel Buildings (ASD)*, June 1, 1989.

Reference 9.8-3   American Institute of Steel Construction, *Seismic Provisions for Structural Steel Buildings*, Part I, 1997, including Supplement 2, November 10, 2000.

Reference 9.8-4   American Iron and Steel Institute (AISI), *Specification for the Design of Cold-Formed Steel Structural Members*, 1996, including Supplement No. 1, July 30, 1999.

Reference 9.8-5   ASCE, *Specification for the Design of Cold-Formed Stainless Steel Structural Members*, ASCE 8-90, 1990.

Reference 9.8-6   Steel Joist Institute, *Standard Specification, Load Tables and Weight Tables for Steel Joists and Joist Girders*, 1994.

Reference 9.8-7   ASCE, *Structural Applications for Steel Cables for Buildings*, ASCE 19-95, 1995.

SECTION 9.9
STRUCTURAL CONCRETE

9.9.1 Reference Documents. The quality and testing of materials and the design and construction of structural concrete components that resist seismic forces shall conform to the requirements of the references listed in this Section except that modifications are necessary to make the reference compatible with the provisions of this document. Appendix A.9.9 provides the supplementary provisions for this compatibility. The load combinations of Section 2.4.1 are not applicable for design of reinforced concrete to resist earthquake loads.

Reference 9.9-1   American Concrete Institute, *Building Code Requirements for Structural Concrete*, ACI 318-02.

SECTION 9.10
COMPOSITE STRUCTURES

9.10.1 Reference Documents. The design, construction, and quality of composite steel and concrete components that resist seismic forces shall conform to the relevant requirements of the following references except as modified by the provisions of this chapter.

Reference 9.10-1   American Institute of Steel Construction (AISC), *Load and Resistance Factor Design Specification for Structural Steel Buildings* (LRFD), 1999.

Reference 9.10-2   American Concete Institute (ACI), *Building Code Requirements for Structural Concrete*, 1999, *excluding*

Appendix A, (Alternate Design Method) and Chapter 22 (Structural Plain Concrete) and making Appendix C (Alternative Load and Strength Reduction Factors) mandatory.

Reference 9.10-3   American Institute of Steel Construction (AISC), *Seismic Provisions for Structural Steel Buildings*, Including Supplement No. 1 (February 15, 1999, July 1997, Parts I and II.

Reference 9.10-4   American Iron and Steel Institute (AISI), *Specification for the Design of Cold-Formed Steel Structural Members*, 1996, including Supplement 2000, excluding ASD provisions.

9.10.1.1 When using ACI [2], Appendix A and Chapter 22 are excluded and Appendix C is mandatory.

SECTION 9.11
MASONRY

9.11.1 Reference Documents. The design, construction, and quality assurance of masonry components that resist seismic forces shall conform to the requirements of the reference listed in this Section except that modifications are necessary to make the reference compatible with the provisions of this document. Appendix A.9.11 provides the supplementary provisions for this compatibility.

Reference 9.11-1   American Concrete Institute, *Building Code Requirements for Masonry Structures*, ACI 530-99/ASCE 5-99/TMS 402-99, 1999 and *Specifications for Masonry Structures, ACI 530.1-99/ASCE 6-99/TMS 602-99*, 1999.

SECTION 9.12
WOOD

9.12.1 Reference Documents. The quality, testing, design, and construction of members and their fastenings in wood systems that resist seismic forces shall conform to the requirements of the reference documents listed in this Section.

9.12.1.1 Consensus Standards.

Reference 9.12-1   National Design Specification for Wood Construction, including supplements, *ANSI/AF&PA NDS-2001* (2001).

Reference 9.12-2   National Institute of Standards and Technology, *American Softwood Lumber Standard, Voluntary Product Standard PS 20-99*, (1999).

Reference 9.12-3   Softwood Plywood–Construction and Industrial *PS 1-95* (1995).

Reference 9.12-4   ANSI, *Wood Particle Board, ANSI A208.1*, (1993).

Reference 9.12-5   *Performance Standard for Wood Based Structural Use Panels PS 2-92* (1992).

Reference 9.12-6   American National Standard for Wood Products–Structural Glued Laminated Timber," *ANSI/AITC A190.1-1992*.

Reference 9.12-7   Wood Poles — Specifications and Dimensions *ANSI 05.1* (1992).

Reference 9.12-8   Standard Specification for Establishing and Monitoring Structural Capacities of Prefabricated Wood I-Joists, *ASTM D 5055-95A* (1995).

Reference 9.12-9   Load and Resistance Factor Design (LRFD), Standard for Engineered Wood Construction, including supplements, ASCE 16-95 (1995).

9.12.1.2 Other References.

Reference 9.12-10   International Code Council, (ICC) *2000 International Residential Code with the 2002 Supplement*, 2000.

SECTION 9.13
PROVISIONS FOR SEISMICALLY ISOLATED STRUCTURES

9.13.1 General. Every seismically isolated structure and every portion thereof shall be designed and constructed in accordance with the requirements of this Section and the applicable requirements of Section 9.1.

The lateral force-resisting system and the isolation system shall be designed to resist the deformations and stresses produced by the effects of seismic ground motions as provided in this Section.

9.13.2 Criteria Selection.

9.13.2.1 Basis for Design. The procedures and limitations for the design of seismically isolated structures shall be determined considering zoning, site characteristics, vertical acceleration, cracked section properties of concrete and masonry members, Seismic Use Group, configuration, structural system, and height in accordance with Section 9.5.2 except as noted below.

9.13.2.2 Stability of the Isolation System. The stability of the vertical load-carrying elements of the isolation system shall be verified by analysis and test, as required, for lateral seismic displacement equal to the total maximum displacement.

9.13.2.3 Seismic Use Group. All portions of the structure, including the structure above the isolation system, shall be assigned a Seismic Use Group in accordance with the requirements of Section 9.1.3. The Occupancy Importance Factor shall be taken as 1.0 for a seismically isolated structure, regardless of its Seismic Use Group categorization.

9.13.2.4 Configuration Requirements. Each structure shall be designated as being regular or irregular on the basis of the structural configuration above the isolation system.

9.13.2.5 Selection of Lateral Response Procedure.

9.13.2.5.1 General. Seismically isolated structures except those defined in Section 9.13.2.5.2 shall be designed using the dynamic lateral response procedure of Section 9.13.4.

9.13.2.5.2 Equivalent Lateral Force Procedure. The equivalent lateral response procedure of Section 9.13.3 is permitted to be used for design of a seismically isolated structure provided that:

1. The structure is located at a site with S_1 less than or equal to 0.60 g.

2. The structure is located on a Class A, B, C, or D site.

3. The structure above the isolation interface is less than or equal to 4 stories or 65 ft (19.8 m) in height.

4. The effective period of the isolated structure at maximum displacement, T_M, is less than or equal to 3.0 sec.

5. The effective period of the isolated structure at the design displacement, T_D, is greater than three times the elastic, fixed-base period of the structure above the isolation system as determined by Eq. 9.5.3.3-1 or 9.5.3.3-2.

6. The structure above the isolation system is of regular configuration.

7. The isolation system meets all of the following criteria:

    a. The effective stiffness of the isolation system at the design displacement is greater than one-third of the effective stiffness at 20% of the design displacement.

    b. The isolation system is capable of producing a restoring force as specified in Section 9.13.6.2.4.

    c. The isolation system has force-deflection properties that are independent of the rate of loading.

    d. The isolation system has force-deflection properties that are independent of vertical load and bilateral load.

    e. The isolation system does not limit maximum considered earthquake displacement to less than S_{M1}/S_{D1} times the total design displacement.

9.13.2.5.3 Dynamic Analysis. The dynamic lateral response procedure of Section 9.13.4 shall be used as specified below.

9.13.2.5.3.1 Response-Spectrum Analysis. Response-spectrum analysis shall not be used for design of a seismically isolated structure unless:

1. The structure is located on a Class A, B, C, or D Site

2. The isolation system meets the criteria of Item 7 of Section 9.13.2.5.2.

9.13.2.5.3.2 Time-History Analysis. Time-history analysis shall be permitted for design of any seismically isolated structure and shall be used for design of all seismically isolated structures not meeting the criteria of Section 9.13.2.5.3.1.

9.13.2.5.3.3 Site-Specific Design Spectra. Site-specific ground-motion spectra of the design earthquake and the maximum considered earthquake developed in accordance with Section 9.13.4.4.1 shall be used for design and analysis of all seismically isolated structures if any one of the following conditions apply:

1. The structure is located on a Class F Site

2. The structure is located at a site with S_1 greater than 0.60 g, as determined in Section 9.4.1.

9.13.3 Equivalent Lateral Force Procedure.

9.13.3.1 General. Except as provided in Section 9.13.4, every seismically isolated structure or portion thereof shall be designed and constructed to resist minimum earthquake displacements and forces as specified by this Section and the applicable requirements of Section 9.5.3.

9.13.3.2 Deformation Characteristics of the Isolation System. Minimum lateral earthquake design displacements and forces on seismically isolated structures shall be based on the deformation characteristics of the isolation system.

The deformation characteristics of the isolation system shall explicitly include the effects of the wind-restraint system if such a system is used to meet the design requirements of this document.

The deformation characteristics of the isolation system shall be based on properly substantiated tests performed in accordance with Section 9.13.9.

9.13.3.3 Minimum Lateral Displacements.

9.13.3.3.1 Design Displacement. The isolation system shall be designed and constructed to withstand minimum lateral earthquake displacements, D_D, that act in the direction of each of the main horizontal axes of the structure in accordance with the following:

$$D_D = \frac{g S_{D1} T_D}{4\pi^2 B_D} \qquad \text{(Eq. 9.13.3.3.1)}$$

where

g = acceleration of gravity. The units of the acceleration of gravity, g, are in./sec^2 (mm/sec^2) if the units of the design displacement, D, are in. (mm)

S_{D1} = design 5% damped spectral acceleration at 1-sec period, in units of g-sec, as determined in Section 9.4.1.1

T_D = effective period of seismically isolated structure in seconds (sec), at the design displacement in the direction under consideration, as prescribed by Eq. 9.13.3.3.2

B_D = numerical coefficient related to the effective damping of the isolation system at the design displacement, D_D, as set forth in Table 9.13.3.3.1

TABLE 9.13.3.3.1
DAMPING COEFFICIENT, B_I

Effective Damping, β_D or β_M (Percentage of Critical)[a,b]	B_D or B_M Factor
≤2%	0.8
5%	1.0
10%	1.2
20%	1.5
30%	1.7
40%	1.9
≥50%	2.0

[a] The damping coefficient shall be based on the effective damping of the isolation system determined in accordance with the requirements of Section 9.13.9.5.2.

[b] The damping coefficient shall be based on linear interpolation for effective damping values other than those given.

9.13.3.3.2 Effective Period at Design Displacement. The effective period of the isolated structure at design displacement, T_D, shall be determined using the deformational characteristics of the isolation system in accordance with the following equation:

$$T_D = 2\pi \sqrt{\frac{W}{k_{Dmin} g}} \qquad \text{(Eq. 9.13.3.3.2)}$$

where

W = total seismic dead load weight of the structure above the isolation interface as defined in Section 9.5.3 (kip or kN)

k_{Dmin} = minimum effective stiffness in kips/in.(kN/mm) of the isolation system at the design displacement in the horizontal direction under consideration as prescribed by Eq. 9.13.9.5.1-2

g = acceleration due to gravity

9.13.3.3.3 Maximum Lateral Displacement. The maximum displacement of the isolation system, D_M, in the most critical direction of horizontal response shall be calculated in accordance with the formula:

$$D_M = \frac{g S_{M1} T_M}{4\pi^2 B_M} \qquad \text{(Eq. 9.13.3.3.3)}$$

where

g = acceleration of gravity

S_{M1} = maximum considered 5% damped spectral acceleration at 1-sec period, in units of g-sec, as determined in Section 9.4.1.2

T_M = effective period, in seconds (sec), of seismic-isolated structure at the maximum displacement in the direction under consideration as prescribed by Eq. 9.13.3.3.4

B_M = numerical coefficient related to the effective damping of the isolation system at the maximum displacement, D_m, as set forth in Table 9.13.3.3.1

9.13.3.3.4 Effective Period at Maximum Displacement. The effective period of the isolated structure at maximum displacement, T_M, shall be determined using the deformational characteristics of the isolation system in accordance with the equation:

$$T_M = 2\pi \sqrt{\frac{W}{k_{Mmin} g}} \qquad \text{(Eq. 9.13.3.3.4)}$$

where

W = total seismic dead load weight of the structure above the isolation interface as

defined in Sections 9.5.3.2 and 9.5.5.3 (kip or kN)

k_{Mmin} = minimum effective stiffness, in kips/in. (kN/mm), of the isolation system at the maximum displacement in the horizontal direction under consideration as prescribed by Eq. 9.13.9.5.1-4

g = acceleration of gravity

9.13.3.3.5 Total Lateral Displacement. The total design displacement, D_{TD}, and the total maximum displacement, D_{TM}, of elements of the isolation system shall include additional displacement due to actual and accidental torsion calculated from the spatial distribution of the lateral stiffness of the isolation system and the most disadvantageous location of mass eccentricity.

The total design displacement, D_{TD}, and the total maximum displacement, D_{TM}, of elements of an isolation system with uniform spatial distribution of lateral stiffness shall not be taken as less than that prescribed by the following equations:

$$D_{TD} = D_D \left[1 + y \frac{12e}{b^2 + d^2} \right] \quad \textbf{(Eq. 9.13.3.3.5-1)}$$

$$D_{TM} = D_M \left[1 + y \frac{12e}{b^2 + d^2} \right] \quad \textbf{(Eq. 9.13.3.3.5-2)}$$

Exception: The total design displacement, D_{TD}, and the total maximum displacement, D_{TM}, are permitted to be taken as less than the value prescribed by Eqs. 9.13.3.3.5-1 and 9.13.3.3.5-2, respectively, but not less than 1.1 times D_D and D_M, respectively, provided the isolation system is shown by calculation to be configured to resist torsion accordingly.

where

D_D = design displacement, in in. (mm), at the center of rigidity of the isolation system in the direction under consideration as prescribed by Eq. 9.13.3.3.1

D_M = maximum displacement, in in. (mm), at the center of rigidity of the isolation system in the direction under consideration as prescribed by Eq. 9.13.3.3.3

y = the distance, in ft (mm), between the centers of rigidity of the isolation system and the element of interest measured perpendicular to the direction of seismic loading under consideration

e = the actual eccentricity, in ft (mm), measured in plan between the center of mass of the structure above the isolation interface and the center of rigidity of the isolation system, plus accidental eccentricity, in ft (mm), taken as 5% of the longest plan dimension of the structure perpendicular to the direction of force under consideration

b = the shortest plan dimension of the structure, in ft (mm), measured perpendicular to d

d = the longest plan dimension of the structure, in ft (mm)

9.13.3.4 Minimum Lateral Forces.

9.13.3.4.1 Isolation System and Structural Elements at or Below the Isolation System.

The isolation system, the foundation, and all structural elements below the isolation system shall be designed and constructed to withstand a minimum lateral seismic force, V_b, using all of the appropriate provisions for a nonisolated structure where

$$V_b = k_{Dmax} D_D \quad \textbf{(Eq. 9.13.3.4.1)}$$

where

V_b = the minimum lateral seismic design force or shear on elements of the isolation system or elements below the isolation system as prescribed by Eq. 9.13.3.4.1

k_{Dmax} = maximum effective stiffness, in kips/in. (kN/mm), of the isolation system at the design displacement in the horizontal direction under consideration

D_D = design displacement, in in. (mm), at the center of rigidity of the isolation system in the direction under consideration as prescribed by Eq. 9.13.3.3.1

V_b shall not be taken as less than the maximum force in the isolation system at any displacement up to and including the design displacement.

9.13.3.4.2 Structural Elements Above the Isolation System. The structure above the isolation system shall be designed and constructed to withstand a minimum shear force, V_s, using all of the appropriate provisions for a nonisolated structure where

$$V_s = \frac{k_{Dmax} D_D}{R_I} \quad \textbf{(Eq. 9.13.3.4.2)}$$

where

k_{Dmax} = maximum effective stiffness, in kips/in. (kN/mm), of the isolation system at the design displacement in the horizontal direction under consideration

D_D = design displacement, in in. (mm), at the center of rigidity of the isolation system in the direction under consideration as prescribed by Eq. 9.13.3.3.1

R_I = numerical coefficient related to the type of lateral force-resisting system above the isolation system

The R_I factor shall be based on the type of lateral-force-resisting system used for the structure above the isolation system and shall be three-eighths of the R value given in Table 9.5.2.2 with an upper-bound value not to exceed 2.0 and a lower bound value not to be less than 1.0.

9.13.3.4.3 Limits on V_s. The value of V_s shall not be taken as less than the following:

1. The lateral seismic force required by Section 9.5.3 for a fixed-base structure of the same weight, W, and a period equal to the isolated period, T_D,

2. The base shear corresponding to the factored design wind load.

3. The lateral seismic force required to fully activate the isolation system (e.g., the yield level of a softening system, the ultimate capacity of a sacrificial wind-restraint system, or the break-away friction level of a sliding system) factored by 1.5.

9.13.3.5 Vertical Distribution of Force. The total force shall be distributed over the height of the structure above the isolation interface in accordance with the following equation:

$$F_x = \frac{V_s w_x h_x}{\sum_{i=1}^{n} w_i h_i} \qquad \textbf{(Eq. 9.13.3.5)}$$

where

V_s = total lateral seismic design force or shear on elements above the isolation system as prescribed by Eq. 9.13.3.4.2

W_x = portion of W that is located at or assigned to Level i, n, or x, respectively

h_x = height above the base Level i, n, or x, respectively

w_i = portion of W that is located at or assigned to Level i, n, or x, respectively

h_i = height above the base Level i, n, or x, respectively

At each level designated as x, the force, F_x, shall be applied over the area of the structure in accordance with

the mass distribution at the level. Stresses in each structural element shall be calculated as the effect of force, F_x, applied at the appropriate levels above the base.

9.13.3.6 Drift Limits. The maximum interstory drift of the structure above the isolation system shall not exceed $0.015h_{sx}$. The drift shall be calculated by Eq. 9.5.3.7.1 with the C_d factor of the isolated structure equal to the R_I factor defined in Section 9.13.3.4.2.

9.13.4 Dynamic Lateral Response Procedure.

9.13.4.1 General. As required by Section 9.13.2, every seismically isolated structure or portion thereof shall be designed and constructed to resist earthquake displacements and forces as specified in this Section and the applicable requirements of Section 9.5.4.

9.13.4.2 Isolation System and Structural Elements Below the Isolation System. The total design displacement of the isolation system shall not be taken as less than 90% of D_{TD} as specified by Section 9.13.3.3.5.

The total maximum displacement of the isolation system shall not be taken as less than 80% of D_{TM} as prescribed by Section 9.13.3.3.5.

The design lateral shear force on the isolation system and structural elements below the isolation system shall not be taken as less than 90% of V_b as prescribed by Eq. 9.13.3.4.1.

The limits on displacements specified by this Section shall be evaluated using values of D_{TD} and D_{TM} determined in accordance with Section 9.13.3.5 except that $D_{D'}$ is permitted to be used in lieu of D_D and $D_{M'}$ is permitted to be used in lieu of D_M where

$$D'_D = \frac{D_D}{\sqrt{1 + (T/T_D)^2}} \qquad \textbf{(Eq. 9.13.4.2-1)}$$

$$D'_M = \frac{D_M}{\sqrt{1 + (T/T_M)^2}} \qquad \textbf{(Eq. 9.13.4.2-2)}$$

where

D_D = design displacement, in in. (mm), at the center of rigidity of the isolation system in the direction under consideration as prescribed by Eq. 9.13.3.3.1

D_M = maximum displacement in in. (mm), at the center of rigidity of the isolation system in the direction under consideration as prescribed by Eq. 9.13.3.3.3

T = elastic, fixed-base period of the structure above the isolation system as determined by Section 9.5.3.3

T_D = effective period of seismically isolated structure in sec, at the design displacement

in the direction under consideration as prescribed by Eq. 9.13.3.3.2

T_M = effective period, in sec, of the seismically isolated structure, at the maximum displacement in the direction under consideration as prescribed by Eq. 9.13.3.3.4

9.13.4.3 Structural Elements Above the Isolation System. The design lateral shear force on the structure above the isolation system, if regular in configuration, shall not be taken as less than 80% of V_s, or less than the limits specified by Section 9.13.3.4.3.

Exception: The design lateral shear force on the structure above the isolation system, if regular in configuration, is permitted to be taken as less than 80%, but shall not be less than 60% of V_s, when time-history analysis is used for design of the structure.

The design lateral shear force on the structure above the isolation system, if irregular in configuration, shall not be taken as less than V_s, or less than the limits specified by Section 9.13.3.4.3.

Exception: The design lateral shear force on the structure above the isolation system, if irregular in configuration, is permitted to be taken as less than 100%, but shall not be less than 80% of V_s, when time-history analysis is used for design of the structure.

9.13.4.4 Ground Motion.

9.13.4.4.1 Design Spectra. Properly substantiated site-specific spectra are required for design of all structures located on Site Class E or F or located at a site with S_1 greater than 0.60 g. Structures that do not require site-specific spectra and for which site-specific spectra have not been calculated shall be designed using the response spectrum shape given in Figure 9.4.1.2.6.

A design spectrum shall be constructed for the design earthquake. This design spectrum shall not be taken as less than the design earthquake response spectrum given in Figure 9.4.1.2.6.

Exception: If a site-specific spectrum is calculated for the design earthquake, the design spectrum is permitted to be taken as less than 100% but shall not be less than 80% of the design earthquake response spectrum given in Figure 9.4.1.2.6.

A design spectrum shall be constructed for the maximum considered earthquake. This design spectrum shall not be taken as less than 1.5 times the design earthquake response spectrum given in

Figure 9.4.1.2.6. This design spectrum shall be used to determine the total maximum displacement and overturning forces for design and testing of the isolation system.

Exception: If a site-specific spectrum is calculated for the maximum considered earthquake, the design spectrum is permitted to be taken as less than 100% but shall not be less than 80% of 1.5 times the design earthquake response spectrum given in Figure 9.4.1.2.6.

9.13.4.4.2 Time Histories. Pairs of horizontal ground-motion time-history components shall be selected and scaled from not less than three recorded events. For each pair of horizontal ground-motion components, the square root sum of the squares (SRSS) of the 5% damped spectrum of the scaled, horizontal components shall be constructed. The motions shall be scaled such that the average value of the SRSS spectra does not fall below 1.3 times the 5% damped spectrum of the design earthquake (or maximum considered earthquake) by more than 10% for periods from $0.5T_D$ seconds to $1.25T_M$ seconds where T_D and T_M are determined in accordance with Sections 9.13.3.3.2 and 9.13.3.3.4, respectively.

9.13.4.5 Mathematical Model.

9.13.4.5.1 General. The mathematical models of the isolated structure including the isolation system, the lateral-force-resisting system, and other structural elements shall conform to Section 9.5.4.2 and to the requirements of Sections 9.13.4.5.2 and 9.13.4.5.3, below.

9.13.4.5.2 Isolation System. The isolation system shall be modeled using deformational characteristics developed and verified by test in accordance with the requirements of Section 9.13.3.2. The isolation system shall be modeled with sufficient detail to:

1. Account for the spatial distribution of isolator units.

2. Calculate translation, in both horizontal directions, and torsion of the structure above the isolation interface considering the most disadvantageous location of mass eccentricity.

3. Assess overturning/uplift forces on individual isolator units.

4. Account for the effects of vertical load, bilateral load, and/or the rate of loading if the force-deflection properties of the isolation system are dependent on one or more of these attributes.

9.13.4.5.3 Isolated Building.

9.13.4.5.3.1 Displacement. The maximum displacement of each floor and the total design displacement and total maximum displacement across the isolation system shall be calculated using a model of the isolated structure that incorporates the force-deflection characteristics of nonlinear elements of the isolation system and the lateral-force-resisting system.

Isolation systems with nonlinear elements include, but are not limited to, systems that do not meet the criteria of Item 7 of Section 9.13.2.5.2.

Lateral-force-resisting systems with nonlinear elements include, but are not limited to, irregular structural systems designed for a lateral force less than V_s and regular structural systems designed for a lateral force less than 80% of V_s.

9.13.4.5.3.2 Forces and Displacements in Key Elements. Design forces and displacements in key elements of the lateral-force-resisting system shall not be calculated using a linear elastic model of the isolated structure unless:

1. Pseudo-elastic properties assumed for nonlinear isolation system components are based on the maximum effective stiffness of the isolation system.

2. All key elements of the lateral-force-resisting system are linear.

9.13.4.6 Description of Analysis Procedures.

9.13.4.6.1 General. Response-spectrum and time-history analyses shall be performed in accordance with Section 9.5.3 and the requirements of this section.

9.13.4.6.2 Input Earthquake. The design earthquake shall be used to calculate the total design displacement of the isolation system and the lateral forces and displacements of the isolated structure. The maximum considered earthquake shall be used to calculate the total maximum displacement of the isolation system.

9.13.4.6.3 Response-Spectrum Analysis. Response-spectrum analysis shall be performed using a damping value equal to the effective damping of the isolation system or 30% of critical, whichever is less.

Response-spectrum analysis used to determine the total design displacement and the total maximum displacement shall include simultaneous excitation of the model by 100% of the most critical direction of ground motion and 30% of the ground motion on the orthogonal axis. The maximum displacement of the isolation system shall be calculated as the vectorial sum of the two orthogonal displacements.

The design shear at any story shall not be less than the story shear obtained using Eq. 9.13.3.5 and a value of V_s taken as that equal to the base shear obtained from the response-spectrum analysis in the direction of interest.

9.13.4.6.4 Time-History Analysis. Time-history analysis shall be performed with at least three appropriate pairs of horizontal time-history components as defined in Section 9.13.4.4.2.

Each pair of time histories shall be applied simultaneously to the model considering the most disadvantageous location of mass eccentricity. The maximum displacement of the isolation system shall be calculated from the vectorial sum of the two orthogonal components at each time step.

The parameter of interest shall be calculated for each time-history analysis. If three time-history analyses are performed, the maximum response of the parameter of interest shall be used for design. If seven or more time-history analyses are performed, the average value of the response parameter of interest shall be used for design.

9.13.4.7 Design Lateral Force.

9.13.4.7.1 Isolation System and Structural Elements at or Below the Isolation System. The isolation system, foundation, and all structural elements below the isolation system shall be designed using all of the appropriate provisions for a nonisolated structure and the forces obtained from the dynamic analysis without reduction.

9.13.4.7.2 Structural Elements Above the Isolation System. Structural elements above the isolation system shall be designed using the appropriate provisions for a nonisolated structure and the forces obtained from the dynamic analysis reduced by a factor of R_I. The R_I factor shall be based on the type of lateral-force-resisting system used for the structure above the isolation system.

9.13.4.7.3 Scaling of Results. When the factored lateral shear force on structural elements, determined using either response-spectrum or time-history analysis, is less than the minimum level prescribed by Sections 9.13.4.2 and 9.13.4.3, all response parameters, including member forces and moments, shall be adjusted upward proportionally.

9.13.4.7.4 Drift Limits. Maximum interstory drift corresponding to the design lateral force including displacement due to vertical deformation of the isolation system shall not exceed the following limits:

1. The maximum interstory drift of the structure above the isolation system calculated by response-spectrum analysis shall not exceed $0.015h_{sx}$.

2. The maximum interstory drift of the structure above the isolation system calculated by time-history analysis based on the force-deflection characteristics of nonlinear elements of the lateral-force-resisting system shall not exceed $0.020h_{sx}$.

Drift shall be calculated using Eq. 9.5.3.7.1 with the C_d factor of the isolated structure equal to the R_I factor defined in Section 9.13.3.4.2.

The secondary effects of the maximum considered earthquake lateral displacement Δ of the structure above the isolation system combined with gravity forces shall be investigated if the interstory drift ratio exceeds $0.010/R_I$.

9.13.5 Lateral Load on Elements of Structures and Nonstructural Components Supported by Buildings.

9.13.5.1 General. Parts or portions of an isolated structure, permanent nonstructural components and the attachments to them, and the attachments for permanent equipment supported by a structure shall be designed to resist seismic forces and displacements as prescribed by this section and the applicable requirements of Section 9.6.

9.13.5.2 Forces and Displacements.

9.13.5.2.1 Components at or Above the Isolation Interface. Elements of seismically isolated structures and nonstructural components, or portions thereof, that are at or above the isolation interface shall be designed to resist a total lateral seismic force equal to the maximum dynamic response of the element or component under consideration.

> **Exception:** Elements of seismically isolated structures and nonstructural components or portions designed to resist total lateral seismic force as prescribed in Sections 9.5.2.6 and 9.6.1.3 as appropriate.

9.13.5.2.2 Components Crossing the Isolation Interface. Elements of seismically isolated structures and nonstructural components, or portions thereof, that cross the isolation interface shall be designed to withstand the total maximum displacement.

9.13.5.2.3 Components Below the Isolation Interface. Elements of seismically isolated structure and nonstructural components, or portions thereof, that are below the isolation interface shall be designed and constructed in accordance with the requirements of Section 9.5.2.

9.13.6 Detailed System Requirements.

9.13.6.1 General. The isolation system and the structural system shall comply with the material requirements of Sections 9.8 through 9.12. In addition, the isolation system shall comply with the detailed system requirements of this section and the structural system shall comply with the detailed system requirements of this section and the applicable portions of Section 9.5.2.

9.13.6.2 Isolation System.

9.13.6.2.1 Environmental Conditions. In addition to the requirements for vertical and lateral loads induced by wind and earthquake, the isolation system shall provide for other environmental conditions including aging effects, creep, fatigue, operating temperature, and exposure to moisture or damaging substances.

9.13.6.2.2 Wind Forces. Isolated structures shall resist design wind loads at all levels above the isolation interface. At the isolation interface, a wind-restraint system shall be provided to limit lateral displacement in the isolation system to a value equal to that required between floors of the structure above the isolation interface.

9.13.6.2.3 Fire Resistance. Fire resistance for the isolation system shall meet that required for the structure columns, walls, or other such gravity-bearing elements in the same area of the structure.

9.13.6.2.4 Lateral Restoring Force. The isolation system shall be configured to produce a restoring force such that the lateral force at the total design displacement is at least $0.025W$ greater than the lateral force at 50% of the total design displacement.

> **Exception:** The isolation system need not be configured to produce a restoring force, as required above, provided the isolation system is capable of remaining stable under full vertical load and accommodating a total maximum displacement equal to the greater of either 3.0 times the total design displacement or $36S_{M1}$ in. (or $915S_{M1}$ mm).

9.13.6.2.5 Displacement Restraint. The isolation system shall not be configured to include a displacement restraint that limits lateral displacement due to the maximum considered earthquake to less than S_{M1}/S_{D1} times the total design displacement unless the seismically isolated structure is designed in accordance with the following criteria when more stringent than the requirements of Section 9.13.2:

1. Maximum considered earthquake response is calculated in accordance with the dynamic analysis requirements of Section 9.13.4 explicitly considering the nonlinear characteristics of the isolation system and the structure above the isolation system.

2. The ultimate capacity of the isolation system and structural elements below the isolation system shall exceed the strength and displacement demands of the maximum considered earthquake.

3. The structure above the isolation system is checked for stability and ductility demand of the maximum considered earthquake.

4. The displacement restraint does not become effective at a displacement less than 0.75 times the total design displacement unless it is demonstrated by analysis that earlier engagement does not result in unsatisfactory performance.

9.13.6.2.6 Vertical-Load Stability. Each element of the isolation system shall be designed to be stable under the design vertical load at a horizontal displacement equal to the total maximum displacement. The design vertical load shall be computed using load combination 5 of Section 2.3.2 for the maximum vertical load and load combination 7 of Section 2.3.2 for the minimum vertical load. The seismic load E is given by Eqs. 9.5.2.7-1 and 9.5.2.7-2 where S_{DS} in these equations is replaced by S_{MS}. The vertical load due to earthquake, Q_E, shall be based on peak response due to the maximum considered earthquake.

9.13.6.2.7 Overturning. The factor of safety against global structural overturning at the isolation interface shall not be less than 1.0 for required load combinations. All gravity and seismic loading conditions shall be investigated. Seismic forces for overturning calculations shall be based on the maximum considered earthquake and W shall be used for the vertical restoring force.

Local uplift of individual elements shall not be allowed unless the resulting deflections do not cause overstress or instability of the isolator units or other structure elements.

9.13.6.2.8 Inspection and Replacement.

1. Access for inspection and replacement of all components of the isolation system shall be provided.

2. A registered design professional shall complete a final series of inspections or observations of structure separation areas and components that cross the isolation interface prior to the issuance of the certificate of occupancy for the seismically isolated structure. Such inspections and observations shall indicate that the conditions allow free and unhindered displacement of the structure to maximum design levels and that all components that cross the isolation interface as installed are able to accommodate the stipulated displacements.

3. Seismically isolated structures shall have a periodic monitoring, inspection, and maintenance program for the isolation system established by the registered design professional responsible for the design of the system.

4. Remodeling, repair, or retrofitting at the isolation system interface, including that of components that cross the isolation interface, shall be performed under the direction of a registered design professional.

9.13.6.2.9 Quality Control. A quality control testing program for isolator units shall be established by the engineer responsible for the structural design.

9.13.6.3 Structural System.

9.13.6.3.1 Horizontal Distribution of Force. A horizontal diaphragm or other structural elements shall provide continuity above the isolation interface and shall have adequate strength and ductility to transmit forces (due to nonuniform ground motion) from one part of the structure to another.

9.13.6.3.2 Building Separations. Minimum separations between the isolated structure and surrounding retaining walls or other fixed obstructions shall not be less than the total maximum displacement.

9.13.6.3.3 Nonbuilded Structures. These shall be designed and constructed in accordance with the requirements of Section 9.14 using design displacements and forces calculated in accordance with Section 9.13.3 or 9.13.4.

9.13.7 Foundations. Foundations shall be designed and constructed in accordance with the requirements of

Section 4 using design forces calculated in accordance with Section 9.13.3 or 9.13.4, as appropriate.

9.13.8 Design and Construction Review.

9.13.8.1 General. A design review of the isolation system and related test programs shall be performed by an independent engineering team including persons licensed in the appropriate disciplines and experienced in seismic analysis methods and the theory and application of seismic isolation.

9.13.8.2 Isolation System. Isolation system design review shall include, but not be limited to, the following:

1. Review of site-specific seismic criteria including the development of site-specific spectra and ground-motion time-histories and all other design criteria developed specifically for the project.
2. Review of the preliminary design including the determination of the total design displacement of the isolation system design displacement and the lateral force design level.
3. Overview and observation of prototype testing (Section 9.13.9).
4. Review of the final design of the entire structural system and all supporting analyses.
5. Review of the isolation system quality control testing program (Section 9.13.6.2.9).

9.13.9 Required Tests of the Isolation System.

9.13.9.1 General. The deformation characteristics and damping values of the isolation system used in the design and analysis of seismically isolated structures shall be based on tests of a selected sample of the components prior to construction as described in this Section.

The isolation system components to be tested shall include the wind-restraint system if such a system is used in the design.

The tests specified in this section are for establishing and validating the design properties of the isolation system and shall not be considered as satisfying the manufacturing quality control tests of Section 9.13.6.2.9.

9.13.9.2 Prototype Tests.

9.13.9.2.1 General. Prototype tests shall be performed separately on two full-size specimens (or sets of specimens, as appropriate) of each predominant type and size of isolator unit of the isolation system. The test specimens shall include the wind-restraint system as well as individual isolator units if such systems are used in the design. Specimens tested shall not be used for construction unless permitted by the registered design professional and authority having jurisdiction.

9.13.9.2.2 Record. For each cycle of tests, the force-deflection and hysteretic behavior of the test specimen shall be recorded.

9.13.9.2.3 Sequence and Cycles. The following sequence of tests shall be performed for the prescribed number of cycles at a vertical load equal to the average dead load plus one-half the effects due to live load on all isolator units of a common type and size:

1. Twenty fully reversed cycles of loading at a lateral force corresponding to the wind design force.
2. Three fully reversed cycles of loading at each of the following increments of the total design displacement—$0.25D_D$, $0.5D_D$, $1.0D_D$, and $1.0D_M$ where D_D and D_M are as determined in Sections 9.13.3.3.1 and 9.13.3.3.3, respectively, or Section 9.13.4 as appropriate.
3. Three fully reversed cycles of loading at the total maximum displacement, $1.0D_{TM}$.
4. $30S_{D1}/S_{DS}B_D$, but not less than 10, fully reversed cycles of loading at 1.0 times the total design displacement, $1.0D_{TD}$.

If an isolator unit is also a vertical-load–carrying element, then Item 2 of the sequence of cyclic tests specified above shall be performed for two additional vertical load cases specified in Section 9.13.6.2.6. The load increment due to earthquake overturning, Q_E, shall be equal to or greater than the peak earthquake vertical force response corresponding to the test displacement being evaluated. In these tests, the combined vertical load shall be taken as the typical or average downward force on all isolator units of a common type and size.

9.13.9.2.4 Units Dependent on Loading Rates. If the force-deflection properties of the isolator units are dependent on the rate of loading, each set of tests specified in Section 9.13.9.2.3 shall be performed dynamically at a frequency, f, equal to the inverse of the effective period, T_D.

If reduced-scale prototype specimens are used to quantify rate-dependent properties of isolators, the reduced-scale prototype specimens shall be of the same type and material and be manufactured with the same processes and quality as full-scale prototypes and shall be tested at a frequency that represents full-scale prototype loading rates.

The force-deflection properties of an isolator unit shall be considered to be dependent on the rate of loading if there is greater than a plus or minus 15% difference in the effective stiffness and the effective damping at the design displacement when tested at a frequency equal to the inverse of the effective period of the isolated structure and when tested at any frequency in the range of 0.1 to 2.0 times the inverse of the effective period of the isolated structure.

9.13.9.2.5 Units Dependent on Bilateral Load. If the force-deflection properties of the isolator units are dependent on bilateral load, the tests specified in Sections 9.13.9.2.3 and 9.13.9.2.4 shall be augmented to include bilateral load at the following increments of the total design displacement: 0.25 and 1.0, 0.50 and 1.0, 0.75 and 1.0, and 1.0 and 1.0.

If reduced-scale prototype specimens are used to quantify bilateral-load–dependent properties, the reduced-scale specimens shall be of the same type and material and manufactured with the same processes and quality as full-scale prototypes.

The force-deflection properties of an isolator unit shall be considered to be dependent on bilateral load if the bilateral and unilateral force-deflection properties have greater than a 15% difference in effective stiffness at the design displacement.

9.13.9.2.6 Maximum and Minimum Vertical Load. Isolator units that carry vertical load shall be statically tested for maximum and minimum downward vertical load at the total maximum displacement. In these tests, the combined vertical loads shall be taken as specified in Section 9.13.6.2.6 on any one isolator of a common type and size. The dead load, D, and live load, L, are specified in Section 9.5.2.7. The seismic load E is given by Eqs. 9.5.2.7-1 and 9.5.2.7-2 where S_{DS} in these equations is replaced by S_{MS} and the vertical load, Q_E, is based on the peak earthquake vertical force response corresponding to the maximum considered earthquake.

9.13.9.2.7 Sacrificial Wind-Restraint Systems. If a sacrificial wind-restraint system is to be utilized, the ultimate capacity shall be established by test.

9.13.9.2.8 Testing Similar Units. The prototype tests are not required if an isolator unit is of similar size and of the same type and material as a prototype isolator unit that has been previously tested using the specified sequence of tests.

9.13.9.3 Determination of Force-Deflection Characteristics. The force-deflection characteristics of the isolation system shall be based on the cyclic load tests of isolator prototypes specified in Section 9.13.9.2.

The effective stiffness of an isolator unit, k_{eff}, shall be calculated for each cycle of loading as follows:

$$k_{eff} = \frac{|F^+| + |F^-|}{|\Delta^+| + |\Delta^-|} \qquad \textbf{(Eq. 9.13.9.3-1)}$$

where F^+ and F^- are the positive and negative forces, at Δ^+ and Δ^-, respectively.

As required, the effective damping, β_{eff}, of an isolator unit shall be calculated for each cycle of loading by the equation:

$$\beta_{eff} = \frac{2}{\pi} \frac{E_{loop}}{k_{eff}(|\Delta^+| + |\Delta^-|)^2} \qquad \textbf{(Eq. 9.13.9.3-2)}$$

where the energy dissipated per cycle of loading, E_{loop}, and the effective stiffness, k_{eff}, shall be based on peak test displacements of Δ^+ and Δ^-.

9.13.9.4 System Adequacy. The performance of the test specimens shall be assessed as adequate if the following conditions are satisfied:

1. For each increment of test displacement specified in Item 2 of Section 9.13.9.2.3 and for each vertical load case specified in Section 9.13.9.2.3, there is no greater than a 15% difference between the effective stiffness at each of the three cycles of test and the average value of effective stiffness for each test specimen.

2. For each increment of test displacement specified in Item 2 of Section 9.13.9.2.3 and for each vertical load case specified in Section 9.13.9.2.3, there is no greater than a 15% difference in the average value of effective stiffness of the two test specimens of a common type and size of the isolator unit over the required three cycles of test.

3. For each specimen, there is no greater than a plus or minus 20% change in the initial effective stiffness of each test specimen over the $15S_{DI}/S_{DS}B_D$, but not less than 10, cycles of test specified in Item 3 of Section 9.13.9.2.3.

4. For each specimen, there is no greater than a 20% decrease in the initial effective damping over for the $15S_{DI}/S_{DS}B_D$, but not less than 10, cycles of test specified in Section 9.13.9.2.3.

5. All specimens of vertical load-carrying elements of the isolation system remain stable up to the total maximum displacement for static load as prescribed in Section 9.13.9.2.6 and shall have a positive incremental force-carrying capacity.

9.13.9.5 Design Properties of the Isolation System.

9.13.9.5.1 Maximum and Minimum Effective Stiffness. At the design displacement, the maximum

and minimum effectiveness stiffness of the isolated system, k_{Dmax} and k_{Dmin}, shall be based on the cyclic tests of Item 2 of Section 9.13.9.2.3 and calculated by the equations:

$$k_{Dmax} = \frac{\sum |F_D^+|_{max} + \sum |F_D^-|_{max}}{2D_D}$$

(Eq. 9.13.9.5.1-1)

$$k_{Dmin} = \frac{\sum |F_D^+|_{min} + \sum |F_D^-|_{min}}{2D_D}$$

(Eq. 9.13.9.5.1-2)

At the maximum displacement, the maximum and minimum effective stiffness of the isolation system, k_{Mmax} and k_{Mmin}, shall be based on the cyclic tests of Item 3 of Section 9.13.9.2.3 and calculated by the equations:

$$k_{Mmax} = \frac{\sum |F_M^+|_{max} + \sum |F_M^-|_{max}}{2D_M}$$

(Eq. 9.13.9.5.1-3)

$$k_{Mmin} = \frac{\sum |F_M^+|_{min} + \sum |F_M^-|_{min}}{2D_M}$$

(Eq. 9.13.9.5.1-4)

The maximum effective stiffness of the isolation system, k_{Dmax} (or k_{Mmax}), shall be based on forces from the cycle of prototype testing at a test displacement equal to D_D (or D_M) that produces the largest value of effective stiffness. Minimum effective stiffness of the isolation system, k_{Dmin} (or k_{Mmin}), shall be based on forces from the cycle of prototype testing at a test displacement equal to D_D (or D_M) that produces the smallest value of effective stiffness.

For isolator units that are found by the tests of Sections 9.13.9.3, 9.13.9.4, and 9.13.9.5 to have force-deflection characteristics that vary with vertical load, rate of loading, or bilateral load, respectively, the values of k_{Dmax} and k_{Mmax} shall be increased and the values of k_{Dmin} and k_{Mmin} shall be decreased, as necessary, to bound the effects of measured variation in effective stiffness.

9.13.9.5.2 Effective Damping. At the design displacement, the effective damping of the isolation system, β_D, shall be based on the cyclic tests of Item 2 of Section 9.13.9.3 and calculated by the equation:

$$\beta_D = \frac{\sum E_D}{2\pi k_{Dmax} D_D^2}$$

(Eq. 9.13.9.5.2-1)

In Eq. 9.13.9.5.2-1, the total energy dissipated per cycle of design displacement response, ΣE_D, shall be taken as the sum of the energy dissipated per cycle in

all isolator units measured at a test displacement equal to D_D. The total energy dissipated per cycle of design displacement response, ΣE_D, shall be based on forces and deflections from the cycle of prototype testing at test displacement D_D that produces the smallest values of effective damping.

At the maximum displacement, the effective damping of the isolation system, β_M, shall be based on the cyclic tests of Item 2 of Section 9.13.9.3 and calculated by the equation:

$$\beta_M = \frac{\sum E_M}{2\pi k_{Mmax} D_M^2}$$

(Eq. 9.13.9.5.2-2)

In Eq. 9.13.9.5.2-2, the total energy dissipated per cycle of design displacement response, ΣE_M, shall be taken as the sum of the energy dissipated per cycle in all isolator units measured at a test displacement equal to D_M. The total energy dissipated per cycle of maximum displacement response, ΣE_M, shall be based on forces and deflections from the cycle of prototype testing at test displacement D_M that produces the smallest value of effective damping.

SECTION 9.14
NONBUILDING STRUCTURES

9.14.1 General.

9.14.1.1 Nonbuilding Structures. Includes all self-supporting structures that carry gravity loads and that may be required to resist the effects of earthquake, with the exception of: buildings, vehicular and railroad bridges, nuclear power generation plants, offshore platforms, dams, and other structures excluded in Section 9.1.2.1. Nonbuilding structures supported by the earth or supported by other structures shall be designed and detailed to resist the minimum lateral forces specified in this section. Design shall conform to the applicable provisions of other sections as modified by this section.

9.14.1.2 Design. The design of nonbuilding structures shall provide sufficient stiffness, strength, and ductility consistent with the requirements specified herein for buildings to resist the effects of seismic ground motions as represented by these design forces:

a. Applicable strength and other design criteria shall be obtained from other portions of Section 9 or its referenced codes and standards.

b. When applicable strength and other design criteria are not contained in or referenced by Section 9, such criteria shall be obtained from approved national standards. Where approved national standards define acceptance criteria in

terms of allowable stresses as opposed to strength, the design seismic forces shall be obtained from this Section and used in combination with other loads as specified in Section 2.4 of this Standard and used directly with allowable stresses specified in the national standards. Detailing shall be in accordance with the approved national standards.

9.14.2 Reference Standards.

9.14.2.1 Consensus Standards. The following references are consensus standards and are to be considered part of the requirements of Section 9 to the extent referred to in Section 9.14:

Reference 9.14-1   ACI. (1995). "Standard Practice for the Design and Construction of Cast-In-Place Reinforced Concrete Chimneys." *ACI 307.*

Reference 9.14-2   ACI. (1997). "Standard Practice for the Design and Construction of Concrete Silos and Stacking Tubes for Storing Granular Materials." *ACI 313.*

Reference 9.14-3   ACI. (2001). "Standard Practice for the Seismic Design of Liquid-Containing Concrete Structures." *ACI 350.3/350.3R.*

Reference 9.14-4   ACI. (1998). "Guide to the Analysis, Design, and Construction of Concrete-Pedestal Water Towers." *ACI 371R.*

Reference 9.14-5   ANSI. "Safety Requirements for the Storage and Handling of Anhydrous Ammonia." *ANSI K61.1.*

Reference 9.14-6   American Petroleum Institute (API). (December 1998). "Design and Construction of Large, Welded, Low Pressure Storage Tanks." *API 620*, 9th Edition, Addendum 3.

Reference 9.14-7   API. (March 2000). "Welded Steel Tanks For Oil Storage." *API 650*, 10th Edition, Addendum 1.

Reference 9.14-8   API. (December 1999). "Tank Inspection, Repair, Alteration, and Reconstruction." *ANSI/API 653*, 2nd Edition, Addendum 4.

Reference 9.14-9   API. (May 1995). "Design and Construction of Liquefied Petroleum Gas Installation." *ANSI/API 2510*, 7th Edition.

Reference 9.14-10   API. (February 1995). "Bolted Tanks for Storage of Production Liquids." *Specification 12B*, 14th Edition.

Reference 9.14-11   ASME. (2000). "Boiler and Pressure Vessel Code." Including Addenda through 2000.

Reference 9.14-12   ASME. (1992). "Steel Stacks." *ASME STS-1.*

Reference 9.14-13   ASME. (1995). "Gas Transmission and Distribution Piping Systems." *ASME B31.8.*

Reference 9.14-14   ASME. (1999). "Welded Aluminum-Alloy Storage Tanks." *ASME B96.1.*

Reference 9.14-15   American Water Works Association (AWWA). (1997). "Welded Steel Tanks for Water Storage." *ANSI/AWWA D100.*

Reference 9.14-16   AWWA. (1997). "Factory-Coated Bolted Steel Tanks for Water Storage." *ANSI/AWWA D103.*

Reference 9.14-17   AWWA. (1995). "Wire- and Strand-Wound Circular for Prestressed Concrete Water Tanks." *ANSI/AWWA D110.*

Reference 9.14-18   AWWA. (1995). "Circular Prestressed Concrete Tanks with Circumferential Tendons." *ANSI/AWWA D115.*

Reference 9.14-19   National Fire Protection Association (NFPA). (2000). "Flammable and Combustible Liquids Code." *ANSI/NFPA 30.*

Reference 9.14-20   NFPA. (2001). "Storage and Handling of Liquefied Petroleum Gas." *ANSI/NFPA 58.*

Reference 9.14-21   NFPA. (2001). "Storage and Handling of Liquefied Petroleum Gases at Utility Gas Plants." *ANSI/NFPA 59.*

Reference 9.14-22   NFPA. (2001). "Production, Storage, and Handling of Liquefied Natural Gas (LNG)." *ANSI/NFPA 59A.*

Reference 9.13-30   NFPA. (2001). "Standard for Bulk Oxygen Systems at Consumer Sites." *ANSI/NFPA 50.*

9.14.2.2 Accepted Standards. The following references are standards developed within the industry and represent acceptable procedures for design and construction:

Reference 9.14-23   ASTM. (1992). "Standard Practice for the Design and Manufacture of Amusement Rides and Devices." *ASTM F1159.*

Reference 9.14-24   ASTM. (1998). "Standard Guide for Design and Construction of Brick Liners for Industrial Chimneys." *ASTM C1298*

Reference 9.14-25   Rack Manufacturers Institute (RMI). (1997). "Specification for the Design, Testing, and Utilization of Industrial Steel Storage Racks."

Reference 9.14-26   U.S. Department of Transportation (DOT). "Pipeline Safety Regulations." *Title 49CFR Part 193*.

Reference 9.14-27   U.S. Naval Facilities Command (NAVFAC). "The Seismic Design of Waterfront Retaining Structures." *NAVFAC R-939*.

Reference 9.14-28   U.S. Naval Facilities Engineering Command. "Piers and Wharves." *NAVFAC DM-25.1*

Reference 9.14-29   U.S. Army Corps of Engineers (USACE). (1992). "Seismic Design for Buildings." *Army TM 5-809-10/ NAVFAC P-355/ Air Force AFM 88-3*, Chapter 13.

Reference 9.14-31   Compressed Gas Association (GGA). (1999). "Guide for Flat-Bottomed LOX/LIN/LAR Storage Tank Systems." 1st Edition.

9.14.3 Industry Design Standards and Recommended Practice.
Table 9.14.3 is a cross-reference of consensus standards/accepted standards and the applicable nonbuilding structures.

9.14.4 Nonbuilding Structures Supported by Other Structures.
If a nonbuilding structure is supported above the base by another structure and the weight of the nonbuilding structure is less than 25% of the combined weight of the nonbuilding structure and the supporting structure, the design seismic forces of the supported nonbuilding structure shall be determined in accordance with the requirements of Section 9.6.1.3.

If the weight of a nonbuilding structure is 25% or more of the combined weight of the nonbuilding structure and the supporting structure, the design seismic forces of the nonbuilding structure shall be determined based on the combined nonbuilding structure and supporting structural system. For supported nonbuilding structures that have nonrigid component dynamic characteristics, the combined system R factor shall be a maximum of 3. For supported nonbuilding structures that have rigid component dynamic characteristics (as defined in Section 9.14.5.2), the combined system R factor shall be the value of the supporting structural system. The supported nonbuilding structure and attachments shall be designed for the forces determined for the nonbuilding structure in a combined systems analysis.

9.14.4.1 Architectural, Mechanical, and Electrical Components.
Architectural, mechanical, and electrical components supported by nonbuilding structures shall be designed in accordance with Section 9.6 of this Standard.

9.14.5 Structural Design Requirements.

9.14.5.1 Design Basis.
Nonbuilding structures having specific seismic design criteria established in approved standards shall be designed using the standards as amended herein. In addition, nonbuilding structures shall be designed in compliance with Sections 9.14.6 and 9.14.7 to resist minimum seismic lateral forces that are not less than the requirements of Section 9.5.5.2 with the following additions and exceptions:

1. The response modification coefficient, R, shall be the lesser of the values given in Table 9.14.5.1.1 or the values in Table 9.5.2.2.

2. For nonbuilding systems that have an R value provided in Table 9.14.5.1.1, the minimum specified value in Eq. 9.5.5.2.1-3 shall be replaced by:

$$C_s = 0.14 S_{DS} I \qquad \textbf{(Eq. 9.14.5.1-1)}$$

and the minimum value specified in Eq. 9.5.5.2.1-4 shall be replaced by:

$$C_s = \frac{0.8 S_1 I}{R} \qquad \textbf{(Eq. 9.14.5.1-2)}$$

3. The overstrength factor, Ω_0, shall be taken from the same table as the R factor.

4. The importance factor, I, shall be as given in Table 9.14.5.1.2.

5. The height limitations shall be as given in Table 9.14.5.1.1 or the values in Table 9.5.2.2 where not specified in Table 9.14.5.1.1.

6. The vertical distribution of the lateral seismic forces in nonbuilding structures covered by this section shall be determined:

    a. Using the requirements of Section 9.5.5.4, or

    b. Using the procedures of Section 9.5.6, or

    c. In accordance with an approved standard applicable to the specific nonbuilding structure.

7. For nonbuilding structural systems containing liquids, gases, and granular solids supported at the base as defined in Section 9.14.7.3.1, the minimum seismic design force shall not be less than that required by the approved standard for the specific system.

8. Irregular structures per Section 9.5.2.3 at sites where the S_{DS} is greater than or equal to 0.50 and that cannot be modeled as a single mass shall use the procedures of Section 9.5.6.

TABLE 9.14.3
STANDARDS, INDUSTRY STANDARDS, AND REFERENCES

Application	Reference
Steel Storage Racks	RMI [25]
Piers and Wharves	NAVFAC R-939 [27], NAVFAC DM-25.1 [28]
Welded Steel Tanks for Water Storage	ACI 371R [4], ANSI/AWWA D100 [15], Army TM 5-809-10/ NAVFAC P-355/ Air Force AFM 88-3 Chapter 13 [29]
Welded Steel and Aluminum Tanks for Petroleum and Petrochemical Storage	API 620, 9th Edition, Addendum 3 [6], API 650, 10th Edition, Addendum 1 [7], ANSI/API 653, 2nd Edition, Addendum 4 [8], ASME B96.1 [14], Army TM 5-809-10/ NAVFAC P-355/ Air Force AFM 88-3 Chapter 13 [29]
Bolted Steel Tanks for Water Storage	ANSI/AWWA D103 [16], Army TM 5-809-10/ NAVFAC P-355/ Air Force AFM 88-3 Chapter 13 [29]
Bolted Steel Tanks for Petroleum and Petrochemical Storage	API Specification 12B, 14th Edition [10], Army TM 5-809-10/ NAVFAC P-355/ Air Force AFM 88-3 Chapter 13 [29]
Concrete Tanks for Water Storage	ACI 350.3/350.3R [3], ANSI/AWWA D110 [17], ANSI/AWWA D115 [18], Army TM 5-809-10/ NAVFAC P-355/ Air Force AFM 88-3 Chapter 13 [29]
Pressure Vessels	ASME [11]

Refrigerated Liquids Storage:	
Liquid Oxygen, Nitrogen, and Argon	ANSI/NFPA 50 [30], GGA [31]
Liquefied Natural Gas (LNG)	ANSI/NFPA 59A [22], DOT Title 49CFR Part 193 [26]
LPG (Propane, Butane, etc.)	ANSI/API 2510, 7th Edition [9], ANSI/NFPA 30 [19], ANSI/NFPA 58 [20], ANSI/NFPA 59 [21]
Ammonia	ANSI K61.1 [5]
Concrete Silos and Stacking Tubes	ACI 313 [2]
Impoundment Dikes and Walls:	
Hazardous Materials	ANSI K61.1 [5]
Flammable Materials	ANSI/NFPA 30 [19]
Liquefied Natural Gas	ANSI/NFPA 59A [22], DOT Title 49CFR Part 193 [26]
Gas Transmission and Distribution Piping Systems	ASME B31.8 [13]
Cast-in-Place Concrete Stacks and Chimneys	ACI 307 [1]
Steel Stacks and Chimneys	ASME STS-1 [12]
Guyed Steel Stacks and Chimneys	ASME STS-1 [12]
Brick Masonry Liners for Stacks and Chimneys	ASTM C1298 [24]
Amusement Structures	ASTM F1159 [23]

TABLE 9.14.5.1.1
SEISMIC COEFFICIENTS FOR NONBUILDING STRUCTURES

Nonbuilding Structure Type	R	Ω_0	C_d	Structural System and Height Limits (ft)[b] Seismic Design Category			
				A & B	C	D	E & F
Nonbuilding frame systems:	See Table 9.5.2.2						
Concentric braced frames of steel				NL	NL	NL	NL
Special concentric braced frames of steel				NL	NL	NL	NL
Moment-resisting frame systems:	See Table 9.5.2.2						
Special moment frames of steel				NL	NL	NL	NL
Ordinary moment frames of steel				NL	NL	50	50
Special moment frames of concrete				NL	NL	NL	NL
Intermediate moment frames of concrete				NL	NL	50[c]	50[c]
Ordinary moment frames of concrete				NL	50	NP	NP
Steel storage racks	4	2	$3\frac{1}{2}$	NL	NL	NL	NL
Elevated tanks, vessels, bins, or hoppers[a]:							
On braced legs	3	2	$2\frac{1}{2}$	NL	NL	NL	NL
On unbraced legs	3	2	$2\frac{1}{2}$	NL	NL	NL	NL
Irregular braced legs single pedestal or skirt-supported	2	2	2	NL	NL	NL	NL
Welded steel	2	2	2	NL	NL	NL	NL
Concrete	2	2	2	NL	NL	NL	NL
Horizontal, saddle supported welded steel vessels	3	2	$2\frac{1}{2}$	NL	NL	NL	NL
Tanks or vessels supported on structural towers similar to buildings	3	2	2	NL	NL	NL	NL
Flat bottom, ground supported tanks, or vessels:							
Anchored (welded or bolted steel)	3	2	$2\frac{1}{2}$	NL	NL	NL	NL
Unanchored (welded or bolted steel)	$2\frac{1}{2}$	2	2	NL	NL	NL	NL
Reinforced or prestressed concrete:							
Tanks with reinforced nonsliding base	2	2	2	NL	NL	NL	NL
Tanks with anchored flexible base	3	2	2	NL	NL	NL	NL
Tanks with unanchored and unconstrained flexible base:	$1\frac{1}{2}$	$1\frac{1}{2}$	$1\frac{1}{2}$	NL	NL	NL	NL
Other material:	$1\frac{1}{2}$	$1\frac{1}{2}$	$1\frac{1}{2}$	NL	NL	NL	NL
Cast-in-place concrete silos, stacks, and chimneys having walls continuous to the foundation	3	$1\frac{3}{4}$	3	NL	NL	NL	NL
All other reinforced masonry structures not similar to buildings	3	2	$2\frac{1}{2}$	NL	NL	50	50
All other nonreinforced masonry structures not similar to buildings	$1\frac{1}{4}$	2	$1\frac{1}{2}$	NL	50	50	50
All other steel and reinforced concrete distributed mass cantilever structures not covered herein including stacks, chimneys, silos, and skirt-supported vertical vessels that are not similar to buildings	3	2	$2\frac{1}{2}$	NL	NL	NL	NL

(continued)

TABLE 9.14.5.1.1 — continued
SEISMIC COEFFICIENTS FOR NONBUILDING STRUCTURES

Nonbuilding Structure Type	R	Ω_0	C_d	Structural System and Height Limits (ft)[b]			
				Seismic Design Category			
				A & B	C	D	E & F
Trussed towers (freestanding or guyed), guyed stacks and chimneys	3	2	$2\frac{1}{2}$	NL	NL	NL	NL
Cooling towers:							
Concrete or steel	$3\frac{1}{2}$	$1\frac{3}{4}$	3	NL	NL	NL	NL
Wood frame	$3\frac{1}{2}$	3	3	NL	NL	50	50
Telecommunication towers							
Truss: Steel	3	$1\frac{1}{2}$	3	NL	NL	NL	NL
Pole: Steel	$1\frac{1}{2}$	$1\frac{1}{2}$	$1\frac{1}{2}$	NL	NL	NL	NL
Wood	$1\frac{1}{2}$	$1\frac{1}{2}$	$1\frac{1}{2}$	NL	NL	NL	NL
Concrete	$1\frac{1}{2}$	$1\frac{1}{2}$	$1\frac{1}{2}$	NL	NL	NL	NL
Frame: Steel	3	$1\frac{1}{2}$	$1\frac{1}{2}$	NL	NL	NL	NL
Wood	$1\frac{1}{2}$	$1\frac{1}{2}$	$1\frac{1}{2}$	NL	NL	NL	NL
Concrete	2	$1\frac{1}{2}$	$1\frac{1}{2}$	NL	NL	NL	NL
Amusement structures and monuments	2	2	2	NL	NL	NL	NL
Inverted pendulum-type structures (except elevated tanks, vessels, bins, and hoppers)	2	2	2	NL	NL	NL	NL
Signs and billboards	$3\frac{1}{2}$	$1\frac{3}{4}$	3	NL	NL	NL	NL
All other self-supporting structures, tanks, or vessels not covered above or by approved standards that are similar to buildings	$1\frac{1}{4}$	2	$2\frac{1}{2}$	NL	50	50	50

[a] Support towers similar to building-type structures, including those with irregularities (see Section 9.5.2.3 of this Standard for definition of irregular structures) shall comply with the requirements of Section 9.5.2.6.
[b] Height shall be measured from the base.
[c] See exception to Section 9.14.6.1.
NL = No limit.
NP = Not permitted.

9. Where an approved national standard provides a basis for the earthquake-resistant design of a particular type of nonbuilding structure covered by Section 9.14, such a standard shall not be used unless the following limitations are met:

  a. The seismic ground acceleration, and seismic coefficient, shall be in conformance with the requirements of Sections 9.4.1 and 9.4.1.2.5, respectively.

  b. The values for total lateral force and total base overturning moment used in design shall not be less than 80% of the base shear value and overturning moment, each adjusted for the effects of soil-structure interaction that is obtained using this Standard.

10. The base shear is permitted to be reduced in accordance with Section 9.5.9.2.1 to account for the effects of soil-structure interaction. In no case shall the reduced base shear, $\tilde{V}$, be less than 0.7 V.

9.14.5.1.1 Seismic Factors.

9.14.5.1.2 Importance Factors and Seismic Use Group Classifications. The importance factor (I) and seismic use group for nonbuilding structures are based on the relative hazard of the contents and the function. The value of I shall be the largest value determined by the approved standards or the largest value as selected from Table 9.14.5.1.2 or as specified elsewhere in Section 9.14.

TABLE 9.14.5.1.2
IMPORTANCE FACTOR (I) AND SEISMIC USE GROUP CLASSIFICATION FOR NONBUILDING STRUCTURES

Importance Factor	I = 1.0	I = 1.25	I = 1.5
Seismic Use Group	I	II	III
Hazard	H-I	H-II	H-III
Function	F-I	F-II	F-III

H-I.   The nonbuilding structures that are not assigned to H-II or H-III.

H-II.   The nonbuilding structures containing hazardous materials and classified as a Category III structure in Table 1-1.

H-III.   The nonbuilding structures containing extremely hazardous materials and classified as a Category IV structure in Table 1-1.

F-I.   Nonbuilding structures not classified as F-II or F-III.

F-II.   Nonbuilding structures classified as Category III structures in Table 1-1.

F-III.   The nonbuilding structures classified as Category IV structures (essential facilities) in Table 1-1.

9.14.5.2 Rigid Nonbuilding Structures. Nonbuilding structures that have a fundamental period, T, less than $0.06\ s$, including their anchorages, shall be designed for the lateral force obtained from the following:

$$V = 0.30 S_{DS} W I \qquad \textbf{(Eq. 9.14.5.2)}$$

where

V = the total design lateral seismic base shear force applied to a nonbuilding structure

S_{DS} = the site design response acceleration as determined from Section 9.4.1.2.5

W = nonbuilding structure operating weight

I = the importance factor as determined from Table 9.14.5.1.2

The force shall be distributed with height in accordance with Section 9.5.5.4.

9.14.5.3 Loads. The weight W for nonbuilding structures shall include all dead load as defined for structures in Section 9.5.5.2. For purposes of calculating design seismic forces in nonbuilding structures, W also shall include all normal operating contents for items such as tanks, vessels, bins, hoppers, and the contents of piping. W shall include snow and ice loads when these loads constitute 25% or more of W or when required by the building official based on local environmental characteristics.

9.14.5.4 Fundamental Period. The fundamental period of the nonbuilding structure shall be determined by methods as prescribed in Section 9.5.5.3 or by other rational methods.

9.14.5.5 Drift Limitations. The drift limitations of Section 9.5.2.8 need not apply to nonbuilding structures if a rational analysis indicates they can be exceeded without adversely affecting structural stability or attached or interconnected components and elements such as walkways and piping. P-delta effects shall be considered when critical to the function or stability of the structure.

9.14.5.6 Materials Requirements. The requirements regarding specific materials in Sections A.9.8, A.9.9, and A.9.11 shall be applicable unless specifically exempted in Section 9.14.

9.14.5.7 Deflection Limits and Structure Separation. Deflection limits and structure separation shall be determined in accordance with this Standard unless specifically amended in Section 9.14.

9.14.5.8 Site-Specific Response Spectra. Where required by an approved standard or the authority having jurisdiction, specific types of nonbuilding structures shall be designed for site-specific criteria that accounts for local seismicity and geology, expected recurrence intervals and magnitudes of events from known seismic hazards (reference section 9.4.1.3 of this Standard). If a longer recurrence interval is defined in the approved standard for the nonbuilding structure such as LNG tanks [Ref. 9.14-22], the recurrence interval required in the approved standard shall be used.

9.14.6 Nonbuilding Structures Similar to Buildings.

9.14.6.1 General. Nonbuilding structures that have structural systems that are designed and constructed in a manner similar to buildings and have a dynamic response similar to building structures shall be designed similar to building structures and in compliance with this Standard with exceptions as contained in this section.

This general category of nonbuilding structures shall be designed in accordance with Sections 9.5 and 9.14.5.

The lateral force design procedure for nonbuilding structures with structural systems similar to building structures (those with structural systems listed in Table 9.5.2.2) shall be selected in accordance with the force and detailing requirements of Section 9.5.2.1. The height limitations shall be as given in Table 9.14.1.1 or the values in Table 9.5.2.2 where not specified in Table 9.14.5.1.1.

Exception: Intermediate moment frames of reinforced concrete shall not be used at sites where the seismic coefficient S_{DS} is greater than or equal to 0.50 unless

1. The nonbuilding structure is less than 50 ft (15.2 m) in height and

2. $R = 3.0$ is used for design.

The combination of load effects, E, shall be determined in accordance with Section 9.5.2.7.

9.14.6.2 Pipe Racks.

9.14.6.2.1 Design Basis. In addition to the provisions of 9.14.6.1, pipe racks supported at the base of the structure shall be designed to meet the force requirements of Sections 9.5.5 or 9.5.6.

Displacements of the pipe rack and potential for interaction effects (pounding of the piping system) shall be considered using the amplified deflections obtained from the following formula:

$$\delta_x = \frac{C_d \delta_{xe}}{I}$$ **(Eq. 9.14.6.2.1)**

where

C_d = deflection amplification factor in Table 9.14.5.1.1

δ_{xe} = deflections determined using the prescribed seismic design forces of this Standard

I = importance factor determined from Table 9.14.5.1.2

Exception: The importance factor, I, shall be determined from Table 9.14.5.1.2 for the calculation of δ_{xe}.

See Section 9.6.3.11 for the design of piping systems and their attachments. Friction resulting from gravity loads shall not be considered to provide resistance to seismic forces.

9.14.6.3 Steel Storage Racks.

In addition to the provisions of 9.14.6.1, steel storage racks shall be designed in accordance with the provisions of Sections 9.14.6.3.1 through 9.14.6.3.4. Alternatively, steel storage racks shall be permitted to be designed in accordance with the method defined in Section 2.7 "Earthquake Forces" of Ref. 9.14-25, where the following changes are included:

1. The values of C_a and C_v used shall equal $S_{DS}/2.5$ and S_{D1}, respectively, where S_{DS} and S_{D1} are determined in accordance with Section 9.4.1.2.5 of this Standard.

2. The value of I_p used shall be determined in accordance with Section 9.6.1.5 of this Standard.

3. For storage racks located at or below grade, the value of C_S used shall not be less than 0.14 S_{DS}. For storage racks located above grade, the value

of C_S used shall not be less than the value for F_p determined in accordance with Section 9.6.2.4.1c of this Standard, where R_p is taken as equal to R from Ref. 9.14-25 and a_p is taken as equal to 2.5.

9.14.6.3.1 General Requirements. Steel storage racks shall satisfy the force requirements of this section.

Exception: Steel storage racks supported at the base are permitted to be designed as structures with an R of 4, provided that the requirements of Section 9.5 are met. Higher values of R are permitted to be used when the detailing requirements of reference documents of Section 9.8 as modified in Section A9.8 are met. The importance factor I shall be taken equal to the I_p values in accordance with Section 9.6.1.5.

9.14.6.3.2 Operating Weight. Steel storage racks shall be designed for each of the following conditions of operating weight, W or W_p.

a. Weight of the rack plus every storage level loaded to 67% of its rated load capacity.

b. Weight of the rack plus the highest storage level only loaded to 100% of its rated load capacity.

The design shall consider the actual height of the center of mass of each storage load component.

9.14.6.3.3 Vertical Distribution of Seismic Forces. For all steel storage racks, the vertical distribution of seismic forces shall be as specified in Section 9.5.5.4 and in accordance with the following:

a. The base shear, V, of the typical structure shall be the base shear of the steel storage rack when loaded in accordance with Section 9.14.6.3.2.

b. The base of the structure shall be the floor supporting the steel storage rack. Each steel storage level of the rack shall be treated as a level of the structure with heights h_i and h_x measured from the base of the structure.

c. The factor k may be taken as 1.0.

d. The factor I shall be in accordance with Section 9.6.1.5.

9.14.6.3.4 Seismic Displacements. Steel storage rack installations shall accommodate the seismic displacement of the storage racks and their contents relative to all adjacent or attached components and elements. The assumed total relative displacement for storage racks shall be not less than 5% of the height above

the base unless a smaller value is justified by test data or analysis in accordance with Section 9.1.2.5.

9.14.6.4 Electrical Power Generating Facilities.

9.14.6.4.1 General. Electrical power generating facilities are power plants that generate electricity by steam turbines, combustion turbines, diesel generators, or similar turbo machinery.

9.14.6.4.2 Design Basis. In addition to the provisions of 9.14.6.1, electrical power generating facilities shall be designed using this Standard and the appropriate factors contained in Section 9.14.5.

9.14.6.5 Structural Towers for Tanks and Vessels.

9.14.6.5.1 General. In addition to the provisions of 9.14.6.1, structural towers which support tanks and vessels shall be designed to meet the provisions of Section 9.14.4 In addition, the following special considerations shall be included:

a. The distribution of the lateral base shear from the tank or vessel onto the supporting structure shall consider the relative stiffness of the tank and resisting structural elements.

b. The distribution of the vertical reactions from the tank or vessel onto the supporting structure shall consider the relative stiffness of the tank and resisting structural elements. When the tank or vessel is supported on grillage beams, the calculated vertical reaction due to weight and overturning shall be increased at least 20% to account for nonuniform support. The grillage beam and vessel attachment shall be designed for this increased design value.

c. Seismic displacements of the tank and vessel shall consider the deformation of the support structure when determining P-delta effects or evaluating required clearances to prevent pounding of the tank on the structure.

9.14.6.6 Piers and Wharves.

9.14.6.6.1 General. Piers and wharves are structures located in waterfront areas that project into a body of water or parallel the shoreline.

9.14.6.6.2 Design Basis. In addition to the provisions of 9.14.6.1, piers and wharves shall be designed to comply with this Standard and approved standards. Seismic forces on elements below the water level shall include the inertial force of the mass of the

displaced water. The additional seismic mass equal to the mass of the displaced water shall be included as a lumped mass on the submerged element, and shall be added to the calculated seismic forces of the pier or wharf structure. Seismic dynamic forces from the soil shall be determined by the registered design professional.

The design shall account for the effects of liquefaction on piers and wharves as required.

9.14.7 Nonbuilding Structures Not Similar to Buildings.

9.14.7.1 General. Nonbuilding structures that have structural systems that are designed and constructed in a manner such that the dynamic response is not similar to buildings shall be designed in compliance with this Standard with exceptions as contained in this section.

This general category of nonbuilding structures shall be designed in accordance with this Standard and the specific applicable approved standards. Loads and load distributions shall not be less than those determined in this Standard.

The combination of load effects, E, shall be determined in accordance with Section 9.5.2.7.

Exception: The redundancy/reliability factor, ρ, per Section. 9.5.2.4 shall be taken as 1.

9.14.7.2 Earth-Retaining Structures.

9.14.7.2.1 General. This section applies to all earth-retaining walls. The applied seismic forces shall be determined in accordance with Section 9.7.5.1 with a geotechnical analysis prepared by a registered design professional.

The seismic use group shall be determined by the proximity of the retaining wall to buildings and other structures. If failure of the retaining wall would affect an adjacent structure, the Seismic Use Group shall not be less than that of the adjacent structure, as determined in Section 9.1.3. Earth-retaining walls are permitted to be designed for seismic loads as either yielding or nonyielding walls. Cantilevered reinforced concrete retaining walls shall be assumed to be yielding walls and shall be designed as simple flexural wall elements.

9.14.7.3 Tanks and Vessels.

9.14.7.3.1 General. This section applies to all tanks, vessels, bins and silos, and similar containers storing liquids, gases, and granular solids supported at the base (hereafter referred to generically as tanks and vessels). Tanks and vessels covered herein include

reinforced concrete, prestressed concrete, steel, aluminum, and fiber-reinforced plastic materials. Tanks supported on elevated levels in buildings shall be designed in accordance with Section 9.6.3.9.

9.14.7.3.2 Design Basis. Tanks and vessels storing liquids, gases, and granular solids shall be designed in accordance with this Standard and shall be designed to meet the requirements of the applicable approved standards shown in Table 9.14.3 and Section 9 of this Standard as defined in this section. Resistance to seismic forces shall be determined from a substantiated analysis based on the approved standards shown in Table 9.14.3.

a. Damping for the convective (sloshing) force component shall be taken as 0.5%.

b. Impulsive and convective components shall be combined by the direct sum or the square root of the sum of the squares (SRSS) method when the modal periods are separated. If significant modal coupling may occur, the complete quadratic combination (CQC) method shall be used.

c. Vertical earthquake forces shall be considered in accordance with the appropriate approved national standard. If the approved national standard permits the user the option of including or excluding the vertical earthquake force, to comply with this Standard, it shall be included. For tanks and vessels not covered by an approved national standard, the vertical earthquake force shall be defined as 67% of the equivalent lateral force.

9.14.7.3.3 Strength and Ductility. Structural components and members that are part of the lateral support system shall be designed to provide the following:

a. Connections and attachments for anchorage and other lateral force-resisting components shall be designed to develop the strength of the anchor (e.g., minimum published yield strength, F_y in direct tension, plastic bending moment), or Ω_o times the calculated element design force.

b. Penetrations, manholes, and openings in shell components shall be designed to maintain the strength and stability of the shell to carry tensile and compressive membrane shell forces.

c. Support towers for tanks and vessels with irregular bracing, unbraced panels, asymmetric bracing, or concentrated masses shall be designed using the provisions of Section 9.5.2.3 for irregular structures. Support towers using chevron or eccentric braced framing shall comply with the requirements of Section 9.5. Support towers using tension only bracing shall be designed such that the full cross-section of the tension element can yield during overload conditions.

d. In support towers for tanks and vessels, compression struts that resist the reaction forces from tension braces shall be designed to resist the lesser of the yield load of the brace ($A_g F_y$), or Ω_o times the calculated tension load in the brace.

e. The vessel stiffness relative to the support system (foundation, support tower, skirt, etc.) shall be considered in determining forces in the vessel, the resisting components, and the connections.

f. For concrete liquid-containing structures, system ductility and energy dissipation under unfactored loads shall not be allowed to be achieved by inelastic deformations to such a degree as to jeopardize the serviceability of the structure. Stiffness degradation and energy dissipation shall be allowed to be obtained either through limited microcracking or by means of lateral force-resistance mechanisms that dissipate energy without damaging the structure.

9.14.7.3.4 Flexibility of Piping Attachments. Piping systems connected to tanks and vessels shall consider the potential movement of the connection points during earthquakes and provide sufficient flexibility to avoid release of the product by failure of the piping system. Forces and deformations imparted by piping systems or other attachments shall not exceed the design limitations at points of attachment to the tank or vessel shell. Mechanical devices which add flexibility such as bellows, expansion joints, and other flexible apparatus may be used when they are designed for seismic loads and displacements.

Unless otherwise calculated, the minimum displacements in Table 9.14.7.3.4 shall be assumed. For attachment points located above the support or foundation elevation, the displacements in Table 9.14.7.3.4 shall be increased to account for drift of the tank or vessel relative to the base of support.

When the elastic deformations are calculated, the minimum design displacements for piping attachments shall be the calculated displacements at the point of attachment increased by the amplification factor C_d.

TABLE 9.14.7.3.4
MINIMUM DISPLACEMENTS FOR PIPING ATTACHMENTS

Anchored Tanks or Vessels	Displacements (in.)
Vertical displacement relative to support or foundation	2
Horizontal (radial and tangential) relative to support or foundation	0.5
Unanchored Tanks or Vessels (at grade)	
Vertical displacement relative to support or foundation	
If designed to meet approved standard	6
If designed for seismic loads per these provisions but not covered by an approved standard	12
For tanks and vessels with a diameter <40 ft, horizontal (radial and tangential) relative to support or foundation	8

The values given in Table 9.14.7.3.4 do not include the influence of relative movements of the foundation and piping anchorage points due to foundation movements (e.g., settlement, seismic displacements). The effects of the foundation movements shall be included in the piping system design including the determination of the mechanical loading on the tank or vessel and the total displacement capacity of the mechanical devices intended to add flexibility.

9.14.7.3.5 Anchorage. Tanks and vessels at grade shall be permitted to be designed without anchorage when they meet the requirements for unanchored tanks in approved standards. Tanks and vessels supported above grade on structural towers or building structures shall be anchored to the supporting structure.

The following special detailing requirements shall apply to steel tank anchor bolts in seismic regions where $S_{DS} > 0.5$, or where the structure is classified as Seismic Use Group III.

a. Hooked anchor bolts (L- or J-shaped embedded bolts) or other anchorage systems based solely on bond or mechanical friction shall not be used when $S_{DS} \geq 0.33$. Postinstalled anchors may be used provided that testing validates their ability to develop yield load in the anchor under cyclic loads in cracked concrete.

b. When anchorage is required, the anchor embedment into the foundation shall be designed to develop the minimum specified yield strength of the anchor.

9.14.7.3.6 Ground-Supported Storage Tanks for Liquids.

9.14.7.3.6.1 General. Ground-supported, flat bottom tanks storing liquids shall be designed to resist the seismic forces calculated using one of the following procedures:

a. The base shear and overturning moment calculated as if tank and the entire contents are a rigid mass system per Section 9.14.5.2 of this Standard, or

b. Tanks or vessels storing liquids in Seismic Use Group III, or with a diameter greater than 20 ft shall be designed to consider the hydrodynamic pressures of the liquid in determining the equivalent lateral forces and lateral force distribution per the approved standards listed in Table 9.14.3 and the requirements of Section 9.14.7.3 of this Standard.

c. The force and displacement provisions of Section 9.14.5 of this Standard.

The design of tanks storing liquids shall consider the impulsive and convective (sloshing) effects and their consequences on the tank, foundation, and attached elements. The impulsive component corresponds to the high-frequency amplified response to the lateral ground motion of the tank roof, shell, and portion of the contents that moves in unison with the shell. The convective component corresponds to the low-frequency amplified response of the contents in the fundamental sloshing mode. Damping for the convective component shall be 0.5% for the sloshing liquid unless otherwise defined by the approved national standard. The following definitions shall apply:

T_c = natural period of the first (convective) mode of sloshing

T_i = fundamental period of the tank structure and impulsive component of the content

V_i = base shear due to impulsive component from weight of tank and contents

V_c = base shear due to the convective component of the effective sloshing mass

The seismic base shear is the combination of the impulsive and convective components:

$$V = V_i + V_c \qquad \textbf{(Eq. 9.14.7.3.6.1-1)}$$

where

$$V_i = \frac{S_{ai} I W_i}{R} \qquad \textbf{(Eq. 9.14.7.3.6.1-2)}$$

$$V_c = \frac{S_{ac} I W_c}{R} \qquad \textbf{(Eq. 9.14.7.3.6.1-3)}$$

S_{ai} = the spectral acceleration as a multiplier of gravity including the site impulsive components at period T_i and 5% damping

For $T_i \leq T_s$

$$S_{ai} = S_{DS} \qquad \textbf{(Eq. 9.14.7.3.6.1-4)}$$

For $T_i > T_s$

$$S_{ai} = \frac{S_{D1}}{T_i} \qquad \textbf{(Eq. 9.14.7.3.6.1-5)}$$

Notes:

a. When an approved national standard is used in which the spectral acceleration for the tank shell, and the impulsive component of the liquid is independent of T_i, then $S_{ai} = S_{DS}$.

b. Eq. 9.14.7.3.6.1-5 shall not be less than $0.14\,S_{DS}\,R$ and in addition in Seismic Design Categories E and F shall not be taken less than $0.8\,S_1$.

S_{ac} = the spectral acceleration of the sloshing liquid based on the sloshing period T_c and 0.5% damping

For $T_c \leq 4.0$ sec

$$S_{ac} = \frac{1.5 S_{D1}}{T_c} \qquad \textbf{(Eq. 9.14.7.3.6.1-6)}$$

For $T_c > 4.0$ sec

$$S_{ac} = \frac{6 S_{D1}}{T_c^2} \qquad \textbf{(Eq. 9.14.7.3.6.1-7)}$$

where

$$T_c = 2\pi \sqrt{\frac{D}{3.68g \tanh\left(\dfrac{3.68H}{D}\right)}}$$

$$\textbf{(Eq. 9.14.7.3.6.1-8)}$$

and
where

D = the tank diameter in ft (or meters), H = liquid height in ft (or meters), and g = acceleration due to gravity in consistent units

W_I = impulsive weight (impulsive component of liquid, roof and equipment, shell, bottom, and internal components)

W_c = the portion of the liquid weight sloshing

The general design response spectra for ground-supported liquid storage tanks is shown in Figure 9.14.7.3.6-1.

9.14.7.3.6.1.1 Distribution of Hydrodynamic and Inertia Forces. Unless otherwise required by the appropriate approved standard in Table 9.14.3, the method given in Ref. 9.14-3 may be used to determine the vertical and horizontal distribution of the hydrodynamic and inertia forces on the walls of circular and rectangular tanks.

9.14.7.3.6.1.2 Freeboard. Sloshing of the liquid within the tank or vessel shall be considered in determining the freeboard required above the top capacity liquid level. A minimum freeboard (δ_s) shall be provided per Table 9.14.7.3.6.1.2. The height of the sloshing wave is permitted to be calculated as:

$$\delta_s = 0.5 D I S_{ac} \qquad \textbf{(Eq. 9.14.7.3.6.1.2)}$$

9.14.7.3.6.1.3 Equipment and Attached Piping. Equipment, piping and walkways or other

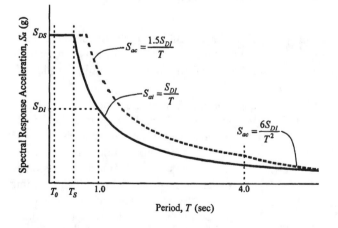

FIGURE 9.14.7.3.6-1
DESIGN RESPONSE SPECTRA FOR GROUND-SUPPORTED LIQUID STORAGE TANKS

TABLE 9.14.7.3.6.1.2
MINIMUM REQUIRED FREEBOARD

Value of S_{DS}	Seismic Use Group		
	I	II	III
$S_{DS} < 0.167$ g	a	a	$\delta_s{}^c$
0.167 g $\leq S_{DS} < 0.33$ g	a	a	$\delta_s{}^c$
0.33 g $\leq S_{DS} < 0.50$ g	a	$0.7\delta_s{}^b$	$\delta_s{}^c$
$S_{DS} \geq 0.50$ g	a	$0.7\delta_s{}^b$	$\delta_s{}^c$

[a] No minimum freeboard is required.

[b] A freeboard equal to 0.7 δ_s is required unless one of the following alternatives is provided:

1. Secondary containment is provided to control the product spill.
2. The roof and supporting structure are designed to contain the sloshing liquid.

[c] Freeboard equal to the calculated wave height, δ_s, is required unless one of the following alternatives is provided:

1. Secondary containment is provided to control the product spill.
2. The roof and supporting structure are designed to contain the sloshing liquid.

appurtenances attached to the structure shall be designed to accommodate the displacements imposed by seismic forces. For piping attachments, see Section 9.14.7.3.4.

9.14.7.3.6.1.4 Internal Components. The attachments of internal equipment and accessories which are attached to the primary liquid or pressure-retaining shell or bottom, or provide structural support for major components (e.g., a column supporting the roof rafters), shall be designed for the lateral loads due to the sloshing liquid in addition to the inertial forces by a substantiated analysis method.

9.14.7.3.6.1.5 Sliding Resistance. The transfer of the total lateral shear force between the tank or vessel and the subgrade shall be considered:

a. For unanchored flat bottom steel tanks, the overall horizontal seismic shear force is permitted to be resisted by friction between the tank bottom and the foundation or subgrade. Unanchored storage tanks shall be designed such that sliding will not occur when the tank is full of stored product. The maximum calculated seismic base shear, V, shall not exceed:

$$V < W \tan 30° \qquad \textbf{(Eq. 9.14.7.3.6.1.5)}$$

W shall be determined using the effective weight of the tank, roof, and contents after reduction for coincident vertical earthquake. Lower values of the friction factor shall be used if the design of the tank bottom to supporting foundation does not justify the friction value above (e.g., leak detection membrane beneath the bottom with a lower friction factor, smooth bottoms, etc.). Alternatively, the friction factor shall be permitted to be determined by testing in accordance with Section 9.1.2.5.

b. No additional lateral anchorage is required for anchored steel tanks designed in accordance with approved standards.

c. The lateral shear transfer behavior for special tank configurations (e.g., shovel bottoms, highly crowned tank bottoms, tanks on grillage) can be unique and are beyond the scope of these provisions.

9.14.7.3.6.1.6 Local Shear Transfer. Local transfer of the shear from the roof to the wall and the wall of the tank into the base shall be considered. For cylindrical tanks and vessels, the peak local tangential shear per unit length shall be calculated by:

$$v_{max} = \frac{2V}{\pi D} \qquad \textbf{(Eq. 9.14.7.3.6.1.6)}$$

a. Tangential shear in flat bottom steel tanks shall be transferred through the welded connection to the steel bottom. This transfer mechanism is deemed acceptable for steel tanks designed in accordance with the approved standards where $S_{DS} < 1.0$ g.

b. For concrete tanks with a sliding base where the lateral shear is resisted by friction between the tank wall and the base, the friction coefficient value used for design shall not exceed tan 30 degrees.

c. Fixed-base or hinged-base concrete tanks, transfer the horizontal seismic base shear shared by membrane (tangential) shear and radial shear into the foundation. For anchored flexible-base concrete tanks, the majority of the base shear is resisted by membrane (tangential) shear through the anchoring system with only insignificant vertical bending in the wall. The connection between the wall and floor

shall be designed to resist the maximum tangential shear.

9.14.7.3.6.1.7 Pressure Stability. For steel tanks, the internal pressure from the stored product stiffens thin cylindrical shell structural elements subjected to membrane compression forces. This stiffening effect may be considered in resisting seismically induced compressive forces if permitted by the approved standard or the authority having jurisdiction.

9.14.7.3.6.1.8 Shell Support. Steel tanks resting on concrete ring walls or slabs shall have a uniformly supported annulus under the shell. Uniform support shall be provided by one of the following methods:

a. Shimming and grouting the annulus.

b. Using fiberboard or other suitable padding.

c. Using butt-welded bottom or annular plates resting directly on the foundation.

d. Using closely spaced shims (without structural grout) provided that the localized bearing loads are considered in the tank wall and foundation to prevent local crippling and spalling.

Anchored tanks shall be shimmed and grouted. Local buckling of the steel shell for the peak compressive force due to operating loads and seismic overturning shall be considered.

9.14.7.3.6.1.9 Repair, Alteration, or Reconstruction. Repairs, modifications, or reconstruction (i.e., cut down and re-erect) of a tank or vessel shall conform to industry standard practice and this Standard. For welded steel tanks storing liquids, see Ref. 9.14-8 and the approved national standard in Table 9.14.3. Tanks that are relocated shall be re-evaluated for the seismic loads for the new site and the requirements of new construction in accordance with the appropriate approved national standard and this Standard.

9.14.7.3.7 Water and Water Treatment Tanks and Vessels.

9.14.7.3.7.1 Welded Steel. Welded steel water storage tanks and vessels shall be designed in accordance with the seismic requirements of Ref. 9.14-15 except that the design input forces shall be modified as follows:

The equations for base shear and overturning moment are defined by the following equations for allowable stress design procedures:

For $T_s < T_c \leq 4.0$ sec

$$V_{ACT} = \frac{S_{DS}I}{(1.4R)} \left[(W_s + W_r + W_f + W_1) + 1.5\frac{T_s}{T_c}W_2 \right]$$

(Eq. 9.14.7.3.7.1-1)

$$M = \frac{S_{DS}I}{(1.4R)} \left[(W_sX_s + W_rH_t + W_1X_1) + 1.5\frac{T_s}{T_c}W_2X_2 \right]$$

(Eq. 9.14.7.3.7.1-2)

For $T_c > 4.0$ sec

$$V_{ACT} = \frac{S_{DS}I}{(1.4R)} \left[(W_s + W_r + W_f + W_1) + 6\frac{T_s}{T_c^2}W_2 \right]$$

(Eq. 9.14.7.3.7.1-3)

$$M = \frac{S_{DS}I}{(1.4R)} \left[(W_sX_s + W_rH_t + W_1X_1) + 6\frac{T_s}{T_c^2}W_2X_2 \right]$$

(Eq. 9.14.7.3.7.1-4)

a. Substitute the above equations for Eqs. 13-4 and 13-8 of Ref. 9.14-15 where S_{DS} and T_S are defined in Section 9.4.1.2.5 and R is defined in Table 9.14.5.1.1.

b. The hydrodynamic seismic hoop tensile stress is defined in Eqs. 13-20 through 13-25 in Ref. 9.14-15. When using these equations, substitute Eq. 9.14.7.3.7.1-5 for $\frac{Z_I}{R_W}$ directly into the equations.

$$\frac{S_{DS}I}{2.5(1.4R)}$$ **(Eq. 9.14.7.3.7.1-5)**

c. Sloshing height shall be calculated per Section 9.14.7.3.7.1.2 instead of Eq. 13-26 of Ref. 9.14-15.

9.14.7.3.7.2 Bolted Steel. Bolted steel water storage structures shall be designed in accordance with the seismic requirements of Ref. 9.14-16 except that the design input forces shall be modified in the same manner shown in Section 9.14.7.3.7.1 of this Standard.

9.14.7.3.7.3 Reinforced and Prestressed Concrete. Reinforced and prestressed concrete tanks shall be designed in accordance with the seismic requirements of Ref. 9.14-3 except that the design input forces for allowable stress design procedures shall be modified as follows:

a. For $T_I < T_o$, and $T_I > T_s$, substitute the term $S_a I/1.4R$, where S_a is defined in Section 9.4.1.2.6, Subsections 1, 2, or 3, for the terms in the appropriate equations as shown below:

For $\dfrac{ZIC_1}{R_1}$ shear and overturning moment equations of Ref. 9.14-17

For $\dfrac{ZIC_1}{R_W}$ shear and overturning moment equations of Ref. 9.14-18

For $\dfrac{ZISC_i}{R_i}$ in the base shear and overturning moment equations of

Ref. 9.14-3

b. For $T_o \leq T_I \leq T_s$, substitute the term $\dfrac{S_{DS}I}{1.4R}$ for terms $\dfrac{ZIC_1}{R_1}$ and $\dfrac{ZISC_i}{R_i}$

c. For all values of T_C (or T_w), $\dfrac{ZIC_c}{R_c}$, $\dfrac{ZIC_c}{R_W}$ and $\dfrac{ZISC_c}{R_c}$ are replaced by

$\dfrac{6S_{D1}}{T_c^2}$ or $\dfrac{6S_{DS}I}{T_c^2}T_s$

where

S_a, S_{D1}, S_{DS}, T_0, and T_s are defined in Section 9.4.1.2.6 of this Standard.

9.14.7.3.8 Petrochemical and Industrial Tanks and Vessels Storing Liquids.

9.14.7.3.8.1 Welded Steel. Welded steel petrochemical and industrial tanks and vessels storing liquids shall be designed in accordance with the seismic requirements of Ref. 9.14-6 and Ref. 9.14-7 except that the design input forces for allowable stress design procedures shall be modified as follows:

a. When using the equations in Section E.3 of Ref. 9.14-7, substitute into the equation for overturning moment M (where S_{DS} and T_s are defined in Section 9.4.1.2.5 of this Standard). Thus,

In the range $T_s < T_c \leq 4.0$ sec,

$$M = S_{DS}I[0.24(W_s X_s + W_t H_t + W_1 X_1) + 0.80 C_2 T_s W_2 X_2]$$

(Eq. 9.14.7.3.8.1-1)

where $C_2 = \dfrac{0.75S}{T_c}$ and $S = 1.0$

In the range $T_w > 4.0$ sec, and

$$M = S_{DS}I[0.24(W_s X_s + W_t H_t + W_1 X_1) + 0.71 C_2 T_s W_2 X_2]$$

(Eq. 9.14.7.3.8.1-2)

where $C_2 = \dfrac{3.375S}{T_c^2}$ and $S = 1.0$

9.14.7.3.8.2 Bolted Steel. Bolted steel tanks used for storage of production liquids. Ref. 9.14-10 covers the material, design, and erection requirements for vertical, cylindrical, aboveground bolted tanks in nominal capacities of 100 to 10,000 barrels for production service. Unless required by the building official having jurisdiction, these temporary structures need not be designed for seismic loads. If design for seismic load is required, the loads may be adjusted for the temporary nature of the anticipated service life.

9.14.7.3.8.3 Reinforced and Prestressed Concrete. Reinforced concrete tanks for the storage of petrochemical and industrial liquids shall be designed in accordance with the force requirements of Section 14.7.3.7.3.

9.14.7.3.9 Ground-Supported Storage Tanks for Granular Materials.

9.14.7.3.9.1 General. The intergranular behavior of the material shall be considered in determining effective mass and load paths, including the following behaviors:

a. Increased lateral pressure (and the resulting hoop stress) due to loss of the intergranular friction of the material during the seismic shaking.

b. Increased hoop stresses generated from temperature changes in the shell after the material has been compacted.

c. Intergranular friction, which can transfer seismic shear directly to the foundation.

9.14.7.3.9.2 Lateral Force Determination. The lateral forces for tanks and vessels storing granular

materials at grade shall be determined by the requirements and accelerations for short period structures (i.e., S_{DS}).

9.14.7.3.9.3 Force Distribution to Shell and Foundation.

9.14.7.3.9.3.1 Increased Lateral Pressure. The increase in lateral pressure on the tank wall shall be added to the static design lateral pressure but shall not be used in the determination of pressure stability effects on the axial buckling strength of the tank shell.

9.14.7.3.9.3.2 Effective Mass. A portion of a stored granular mass will act with the shell (the effective mass). The effective mass is related to the physical characteristics of the product, the height-to-diameter (H/D) ratio of the tank, and the intensity of the seismic event. The effective mass shall be used to determine the shear and overturning loads resisted by the tank.

9.14.7.3.9.3.3 Effective Density. The effective density factor (that part of the total stored mass of product which is accelerated by the seismic event) shall be determined in accordance with Ref. 9.14-2.

9.14.7.3.9.3.4 Lateral Sliding. For granular storage tanks that have a steel bottom and are supported such that friction at the bottom to foundation interface can resist lateral shear loads, no additional anchorage to prevent sliding is required. For tanks without steel bottoms (i.e., the material rests directly on the foundation), shear anchorage shall be provided to prevent sliding.

9.14.7.3.9.3.5 Combined Anchorage Systems. If separate anchorage systems are used to prevent overturning and sliding, the relative stiffness of the systems shall be considered in determining the load distribution.

9.14.7.3.9.4 Welded Steel Structures. Welded steel granular storage structures shall be designed in accordance with Section 9 of this Standard. Component allowable stresses and materials shall be per Ref. 9.14-15, except the allowable circumferential membrane stresses and material requirements in Ref. 9.14-7 shall apply.

9.14.7.3.9.5 Bolted Steel Structures. Bolted steel granular storage structures shall be designed in accordance with Section 9. Component allowable stresses and materials shall be per Ref. 9.14-16.

9.14.7.3.9.6 Reinforced Concrete Structures. Reinforced concrete structures for the storage of granular materials shall be designed in accordance with the force requirements of Section 9 and the requirements of Ref. 9.14-2.

9.14.7.3.9.7 Prestressed Concrete Structures. Prestressed concrete structures for the storage of granular materials shall be designed in accordance with the force provisions of Section 9 of this Standard and the requirements of Ref. 9.14-2.

9.14.7.3.10 Elevated Tanks and Vessels for Liquids and Granular Materials.

9.14.7.3.10.1 General. This section applies to tanks, vessels, bins, and hoppers that are elevated above grade where the supporting tower is an integral part of the structure, or where the primary function of the tower is to support the tank or vessel. Tanks and vessels that are supported within buildings, or are incidental to the primary function of the tower, are considered mechanical equipment and shall be designed in accordance with Section 9.6 of this Standard.

Elevated tanks shall be designed for the force and displacement requirements of the applicable approved standard, or Section 9.14.5.

9.14.7.3.10.2 Effective Mass. The design of the supporting tower or pedestal, anchorage, and foundation for seismic overturning shall assume the material stored is a rigid mass acting at the volumetric center of gravity. The effects of fluid-structure interaction may be considered in determining the forces, effective period, and mass centroids of the system if the following requirements are met:

a. The sloshing period, T_c, is greater than $3T$ where T = natural period of the tank with confined liquid (rigid mass) and supporting structure.
b. The sloshing mechanism (i.e., the percentage of convective mass and centroid) is determined for the specific configuration of the container by detailed fluid-structure interaction analysis or testing.

Soil-structure interaction may be included in determining T providing the provisions of Section 9.5.9 are met.

9.14.7.3.10.3 *P*-Delta Effects. The lateral drift of the elevated tank shall be considered as follows:

a. The design drift, the elastic lateral displacement of the stored mass center of gravity shall be increased by the factor, C_d for evaluating the additional load in the support structure.

b. The base of the tank shall be assumed to be fixed rotationally and laterally.

c. Deflections due to bending, axial tension, or compression shall be considered. For pedestal tanks with a height-to-diameter ratio less than 5, shear deformations of the pedestal shall be considered.

d. The dead load effects of roof-mounted equipment or platforms shall be included in the analysis.

e. If constructed within the plumbness tolerances specified by the approved standard, initial tilt need not be considered in the *P*-delta analysis.

9.14.7.3.10.4 Transfer of Lateral Forces into Support Tower. For post-supported tanks and vessels which are cross-braced:

a. The bracing shall be installed in such a manner as to provide uniform resistance to the lateral load (e.g., pretensioning or tuning to attain equal sag).

b. The additional load in the brace due to the eccentricity between the post to tank attachment and the line of action of the bracing shall be included.

c. Eccentricity of compression strut line of action (elements that resist the tensile pull from the bracing rods in the lateral-force-resisting systems) with their attachment points shall be considered.

d. The connection of the post or leg with the foundation shall be designed to resist both the vertical and lateral resultant from the yield load in the bracing assuming the direction of the lateral load is oriented to produce the maximum lateral shear at the post to foundation interface. Where multiple rods are connected to the same location, the anchorage shall be designed to resist the concurrent tensile loads in the braces.

9.14.7.3.10.5 Evaluation of Structures Sensitive to Buckling Failure. Shell structures that support substantial loads may exhibit a primary mode of failure from localized or general buckling of the support pedestal or skirt during seismic loads. Such structures may include single-pedestal water towers, skirt-supported process vessels, and similar single-member towers. Where the structural assessment concludes that buckling of the support is the governing primary mode of failure, structures and components in Seismic Use Group III shall be designed to resist the seismic forces as follows:

a. The seismic response coefficient for this evaluation shall be per Section 9.5.3.2.1 of this Standard with I/R set equal to 1.0. Soil-structure and fluid-structure interaction may be utilized in determining the structural response. Vertical or orthogonal combinations need not be considered.

b. The resistance of the structure or component shall be defined as the critical buckling resistance of the element; i.e., a factor of safety set equal to 1.0

c. The anchorage and foundation shall be designed to resist the load determined in (a). The foundation shall be proportioned to provide a stability ratio of at least 1.2 for the overturning moment. The maximum toe pressure under the foundation shall not exceed the lesser of the ultimate bearing capacity or 3 times the allowable bearing capacity. All structural components and elements of the foundation shall be designed to resist the combined loads with a load factor of 1.0 on all loads including dead load, live load, and earthquake load. Anchors shall be permitted to yield.

9.14.7.3.10.6 Welded Steel Water Storage Structures. Welded steel elevated water storage structures shall be designed and detailed in accordance with the seismic requirements of Ref. 9.14-15 and this Standard except that the design input forces for allowable stress design procedures shall be modified by substituting the following terms for ZIC/R_W into Eqs. 13-1 and 13-3 of Ref. 9.14-15 and set the value for $S = 1.0$.

For $T \leq Ts$, substitute the term

$$\frac{S_{DS}I}{1.4R} \qquad \textbf{(Eq. 9.14.7.3.10.6-1)}$$

For $Ts < T \leq 4.0$ second, substitute the term

$$\frac{S_{D1}I}{T(1.4R)} \qquad \textbf{(Eq. 9.14.7.3.10.6-2)}$$

For $T > 4.0$ second, substitute the term

$$\frac{S_{D1}I}{T^2(1.4R)} \qquad \textbf{(Eq. 9.14.7.3.10.6-3)}$$

9.14.7.3.10.6.1 Analysis Procedures. The equivalent lateral force procedure shall be permitted. A more rigorous analysis shall also be permitted. Analysis of single-pedestal structures shall be based on a fixed-base, single degree-of-freedom model. All mass, including the liquid, shall be considered rigid unless the sloshing mechanism (i.e., the percentage of convective mass and centroid) is determined for the specific configuration of the container by detailed fluid-structure interaction analysis or testing. The inclusion of soil-structure interaction shall be permitted.

9.14.7.3.10.6.2 Structure Period. The fundamental period of vibration of the structure shall be established using the structural properties and deformational characteristics of the resisting elements in a substantiated analysis. The period used to calculate the seismic response coefficient shall not exceed 4.0 secs. See Ref. 9.14-15 for guidance on computing the fundamental period of cross-braced structures.

9.14.7.3.10.7 Concrete Pedestal (Composite) Tanks. Concrete pedestal (composite) elevated water storage structures shall be designed in accordance with the requirements of Ref. 9.14-4 except that the design input forces shall be modified as follows:

In Eq. 4-8a of Ref. 9.14-4,
For $Ts < T \le 4.0$ sec, replace the term $1.2C_V/RT^{2/3}$ with

$$\frac{S_{D1}I}{TR} \qquad \textbf{(Eq. 9.14.7.3.10.7-1)}$$

For $T > 4.0$ sec, replace the term $1.2C_V/RT^{2/3}$ with

$$\frac{4S_{D1}I}{T^2R} \qquad \textbf{(Eq. 9.14.7.3.10.7-2)}$$

In Eq. 4-8b of Ref. 9.14-4, replace the term $2.5C_a/R$ with

$$\frac{S_{DS}I}{R} \qquad \textbf{(Eq. 9.14.7.3.10.7-3)}$$

In Eq. 4-9 of Ref. 9.14-4, replace the term $0.5C_a$ with

$$0.2S_{DS} \qquad \textbf{(Eq. 9.14.7.3.10.7-4)}$$

9.14.7.3.10.7.1 Analysis Procedures. The equivalent lateral force procedure shall be permitted for all concrete pedestal tanks and shall be based on a fixed-base, single degree-of-freedom model. All mass, including the liquid, shall be considered rigid unless the sloshing mechanism (i.e., the percentage of convective mass and centroid) is determined for the specific configuration of the container by detailed fluid-structure interaction analysis or testing. Soil-structure interaction may be included. A more rigorous analysis shall be permitted.

9.14.7.3.10.7.2 Structure Period. The fundamental period of vibration of the structure shall be established using the uncracked structural properties and deformational characteristics of the resisting elements in a properly substantiated analysis. The period used to calculate the seismic response coefficient shall not exceed 2.5 sec.

9.14.7.3.11 Boilers and Pressure Vessels.

9.14.7.3.11.1 General. Attachments to the pressure boundary, supports, and lateral-force-resisting anchorage systems for boilers and pressure vessels shall be designed to meet the force and displacement requirements of Sections 9.5.5 and 9.5.6 and the additional requirements of this section. Boilers and pressure vessels categorized as Seismic Use Group III or III shall be designed to meet the force and displacement requirements of Sections 9.5.5 and 9.5.6.

9.14.7.3.11.2 ASME Boilers and Pressure Vessels. Boilers or pressure vessels designed and constructed in accordance with Ref. 9.14-11 shall be deemed to meet the requirements of this section provided that the displacement requirements of Sections 9.5.5 and 9.5.6 are used with appropriate scaling of the force and displacement requirements to the working stress design basis.

9.14.7.3.11.3 Attachments of Internal Equipment and Refractory. Attachments to the pressure boundary for internal and external ancillary components (refractory, cyclones, trays, etc.) shall be designed to resist the seismic forces specified in this Standard to safeguard against rupture of the pressure boundary. Alternatively, the element attached may be designed to fail prior to damaging the pressure boundary provided that the consequences of the failure do not place the pressure boundary in jeopardy. For boilers or vessels

containing liquids, the effect of sloshing on the internal equipment shall be considered if the equipment can damage the integrity of the pressure boundary.

9.14.7.3.11.4 Coupling of Vessel and Support Structure. Where the mass of the operating vessel or vessels supported is greater than 25% of the total mass of the combined structure, the structure and vessel designs shall consider the effects of dynamic coupling between each other. Coupling with adjacent, connected structures such as multiple towers shall be considered if the structures are interconnected with elements that will transfer loads from one structure to the other.

9.14.7.3.11.5 Effective Mass. Fluid-structure interaction (sloshing) shall be considered in determining the effective mass of the stored material providing sufficient liquid surface exists for sloshing to occur and the T_c is greater than $3T$. Changes to or variations in material density with pressure and temperature shall be considered.

9.14.7.3.11.6 Other Boilers and Pressure Vessels. Boilers and pressure vessels designated Seismic Use Group III but not designed and constructed in accordance with the requirements of Ref. 9.14-11 shall meet the following requirements:

The seismic loads in combination with other service loads and appropriate environmental effects shall not exceed the material strength shown in Table 9.14.7.3.11.6.

Consideration shall be made to mitigate seismic impact loads for boiler or vessel components constructed of nonductile materials or vessels operated in such a way that material ductility is reduced (e.g., low-temperature applications).

9.14.7.3.11.7 Supports and Attachments for Boilers and Pressure Vessels. Attachments to the pressure boundary and support for boilers and pressure vessels shall meet the following requirements:

a. Attachments and supports transferring seismic loads shall be constructed of ductile materials suitable for the intended application and environmental conditions.

b. Seismic anchorages embedded in concrete shall be ductile and detailed for cyclic loads.

c. Seismic supports and attachments to structures shall be designed and constructed so that the support or attachment remains ductile throughout the range of reversing seismic lateral loads and displacements.

d. Vessel attachments shall consider the potential effect on the vessel and the support for uneven vertical reactions based on variations in relative stiffness of the support members, dissimilar details, nonuniform shimming, or irregular supports. Uneven distribution of lateral forces shall consider the relative distribution of the resisting elements, the behavior of the connection details, and vessel shear distribution.

The requirements of Sections 9.14.5 and 9.14.7.3.10.5 shall also be applicable to this section.

9.14.7.3.12 Liquid and Gas Spheres.

9.14.7.3.12.1 General. Attachments to the pressure or liquid boundary, supports, and lateral-force-resisting anchorage systems for liquid and gas spheres shall be designed to meet the force and displacement requirements of Sections 9.5.5 and 9.5.6 and the additional requirements of this section. Spheres categorized as Seismic Use Group III or III shall themselves be designed to meet the force and displacement requirements of Sections 9.5.5 and 9.5.6.

9.14.7.3.12.2 ASME Spheres. Spheres designed and constructed in accordance with Section VIII of Ref. 9.14-11 shall be deemed to meet the requirements of this section providing the displacement requirements of Sections 9.5.5 and 9.5.6 are used with appropriate scaling of the force and displacement requirements to the working stress design basis.

TABLE 9.14.7.3.11.6
MAXIMUM MATERIAL STRENGTH

Material	Minimum Ratio F_u/F_y	Max Material Strength Vessel Material	Max Material Strength Threaded Material[a]
Ductile (e.g., steel, aluminum, copper)	1.33[b]	90%[d]	70%[d]
Semiductile	1.2[c]	70%[d]	50%[d]
Nonductile (e.g., cast iron, ceramics, fiberglass)	NA	25%[e]	20%[e]

[a] Threaded connection to vessel or support system.
[b] Minimum 20% elongation per the ASTM material specification.
[c] Minimum 15% elongation per the ASTM material specification.
[d] Based on material minimum specified yield strength.
[e] Based on material minimum specified tensile strength.

9.14.7.3.12.3 Attachments of Internal Equipment and Refractory. Attachments to the pressure or liquid boundary for internal and external ancillary components (refractory, cyclones, trays, and so on) shall be designed to resist the seismic forces specified in this Standard to safeguard against rupture of the pressure boundary. Alternatively, the element attached to the sphere could be designed to fail prior to damaging the pressure or liquid boundary providing the consequences of the failure do not place the pressure boundary in jeopardy. For spheres containing liquids, the effect of sloshing on the internal equipment shall be considered if the equipment can damage the pressure boundary.

9.14.7.3.12.4 Effective Mass. Fluid-structure interaction (sloshing) shall be considered in determining the effective mass of the stored material providing sufficient liquid surface exists for sloshing to occur and the T_c is greater than $3T$. Changes to or variations in fluid density shall be considered.

9.14.7.3.12.5 Post and Rod Supported. For post-supported spheres that are cross-braced:

a. The requirements of Section 9.14.7.3.10.4 shall also be applicable to this section.

b. The stiffening effect of (reduction in lateral drift) from pre-tensioning of the bracing shall be considered in determining the natural period.

c. The slenderness and local buckling of the posts shall be considered.

d. Local buckling of the sphere shell at the post attachment shall be considered.

e. For spheres storing liquids, bracing connections shall be designed and constructed to develop the minimum published yield strength of the brace. For spheres storing gas vapors only, bracing connection shall be designed for Ω_o times the maximum design load in the brace. Lateral bracing connections directly attached to the pressure or liquid boundary are prohibited.

9.14.7.3.12.6 Skirt Supported. For skirt-supported spheres, the following requirements shall apply:

a. The provisions of Section 9.14.7.3.10.5 shall also apply.

b. The local buckling of the skirt under compressive membrane forces due to axial load and bending moments shall be considered.

c. Penetration of the skirt support (manholes, piping, and so on) shall be designed and constructed to maintain the strength of the skirt without penetrations.

9.14.7.3.13 Refrigerated Gas Liquid Storage Tanks and Vessels.

9.14.7.3.13.1 General. The seismic design of the tanks and facilities for the storage of liquefied hydrocarbons and refrigerated liquids is beyond the scope of this section. The design of such tanks is addressed in part by various approved standards as listed in Table 9.14.3.

Exception: Low-pressure, welded steel storage tanks for liquefied hydrocarbon gas (e.g., LPG, butane, and so on) and refrigerated liquids (e.g., ammonia) shall be designed in accordance with the requirements of Section 9.14.7.3.8 and Ref. 9.14-6.

9.14.7.3.14 Horizontal, Saddle-Supported Vessels for Liquid or Vapor Storage.

9.14.7.3.14.1 General. Horizontal vessels supported on saddles (sometimes referred to as blimps) shall be designed to meet the force and displacement requirements of Sections 9.5.5 and 9.5.6.

9.14.7.3.14.2 Effective Mass. Changes to or variations in material density shall be considered. The design of the supports, saddles, anchorage, and foundation for seismic overturning shall assume the material stored is a rigid mass acting at the volumetric center of gravity.

9.14.7.3.14.3 Vessel Design. Unless a more rigorous analysis is performed:

a. Horizontal vessels with a length-to-diameter ratio of 6 or more may be assumed to be a simply supported beam spanning between the saddles for determining the natural period of vibration and global bending moment.

b. Horizontal vessels with a length-to-diameter ratio of less than 6, the effects of "deep beam shear" shall be considered when determining the fundamental period and stress distribution.

c. Local bending and buckling of the vessel shell at the saddle supports due to seismic load shall be considered. The stabilizing

effects of internal pressure shall not be considered to increase the buckling resistance of the vessel shell.

d. If the vessel is a combination of liquid and gas storage, the vessel and supports shall be designed both with and without gas pressure acting (assume piping has ruptured and pressure does not exist).

9.14.7.4 Stacks and Chimneys.

9.14.7.4.1 General. Stacks and chimneys are permitted to be either lined or unlined, and shall be constructed from concrete, steel, or masonry.

9.14.7.4.2 Design Basis. Steel stacks, concrete stacks, steel chimneys, concrete chimneys, and liners shall be designed to resist seismic lateral forces determined from a substantiated analysis using approved standards. Interaction of the stack or chimney with the liners shall be considered. A minimum separation shall be provided between the liner and chimney equal to C_d times the calculated differential lateral drift.

9.14.7.5 Amusement Structures.

9.14.7.5.1 General. Amusement structures are permanently fixed structures constructed primarily for the conveyance and entertainment of people.

9.14.7.5.2 Design Basis. Amusement structures shall be designed to resist seismic lateral forces determined from a substantiated analysis using approved standards.

9.14.7.6 Special Hydraulic Structures.

9.14.7.6.1 General. Special hydraulic structures are structures that are contained inside liquid-containing structures. These structures are exposed to liquids on both wall surfaces at the same head elevation under normal operating conditions. Special hydraulic structures are subjected to out-of-plane forces only during an earthquake when the structure is subjected to differential hydrodynamic fluid forces. Examples of special hydraulic structures include separation walls, baffle walls, weirs, and other similar structures.

9.14.7.6.2 Design Basis. Special hydraulic structures shall be designed for out-of-phase movement of the fluid. Unbalanced forces from the motion of the liquid must be applied simultaneously "in front of" and "behind" these elements.

Structures subject to hydrodynamic pressures induced by earthquakes shall be designed for rigid body and sloshing liquid forces and their own inertia force. The height of sloshing shall be determined and compared to the freeboard height of the structure.

Interior elements, such as baffles or roof supports, also shall be designed for the effects of unbalanced forces and sloshing.

9.14.7.7 Secondary Containment Systems. Secondary containment systems such as impoundment dikes and walls shall meet the requirements of the applicable standards for tanks and vessels and the authority having jurisdiction.

Secondary containment systems shall be designed to withstand the effects of the maximum considered earthquake ground motion when empty and two-thirds of the maximum considered earthquake ground motion when full including all hydrodynamic forces as determined in accordance with the procedures of Section 9.4.1. When determined by the risk assessment required by Section 1.5.2 or by the authority having jurisdiction that the site may be subject to aftershocks of the same magnitude as the maximum considered ground motion, secondary containment systems shall be designed to withstand the effects of the maximum considered earthquake ground motion when full including all hydrodynamic forces as determined in accordance with the procedures of Section 9.4.1.

9.14.7.7.2 Freeboard. Sloshing of the liquid within the secondary containment area shall be considered in determining the height of the impound. Where the primary containment has not been designed with a reduction in the structure category (i.e., no reduction in importance factor I) as permitted by Section 1.5.2, no freeboard provision is required. Where the primary containment has been designed for a reduced structure category (i.e., importance factor I reduced) as permitted by Section 1.5.2, a minimum freeboard, δ_S, shall be provided where

$$\delta_S = 0.50 D S_{ac} \qquad \textbf{(Eq. 9.14.7.7.2)}$$

where S_{ac} is the spectral acceleration of the convective component and is determined according to the procedures of Section 9.4.1 using 0.5% damping. For circular impoundment dikes, D shall be taken as the diameter of the impoundment dike. For rectangular impoundment dikes, D shall be taken as the longer plan dimension of the impoundment dike.

9.14.7.8 Telecommunication Towers. Self-supporting and guyed telecommunication towers shall be designed to resist seismic lateral forces determined from a substantiated analysis using approved standards.

SECTION 10.0
ICE LOADS — ATMOSPHERIC ICING

SECTION 10.1
GENERAL

Atmospheric ice loads due to freezing rain, snow, and in-cloud icing shall be considered in the design of ice-sensitive structures. In areas where records or experience indicate that snow or in-cloud icing produces larger loads than freezing rain, site-specific studies shall be used. Structural loads due to hoarfrost are not a design consideration. Roof snow loads are covered in Section 7.

10.1.1 Site-Specific Studies. Mountainous terrain and gorges shall be examined for unusual icing conditions. Site-specific studies shall be used to determine the 50-year mean recurrence interval ice thickness and concurrent wind speed in:

1. Alaska.
2. areas where records or experience indicate that snow or in-cloud icing produces larger loads than freezing rain.
3. special icing regions shown in Figure 10-2.
4. mountainous terrain and gorges where examination indicates unusual icing conditions exist.

Site-specific studies shall be subject to review and approval by the authority having jurisdiction.

In lieu of using the mapped values, it shall be permitted to determine the ice thickness and the concurrent wind speed for a structure from local meteorological data based on a 50-year mean recurrence interval provided that:

1. the quality of the data for wind and type and amount of precipitation has been taken into account,
2. a robust ice accretion algorithm has been used to estimate uniform ice thicknesses and concurrent wind speeds from these data,
3. extreme-value statistical analysis procedures acceptable to the authority having jurisdiction have been employed in analyzing the ice thickness and concurrent wind speed data, and
4. the length of record and sampling error have been taken into account.

10.1.2 Dynamic Loads. Dynamic loads, such as those resulting from galloping, ice shedding, and aeolian vibrations, that are caused or enhanced by an ice accretion on a flexible structural member, component, or appurtenance are not covered in this Section.

10.1.3 Exclusions. Electric transmission systems, communications towers and masts, and other structures for which national standards exist are excluded from the requirements of this section. Applicable standards and guidelines include Refs. 10-1, 10-2, and 10-3.

SECTION 10.2
DEFINITIONS

The following definitions apply only to the provisions of Section 10.

COMPONENTS AND APPURTENANCES. Nonstructural elements that may be exposed to atmospheric icing. Examples are ladders, handrails, antennas, waveguides, radio frequency (RF) transmission lines, pipes, electrical conduits, and cable trays.

FREEZING RAIN. Rain or drizzle that falls into a layer of subfreezing air at the earth's surface and freezes on contact with the ground or an object to form glaze ice.

GLAZE. Clear high-density ice.

HOARFROST. An accumulation of ice crystals formed by direct deposition of water vapor from the air onto an object.

ICE-SENSITIVE STRUCTURES. Structures for which the effect of an atmospheric icing load governs the design of part or all of the structure. This includes, but is not limited to, lattice structures, guyed masts, overhead lines, light suspension and cable-stayed bridges, aerial cable systems (e.g., for ski lifts and logging operations), amusement rides, open catwalks and platforms, flagpoles, and signs.

IN-CLOUD ICING. Occurs when supercooled cloud or fog droplets carried by the wind freeze on impact with objects. In-cloud icing usually forms rime but may also form glaze.

RIME. White or opaque ice with entrapped air.

SNOW. Snow that adheres to objects by some combination of capillary forces, freezing, and sintering.

SECTION 10.3
SYMBOLS AND NOTATION

A_s = surface area of one side of a flat plate or the projected area of complex shapes

A_i = cross-sectional area of ice

D = diameter of a circular structure or member as defined in Section 6, in ft (m)

D_c = diameter of the cylinder circumscribing an object

f_z = factor to account for the increase in ice thickness with height

I_i = importance factor and multiplier on ice thickness based on the structure category as defined in Section 1

I_w = importance factor and multiplier on the concurrent wind pressure based on the structure category as defined in Section 1

K_{zt} = topographic factor as defined in Section 6

q_z = velocity pressure evaluated at height z above ground, in lb/ft^2 (N/m^2) as defined in Section 6

r = radius of the maximum cross-section of a dome or radius of a sphere

t = nominal ice thickness due to freezing rain at a height of 33 ft (10 m) from Figure 10-2, 10-3, or 10-4 in in. (mm)

t_d = design ice thickness in in. (mm) from Eq. 10-5

V_c = concurrent wind speed mph (m/s) from Figures 10-2, 10-3, and 10-4

V_i = volume of ice

z = height above ground in ft (m)

$\in$ = solidity ratio as defined in Section 6

SECTION 10.4
ICE LOADS DUE TO FREEZING RAIN

10.4.1 Ice Weight. The ice load shall be determined using the weight of glaze ice formed on all exposed surfaces of structural members, guys, components, appurtenances, and cable systems. On structural shapes, prismatic members, and other similar shapes, the cross-sectional area of ice shall be determined by:

$$A_i = \pi t_d (D_c + t_d) \qquad \textbf{(Eq. 10-1)}$$

D_c is shown for a variety of cross-sectional shapes in Figure 10-1.

On flat plates and large three-dimensional objects such as domes and spheres, the volume of ice shall be determined by:

$$V_i = \pi t_d A_s \qquad \textbf{(Eq. 10-2)}$$

For a flat plate A_s shall be the area of one side of the plate, for domes and spheres A_s shall be determined by:

$$A_s = \pi r^2 \qquad \textbf{(Eq. 10-3)}$$

It is acceptable to multiply V_i by 0.8 for vertical plates and 0.6 for horizontal plates.

The ice density shall be not less than 56 pcf (900 kg/m^3).

10.4.2 Nominal Ice Thickness. Figures 10-2, 10-3, and 10-4 show the equivalent uniform radial thicknesses t of ice due to freezing rain at a height of 33 ft (10 m) over the contiguous United States for a 50-year mean recurrence interval. Also shown are concurrent 3-second gust wind speeds. Thicknesses for Alaska and Hawaii, and for ice accretions due to other sources in all regions, shall be obtained from local meteorological studies.

10.4.3 Height Factor. The height factor f_z used to increase the radial thickness of ice for height above ground z shall be determined by:

$$f_z = \left(\frac{z}{33}\right)^{0.10} \quad \text{for} \quad 0 \text{ ft} < z \leq 900 \text{ ft} \quad \textbf{(Eq. 10-4)}$$

$$f_z = 1.4 \text{ for} \quad z > 900 \text{ ft}$$

In SI

$$f_z = \left(\frac{z}{10}\right)^{0.10} \quad \text{for} \quad 0 \text{ m} < z \leq 275 \text{ m}$$

$$f_z = 1.4 \text{ for} \quad z > 275 \text{ m}$$

10.4.4 Importance Factors. Importance factors to be applied to the radial ice thickness and wind pressure according to the structure classifications (as defined in Table 1-1) are listed in Table 10-1. The importance factor multiplier I_i must be on the ice thickness, not the ice weight because the ice weight is not a linear function of thickness.

10.4.5 Topographic Factor. Both the ice thickness and concurrent wind speed for structures on hills, ridges, and escarpments are higher than those on level terrain because of wind speed-up effects. The topographic factor for the concurrent wind pressure is K_{zt} and the topographic factor for ice thickness is $(K_{zt})^{0.35}$, where K_{zt} is obtained from Eq. 6-1.

10.4.6 Design Ice Thickness for Freezing Rain. The design ice thickness t_d shall be calculated from Eq. 10-5.

$$t_d = 2.0 t I_i f_z (K_{zt})^{0.35} \qquad \textbf{(Eq. 10-5)}$$

SECTION 10.5
WIND ON ICE-COVERED STRUCTURES

Ice accreted on structural members, components, and appurtenances increases the projected area of the structure exposed to wind. The projected area shall be increased by adding t_d to all free edges of the projected area. Wind loads on this increased projected area shall be used in the design of ice-sensitive structures. Figures 10-2, 10-3, and 10-4 include 3-second gust wind speeds at 33 ft (10 m)

above grade that are concurrent with the ice loads due to freezing rain. Wind loads shall be calculated in accordance with Section 6 as modified by Sections 10.5.1 to 10.5.4 below.

10.5.1 Wind on Ice-Covered Chimneys, Tanks, and Similar Structures. Force coefficients for structures with square, hexagonal, and octagonal cross-sections shall be as given in Table 6-19. Force coefficients for structures with round cross-sections shall be as given in Table 6-19 for round cross-sections with $D\sqrt{q_z} \leq 2.5$ for all ice thicknesses, wind speeds, and structure diameters.

10.5.2 Wind on Ice-Covered Solid Freestanding Walls and Solid Signs. Force coefficients shall be as given in Table 6-10 based on the dimensions of the wall or sign including ice.

10.5.3 Wind on Ice-Covered Open Signs and Lattice Frameworks. The solidity ratio $\in$ shall be based on the projected area including ice. The force coefficient for the projected area of flat members shall be as given in Table 6-21. The force coefficient for rounded members and for the additional projected area due to ice on both flat and rounded members shall be as given in Table 6-21 for rounded members with $D\sqrt{q_z} \leq 2.5$ for all ice thicknesses, wind speeds and member diameters.

10.5.4 Wind on Ice-Covered Trussed Towers. The solidity ratio $\in$ shall be based on the projected area including ice. The force coefficients shall be as given in Table 6-22. It is acceptable to reduce the force coefficients for the additional projected area due to ice on both round and flat members by the factor for rounded members in Note 3 of Table 6-22.

SECTION 10.6
PARTIAL LOADING

The effects of a partial ice load shall be considered when this condition is critical for the type of structure under consideration. It is permitted to consider this to be a static load.

SECTION 10.7
DESIGN PROCEDURE

1. The nominal ice thickness, t, and the concurrent wind speed, V_c, for the site shall be determined from Figures 10-2, 10-3, 10-4, or a site-specific study.

2. The topographic factor for the site, K_{zt}, shall be determined in accordance with Section 10.4.5.

3. The importance factor, I_i, shall be determined in accordance with Section 10.4.4.

4. The height factor, f_z, shall be determined in accordance with Section 10.4.3 for each design segment of the structure.

5. The design ice thickness, t_d, shall be determined in accordance with Section 10.4.6, Eq. 10-5.

6. The weight of ice shall be calculated for the design ice thickness, t_d in accordance with Section 10.4.1.

7. The velocity pressure q_z for wind speed V_c shall be determined in accordance with Section 6.5.10 using I_w for the importance factor I.

8. The wind force coefficients shall be determined in accordance with Section 10.5.

9. The gust effect factor shall be determined in accordance with Section 6.5.8.

10. The design wind force shall be determined in accordance with Section 6.5.13.

11. The iced structure shall be analyzed for the load combinations in either Section 2.3 or Section 2.4.

SECTION 10.8
REFERENCES

Ref. 10-1 ANSI. (2002). "National Electrical Safety Code." (NESC). *ANSI C2.*

Ref. 10-2 ASCE. (1991). "Guidelines for Electrical Transmission Line Structural Loading." *ASCE 74.*

Ref. 10-3 ANSI/EIA/TIA. (1996). "Structural Standards for Steel Antenna Towers and Antenna Supporting Structures." *EIA/TIA-222.*

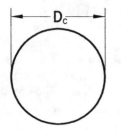

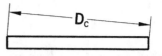

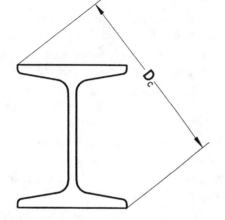

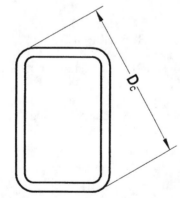

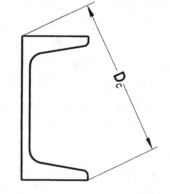

FIGURE 10-1
CHARACTERISTIC DIMENSION "D_C" FOR CALCULATING THE ICE AREA FOR A VARIETY OF CROSS-SECTIONAL SHAPES

This page intentionally left blank.

See Fig. 10-4

.25"

.25"

60 mph

.5"

50 mph

0"

40 mph

.5"

.75"

.25"

40 mph

30 mph

.75"

.5"

.25"

——————— Ice Load Zones
– – – – – Wind Speed Zones

Notes

1. In the Appalachian Mountains, indicated by the gray fill, ice loads may vary significantly over short distances.

2. Ice loads on structures in exposed locations at elevations higher than the surrounding terrain and in valleys and gorges may be higher than the mapped loads.

FIGURE 10-2a
50-YR MEAN RECURRENCE INTERVAL UNIFORM ICE THICKNESSES DUE TO FREEZING RAIN WITH CONCURRENT 3-s GUST WIND SPEEDS: CONTIGUOUS 48 STATES

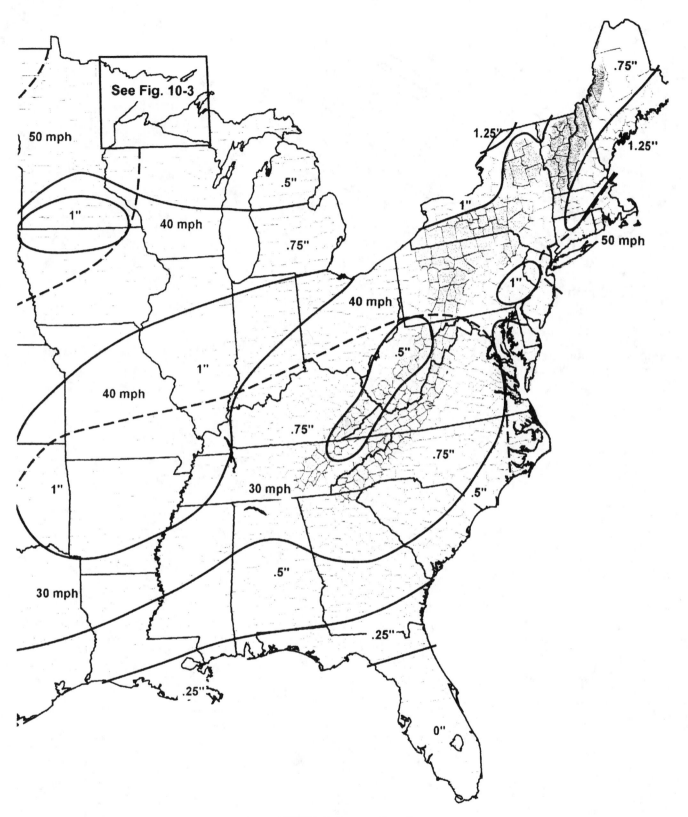

FIGURE 10-2b — continued

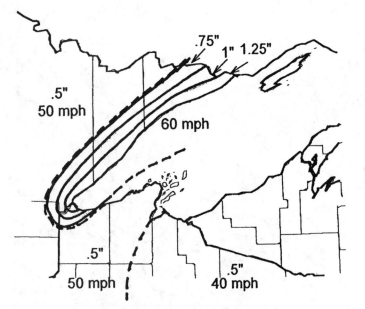

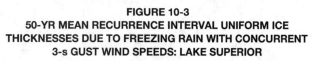

FIGURE 10-3
50-YR MEAN RECURRENCE INTERVAL UNIFORM ICE
THICKNESSES DUE TO FREEZING RAIN WITH CONCURRENT
3-s GUST WIND SPEEDS: LAKE SUPERIOR

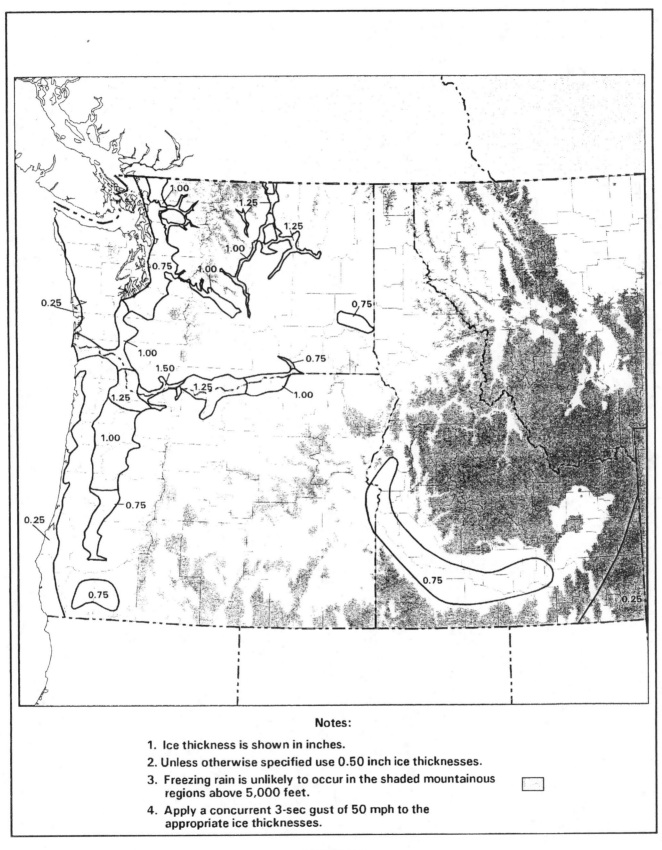

Notes:

1. Ice thickness is shown in inches.
2. Unless otherwise specified use 0.50 inch ice thicknesses.
3. Freezing rain is unlikely to occur in the shaded mountainous regions above 5,000 feet.
4. Apply a concurrent 3-sec gust of 50 mph to the appropriate ice thicknesses.

FIGURE 10-4

50-YR MEAN RECURRENCE INTERVAL UNIFORM ICE THICKNESSES DUE TO FREEZING RAIN WITH CONCURRENT 3-s GUST WIND SPEEDS: PACIFIC NORTHWEST

TABLE 10-1
IMPORTANCE FACTOR I_i AND I_w

Structure Category	I_i (Multiplier on Ice Thickness)	I_w (Multiplier on Concurrent Wind Pressure)
I	0.80	1.0
II	1.00	1.0
III	1.25	1.0
IV	1.25	1.0

APPENDIX A.9
SUPPLEMENTAL PROVISIONS

SECTION A.9.1
PURPOSE

These provisions are not directly related to computation of earthquake loads, but they are deemed essential for satisfactory performance in an earthquake when designing with the loads determined from Section 9, due to the substantial cyclic inelastic strain capacity assumed to exist by the load procedures in Section 9. These supplemental provisions form an integral part of Section 9.

SECTION A.9.3
QUALITY ASSURANCE

This section provides minimum requirements for quality assurance for seismic force-resisting systems and other designated seismic systems. These requirements supplement the testing and inspection requirements contained in the reference standards given in Sections 9.6 through 9.12.

A.9.3.1 Scope. As a minimum, the quality assurance provisions apply to the following:

1. The seismic force-resisting systems in structures assigned to Seismic Design Categories C, D, E, and F.
2. Other designated seismic systems in structures assigned to Seismic Design Categories D, E, and F that are required in Table 9.6.1.7.

> **Exception:** Structures that comply with the following criteria are exempt from the preparation of a quality assurance plan but those structures are not exempt from special inspection(s) or testing requirements:
>
> a. The structure is constructed of light wood framing or light-gauge cold-formed steel framing, S_{DS} does not exceed 0.50 g, the height of the structure does not exceed 35 ft above grade, and the structure meets the requirements in Items c and d below,
> or
> b. The structure is constructed using a reinforced masonry structural system or reinforced concrete structural system, S_{DS} does not exceed 0.50 g, the height of the structure does not exceed 25 ft above grade, and the structure meets the requirements in Items c and d below.

c. The structure is classified as Seismic Use Group I.
d. The structure does not have any of the following plan irregularities as defined in Table 9.5.2.3.2 or any of the following vertical irregularities as defined in Table 9.5.2.3.3:
   1. Torsional irregularity
   2. Extreme torsional irregularity
   3. Nonparallel systems
   4. Stiffness irregularity/soft story
   5. Stiffness irregularity/extreme soft story
   6. Discontinuity in capacity/weak story

The following standards are referenced in the provisions for inspection and testing:

Ref. 9.1.6-1   ANSI. (1998). "Structural Welding Code-Steel." ANSI/AWS D1.1-98.
Ref. 9.1.6-2   ASTM. (1990). "Specification for Straight Beam Ultrasound Examination of Steel Plates." ASTM A435-90.
Ref. 9.1.6-3   ASTM. (1991). "Specification for Straight Beam Ultrasound Examination for Rolled Steel Shapes." ASTM A898-91.

A.9.3.2 Quality Assurance Plan. A quality assurance plan shall be submitted to the authority having jurisdiction.

A.9.3.2.1 Details of Quality Assurance Plan. The quality assurance plan shall specify the designated seismic systems or seismic force-resisting system in accordance with Section A.9.3 that are subject to quality assurance. The registered design professional in responsible charge of the design of a seismic force-resisting system and a designated seismic system shall be responsible for the portion of the quality assurance plan applicable to that system. The special inspections and special tests needed to establish that the construction is in conformance with these provisions shall be included in the portion of the quality assurance plan applicable to the designated seismic system. The quality assurance plan shall include:

a. The seismic force-resisting systems and designated seismic systems in accordance with this chapter that are subject to quality assurance.
b. The special inspections and testing to be provided as required by these provisions and the reference standards in Section 9.

c. The type and frequency of testing.

d. The type and frequency of special inspections.

e. The frequency and distribution of testing and special inspection reports.

f. The structural observations to be performed.

g. The frequency and distribution of structural observation reports.

A.9.3.2.2 Contractor Responsibility. Each contractor responsible for the construction of a seismic force-resisting system, designated seismic system, or component listed in the quality assurance plan shall submit a written contractor's statement of responsibility to the regulatory authority having jurisdiction and to the owner prior to the commencement of work on the system or component. The contractor's statement of responsibility shall contain the following:

1. Acknowledgment of awareness of the special requirements contained in the quality assurance plan.

2. Acknowledgment that control will be exercised to obtain conformance with the design documents approved by the authority having jurisdiction.

3. Procedures for exercising control within the contractor's organization, the method and frequency of reporting, and the distribution of the reports.

4. Identification and qualifications of the person(s) exercising such control and their position(s) in the organization.

A.9.3.3 Special Inspection. The building owner shall employ a special inspector(s) to observe the construction of all designated seismic systems in accordance with the quality assurance plan for the following construction work:

A.9.3.3.1 Foundations. Continuous special inspection is required during driving of piles and placement of concrete in piers or piles. Periodic special inspection is required during construction of drilled piles, piers, and caisson work, the placement of concrete in shallow foundations, and the placement of reinforcing steel.

A.9.3.3.2 Reinforcing Steel.

A.9.3.3.2.1 Periodic special inspection during and on completion of the placement of reinforcing steel in intermediate and special moment frames of concrete and concrete shear walls.

A.9.3.3.2.2 Continuous special inspection during the welding of reinforcing steel resisting flexural and axial forces in intermediate and special moment frames of concrete, in boundary members of concrete shear walls, and welding of shear reinforcement.

A.9.3.3.3 Structural Concrete. Periodic special inspection during and on completion of the placement of concrete in intermediate and special moment frames, and in boundary members of concrete shear walls.

A.9.3.3.4 Prestressed Concrete. Periodic special inspection during the placement and after the completion of placement of prestressing steel and continuous special inspection is required during all stressing and grouting operations and during the placement of concrete.

A.9.3.3.5 Structural Masonry.

A.9.3.3.5.1 Periodic special inspection during the preparation of mortar, the laying of masonry units, and placement of reinforcement; and prior to placement of grout.

A.9.3.3.5.2 Continuous special inspection during welding of reinforcement, grouting, consolidation and reconsolidation, and placement of bent-bar anchors as required by Section A.9.11.

A.9.3.3.6 Structural Steel.

A.9.3.3.6.1 Continuous special inspection is required for all structural welding.

Exception: Periodic special inspection for single-pass fillet or resistance welds and welds loaded to less than 50% of their design strength shall be the minimum requirement, provided the qualifications of the welder and the welding electrodes are inspected at the beginning of the work and all welds are inspected for compliance with the approved construction documents at the completion of welding.

A.9.3.3.6.2 Periodic special inspection is required in accordance with Ref. 9.8-1 or 9.8-2 for installation and tightening of fully tensioned high-strength bolts in slip-critical connections and in connections subject to direct tension. Bolts in connections identified as not being slip-critical or subject to direct tension need not be inspected for bolt tension other than to ensure that the plies of the connected elements have been brought into snug contact.

A.9.3.3.7 Structural Wood.

A.9.3.3.7.1 Continuous special inspection during all field gluing operations of elements of the seismic force-resisting system.

A.9.3.3.7.2 Periodic special inspection is required for nailing, bolting, anchoring, and other fastening of components within the seismic force-resisting system including drag struts, braces, and hold-downs.

A.9.3.3.7.3 Periodic special inspections for nailing and other fastening of wood sheathing used for wood shear walls, shear panels, and diaphragms where the required fastener spacing is 4 in. or less, and which are included in the seismic force-resisting system.

A.9.3.3.8 Cold-Formed Steel Framing.

A.9.3.3.8.1 Periodic special inspection is required during all welding operations of elements of the seismic force-resisting system.

A.9.3.3.8.2 Periodic special inspection is required for screw attachment, bolting, anchoring, and other fastening of components within the seismic force-resisting system including struts, braces, and hold-downs.

A.9.3.3.9 Architectural Components. Special inspection for architectural components shall be as follows:

1. Periodic special inspection during the erection and fastening of exterior cladding, interior and exterior nonbearing walls, and interior and exterior veneer in Seismic Design Categories D, E, and F.

Exceptions:

   a. Architectural components less than 30 ft (9 m) above grade or walking surface.
   b. Cladding and veneer weighing 5 lb/ft^2 (239 N/m^2) or less.
   c. Interior nonbearing walls weighing 15 lbs/ft^2 (718 N/m^2) or less.

2. Periodic special inspection during the anchorage of access floors, suspended ceilings, and storage racks 8 ft (2.5 m) or greater in height in Seismic Design Categories D, E, and F.

3. Periodic special inspection during erection of glass 30 ft (9 m) or more above an adjacent grade or walking surface in glazed curtain walls, glazed storefronts, and interior glazed partitions in Seismic Design Categories D, E, and F.

A.9.3.3.10 Mechanical and Electrical Components. Special inspection for mechanical and electrical components shall be as follows:

1. Periodic special inspection during the anchorage of electrical equipment for emergency or standby power systems in Seismic Design Categories C, D, E, and F.

2. Periodic special inspection during the installation of anchorage of all other electrical equipment in Seismic Design Categories E and F.

3. Periodic special inspection during the installation for flammable, combustible, or highly toxic piping systems and their associated mechanical units in Seismic Design Categories C, D, E, and F.

4. Periodic special inspection during the installation of HVAC ductwork that will contain hazardous materials in Seismic Design Categories C, D, E, and F.

5. Periodic special inspection during the installation of vibration isolation systems when the construction documents indicate a maximum clearance (air gap) between the equipment support frame and restraint less than or equal to 1/4 in.

A.9.3.3.11 Seismic Isolation System. Periodic special inspection is required during the fabrication and installation of isolator units and energy-dissipation devices if used as part of the seismic isolation system.

A.9.3.4 Testing. The special inspector(s) shall be responsible for verifying that the special test requirements are performed by an approved testing agency for the types of work in designated seismic systems listed below.

A.9.3.4.1 Reinforcing and Prestressing Steel. Special testing of reinforcing and prestressing steel shall be as follows:

A.9.3.4.1.1 Examine certified mill test reports for each shipment of reinforcing steel used to resist flexural and axial forces in reinforced concrete intermediate and special moment frames and boundary members of reinforced concrete shear walls or reinforced masonry shear walls and determine conformance with construction documents.

A.9.3.4.1.2 Where ASTM A615 reinforcing steel is used to resist earthquake-induced flexural and axial forces in special moment frames and in wall boundary elements of shear walls in structures of Seismic Design Categories D, E, and F, verify that the requirements of Section 21.2.5.1 of Ref. 9.9-1 have been satisfied.

A.9.3.4.1.3 Where ASTM A615 reinforcing steel is to be welded, verify that chemical tests have been

performed to determine weldability in accordance with Section 3.5.2 of Ref. 9.9-1.

A.9.3.4.2 Structural Concrete. Samples of structural concrete shall be obtained at the project site and tested in accordance with the requirements of Ref. 9.9-1.

A.9.3.4.3 Structural Masonry. Quality assurance testing of structural masonry shall be in accordance with the requirements of Ref. 9.11-1.

A.9.3.4.4 Structural Steel. The testing needed to establish that the construction is in conformance with these provisions shall be included in a quality assurance plan. The minimum testing contained in the quality assurance plan shall be as required in Ref. 9.8-3 and the following requirements:

A.9.3.4.4.1 Base Metal Testing. Base metal thicker than 1.5 in. (38 mm), when subject to through-thickness weld shrinkage strains, shall be ultrasonically tested for discontinuities behind and adjacent to such welds after joint completion. Any material discontinuities shall be accepted or rejected on the basis of ASTM A435, *Specification for Straight Beam Ultrasound Examination of Steel Plates*, or ASTM A898, *Specification for Straight Beam Ultrasound Examination for Rolled Steel Shapes (Level 1 Criteria)*, and criteria as established by the registered design professional(s) in responsible charge and the construction documents.

A.9.3.4.5 Mechanical and Electrical Equipment. As required to ensure compliance with the seismic design provisions herein, the registered design professional in responsible charge shall clearly state the applicable requirements on the construction documents. Each manufacturer of these designated seismic system components shall test or analyze the component and its mounting system or anchorage as required and shall submit a certificate of compliance for review and acceptance by the registered design professional in responsible charge of the design of the designated seismic system and for approval by the authority having jurisdiction. The basis of certification shall be by actual test on a shaking table, by three-dimensional shock tests, by an analytical method using dynamic characteristics and forces, by the use of experience data (i.e., historical data demonstrating acceptable seismic performance), or by more rigorous analysis providing for equivalent safety. The special inspector shall examine the designated seismic system and shall determine whether its anchorages and label conform with the certificate of compliance.

A.9.3.4.6 Seismic-Isolated Structures. For required system tests, see Section 9.13.9

A.9.3.5 Structural Observations. Structural observations shall be provided for those structures included in Seismic Design Categories D, E, and F when one or more of the following conditions exist:

1. The structure is included in Seismic Use Group II or Seismic Use Group III, or
2. The height of the structure is greater than 75 ft above the base, or
3. The structure is in Seismic Design Category E or F and Seismic Use Group I and is greater than two stories in height.

Observed deficiencies shall be reported in writing to the owner and the authority having jurisdiction.

A.9.3.6 Reporting and Compliance Procedures. Each special inspector shall furnish to the authority having jurisdiction, the registered design professional in responsible charge, the owner, the persons preparing the quality assurance plan, and the contractor copies of regular weekly progress reports of his observations, noting therein any uncorrected deficiencies and corrections of previously reported deficiencies. All deficiencies shall be brought to the immediate attention of the contractor for correction. At completion of construction, each special inspector shall submit a final report to the authority having jurisdiction certifying that all inspected work was completed substantially in accordance with approved construction documents. Work not in compliance shall be described in the final report. At completion of construction, the building contractor shall submit a final report to the authority having jurisdiction certifying that all construction work incorporated into the seismic force-resisting system and other designated seismic systems was constructed substantially in accordance with the approved construction documents and applicable workmanship requirements. Work not in compliance shall be described in the final report. The contractor shall correct all deficiencies as required.

SECTION A.9.7
SUPPLEMENTARY FOUNDATION
REQUIREMENTS

A.9.7.4.4 Special Pile Requirements for Category C. All concrete piles and concrete-filled pipe piles shall be connected to the pile cap by embedding the pile reinforcement in the pile cap for a distance equal to the development length as specified in Ref. 9.9-1 as modified by Section A9.9 of this Standard or by the use of field-placed dowels anchored in the concrete pile. For deformed bars, the development length is the full development length for compression or tension, in the case of uplift, without reduction in length for excess area.

Hoops, spirals, and ties shall be terminated with seismic hooks as defined in Section 21.1 of Ref. 9.9-1.

Where required for resistance to uplift forces, anchorage of steel pipe (round HSS sections), concrete-filled steel pipe or H piles to the pile cap shall be made by means other than concrete bond to the bare steel section.

Exception: Anchorage of concrete-filled steel pipe piles is permitted to be accomplished using deformed bars developed into the concrete portion of the pile.

Where a minimum length for reinforcement or the extent of closely spaced confinement reinforcement is specified at the top of the pile, provisions shall be made so that those specified lengths or extents are maintained after pile cutoff.

A.9.7.4.4.1 Uncased Concrete Piles. Reinforcement shall be provided where required by analysis. As a minimum, longitudinal reinforcement ratio of 0.0025 shall be provided for uncased cast-in-place concrete drilled or augered piles, piers, or caissons in the top one-third of the pile length or a minimum length of 10 ft (3 m) below the ground. There shall be a minimum of four longitudinal bars with closed ties (or equivalent spirals) of a minimum of a 3/8 in. (9 mm) diameter provided at 16 longitudinal-bar-diameter maximum spacing. Transverse confinement reinforcing with a maximum spacing of 6 in. (152 mm) or 8 longitudinal-bar-diameters, whichever is less, shall be provided in the pile within three pile diameters of the bottom of the pile cap.

A.9.7.4.4.2 Metal-Cased Concrete Piles. Reinforcement requirements are the same as for uncased concrete piles.

Exception: Spiral welded metal casing of a thickness not less than No. 14 gauge can be considered as providing concrete confinement equivalent to the closed ties or equivalent spirals required in an uncased concrete pile, provided that the metal casing is adequately protected against possible deleterious action due to soil constituents, changing water levels, or other factors indicated by boring records of site conditions.

A.9.7.4.4.3 Concrete-Filled Pipe. Minimum reinforcement 0.01 times the cross-sectional area of the pile concrete shall be provided in the top of the pile with a length equal to two times the required cap embedment anchorage into the pile cap.

A.9.7.4.4.4 Precast Nonprestressed Concrete Piles. A minimum longitudinal steel reinforcement ratio of 0.01 shall be provided for precast nonprestressed concrete piles. The longitudinal reinforcing shall be confined with closed ties or equivalent spirals of a minimum 3/8 in. (10 mm) diameter. Transverse confinement reinforcing shall be provided at a maximum

spacing of eight times the diameter of the smallest longitudinal bar, but not to exceed 6 in. (152 mm), within three pile diameters of the bottom of the pile cap. Outside of the confinement region, closed ties or equivalent spirals shall be provided at a 16 longitudinal-bar-diameter maximum spacing, but not greater than 8 in. (200 mm). Reinforcement shall be full length.

A.9.7.4.4.5 Precast Prestressed Piles. For the upper 20 ft (6 m) of precast prestressed piles, the minimum volumetric ratio of spiral reinforcement shall not be less than 0.007 or the amount required by the following formula:

$$\rho_s = \frac{0.12 f_c'}{f_{yh}} \qquad \textbf{(Eq. A.9.7.4.4.5-1)}$$

where

ρ_s = volumetric ratio (vol. spiral/vol. core)
f_c' = specified compressive strength of concrete, psi (MPa)
f_{yh} = specified yield strength of spiral reinforcement, which shall not be taken greater than 85,000 psi (586 MPa)

A minimum of one-half of the volumetric ratio of spiral reinforcement required by Eq. A.9.7.4.4.5-1 shall be provided for the remaining length of the pile.

A.9.7.5 Special Pile Requirements for Categories D, E, and F.

A.9.7.5.4.1 Uncased Concrete Piles. A minimum longitudinal reinforcement ratio of 0.005 shall be provided for uncased cast-in-place drilled or augered concrete piles, piers, or caissons in the top one-half of the pile length, or a minimum length of 10 ft (3m) below ground, or throughout the flexural length of the pile, whichever length is greatest. The flexural length shall be taken as the length of pile to a point where the concrete section cracking moment multiplied by the resistance factor 0.4 exceeds the required factored moment at that point. There shall be a minimum of four longitudinal bars with transverse confinement reinforcement in the pile in accordance with Sections 21.4.4.1, 21.4.4.2, and 21.4.4.3 of Ref. 9.9-1. Such transverse confinement reinforcement shall extend the full length of the pile in Site Classes E or F, a minimum of seven times the least pile dimension above and below the interfaces of soft to medium stiff clay or liquefiable strata, and three times the least pile dimension below the bottom of the pile cap in Site Classes other than E or F.

In other than Site Classes E or F, it shall be permitted to use a transverse spiral reinforcement

ratio of not less than one-half of that required in Section 21.4.4.1(a) of Ref. 9.9-1 throughout the remainder of the pile length. Tie spacing throughout the remainder of the pile length shall not exceed 12 longitudinal-bar-diameters, one-half the diameter of the section, or 12 in. (305 mm). Ties shall be a minimum of No. 3 bars for up to 20-in.-diameter (500 mm) piles and No. 4 bars for piles of larger diameter.

A.9.7.5.4.2 Metal-Cased Concrete Piles. Reinforcement requirements are the same as for uncased concrete piles.

> **Exception:** Spiral welded metal-casing of a thickness not less than No. 14 gauge can be considered as providing concrete confinement equivalent to the closed ties or equivalent spirals required in an uncased concrete pile, provided that the metal casing is adequately protected against possible deleterious action due to soil constituents, changing water levels, or other factors indicated by boring records of site conditions.

A.9.7.5.4.3 Precast Concrete Piles. Transverse confinement reinforcement consisting of closed ties or equivalent spirals shall be provided in accordance with Sections 21.4.4.1, 21.4.4.2, and 21.4.4.3 of Ref. 9.9-1 for the full length of the pile.

> **Exception:** In other than Site Classes E or F, the specified transverse confinement reinforcement shall be provided within three pile diameters below the bottom of the pile cap, but it shall be permitted to use a transverse reinforcing ratio of not less than one-half of that required in Section 21.4.4.1(a) of Ref. 9.9-1 throughout the remainder of the pile length.

A.9.7.5.4.4 Precast Prestressed Piles. In addition to the requirements for Seismic Design Category C, the following requirements shall be met:

1. Requirements of Ref. 9.9-1, Chapter 21, need not apply.

2. Where the total pile length in the soil is 35 ft (10,668 mm) or less, the lateral transverse reinforcement in the ductile region shall occur through the length of the pile. Where the pile length exceeds 35 ft (10,668 mm), the ductile pile region shall be taken as the greater of 35 ft (10,668 mm) or the distance from the underside of the pile cap to the point of zero curvature plus three times the least pile dimension.

3. In the ductile region, the center-to-center spacing of the spirals or hoop reinforcement shall

not exceed one-fifth of the least pile dimension, six times the diameter of the longitudinal strand, or 8 in. (203 mm), whichever is smaller.

4. Spiral reinforcement shall be spliced by lapping one full turn by welding or by the use of a mechanical connector. Where spiral reinforcement is lap spliced, the ends of the spiral shall terminate in a seismic hook in accordance with Ref. 9.9-1, except that the bend shall be not less than 135 degrees. Welded splices and mechanical connectors shall comply with Section 12.14.3 of Ref. 9.9-1.

5. Where the transverse reinforcement consists of circular spirals, the volumetric ratio of spiral transverse reinforcement in the ductile region shall comply with:

$$\rho_s = 0.25 \frac{f_c'}{f_{yh}} \left[\frac{A_g}{A_{ch}} - 1.0 \right]$$
$$\times \left[0.5 + \frac{1.4P}{f_c' A_g} \right]$$

but not less than

$$\rho_s = 0.12 \frac{f_c'}{f_{yh}} \left[0.5 + \frac{1.4P}{f_c' A_g} \right]$$

and not to exceed $\rho_s = 0.021$
where

ρ_s = volumetric ratio (vol. spiral/vol. core)
$f_c' \leq 6000$ psi (41.4 MPa)
f_{yh} = yield strength of spiral reinforcement ≤ 85 ksi (586 MPa)
A_g = pile cross-sectional area, in.^{2} (mm^2)
A_{ch} = core area defined by spiral outside diameter, in.^{2} (mm^2)
P = axial load on pile resulting from the load combination $1.2D + 0.5L + 1.0E$, pounds (kN)

This required amount of spiral reinforcement is permitted to be obtained by providing an inner and outer spiral.

6. When transverse reinforcement consists of rectangular hoops and cross ties, the total cross-sectional area of lateral transverse reinforcement in the ductile region with spacings, and perpendicular to dimension, h_c, shall conform to:

$$A_{sh} = 0.3 s h_c \frac{f_c'}{f_{yh}} \left[\frac{A_g}{A_{ch}} - 1.0 \right]$$
$$\times \left[0.5 + \frac{1.4P}{f_c' A_g} \right]$$

but not less than

$$A_{sh} = 0.12 s h_c \frac{f'_c}{f_{yh}} \left[0.5 + \frac{1.4 P}{f'_c A_g} \right]$$

where

s = spacing of transverse reinforcement measured along length of pile, in (mm)
h_c = cross-sectional dimension of pile core measured center-to-center of hoop reinforcement, in (mm)
$f_{yh} \leq 70$ ksi (483 MPa)

The hoops and cross ties shall be equivalent to deformed bars not less than No. 3 in size. Rectangular hoop ends shall terminate at a corner with seismic hooks.

Outside of the length of the pile requiring transverse reinforcement, the spiral or hoop reinforcement with a volumetric ratio not less than one-half of that required for transverse confinement reinforcement shall be provided.

A.9.7.5.4.5 Steel Piles. The connection between the pile cap and steel piles or unfilled steel pipe piles shall be designed for a tensile force equal to 10% of the pile compression capacity.

> **Exceptions:** Connection tensile capacity need not exceed the strength required to resist the special seismic loads of Section 9.5.2.7.1. Connections need not be provided where the foundation or supported structure does not rely on the tensile capacity of the piles for stability under the design seismic forces.

SECTION A.9.8
SUPPLEMENTARY PROVISIONS FOR STEEL

A.9.8.1 General. The design, construction, and quality of steel components that resist seismic forces shall conform to the requirements of the references listed in Section 9.8 except as modified by the requirements of this section.

A.9.8.2 Seismic Requirements for Steel Structures. Steel structures and structural elements therein that resist seismic forces shall be designed in accordance with the requirements of Sections A.9.8.3 and A.9.8.4 for the appropriate Seismic Design Category.

A.9.8.3 Seismic Design Categories A, B, and C. Steel structures assigned to Seismic Design Categories A, B, and C shall be of any construction permitted by the references in Section A.9.8.1.1. An R factor as set forth in Table 9.5.2.2 shall be permitted when the structure is designed and detailed in accordance with the requirements of Ref. 9.8-3

for structural steel buildings as modified by this chapter and Section A.9.8.7 for light-framed walls. Systems not detailed in accordance with Ref. 9.8-3 shall use the R factor designated for "Systems not detailed for seismic."

A.9.8.4 Seismic Design Categories D, E, and F. Steel structures assigned to Seismic Design Categories D, E, and F shall be designed and detailed in accordance with Ref. 9.8-3 Part I or Part III or Section A.9.8.6 for light-framed cold-formed steel wall systems.

A.9.8.5 Cold-Formed Steel Seismic Requirements. The design of cold-formed carbon or low-alloy steel to resist seismic loads shall be in accordance with the provisions of Ref. 9.8-4, and the design of cold-formed stainless steel structural members to resist seismic loads shall be in accordance with the provisions of Ref. 9.8-5, except as modified by this section. The references to section and paragraph numbers are to those of the particular specification modified.

A.9.8.5.1 Revise Section A5.1.3 of Ref. 9.8-4 by deleting the reference to earthquake or seismic loads in the sentence permitting the 0.75 factor. Seismic load combinations shall be as determined by ASCE 7.

A.9.8.6 Light-Framed Wall Requirements. Cold-formed steel stud wall systems designed in accordance with Ref. 9.8-4 or 9.8-5 shall, when required by the provisions of Sections A.9.8.3 or A.9.8.4, also comply with the requirements of this section.

A.9.8.6.1 Boundary Members. All boundary members, chords, and collectors shall be designed to transmit the axial force induced by the specified loads of Section 9.

A.9.8.6.2 Connections. Connections of diagonal bracing members, top chord splices, boundary members, and collectors shall have a design strength equal to or greater than the nominal tensile strength of the members being connected or Ω_o times the design seismic forces. The pullout resistance of screws shall not be used to resist seismic forces.

A.9.8.6.3 Braced Bay Members. In stud systems where the lateral forces are resisted by braced frames, the vertical and diagonal members of braced bays shall be anchored such that the bottom tracks are not required to resist tensile forces by bending of the track or track web. Both flanges of studs in a bracing bay shall be braced to prevent lateral torsional buckling. In braced shear walls, the vertical boundary members shall be anchored so the bottom track is not required to resist uplift forces by bending of the track web.

A.9.8.6.4 Diagonal Braces. Provision shall be made for pre-tensioning or other methods of installation of tension-only bracing to prevent loose diagonal straps.

A.9.8.6.5 Shear Walls. Nominal shear values for wall sheathing materials are given in Table A.9.8.6.5. Design shear values shall be determined by multiplying the nominal values therein by a factor of 0.55. In structures over 1 story in height, the assemblies in Table A.9.8.6.5 shall not be used to resist horizontal loads contributed by forces imposed by masonry or concrete construction.

Panel thicknesses shown in Table A.9.8.6.5 shall be considered to be minimums. No panels less than 2-in. wide shall be used. Plywood or oriented strand board structural panels shall be of a type that is manufactured using exterior glue. Framing members, blocking, or strapping shall be provided at the edges of all sheets. Fasteners along the edges in shear panels shall be placed not less than 3/8 in. (9.5 mm) in from panel edges. Screws shall be of sufficient length to ensure penetration into the steel stud by at least two full diameter threads.

The height-to-length ratio of wall systems listed in Table A.9.8.6.5 shall not exceed 2:1.

Perimeter members at openings shall be provided and shall be detailed to distribute the shearing stresses. Wood sheathing shall not be used to splice these members.

Wall studs and track shall have a minimum uncoated base thickness of not less than 0.033 in. (0.84 mm) and shall not have an uncoated base metal thickness greater than 0.048 in. (1.22 mm). Panel end studs and their uplift anchorage shall have the design strength to resist the forces determined by the seismic loads determined by Eqs. 9.5.2.7.1-1 and 9.5.2.7.1-2.

A.9.8.7 Seismic Requirements for Steel Deck Diaphragms. Steel deck diaphragms shall be made from materials conforming to the requirements of Ref. 9.8-4 or 9.8-5. Nominal strengths shall be determined in accordance with approved analytical procedures or with test procedures prepared by a registered design professional experienced in testing of cold-formed steel assemblies and approved by the authority having jurisdiction. Design strengths shall be determined by multiplying the nominal strength by a resistance factor, ϕ equal to 0.60 for mechanically connected diaphragms and equal to 0.50 for welded diaphragms. The steel deck installation for the building, including fasteners, shall comply with the test assembly arrangement. Quality standards established for the nominal strength test shall be the minimum standards required for the steel deck installation, including fasteners.

A.9.8.8 Steel Cables. The design strength of steel cables shall be determined by the provisions of Ref. 9.8-7 except as modified by this section. Ref. 9.8-7, Section 5d, shall be modified by substituting $1.5(T_4)$ when T_4 is the net tension in cable due to dead load, prestress, live load, and seismic load. A load factor of 1.1 shall be applied to the prestress force to be added to the load combination of Section 3.1.2 of Ref. 9.5-7.

SECTION A.9.9
SUPPLEMENTAL PROVISIONS FOR CONCRETE

A.9.9.1 Modifications to Ref. 9.9-1 (ACI 318-02). The text of Ref. 9.9-1 (ACI 318-02) shall be modified as indicated in Sections A.9.9.1.1 through A.9.9.1.6. Italic type is used for text within Sections A.9.9.1.1 through A.9.9.1.6 to indicate provisions which differ from Ref. 9.9-1 (ACI 318-02).

A.9.9.1.1 ACI 318, Section 21.0. Add the following notations to Section 21.0:

h = overall *dimension of member in the direction of action considered*

A.9.9.1.2 ACI 318, Section 21.1. Add the following definition to Section 21.1.

Wall Pier. *A wall segment with a horizontal length-to-thickness ratio of at least 2.5, but not exceeding 6, whose clear height is at least two times its horizontal length*

A.9.9.1.3 ACI 318, Section 21.2.5. Modify Section 21.2.5 by renumbering as Section 21.2.5.1 and adding new Sections 21.2.5.2, 21.2.5.3, and 21.2.5.4 to read as follows:

21.2.5 Reinforcement in Members Resisting Earthquake-Induced Forces.

21.2.5.1 *Except as permitted in Sections 21.2.5.2 and 21.2.5.3,* reinforcement resisting earthquake-induced flexural and axial forces in frame members and in wall boundary elements shall comply with ASTM A706. ASTM A615 Grades 40 and 60 reinforcement shall be permitted in these members if (a) the actual yield strength based on mill tests does not exceed the specified yield strength by more than 18,000 psi (retests shall not exceed this value by more than an additional 3000 psi), and (b) the ratio of the actual ultimate tensile strength to the actual tensile yield strength is not less than 1.25.

21.2.5.2 *Prestressing tendons shall be permitted in flexural members of frames provided the average prestress, f_{pc}, calculated for an area equal to the member's shortest cross-sectional dimension multiplied by the perpendicular dimension shall be the lesser of 700 psi*

(4.83 MPa) or *$f_c'/6$ at locations of nonlinear action where prestressing tendons are used in members of frames.*

21.2.5.3 *Unless the seismic force-resisting frame is qualified for use through structural testing as required by the ACI Provisional Standard ACI T1.1-01, Acceptance Criteria for Moment Frames Based on Structural Testing, for members in which prestressing tendons are used together with mild reinforcement to resist earthquake-induced forces, prestressing tendons shall not provide more than one-quarter of the strength for either positive or negative moments at the nonlinear action location and shall be anchored at the exterior face of the joint or beyond.*

21.2.5.4 *Anchorages for tendons shall be demonstrated to perform satisfactorily for seismic loadings. Anchorage assemblies shall withstand, without failure, a minimum of 50 cycles of loading ranging between 40% and 85% of the minimum specified tensile strength of the prestressing steel.*

A.9.9.1.4 ACI 318, Section 21.7. Modify Section 21.7 by adding a new Section 21.7.10 to read as follows:

21.7.10 Wall Piers and Wall Segments.

21.7.10.1 *Wall piers not designed as a part of a special moment-resisting frame shall have transverse reinforcement designed to satisfy the requirements in Section 21.7.10.2.*

Exceptions:

1. Wall piers that satisfy Section 21.11.
2. Wall piers along a wall line within a story where other shear wall segments provide lateral support to the wall piers, and such segments have a total stiffness of at least six times the sum of the stiffnesses of all the wall piers.

21.7.10.2 Transverse reinforcement shall be designed to resist the shear forces determined from Sections 21.4.5.1 and 21.3.4.2. Where the axial compressive force, including earthquake effects, is less than $A_g f_c'/20$, transverse reinforcement in wall piers is permitted to have standard hooks at each end in lieu of hoops. Spacing of transverse reinforcement shall not exceed 6 in. (152 mm). Transverse reinforcement shall be extended beyond the pier clear height for at least the development length of the largest longitudinal reinforcement in the wall pier.

21.7.10.3 Wall segments with a horizontal length-to-thickness ratio less than 2 1/2, shall be designed as columns.

A.9.9.1.5 ACI 318, Section 21.10. Modify Section 21.10.1.1 to read as follows:

21.8.1 Foundations resisting earthquake-induced forces or transferring earthquake-induced forces between structure and ground shall comply with requirements of Section 21.10 and other applicable code provisions *unless modified by Section 9.7 of ASCE 7-02.*

A.9.9.1.6 ACI 318, Section 21.11. Modify Section 21.11.2.2 to read as follows:

21.11.2.2 Members with factored gravity axial forces exceeding $(A_g f_c'/10)$ shall satisfy Sections 21.4.3.1, 21.4.4.1 (c), 21.4.4.3, and 21.4.5. The maximum longitudinal spacing of ties shall be s_o for the full column height. The spacing s_o shall not be more than six diameters of the smallest longitudinal bar enclosed or 6 in. (152 mm), whichever is smaller.

A.9.9.1.7 ACI 318 Section D.4.2. Modify Appendix Section D.4 by adding a new Exception at the end of Section D.4.2.2 to read as follows:

Exception: If N_b is determined using Eq. D-6a, the concrete breakout strength of D.4.2 shall be considered satisfied by the design procedure of D.5.2 and D.6.2 without the need for testing regardless of anchor bolt diameter and tensile embedment.

A.9.9.2 Classification of Shear Walls. Structural concrete shear walls which resist seismic forces shall be classified in accordance with Sections A.9.9.2.1 through A.9.9.2.4.

A.9.9.2.1 Ordinary Plain Concrete Shear Walls. Ordinary plain concrete shear walls are walls conforming to the requirements of Ref. 9.9-1 for ordinary structural plain concrete walls.

A.9.9.2.2 Detailed Plain Concrete Shear Walls. Detailed plain concrete shear walls are walls conforming to the requirements for ordinary plain concrete shear walls and shall have reinforcement as follows. Vertical reinforcement of at least 0.20 in.^{2} (129 mm^2) in cross-sectional area shall be provided continuously from support to support at each corner, at each side of each opening, and at the ends of walls. The continuous vertical bar required beside an opening is permitted to substitute for one of the two No. 5 bars required by Section 22.6.6.5 of Ref. 9.9-1. Horizontal reinforcement

at least 0.20 in.^{2} (129 mm^2) in cross-sectional area shall be provided:

1. Continuously at structurally connected roof and floor levels and at the top of walls.
2. At the bottom of load-bearing walls or in the top of foundations where doweled to the wall.
3. At a maximum spacing of 120 in. (3048 mm).

Reinforcement at the top and bottom of openings, where used in determining the maximum spacing specified in Item 3 above, shall be continuous in the wall.

A.9.9.2.3 Ordinary Reinforced Concrete Shear Walls.
Ordinary reinforced concrete shear walls are walls conforming to the requirements of Ref. 9.9-1 for ordinary reinforced concrete structural walls.

A.9.9.2.4 Special Reinforced Concrete Shear Walls.
Special reinforced concrete shear walls are walls conforming to the requirements of Ref. 9.9-1 for special reinforced concrete structural walls or special precast structural walls.

A.9.9.3 Seismic Design Category B.
Structures assigned to Seismic Design Category B, as determined in Section 9.4.2, shall conform to the requirements for Seismic Design Category A and to the additional requirements for Seismic Design Category B of this section.

A.9.9.3.1 Ordinary Moment Frames.
In flexural members of ordinary moment frames forming part of the seismic force-resisting system, at least two main flexural reinforcing bars shall be provided continuously top and bottom throughout the beams, through or developed within exterior columns or boundary elements.

Columns of ordinary moment frames having a clear height to maximum plan dimension ratio of five or less shall be designed for shear in accordance with Section 21.12.3 of Ref. 9.9-1.

A.9.9.4 Seismic Design Category C.
Structures assigned to Seismic Design Category C, as determined in Section 9.4.2, shall conform to the requirements for Seismic Design Category B and to the additional requirements for Seismic Design Category C of this section.

A.9.9.4.1 Seismic Force-Resisting Systems.
Moment frames used to resist seismic forces shall be "intermediate moment frames" or "special moment frames." Shear walls used to resist seismic forces shall be ordinary reinforced concrete shear walls, or special reinforced concrete shear walls. Ordinary reinforced concrete shear walls constructed of precast concrete elements shall comply with the additional requirements of Section 21.13 of Ref. 9.9-1 (ACI 318) for intermediate precast concrete structural walls.

A.9.9.4.2 Discontinuous Members.
Columns supporting reactions from discontinuous stiff members such as walls shall be designed for the special load combinations in Section 9.5.2.7 and shall be provided with transverse reinforcement at the spacing s_o as defined in Section 21.12.5.1 of Ref. 9.9-1 over their full height beneath the level at which the discontinuity occurs. This transverse reinforcement shall be extended above and below the column as required in Section 21.4.4.5 of Ref. 9.9-1.

A.9.9.4.3 Anchor Bolts in the Top of Columns.
Anchor bolts set in the top of a column shall be provided with ties that enclose at least four longitudinal column bars. There shall be at least two No. 4, or three No. 3 ties within 5 in. of the top of the column. The ties shall have hooks on each free end that comply with Section 7.1.3(c) of Ref. 9.9-1.

A.9.9.4.4 Plain Concrete.
Structural plain concrete members in structures assigned to Seismic Design Category C shall conform to Ref. 9.9-1 and Sections A.9.9.4.4.1 through A.9.9.4.4.3.

A.9.9.4.4.1 Walls.
Structural plain concrete walls are not permitted in structures assigned to Seismic Design Category C.

Exception: Structural plain concrete basement, foundation, or other walls below the base are permitted in detached one-and two-family dwellings constructed with stud-bearing walls. Such walls shall have reinforcement in accordance with Section 22.6.6.5 of Ref. 9.9-1.

A.9.9.4.4.2 Footings.
Isolated footings of plain concrete supporting pedestals or columns are permitted provided the projection of the footing beyond the face of the supported member does not exceed the footing thickness.

Exception: In detached one- and two-family dwellings, three stories or less in height, the projection of the footing beyond the face of the supported member is permitted to exceed the footing thickness.

Plain concrete footings supporting walls shall be provided with not less than two continuous longitudinal reinforcing bars. Bars shall not be smaller than No. 4 and shall have a total area of not less than 0.002

times the gross cross-sectional area of the footing. For footings that exceed 8 in. (203 mm) in thickness, a minimum of one bar shall be provided at the top and bottom of the footing. For foundation systems consisting of a plain concrete footing and a plain concrete stemwall, a minimum of one bar shall be provided at the top of the stemwall and at the bottom of the footing. Continuity of reinforcement shall be provided at corners and intersections.

Exceptions:

1. In detached one- and two-family dwellings, three stories or less in height and constructed with stud-bearing walls, plain concrete footings supporting walls are permitted without longitudinal reinforcement.
2. Where a slab-on-ground is cast monolithically with the footing, one No. 5 bar is permitted to be located at either the top or bottom of the footing.

A.9.9.4.4.3 Pedestals. Plain concrete pedestals shall not be used to resist lateral seismic forces.

A.9.9.5 Seismic Design Categories D, E, and F. Structures assigned to Seismic Design Categories D, E, or F, as determined in Section 9.4.2, shall conform to the requirements for Seismic Design Category C and to the additional requirements of this section.

A.9.9.5.1 Seismic Force-Resisting Systems. Moment frames used to resist seismic forces shall be special moment frames. Shear walls used to resist seismic forces shall be special reinforced concrete shear walls.

A.9.9.5.2 Frame Members Not Proportioned to Resist Forces Induced by Earthquake Motions. Frame components assumed not to contribute to lateral force resistance shall conform to Section 21.11 of Ref. 9.9-1.

SECTION A.9.11
SUPPLEMENTARY PROVISIONS FOR
MASONRY

A.9.11.1 For the purposes of design of masonry structures using the earthquake loads given in ASCE 7, several amendments to the reference standard are necessary.

A.9.11.2 The references to "Seismic Performance Category" in Section 1.13 and elsewhere of the reference standard shall be replaced by "Seismic Design Category," and the rules given for Category E also apply to Category F.

A.9.11.3 The anchorage forces given in Section 1.13.3.2 of the reference standard shall not be interpreted to replace the anchorage forces given in Section 9 of this standard.

A.9.11.4 To qualify for the R factors given in Section 9 of this standard, the requirements of the reference standard shall be satisfied and amended as follows:

Ordinary and detailed plain masonry shear walls shall be designed according to Sections 2.1 and 2.2 of the reference standard. Detailed plain masonry shear walls shall be reinforced as a minimum as required in Section 1.13.5.3.3 of the reference standard.

Ordinary, intermediate, and special reinforced masonry shear walls shall be designed according to Sections 2.1 and 2.3 of the reference standard. Ordinary and intermediate reinforced masonry shear walls shall be reinforced as a minimum as required in Section 1.13.5.3.3 of the reference standard. In addition, intermediate reinforced masonry shear walls shall have vertical reinforcing bars spaced no farther apart than 48 in. Special reinforced masonry shear walls shall be reinforced as a minimum as required in Sections 1.13.6.3 and 1.13.6.4 of the reference standard.

Special reinforced masonry shear walls shall not have a ratio of reinforcement, ρ, that exceeds that given by either Method A or B below:

A.9.11.4.1 Method A. Method A is permitted to be used where the story drift does not exceed $0.010h_{sx}$ as given in Table 9.5.2.8 and if the extreme compressive fiber strains are less than 0.0035 in./in. (mm/mm) for clay masonry and 0.0025 in./in./ (mm/mm) for concrete masonry.

1. When walls are subjected to in-plane forces, and for columns and beams, the critical strain condition corresponds to a strain in the extreme tension reinforcement equal to five times the strain associated with the reinforcement yield stress, f_y.
2. When walls are subjected to out-of-plane forces, the critical strain condition corresponds to a strain in the reinforcement equal to 1.3 times the strain associated with reinforcement yield stress, f_y.

The strain at the extreme compression fiber shall be assumed to be 0.0035 in./in. (mm/mm) for clay masonry and 0.0025 in./in. (mm/mm) for concrete masonry.

The calculation of the maximum reinforcement ratio shall include factored gravity axial loads. The stress in the tension reinforcement shall be assumed to be $1.25f_y$. Tension in the masonry shall be neglected. The strength of the compressive zone shall be calculated as 80% of the area of the compressive zone. Stress in reinforcement in the compression zones shall be based on a linear strain distribution.

A.9.11.4.2 Method B. Method B is permitted to be used where story drift does not exceed $0.013h_{sx}$ as given in Table 9.5.2.8.

1. Boundary members shall be provided at the boundaries of shear walls when the compressive strains in the wall exceed 0.002. The strain shall be determined using factored forces and R equal to 1.5.

2. The minimum length of the boundary member shall be three times the thickness of the wall, but shall include all areas where the compressive strain per Section A.9.11.4.2 item 1, is greater than 0.002.

3. Lateral reinforcement shall be provided for the boundary elements. The lateral reinforcement shall be a minimum of No. 3 closed ties at a maximum spacing of 8 in. (203 mm) on center within the grouted core, or equivalent approved confinement, to develop an ultimate compressive strain of at least 0.006.

4. The maximum longitudinal reinforcement ratio shall not exceed $0.15 f'_m / f_y$.

A.9.11.5 Where allowable stress design is used for load combinations including earthquake, load combinations 3 and 5 of Section 2.4.1 of this standard shall replace combinations (c) and (e) of Section 2.1.1.1.1 of the reference standard. If the increase in stress given in 2.1.1.1.3 of the reference standard is used, the restriction on load reduction in Section 2.4.3 of this standard shall be observed.

TABLE A.9.8.6.5
NOMINAL SHEAR VALUES FOR SEISMIC FORCES FOR SHEAR WALLS FRAMED WITH
COLD-FORMED STEEL STUDS (IN LBS/FT)[a,b]

Assembly Description	Fastener Spacing at Panel Edges[c] (in.)				Framing Spacing (in. o.c.)
	6	4	3	2	
15/32 in. rated structural I sheathing (4-ply) plywood one side[d]	780	990	1465	1625	24
7/16 in. oriented strand board one side[d]	700	915	1275	1625	24

Note: For fastener and framing spacing, multiply inches by 25.4 to obtain mm.

[a] Nominal shear values shall be multiplied by the appropriate strength reduction factor to determine design strength as set forth in Section A.9.8.6.5.

[b] Studs shall be a minimum 1 5/8 in. × 3 1/2 in. with a 3/8 in. return lip. Track shall be a minimum 1 1/4 in. × 3 1/2 in. Both studs and track shall have a minimum uncoated base metal thickness of 0.033 in. and shall be ASTM A653 SS Grade 33, ASTM A792 SS Grade 33, or ASTM A875 SS Grade 33. Framing screws shall be No. 8 × 5/8 in. wafer head self-drilling. Plywood and OSB screws shall be a minimum No. 8 × 1 in. bugle head. Where horizontal straps are used to provide blocking, they shall be a minimum 1 1/2 in. wide and of the same material and thickness as the stud and track.

[c] Screws in the field of the panel shall be installed 12 in. on center unless otherwise shown.

[d] Both flanges of the studs shall be braced in accordance with Section A.9.8.6.3.

APPENDIX B.0
SERVICEABILITY CONSIDERATIONS

This Appendix is not a mandatory part of the standard, but provides guidance for design for serviceability in order to maintain the function of a building and the comfort of its occupants during normal usage. Serviceability limits (e.g., maximum static deformations, accelerations, and so on) shall be chosen with due regard to the intended function of the structure.

Serviceability shall be checked using appropriate loads for the limit state being considered.

SECTION B.1
DEFLECTION, VIBRATION, AND DRIFT

B.1.1 Vertical Deflections. Deformations of floor and roof members and systems due to service loads shall not impair the serviceability of the structure.

B.1.2 Drift of Walls and Frames. Lateral deflection or drift of structures and deformation of horizontal diaphragms and bracing systems due to wind effects shall not impair the serviceability of the structure.

B.1.3 Vibrations. Floor systems supporting large open areas free of partitions or other sources of damping, where vibration due to pedestrian traffic might be objectionable, shall be designed with due regard for such vibration.

Mechanical equipment that can produce objectionable vibrations in any portion of an inhabited structure shall be isolated to minimize the transmission of such vibrations to the structure.

Building structural systems shall be designed so that wind-induced vibrations do not cause occupant discomfort or damage to the building, its appurtenances, or its contents.

SECTION B.2
DESIGN FOR LONG-TERM DEFLECTION

Where required for acceptable building performance, members and systems shall be designed to accommodate long-term irreversible deflections under sustained load.

SECTION B.3
CAMBER

Special camber requirements that are necessary to bring a loaded member into proper relations with the work of other trades shall be set forth in the design documents.

Beams detailed without specified camber shall be positioned during erection so that any minor camber is upward. If camber involves the erection of any member under preload, this shall be noted in the design documents.

SECTION B.4
EXPANSION AND CONTRACTION

Dimensional changes in a structure and its elements due to variations in temperature, relative humidity, or other effects shall not impair the serviceability of the structure.

Provision shall be made either to control crack widths or to limit cracking by providing relief joints.

SECTION B.5
DURABILITY

Buildings and other structures shall be designed to tolerate long-term environmental effects or shall be protected against such effects.

COMMENTARY TO AMERICAN SOCIETY OF CIVIL ENGINEERS STANDARD 7-02

(This Commentary is not a part of the ASCE Standard *Minimum Design Loads for Buildings and Other Structures*. It is included for information purposes.)

This Commentary consists of explanatory and supplementary material designed to assist local building code committees and regulatory authorities in applying the recommended requirements. In some cases, it will be necessary to adjust specific values in the standard to local conditions; in others, a considerable amount of detailed information is needed to put the provisions into effect. This Commentary provides a place for supplying material that can be used in these situations and is intended to create a better understanding of the recommended requirements through brief explanations of the reasoning employed in arriving at them.

The sections of the Commentary are numbered to correspond to the sections of the standard to which they refer. Since it is not necessary to have supplementary material for every section in the standard, there are gaps in the numbering in the Commentary.

SECTION C1.0
GENERAL

SECTION C1.1
SCOPE

The minimum load requirements contained in this standard are derived from research and service performance of buildings and other structures. The user of this standard, however, must exercise judgment when applying the requirements to "other structures." Loads for some structures other than buildings may be found in Sections 3 through 10 of this Standard and additional guidance may be found in the Commentary.

Both loads and load combinations are set forth in this document with the intent that they be used together. If one were to use loads from some other source with the load combinations set forth herein, or vice versa, the reliability of the resulting design may be affected.

Earthquake loads contained herein are developed for structures that possess certain qualities of ductility and postelastic energy-dissipation capability. For this reason, provisions for design, detailing, and construction are provided in Appendix A. In some cases, these provisions modify or add to provisions contained in design specifications.

SECTION C1.3
BASIC REQUIREMENTS

C1.3.1 Strength. Buildings and other structures must satisfy strength limit states in which members are proportioned to carry the design loads safely to resist buckling, yielding, fracture, and so on. It is expected that other standards produced under consensus procedures and intended for use in connection with building code requirements will contain recommendations for resistance factors for strength design methods or allowable stresses (or safety factors) for allowable stress design methods.

C1.3.2 Serviceability. In addition to strength limit states, buildings and other structures must also satisfy serviceability limit states that define functional performance and behavior under load and include such items as deflection and vibration. In the United States, strength limit states have traditionally been specified in building codes because they control the safety of the structure. Serviceability limit states, on the other hand, are usually noncatastrophic, define a level of quality of the structure or element, and are a matter of judgment as to their application. Serviceability limit states involve the perceptions and expectations of the owner or user and are a contractual matter between the owner or user and the designer and builder. It is for

these reasons, and because the benefits themselves are often subjective and difficult to define or quantify, that serviceability limit states for the most part are not included within the three model U.S. Building Codes. The fact that serviceability limit states are usually not codified should not diminish their importance. Exceeding a serviceability limit state in a building or other structure usually means that its function is disrupted or impaired because of local minor damage or deterioration or because of occupant discomfort or annoyance.

C1.3.3 Self-Straining Forces. Constrained structures that experience dimensional changes develop self-straining forces. Examples include moments in rigid frames that undergo differential foundation settlements and shears in bearing wall that support concrete slabs that shrink. Unless provisions are made for self-straining forces, stresses in structural elements, either alone or in combination with stresses from external loads, can be high enough to cause structural distress.

In many cases, the magnitude of self-straining forces can be anticipated by analyses of expected shrinkage, temperature fluctuations, foundation movement, and so on. However, it is not always practical to calculate the magnitude of self-straining forces. Designers often provide for self-straining forces by specifying relief joints, suitable framing systems, or other details to minimize the effects of self-straining forces.

This section of this Standard is not intended to require the designer to provide for self-straining forces that cannot be anticipated during design. An example is settlement resulting from future adjacent excavation.

SECTION C1.4
GENERAL STRUCTURAL INTEGRITY

Through accident, misuse, or sabotage, properly designed structures may be subject to conditions that could lead to general or local collapse. Except for specially designed protective systems, it is usually impractical for a structure to be designed to resist general collapse caused by gross misuse of a large part of the system or severe abnormal loads acting directly on a large portion of it. However, precautions can be taken in the design of structures to limit the effects of local collapse and to prevent or minimize progressive collapse. Progressive collapse is defined as the spread of an initial local failure from element to element, eventually resulting in the collapse of an entire structure or a disproportionately large part of it.

Some authors have defined resistance to progressive collapse to be the ability of a structure to accommodate, with only local failure, the notional removal of any single structural member. Aside from the possibility of further damage that uncontrolled debris from the failed member may cause, it appears prudent to consider whether the abnormal event will fail only a single member.

Since accidents, misuse, and sabotage are normally unforeseeable events, they cannot be defined precisely. Likewise, general structural integrity is a quality that cannot be stated in simple terms. It is the purpose of Section 1.4 and this Commentary to direct attention to the problem of local collapse, present guidelines for handling it that will aid the design engineer, and promote consistency of treatment in all types of structures and in all construction materials. ASCE does not intend, at this time, for this standard to establish specific events to be considered during design, or for this standard to provide specific design criteria to minimize the risk of progressive collapse.

Accidents, Misuse, Sabotage, and Their Consequences.
In addition to unintentional or willful misuse, some of the incidents that may cause local collapse are [C1-1][1]: explosions due to ignition of gas or industrial liquids; boiler failures; vehicle impact; impact of falling objects; effects of adjacent excavations; gross construction errors; very high winds such as tornadoes; and sabotage. Generally, such abnormal events would not be a part of normal design considerations. The distinction between general collapse and limited local collapse can best be made by example as follows.

General Collapse.
The immediate, deliberate demolition of an entire structure by phased explosives is an obvious instance of general collapse. Also, the failure of one column in a 1-, 2-, 3-, or possibly even 4-column structure could precipitate general collapse, because the local failed column is a significant part of the total structural system at that level. Similarly, the failure of a major bearing element in the bottom story of a 2- or 3-story structure might cause general collapse of the whole structure. Such collapses are beyond the scope of the provisions discussed herein. There have been numerous instances of general collapse that have occurred as the result of such events as bombin landslides, and floods.

Limited Local Collapse.
An example of limited local collapse would be the containment of damage to adjacent bays and stories following the destruction of one or two neighboring columns in a multibay structure. The restriction of damage to portions of two or three stories of a higher

structure following the failure of a section of bearing wall in one story is another example.

Examples of General Collapse.

Ronan Point. A prominent case of local collapse that progressed to a disproportionate part of the whole building (and is thus an example of the type of failure of concern here) was the Ronan Point disaster, which brought the attention of the profession to the matter of general structural integrity in buildings. Ronan Point was a 22-story apartment building of large, precast concrete, load-bearing panels in Canning Town, England. In March 1968, a gas explosion in an 18th-story apartment blew out a living room wall. The loss of the wall led to the collapse of the whole corner of the building. The apartments above the 18th story, suddenly losing support from below and being insufficiently tied and reinforced, collapsed one after the other. The falling debris ruptured successive floors and walls below the 18th story, and the failure progressed to the ground. Better continuity and ductility might have reduced the amount of damage at Ronan Point. Another example is the failure of a 1-story parking garage reported in [C1-2]. Collapse of one transverse frame under a concentration of snow led to the later progressive collapse of the whole roof, which was supported by 20 transverse frames of the same type. Similar progressive collapses are mentioned in [C1-3].

Alfred P. Murrah Federal Building. [C1-4, C1-5, C1-6, C1-7, C1-8] On April 19, 1995, a truck containing approximately 4000 pounds of fertilizer-based explosive (ANFO) was parked near the sidewalk next to the 9-story reinforced concrete office building. The side facing the blast had corner columns and four other perimeter columns. The blast shock wave disintegrated one of the 20 in. × 36 in. perimeter columns and caused brittle failures of two others. The transfer girder at the third level above these columns failed, and the upper-story floors collapsed in a progressive fashion. Approximately 70% of the building experienced dramatic collapse, and 168 people died, many of them as a direct result of progressive collapse. There might have been less damage had this structure not relied on transfer girders for support of upper floors, if there had been better detailing for ductility and greater redundancy, and if there had been better resistance for uplift loads on floor slabs.

Following are a number of factors that contribute to the risk of damage propagation in modern structures [C1-9]:

1. There is an apparent lack of general awareness among engineers that structural integrity against collapse is important enough to be regularly considered in design.

2. In order to have more flexibility in floor plans and to keep costs down, interior walls and partitions are

---

[1] Numbers in brackets refer to references listed at the ends of the major sections in which they appear (i.e., Section 1, Section 2, and so on).

often non–load-bearing and hence may be unable to assist in containing damage.

3. In attempting to achieve economy in structure through greater speed of erection and less site labor, systems may be built with minimum continuity, ties between elements, and joint rigidity.

4. Unreinforced or lightly reinforced load-bearing walls in multistory structures may also have inadequate continuity, ties, and joint rigidity.

5. In roof trusses and arches, there may not be sufficient strength to carry the extra loads or sufficient diaphragm action to maintain lateral stability of the adjacent members if one collapses.

6. In eliminating excessively large safety factors, code changes over the past several decades have reduced the large margin of safety inherent in many older structures. The use of higher-strength materials permitting more slender sections compounds the problem in that modern structures may be more flexible and sensitive to load variations and, in addition, may be more sensitive to construction errors.

Experience has demonstrated that the principle of taking precautions in design to limit the effects of local collapse is realistic and can be satisfied economically. From a public safety viewpoint, it is reasonable to expect all multistory structures to possess general structural integrity comparable to that of properly designed, conventional framed structures [C1-9, C1-10].

Design Alternatives. There are a number of ways to obtain resistance to progressive collapse. In [C1-11], a distinction is made between direct and indirect design, and the following approaches are defined:

Direct Design. Explicit consideration of resistance to progressive collapse during the design process through either:

Alternate Path Method. a method that allows local failure to occur, but seeks to provide alternate load paths so that the damage is absorbed and major collapse is averted, or

Specific Local Resistance Method. A method that seeks to provide sufficient strength to resist failure from accidents or misuse.

Indirect Design. Implicit consideration of resistance to progressive collapse during the design process through the provision of minimum levels of strength, continuity, and ductility.

The general structural integrity of a structure may be tested by analysis to ascertain whether alternate paths around hypothetically collapsed regions exist. Alternatively, alternate path studies may be used as guides for developing rules for the minimum levels of continuity and ductility needed to apply the indirect design approach to enhance general structural integrity. Specific local resistance may be provided in regions of high risk, since it may be necessary for some element to have sufficient strength to resist abnormal loads in order for the structure as a whole to develop alternate paths. Specific suggestions for the implementation of each of the defined methods are contained in [C1-11].

Guidelines for the Provision of General Structural Integrity. Generally, connections between structural components should be ductile and have a capacity for relatively large deformations and energy absorption under the effect of abnormal conditions. These criteria are met in many different ways, depending on the structural system used. Details that are appropriate for resistance to moderate wind loads and seismic loads often provide sufficient ductility. In 1999, ASCE issued a state of practice report that is a good introduction to the complex field of blast-resistant design [C1-12].

Work with large precast panel structures [C1-13, C1-14, C1-15] provides an example of how to cope with the problem of general structural integrity in a building system that is inherently discontinuous. The provision of ties combined with careful detailing of connections can overcome difficulties associated with such a system. The same kind of methodology and design philosophy can be applied to other systems [C1-16]. The ACI Building Code Requirements for Structural Concrete [C1-17] includes such requirements in Section 7.13.

There are a number of ways of designing for the required integrity to carry loads around severely damaged walls, trusses, beams, columns, and floors. A few examples of design concepts and details follow:

1. Good plan layout. An important factor in achieving integrity is the proper plan layout of walls and columns. In bearing-wall structures there should be an arrangement of interior longitudinal walls to support and reduce the span of long sections of crosswall, thus enhancing the stability of individual walls and of the structures as a whole. In the case of local failure, this will also decrease the length of wall likely to be affected.

2. Provide an integrated system of ties among the principal elements of the structural system. These ties may be designed specifically as components of secondary load-carrying systems, which often must sustain very large deformations during catastrophic events.

3. Returns on walls. Returns on interior and exterior walls will make them more stable.

4. Changing directions of span of floor slab. Where a floor slab is reinforced in order that it can, with a low safety factor, span in another direction if a load-bearing wall is removed, the collapse of the slab will be prevented and the debris loading of other parts of the structure will be minimized. Often, shrinkage and temperature steel will be enough to enable the slab to span in a new direction.

5. Load-bearing interior partitions. The interior walls must be capable of carrying enough load to achieve the change of span direction in the floor slabs.

6. Catenary action of floor slab. Where the slab cannot change span direction, the span will increase if an intermediate supporting wall is removed. In this case, if there is enough reinforcement throughout the slab and enough continuity and restraint, the slab may be capable of carrying the loads by catenary action, though very large deflections will result.

7. Beam action of walls. Walls may be assumed to be capable of spanning an opening if sufficient tying steel at the top and bottom of the walls allows them to act as the web of a beam with the slabs above and below acting as flanges (see [C1-13]).

8. Redundant structural systems. Provide a secondary load path (e.g., an upper-level truss or transfer girder system that allows the lower floors of a multistory building to hang from the upper floors in an emergency) that allows framing to survive removal of key support elements.

9. Ductile detailing. Avoid low ductility detailing in elements that might be subject to dynamic loads or very large distortions during localized failures (e.g., consider the implications of shear failures in beams or supported slabs under the influence of building weights falling from above).

10. Provide additional reinforcement to resist blast and load reversal when blast loads are considered in design [C1-18].

11. Consider the use of compartmentalized construction in combination with special moment-resisting frames [C1-19] in the design of new buildings when considering blast protection.

While not directly adding structural integrity for the prevention of progressive collapse, the use of special, nonfrangible glass for fenestration can greatly reduce risk to occupants during exterior blasts [C1-18]. To the extent that nonfrangible glass isolates a building's interior from blast shock waves, it can also reduce damage to interior framing elements (e.g., supported floor slabs could be made to be less likely to fail due to uplift forces) for exterior blasts.

SECTION C1.5
CLASSIFICATION OF BUILDINGS AND OTHER STRUCTURES

C1.5.1 Nature of Occupancy. The categories in Table 1-1 are used to relate the criteria for maximum environmental loads or distortions specified in this standard to the consequence of the loads being exceeded for the structure and its occupants. The category numbering is unchanged from that in the previous edition of the standard (ASCE 7-98). Classification continues to reflect a progression of the anticipated seriousness of the consequence of failure from lowest hazard to human life (Category I) to highest (Category IV).

In Sections 6, 7, 9, and 10, importance factors are presented for the four categories identified. The specific importance factors differ according to the statistical characteristics of the environmental loads and the manner in which the structure responds to the loads. The principle of requiring more stringent loading criteria for situations in which the consequence of failure may be severe has been recognized in previous versions of this standard by the specification of mean recurrence interval maps for wind speed and ground snow load.

This section now recognizes that there may be situations when it is acceptable to assign multiple categories to a structure based on use and the type of load condition being evaluated. For instance, there are circumstances when a structure should appropriately be designed for wind loads with importance factors greater than one, but would be penalized unnecessarily if designed for seismic loads with importance factors greater than one. An example would be a hurricane shelter in a low seismic area. The structure would be classified in Category IV for wind design and in Category II for seismic design.

Category I contains buildings and other structures that represent a low hazard to human life in the event of failure either because they have a small number of occupants or have a limited period of exposure to extreme environmental loadings. Examples of agricultural structures that fall under Category I are farm storage structures used exclusively for the storage of farm machinery and equipment, grain bins, corn cribs, and general purpose barns for the temporary feeding of livestock [C1-20]. Category II contains all occupancies other than those in Categories I, III, and IV and are sometimes referred to as "ordinary" for the purpose of risk exposure. Category III contains those buildings and other structures that have large numbers of occupants, are designed for public assembly, or are designed for occupants who are restrained or otherwise restricted from movement, or are designed for evacuation. Buildings and other structures in Category III, therefore, represent a substantial hazard to human life in the event of failure.

Category IV contains buildings and other structures that are designated as essential facilities and are intended to remain operational in the event of extreme environmental loadings. Such occupancies include, but are not limited to,

hospitals and fire, rescue, and other emergency response facilities. Ancillary structures required for the operation of Category IV facilities during an emergency are also included in this category. In addition to essential facilities, buildings and other structures containing extremely hazardous materials have been added to Category IV to recognize the potential devastating effect a release of extremely hazardous materials may have on a population.

C1.5.2 Hazardous Materials and Extremely Hazardous Materials.
A common method of categorizing structures storing hazardous materials or extremely hazardous materials is by the use of a table of exempt amounts of these materials [C1-21, C1-22]. These and other references are sources of guidance on the identification of materials of these general classifications. A drawback to the use of tables of exempt amounts is the fact that the method cannot handle the interaction of multiple materials. Two materials may be exempt because alone neither poses a risk to the public, but may be deadly in a combined release. Therefore, an alternate and superior method of evaluating the risk to the public of a release of a material is by a hazard assessment as part of an overall risk management plan (RMP).

Buildings and other structures containing hazardous materials or extremely hazardous materials may be classified as Category II structures if it can be demonstrated that the risk to the public from a release of these materials is minimal. Companies that operate industrial facilities typically perform HAZOP (Hazard and Operability) studies, conduct quantitative risk assessments, and develop risk management and emergency response plans. Federal regulations and local laws mandate many of these studies and plans [C1-23]. Additionally, many industrial facilities are located in areas remote from the public and have restricted access, which further reduces the risk to the public.

The risk management plan generally deals with mitigating the risk to the general public. Risk to individuals outside the facility storing hazardous or extremely hazardous materials is emphasized because plant personnel are not placed at as high a risk as the general public due to the plant personnel's training in the handling of the hazardous or extremely hazardous materials and due to the safety procedures implemented inside the facilities. When these elements (trained personnel and safety procedures) are not present in a facility, then the risk management plan must mitigate the risk to the plant personnel in the same manner as it mitigates the risk to the general public.

As the result of the prevention program portion of a risk management plan, buildings and other structures normally falling into Category III may be classified into Category II if means (e.g., secondary containment) are provided to contain the hazardous materials or extremely hazardous materials in the case of a release. To qualify, secondary containment systems must be designed, installed, and operated to prevent migration of harmful quantities of toxic or explosive substances out of the system to the air,

soil, ground water, or surface water at any time during the use of the structure. This requirement is not to be construed as requiring a secondary containment system to prevent a release of any toxic or explosive substance into the air. By recognizing that secondary containment shall not allow releases of "harmful" quantities of contaminants, this standard acknowledges that there are substances that might contaminate ground water but do not produce a sufficient concentration of hazardous materials or extremely hazardous materials during a vapor release to constitute a health or safety risk to the public. Because it represents the "last line of defense," secondary containment does not qualify for the reduced classification.

If the beneficial effect of secondary containment can be negated by external forces, such as the overtopping of dike walls by floodwaters or the loss of liquid containment of an earthen dike due to excessive ground displacement during a seismic event, then the buildings or other structures in question may not be classified into Category II. If the secondary containment is to contain a flammable substance, then implementation of a program of emergency response and preparedness combined with an appropriate fire suppression system would be a prudent action associated with a Category II classification. In many jurisdictions, such actions are required by local fire codes.

Also as the result of the prevention program portion of a risk management plan, buildings and other structures containing hazardous materials or extremely hazardous materials also could be classified as Category II for hurricane wind loads when mandatory procedures are used to reduce the risk of release of hazardous substances during and immediately after these predictable extreme loadings. Examples of such procedures include draining hazardous fluids from a tank when a hurricane is predicted or, conversely, filling a tank with fluid to increase its buckling and overturning resistance. As appropriate to minimize the risk of damage to structures containing hazardous materials or extremely hazardous materials, mandatory procedures necessary for the Category II classification should include preventative measures such as the removal of objects that might become air-borne missiles in the vicinity of the structure.

SECTION C1.7
LOAD TESTS

No specific method of test for completed construction has been given in this standard since it may be found advisable to vary the procedure according to conditions. Some codes require the construction to sustain a superimposed load equal to a stated multiple of the design load without evidence of serious damage. Others specify that the superimposed load shall be equal to a stated multiple of the live load plus a portion of the dead load. Limits are set on maximum deflection under load and after removal of the

load. Recovery of at least three-quarters of the maximum deflection, within 24 hours after the load is removed, is a common requirement [C1-17].

REFERENCES

[C1-1] Leyendecker, E.V., Breen, J.E., Somes, N.F., and Swatta, M. Abnormal loading on buildings and progressive collapse—An annotated bibliography. Washington, D.C.: U.S. Dept. of Commerce, National Bureau of Standards. NBS BSS 67, Jan. 1976.

[C1-2] Granstrom, S. and Carlsson, M. "Byggfurskningen T3: Byggnaders beteende vid overpaverkningar" [The behavior of buildings at excessive loadings]. Stockholm, Sweden: Swedish Institute of Building Research, 1974.

[C1-3] Seltz-Petrash, A. "Winter roof collapses: bad luck, bad construction, or bad design?" *Civil Engineering*, 42–45, Dec. 1979.

[C1-4] *Engineering News Record.* (May 1, 1995).

[C1-5] Federal Emergency Management Agency (FEMA). "The Oklahoma City Bombing: Improving Building Performance through Multi-Hazard Mitigation," FEMA 277, American Society of Civil Engineers and Federal Emergency Management Agency, 1996.

[C1-6] Weidlinger, P. "Civilian structures: taking the defensive," *Civil Engineering*, Nov. 1994.

[C1-7] Longinow, A. "The threat of terrorism: can buildings be protected?" *Building Operating Management*, July 1995.

[C1-8] FEMA. "The Oklahoma City Bombing: Improving Building Performance through Multi-Hazard Mitigation," FEMA 277, American Society of Civil Engineers and Federal Emergency Management Agency, 1996.

[C1-9] Breen, J.E., ed. Progressive collapse of building structures [summary report of a workshop held at the University of Texas at Austin, Oct. 1975]. Washington, D.C.: U.S. Department of Housing and Urban Development. Rep. PDR-182, Sept. 1976.

[C1-10] Burnett, E.F.P. The avoidance of progressive collapse: Regulatory approaches to the problem. Washington, D.C.: U.S. Department of Commerce, National Bureau of Standards. NBS GCR 75-48, Oct. 1975. [Available from: National Technical Information Service, Springfield, VA.]

[C1-11] Ellingwood, B.R. and Leyendecker, E.V. "Approaches for design against progressive collapse." *J. Struct. Div.*, 104(3), 413–423, 1978.

[C1-12] Mlakar, P. S. *Structural Design for Physical Security: State of the Practice*, Reston, VA: ASCE, 1999.

[C1-13] Schultz, D.M., Burnett, E.F.P., and Fintel, M. A design approach to general structural integrity, design and construction of large-panel concrete structures. Washington, D.C.: U.S. Department of Housing and Urban Development, 1977.

[C1-14] PCI Committee on Precast Bearing Walls. "Considerations for the design of precast bearing-wall buildings to withstand abnormal loads." *J. Prestressed Concrete Inst.*, 21(2): 46–69, March/April 1976.

[C1-15] Fintel, M. and Schultz, D.M. "Structural integrity of large-panel buildings." *J. Am. Concrete Inst.*, 76(5), 583–622, May 1979.

[C1-16] Fintel, M. and Annamalai, G. "Philosophy of structural integrity of multistory load-bearing concrete masonry structures." *J. Am. Concrete Inst.*, 1(5), 27–35, May 1979.

[C1-17] American Concrete Institute (ACI). "Building Code Requirements for Structural Concrete," ACI Standard 318-02, American Concrete Institute Detroit, MI, 2002.

[C1-18] ASCE. "Design of Blast Resistant Buildings in Petrochemical Facilities," ASCE Petrochemical Energy Committee, Task Committee on Blast Resistant Design, 1977.

[C1-19] FEMA. "NEHRP Recommended Provisions for Seismic Regulations for New Buildings and Other Structures," 1997 Edition, FEMA 302/Feb. 1998, Part 1–Provisions, 1998.

[C1-20] FEMA. "Wet Floodproofing Requirements for Structures Located in Special Flood Hazard Areas in Accordance with the National Flood Insurance Program," Technical Bulletin 7-93, Federal Emergency Management Agency, Mitigation Directorate, Federal Insurance Administration, Washington, D.C., 1993.

[C1-21] International Code Council. "2000 International Building Code," Tables 307.7(1) and 307.7(2), International Code Council, Falls Church, VA, 2000.

[C1-22] Environmental Protection Agency (EPA). "Emergency Planning and Notification—The List of Extremely Hazardous Substances and Their Threshold Planning Quantities," 40 CFR Part 355 Appendix A, Environmental Protection Agency, Washington, D.C., July 1999.

[C1-23] EPA. "Chemical Accident Prevention Provisions," 40 CFR Part 68, Environmental Protection Agency, Washington, D.C., July 1999.

SECTION C2.0
COMBINATIONS OF LOADS

Loads in this standard are intended for use with design specifications for conventional structural materials including steel, concrete, masonry, and timber. Some of these specifications are based on allowable stress design, while others employ strength design. In the case of allowable stress design, design specifications define allowable stresses that may not be exceeded by load effects due to unfactored loads, that is, allowable stresses contain a factor of safety. In strength design, design specifications provide load factors and, in some instances, resistance factors. Structural design specifications based on limit states design have been adopted by a number of specification-writing groups. Therefore, it is desirable to include herein common load factors that are applicable to these new specifications. It is intended that these load factors be used by all material-based design specifications that adopt a strength design philosophy in conjunction with nominal resistances and resistance factors developed by individual material-specification-writing groups. Load factors given herein were developed using a first-order probabilistic analysis and a broad survey of the reliabilities inherent in contemporary design practice. References [C2-9, C2-10, and C2-19] also provide guidelines for materials-specification-writing groups to aid them in developing resistance factors that are compatible, in terms of inherent reliability, with load factors and statistical information specific to each structural material.

SECTION C2.2
SYMBOLS AND NOTATION

Self-straining forces can be caused by differential settlement foundations, creep in concrete members, shrinkage in members after placement, expansion of shrinkage-compensating concrete, and changes in temperature of members during the service life of the structure. In some cases, these forces may be a significant design consideration. In concrete or masonry structures, the reduction in stiffness that occurs upon cracking may relieve these self-straining forces, and the assessment of loads should consider this reduced stiffness.

Some permanent loads, such as landscaping loads on plaza areas, may be more appropriately considered as live loads for purposes of design.

SECTION C2.3
COMBINING LOADS USING STRENGTH DESIGN

C2.3.1 Applicability. Load factors and load combinations given in this section apply to limit states or strength design criteria (referred to as "Load and Resistance Factor Design" by the steel and wood industries), and they should not be used with allowable stress design specifications.

C2.3.2 Basic Combinations. Unfactored loads to be used with these load factors are the nominal loads of Sections 3 through 9 of this standard. Load factors are from NBS SP 577 with the exception of the factor 1.0 for E, which is based on the more recent NEHRP research on seismic-resistant design [C2-21]. The basic idea of the load combination scheme is that in addition to dead load, which is considered to be permanent, one of the variable loads takes on its maximum lifetime value while the other variable loads assume "arbitrary point-in-time" values, the latter being loads that would be measured at any instant of time [C2-23]. This is consistent with the manner in which loads actually combine in situations in which strength limit states may be approached. However, nominal loads in Sections 3 through 9 are substantially in excess of the arbitrary point-in-time values. To avoid having to specify both a maximum and an arbitrary point-in-time value for each load type, some of the specified load factors are less than unity in combinations (2) through (6).

Load factors in Section 2.3.2 are based on a survey of reliabilities inherent in existing design practice. The load factor on wind load in combinations (4) and (6) was increased to 1.6 in ASCE 7-98 from the value of 1.3 appearing in ASCE 7-95. The reasons for this increase are twofold.

First, the previous wind load factor, 1.3, incorporated a factor of 0.85 to account for wind directionality, that is, the reduced likelihood that the maximum wind speed occurs in a direction that is most unfavorable for building response [C2-12]. This directionality effect was not taken into account in ASD. Recent wind-engineering research has made it possible to identify wind directionality factors explicitly for a number of common structures. Accordingly, new wind directionality factors, K_d, are presented in Table 6-6 of this standard; these factors now are reflected in the nominal wind forces, W, used in both strength design and allowable stress design. This change alone mandates an increase in the wind load factor to approximately 1.53.

Second, the value in ASCE 7-95 of 1.3 was based on a statistical analysis of wind forces on buildings at sites not exposed to hurricane winds [C2-12]. Studies have shown that, owing to differences between statistical characteristics of wind forces in hurricane-prone coastal areas of the United States [C2-15, C2-16, and C2-17], the probability of exceeding the factored (or design-basis) wind force $1.3W$ is

higher in hurricane-prone coastal areas than in the interior regions. Two recent studies [C2-13 and C2-14] have shown that the wind load factor in hurricane-prone areas should be increased to approximately 1.5 to 1.8 (depending on site) to maintain comparable reliability.

To move toward uniform risk in coastal and interior areas across the country, two steps were taken. First, the wind speed contours in hurricane-prone areas were adjusted to take the differences in extreme hurricane wind speed probability distributions into account (as explained in Section C6.5.4); these differences previously were accounted for in ASCE 7-95 by the "importance factor." Second, the wind load factor was increased from 1.3 to 1.6. This approach (a) reflects the removal of the directionality factor, and (b) avoids having to specify separate load criteria for coastal and interior areas.

Load combinations (6) and (7) apply specifically to cases in which the structural actions due to lateral forces and gravity loads counteract one another.

Load factors given herein relate only to strength limit states. Serviceability limit states and associated load factors are covered in Appendix B of this standard.

This Standard historically has provided specific procedures for determining magnitudes of dead, occupancy, live, wind, snow, and earthquake loads. Other loads not traditionally considered by this standard may also require consideration in design. Some of these loads may be important in certain material specifications and are included in the load criteria to enable uniformity to be achieved in the load criteria for different materials. However, statistical data on these loads are limited or nonexistent, and the same procedures used to obtain load factors and load combinations in Section 2.3.2 cannot be applied at the present time. Accordingly, load factors for fluid load (F), lateral pressure due to soil and water in soil (H), and self-straining forces and effects (T) have been chosen to yield designs that would be similar to those obtained with existing specifications if appropriate adjustments consistent with the load combinations in Section 2.3.2 were made to the resistance factors. Further research is needed to develop more accurate load factors because the load factors selected for H and F_a, are probably conservative.

Fluid load, F, defines structural actions in structural supports, framework, or foundations of a storage tank, vessel, or similar container due to stored liquid products. The product in a storage tank shares characteristics of both dead and live load. It is similar to a dead load in that its weight has a maximum calculated value, and the magnitude of the actual load may have a relatively small dispersion. However, it is not permanent; emptying and filling causes fluctuating forces in the structure, the maximum load may be exceeded by overfilling; and densities of stored products in a specific tank may vary. Adding F to combination 1 provides additional conservatism for situations in which F is the dominant load.

It should be emphasized that uncertainties in lateral forces from bulk materials, included in H, are higher than those in fluids, particularly when dynamic effects are introduced as the bulk material is set in motion by filling or emptying operations. Accordingly, the load factor for such loads is set equal to 1.6.

C2.3.3 Load Combinations Including Flood Load. The nominal flood load, F_a, is based on the 100-year flood (Section 5.3.3.1). The recommended flood load factor of 2.0 in V Zones and coastal A Zones is based on a statistical analysis of flood loads associated with hydrostatic pressures, pressures due to steady overland flow, and hydrodynamic pressures due to waves, as specified in Section 5.3.3.

The flood load criteria were derived from an analysis of hurricane-generated storm tides produced along the U.S. East and Gulf coasts [C2-14], where storm tide is defined as the water level above mean sea level resulting from wind-generated storm surge added to randomly phased astronomical tides. Hurricane wind speeds and storm tides were simulated at 11 coastal sites based on historical storm climatology and on accepted wind speed and storm surge models. The resulting wind speed and storm tide data were then used to define probability distributions of wind loads and flood loads using wind and flood load equations specified in Sections 6.5, 5.3.3 and in other publications (U.S. Army Corps of Engineers). Load factors for these loads were then obtained using established reliability methods [C2-10] and achieve approximately the same level of reliability as do combinations involving wind loads acting without floods. The relatively high flood load factor stems from the high variability in floods relative to other environmental loads. The presence of 2.0 F_a in both Eqs. 4 and 6 in V-Zones and Coastal A-Zones is the result of high stochastic dependence between extreme wind and flood in hurricane-prone coastal zones. The 2.0 F_a also applies in coastal areas subject to northeasters, extra tropical storms, or coastal storms other than hurricanes where a high correlation exists between extreme wind and flood.

Flood loads are unique in that they are initiated only after the water level exceeds the local ground elevation. As a result, the statistical characteristics of flood loads vary with ground elevation. The load factor 2.0 is based on calculations (including hydrostatic, steady flow, and wave forces) with still-water flood depths ranging from approximately 4 to 9 ft (average still-water flood depth of approximately 6 ft), and applies to a wide variety of flood conditions. For lesser flood depths, load factors exceed 2.0 because of the wide dispersion in flood loads relative to the nominal flood load. As an example, load factors appropriate to water depths slightly less than 4 ft equal 2.8 [C2-14]. However, in such circumstances, the flood load itself generally is small. Thus, the load factor 2.0 is based on the recognition that flood loads of most importance to structural design occur in situations where the depth of flooding is greatest.

C2.3.4 Load Combinations Including Atmospheric Ice Loads. Load combinations 1 and 2 in Sections 2.3.4 and 2.4.3 include the simultaneous effects of snow loads as defined in Section 7 and Atmospheric Ice Loads as defined in Section 10. Load combinations 2 and 3 in Sections 2.3.4 and 2.4.3 introduce the simultaneous effect of wind on the atmospheric ice. The wind load on the atmospheric ice, W_i, corresponds to approximately a 500-year MRI. Accordingly, the load factors on W_i and D_i are set equal to 1.0 and 0.7 in Sections 2.3.4 and 2.4.3, respectively: The rationale is exactly the same as that used to specify the earthquake force as $0.7E$ in the load combinations applied in working stress design. The snow loads defined in Section 7 are based on measurements of frozen precipitation accumulated on the ground, which includes snow, ice due to freezing rain, and rain that falls onto snow and later freezes. Thus the effects of freezing rain are included in the snow loads for roofs, catwalks, and other surfaces to which snow loads are normally applied. The atmospheric ice loads defined in Section 10 are applied simultaneously to those portions of the structure on which ice due to freezing rain, in-cloud icing, or snow accrete that are not subject to the snow loads in Section 7. A trussed tower installed on the roof of a building is one example. The snow loads from Section 7 would be applied to the roof with the atmospheric ice loads from Section 10 applied to the trussed tower. If a trussed tower has working platforms, the snow loads would be applied to the surface of the platforms with the atmospheric ice loads applied to the tower. If a sign is mounted on a roof, the snow loads would be applied to the roof and the atmospheric ice loads to the sign.

SECTION C2.4
COMBINING LOADS USING ALLOWABLE STRESS DESIGN

C2.4.1 Basic Combinations. The load combinations listed cover those loads for which specific values are given in other parts of this standard. However, these combinations are not all-inclusive, and designers will need to exercise judgment in some situations. Design should be based on the load combination causing the most unfavorable effect. In some cases, this may occur when one or more loads are not acting. No safety factors have been applied to these loads since such factors depend on the design philosophy adopted by the particular material specification.

Wind and earthquake loads need not be assumed to act simultaneously. However, the most unfavorable effects of each should be considered separately in design, where appropriate. In some instances, forces due to wind might exceed those due to earthquake, while ductility requirements might be determined by earthquake loads.

Load combinations (7) and (8) were new to the 1998 edition of ASCE 7. They address the situation in which the effects of lateral or uplift forces counteract the effect of gravity loads. This eliminates an inconsistency in the treatment of counteracting loads in allowable stress design and strength design, and emphasizes the importance of checking stability. The earthquake load effect is multiplied by 0.7 to align allowable stress design for earthquake effects with the definition of E in Section 9.2.2.6, which is based on strength principles.

Most loads, other than dead loads, vary significantly with time. When these variable loads are combined with dead loads, their combined effect should be sufficient to reduce the risk of unsatisfactory performance to an acceptably low level. However, when more than one variable load is considered, it is extremely unlikely that they will all attain their maximum value at the same time. Accordingly, some reduction in the total of the combined load effects is appropriate. This reduction is accomplished through the 0.75 load combination factor. The 0.75 factor applies only to the variable loads, not to the dead load.

Some material design standards that permit a one-third increase in allowable stress for certain load combinations have justified that increase by this same concept. Where that is the case, simultaneous use of both the one-third increase in allowable stress and the 25% reduction in combined loads is unsafe and is not permitted. In contrast, allowable stress increases that are based on duration of load or loading rate effects, which are independent concepts, may be combined with the reduction factor for combining multiple variable loads. Effects apply to the total stress; that is, the stress resulting from the combination of all loads. Load combination reduction factors for combined variable loads are different in that they apply only to the variable loads, and they do not affect the permanent loads or the stresses caused by permanent loads. This explains why the 0.75 factor applied to the sum of all loads, to this edition, in which the 0.75 factor applies only to the sum of the variable loads, not the dead load.

Certain material design standards permit a one-third increase in allowable stress for load combinations with one variable load where that variable is earthquake load. This standard handles allowable stress design for earthquake loads in a fashion to give results comparable to the strength design basis for earthquake loads as explained in the C.9 Commentary section titled "Use of Allowable Stress Design Standards."

C2.4.2 Load Combinations Including Flood Load. The basis for the load combinations involving flood load is presented in detail in Section C2.3.3 on strength design. Consistent with the treatment of flood loads for strength design, F_a has been added to load combinations (3) and (4); the multiplier on F_a aligns allowable stress design for flood load with strength design.

C2.4.3 Load Combinations Including Atmospheric Ice Loads. See Section C2.3.4.

SECTION C2.5
LOAD COMBINATIONS FOR
EXTRAORDINARY EVENTS

ASCE Standard 7 Commentary C1.4 recommends approaches to providing general structural integrity in building design and construction. Commentary C2.5 explains the basis for the load combinations that the designer should use if the "Direct Design" alternative in Commentary C1.4 is selected. If the authority having jurisdiction requires the "Indirect Design" alternative, that authority may use these load requirements as one basis for determining minimum required levels of strength, continuity, and ductility. Generally, extraordinary events with a probability of occurrence in the range 10^{-6} to 10^{-4}/yr or greater should be identified, and measures should be taken to ensure that the performance of key loading-bearing structural systems and components is sufficient to withstand such events.

Extraordinary events arise from extraordinary service or environmental conditions that traditionally are not considered explicitly in design of ordinary buildings and structures. Such events are characterized by a low probability of occurrence and usually a short duration. Few buildings are ever exposed to such events, and statistical data to describe their magnitude and structural effects are rarely available. Included in the category of extraordinary events would be fire, explosions of volatile liquids or natural gas in building service systems, sabotage, vehicular impact, misuse by building occupants, subsidence (not settlement) of subsoil, and tornadoes. The occurrence of any of these events is likely to lead to structural damage or failure. If the structure is not properly designed and detailed, this local failure may initiate a chain reaction of failures that propagates throughout a major portion of the structure and leads to a potentially catastrophic collapse. Approximately 15% to 20% of building collapses occur in this way [C2-1]. Although all buildings are susceptible to progressive failures in varying degrees, types of construction that lack inherent continuity and ductility are particularly vulnerable [C2-2, C2-22].

Good design practice requires that structures be robust and that their safety and performance not be sensitive to uncertainties in loads, environmental influences, and other situations not explicitly considered in design. The structural system should be designed in such a way that if an extraordinary event occurs, the probability of damage disproportionate to the original event is sufficiently small [C2-7]. The philosophy of designing to limit the spread of damage rather than to prevent damage entirely is different from the traditional approach to designing to withstand dead, live, snow, and wind loads, but is similar to the philosophy adopted in modern earthquake-resistant design [C2-21].

In general, structural systems should be designed with sufficient continuity and ductility that alternate load paths can develop following individual member failure so that failure of the structure as a whole does not ensue. At a simple level, continuity can be achieved by requiring development of a minimum tie force — say 20 kN/m — between structural elements [C2-3]. Member failures may be controlled by protective measures that ensure that no essential load-bearing member is made ineffective as a result of an accident, although this approach may be more difficult to implement. Where member failure would inevitably result in a disproportionate collapse, the member should be designed for a higher degree of reliability [C2-20]. In either approach, an enhanced quality assurance and maintenance program may be required.

Design limit states include loss of equilibrium as a rigid body, large deformations leading to significant second-order effects, yielding or rupture of members of connections, formation of a mechanism, instability of members, or the structure as a whole. These limit states are the same as those considered for other load events, but the load-resisting mechanisms in a damaged structure may be different, and sources of load-carrying capacity that normally would not be considered in ordinary ultimate limit states design, such as a membrane or catenary action, may be included. The use of elastic analysis vastly underestimates the load-carrying capacity of the structure. Materially or geometrically nonlinear or plastic analyses may be used, depending on the response of the structure to the actions.

Specific design provisions to control the effect of extraordinary loads and risk of progressive failure can be developed with a probabilistic basis [C2-8 and C2-11]. One can either attempt to reduce the likelihood of the extraordinary event or design the structure to withstand or absorb damage from the event if it occurs. Let F be the event of failure and A be the event that a structurally damaging event occurs. The probability of failure due to event A is

$$P_f = P(F \mid A)P(A) \qquad \textbf{(Eq. C2-1)}$$

in which $P(F \mid A)$ is the conditional probability of failure of a damaged structure and $P(A)$ is the probability of occurrence of event A. The separation of $P(F \mid A)$ and $P(A)$ allows one to focus on strategies for reducing risk. $P(A)$ depends on siting, controlling the use of hazardous substances, limiting access, and other actions that are essentially independent of structural design. In contrast, $P(F \mid A)$ depends on structural design measures ranging from minimum provisions for continuity to a complete post-damage structural evaluation.

The probability, $P(A)$, depends on the specific hazard. Limited data for severe fires, gas explosions, bomb explosions, and vehicular collisions indicate that the event probability depends on building size, measured in dwelling units or square footage, and ranges from about 0.23×10^{-6}/dwelling unit/year to about 7.8×10^{-6}/dwelling unit/year [C2-6 and C2-8]. Thus, the probability that a building structure is affected depends on the number of dwelling units (or square footage) in the building. If one were

to set the conditional limit state probability, $P(F \mid A) = 0.1 - 0.2$/yr, however, the annual probability of structural failure from Eq. C2.5.1 would be on the order of 10^{-7} to 10^{-6}, placing the risk in the low-magnitude background along with risks from rare accidents [C2-24].

Design requirements corresponding to this desired $P(F \mid A) = 0.1 - 0.2$ can be developed using first-order reliability analysis if the limit state function describing structural behavior is available [C2-10 and C2-19]. As an alternative, one can leave material and structural behavior considerations to the responsible material specifications and consider only the load combination aspect of the safety check, which is more straightforward.

For checking a structure to determine its residual load-carrying capacity following occurrence of a damaging extraordinary event, selected load-bearing elements should be notionally removed and the capacity of the remaining structure evaluated using the following load combination:

$$(0.9 \text{ or } 1.2)D + (0.5L \text{ or } 0.2S) + 0.2W \quad \textbf{(Eq. C2-2)}$$

For checking the capacity of a structure or structural element to withstand the effect of an extraordinary event, the following load combinations should be used:

$$1.2D + A_k + (0.5L \text{ or } 0.2S) \quad \textbf{(Eq. C2-3)}$$

$$(0.9 \text{ or } 1.2)D + A_k + 0.2W \quad \textbf{(Eq. C2-4)}$$

The value of the load or load effect resulting from extraordinary event A used in design is denoted A_k. Only limited data are available to define the frequency distribution of the load, and A_k must be specified by the authority having jurisdiction [C2-4]. The uncertainty in the load due to the extraordinary event is encompassed in the selection of a conservative A_k and thus the load factor on A_k is set equal to 1.0, as is done in the earthquake load combinations in Section 2.3. Load factors less than 1.0 on the companion actions reflect the small probability of a joint occurrence of the extraordinary load and the design live, snow, or wind load. The companion action $0.5L$ corresponds, approximately, to the mean of the yearly maximum load [C2-5]. Companion actions $0.2S$ and $0.2W$ are interpreted similarly. A similar set of load combinations for extraordinary events appears in [C2-18].

REFERENCES

[C2-1] Allen, D.E. and Schriever, W. "Progressive collapse, abnormal loads and building codes." *Structural Failures: Modes, Causes, Responsibilities*, ASCE, New York, 21–48, 1973.

[C2-2] Breen, J.E. and Siess, C.P. "Progressive collapse — symposium summary." *J. Am. Concrete Inst.*, 76(9), 997–1004, 1979.

[C2-3] British Standards Institution. "British standard structural use of steelwork in building." (BS 5950: Part 8), London, U.K., 1990.

[C2-4] Burnett, E.F.P. "The avoidance of progressive collapse: Regulatory approaches to the problem." National Bureau of Standards Report GCR 75-48, Washington, D.C. (Available from National Technical Information Service, Springfield, Va.), 1975.

[C2-5] Chalk, P.L. and Corotis, R.B. "Probability models for design live loads." *J. Struct. Div.*, ASCE, 106(10), 2017–2033, 1980.

[C2-6] CIB W14. "A conceptual approach towards a probability based design guide on structural fire safety." *Fire Safety Journal*, 6(1) 1–79, 1983.

[C2-7] "Commentary C Structural Integrity." National Building Code of Canada, National Research Council of Canada, Ottawa, Ontario, 1990.

[C2-8] Ellingwood, B. and Leyendecker, E.V. "Approaches for design against progressive collapse." *J. Struct. Div.*, ASCE, 104(3), 413–423, 1978.

[C2-9] Ellingwood, B., Galambos, T.V., MacGregor, J.G., and Cornell, C.A. "Development of a probability-based load criterion for American National Standard A58." Washington, D.C.: U.S. Dept. of Commerce, National Bureau of Standards, NBS SP 577, 1980.

[C2-10] Ellingwood, B., Galambos, T.V., MacGregor, J.G., and Cornell, C.A. "Probability-based load criteria: load factors and load combinations." *J. Struct. Div.*, ASCE, 108(5), 978–997, 1982.

[C2-11] Ellingwood, B. and Corotis, R.B. "Load combinations for buildings exposed to fires." *Engrg. J.*, ASIC, 28(1), 37–44, 1991.

[C2-12] Ellingwood, B. "Wind and snow load statistics for probabilistic design." *J. Struct. Engrg.*, ASCE, 107(7), 1345–1349, 1981.

[C2-13] Ellingwood, B. And Tekie, P.B. "Wind load statistics for probability-based structural design." *J. Struct. Engrg.*, ASCE, 125(4), 453–464, 1999.

[C2-14] Mehta, K.C., et al. "An investigation of load factors for flood and combined wind and flood." Report prepared for Federal Emergency Management Agency, Washington, D.C., 1998.

[C2-15] Peterka, J.A. and Shahid, S. "Design gust wind speeds in the United States." *J. Struct. Engrg.*, ASCE, 124(2), 207–214, 1998.

[C2-16] Vickery, P.J. and Twisdale, L.A. "Prediction of hurricane windspeeds in the United States." *J. Struct. Engrg.*, ASCE, 121(11), 1691–1699, 1995.

[C2-17] Whalen, T. "Probabilistic estimates of design load factors for wind-sensitive structures using the 'peaks over threshold' approach." National Institute of Standards and Technology, NIST TN 1418, Gaithersburg, Md., 1996.

[C2-18] Eurocode 1. "Common unified rules for different types of construction and material." Commission of the European Communities, Brussels, Belgium, 1990.

[C2-19] Galambos, T.V., Ellingwood, B., MacGregor, J.G., and Cornell, C.A. "Probability-based load criteria: assessment of current design practice." *J. Struct. Div.*, ASCE, 108(5), 959–977, 1982.

[C2-20] Nordic Committee on Building Regulations (NKB). "Guidelines for loading and safety regulations for structural design." NKB Report No. 55E, Copenhagen, Denmark, 1987.

[C2-21] Federal Emergency Management Agency (FEMA). "NEHRP recommended provisions for the development of seismic regulations for buildings." FEMA Report 222, Washington, D.C., 1992.

[C2-22] Taylor, D.A. "Progressive collapse." *Can. J. Civ. Engrg.*, 2(4), 517–529, 1975.

[C2-23] Turkstra, C.J. and Madsen, H. "Load combinations for codified structural design." *J. Struct. Div.*, ASCE, 106(12), 2527–2543, 1980.

[C2-24] Wilson, R. and Crouch, E.A. "Risk assessment and comparisons: an introduction." *Science*, 236, 267–270.1, 1987.

SECTION C3.0
DEAD LOADS

SECTION C3.2
WEIGHTS OF MATERIALS AND CONSTRUCTIONS

To establish uniform practice among designers, it is desirable to present a list of materials generally used in building construction, together with their proper weights. Many building codes prescribe the minimum weights for only a few building materials and, in other instances, no guide whatsoever is furnished on this subject. In some cases, the codes are so drawn up as to leave the question of what weights to use to the discretion of the building official, without providing any authoritative guide. This practice, as well as the use of incomplete lists, has been subjected to much criticism. The solution chosen has been to present, in this commentary, an extended list that will be useful to designer and official alike. However, special cases will unavoidably arise, and authority is therefore granted in the standard for the building official to deal with them.

For ease of computation, most values are given in terms of pounds per square foot (lb/ft^2) (kN/m^2) of given thickness (see Table C3-1). Pounds per cubic foot (lb/ft^3) (kN/m^3) values, consistent with the pounds per square foot (kilonewtons per square meter) values, are also presented in some cases (see Table C3-2). Some constructions for which a single figure is given actually have a considerable range in weight. The average figure given is suitable for general use, but when there is reason to suspect a considerable deviation from this, the actual weight should be determined.

Engineers, architects, to consider factors that res and calculated loads.

Engineers and architects ca cumstances beyond their contr however, that conditions are encountered which, if not sidered in design, may reduce the future utility of a building or reduce its margin of safety. Among them are:

1. Dead Loads. There have been numerous instances in which the actual weights of members and construction materials have exceeded the values used in design. Care is advised in the use of tabular values. Also, allowances should be made for such factors as the influence of formwork and support deflections on the actual thickness of a concrete slab of prescribed nominal thickness.

2. Future Installations. Allowance should be made for the weight of future wearing or protective surfaces where there is a good possibility that such may be applied. Special consideration should be given to the likely types and position of partitions, as insufficient provision for partitioning may reduce the future utility of the building.

Attention is directed also to the possibility of temporary changes in the use of a building, as in the case of clearing a dormitory for a dance or other recreational purpose.

TABLE C3-1
MINIMUM DESIGN DEAD LOADS*

Component	Load (psf)	Component	Load (psf)
CEILINGS		Decking, 2-in. wood (Douglas fir)	5
Acoustical Fiber Board	1	Decking, 3-in. wood (Douglas fir)	8
Gypsum board (per mm thickness)	0.55	Fiberboard, 1/2-in.	0.75
Mechanical duct allowance	4	Gypsum sheathing, 1/2-in.	2
Plaster on tile or concrete	5	Insulation, roof boards (per inch thickness)	
Plaster on wood lath	8	Cellular glass	0.7
Suspended steel channel system	2	Fibrous glass	1.1
Suspended metal lath and cement plaster	15	Fiberboard	1.5
Suspended metal lath and gypsum plaster	10	Perlite	0.8
Wood furring suspension system	2.5	Polystyrene foam	0.2
COVERINGS, ROOF, AND WALL		Urethane foam with skin	0.5
Asbestos-cement shingles	4	Plywood (per 1/8-in. thickness)	0.4
Asphalt shingles	2	Rigid insulation, 1/2-in.	0.75
Cement tile	16	Skylight, metal frame, 3/8-in. wire glass	8
Clay tile (for mortar add 10 psf)		Slate, 3/16-in.	7
Book tile, 2-in.	12	Slate, 1/4-in.	10
Book tile, 3-in.	20	Waterproofing membranes:	
Ludowici	10	Bituminous, gravel-covered	5.5
Roman	12	Bituminous, smooth surface	1.5
Spanish	19	Liquid applied	1
Composition:		Single-ply, sheet	0.7
Three-ply ready roofing	1	Wood sheathing (per inch thickness)	3
Four-ply felt and gravel	5.5	Wood shingles	3
Five-ply felt and gravel	6	FLOOR FILL	
Copper or tin	1	Cinder concrete, per inch	9
Corrugated asbestos-cement roofing	4	Lightweight concrete, per inch	8
Deck, metal, 20 gage	2.5	Sand, per inch	8
Deck, metal, 18 gage	3	Stone concrete, per inch	12

(continued)

FLOORS AND FLOOR FINISHES

Component	Load (psf)
Asphalt block (2-in.), 1/2-in. mortar	30
Cement finish (1-in.) on stone-concrete fill	32
Ceramic or quarry tile (3/4-in.) on 1/2-in. mortar bed	16
Ceramic or quarry tile (3/4-in.) on 1-in. mortar bed	23
Concrete fill finish (per inch thickness)	12
Hardwood flooring, 7/8-in.	4
Linoleum or asphalt tile, 1/4-in.	1
Marble and mortar on stone-concrete fill	33
Slate (per mm thickness)	15
Solid flat tile on 1-in. mortar base	23
Subflooring, 3/4-in.	3
Terrazzo (1-1/2-in.) directly on slab	19
Terrazzo (1-in.) on stone-concrete fill	32
Terrazzo (1-in.), 2-in. stone concrete	32
Wood block (3-in.) on mastic, no fill	10
Wood block (3-in.) on 1/2-in. mortar base	16

FLOORS, WOOD-JOIST (NO PLASTER) DOUBLE WOOD FLOOR

Joist sizes (inches):	12-in. spacing (lb/ft²)	16-in. spacing (lb/ft²)	24-in. spacing (lb/ft²)
2 × 6	6	5	5
2 × 8	6	6	5
2 × 10	7	6	6
2 × 12	8	7	6

FRAME PARTITIONS

Component	Load (psf)
Movable steel partitions	4
Wood or steel studs, 1/2-in. gypsum board each side	8
Wood studs, 2 × 4, unplastered	4
Wood studs, 2 × 4, plastered one side	12
Wood studs, 2 × 4, plastered two sides	20

FRAME WALLS

Component	Load (psf)
Exterior stud walls:	
2 × 4 @ 16-in., 5/8-in. gypsum, insulated, 3/8-in. siding	11
2 × 6 @ 16-in., 5/8-in. gypsum, insulated, 3/8-in. siding	12
Exterior stud walls with brick veneer	48
Windows, glass, frame and sash	8

Clay brick wythes:

Component	Load (psf)
4 in.	39
8 in.	79
12 in.	115
16 in.	155

Hollow concrete masonry unit wythes:

Wythe thickness (in inches)	4	6	8	10	12
Density of unit (16.49 kN/m³)					
No grout	22	24	31	37	43
48″ o.c.		29	38	47	55
40″ o.c. grout		30	40	49	57
32″ o.c. spacing		32	42	52	61
24″ o.c.		34	46	57	67
16″ o.c.		40	53	66	79
Full Grout		55	75	95	115
Density of unit (125 pcf):					
No grout	26	28	36	44	50
48″ o.c.		33	44	54	62
40″ o.c. grout		34	45	56	65
32″ o.c. spacing		36	47	58	68
24″ o.c.		39	51	63	75
16″ o.c.		44	59	73	87
Full Grout		59	81	102	123
Density of unit (21.21 kN/m³)					
No grout	29	30	39	47	54
48″ o.c.		36	47	57	66
40″ o.c. grout		37	48	59	69
32″ o.c. spacing		38	50	62	72
24″ o.c.		41	54	67	78
16″ o.c.		46	61	76	90
Full Grout		62	83	105	127

Solid concrete masonry unit wythes (incl. Wythe thickness (in mm)

Component	4	6	8	10	12
Density of unit (105 pcf):	32	51	69	87	105
Density of unit (125 pcf):	38	60	81	102	124
Density of unit (135 pcf):	41	64	87	110	133

* Weights of masonry include mortar but not plaster. For plaster, add 5 lb/ft² for each face plastered. Values given represent averages. In some cases, there is a considerable range of weight for the same construction.

TABLE C3-1

MINIMUM DESIGN DEAD LOADS*

Component	Load (kN/m²)	Component	Load (kN/m²)
CEILINGS		Decking, 51 mm wood (Douglas fir)	0.24
Acoustical fiberboard	0.05	Decking, 76 mm wood (Douglas fir)	0.38
Gypsum board (per mm thickness)	0.008	Fiberboard, 13 mm	0.04
Mechanical duct allowance	0.19	Gypsum sheathing, 13 mm	0.10
Plaster on tile or concrete	0.24	Insulation, roof boards (per mm thickness)	
Plaster on wood lath	0.38	Cellular glass	0.0013
Suspended steel channel system	0.10	Fibrous glass	0.0021
Suspended metal lath and cement plaster	0.72	Fiberboard	0.0028
Suspended metal lath and gypsum plaster	0.48	Perlite	0.0015
Wood furring suspension system	0.12	Polystyrene foam	0.0004
COVERINGS, ROOF, AND WALL		Urethane foam with skin	0.0009
Asbestos-cement shingles	0.19	Plywood (per mm thickness)	0.006
Asphalt shingles	0.10	Rigid insulation, 13 mm	0.04
Cement tile	0.77	Skylight, metal frame, 10 mm wire glass	0.38
Clay tile (for mortar add 0.48 kN/m²)		Slate, 5 mm	0.34
Book tile, 51 mm	0.57	Slate, 6 mm	0.48
Book tile, 76 mm	0.96	Waterproofing membranes:	
Ludowici	0.48	Bituminous, gravel-covered	0.26
Roman	0.57	Bituminous, smooth surface	0.07
Spanish	0.91	Liquid applied	0.05
Composition:		Single-ply, sheet	0.03
Three-ply ready roofing	0.05	Wood sheathing (per mm thickness)	0.0057
Four-ply felt and gravel	0.26	Wood shingles	0.14
Five-ply felt and gravel	0.29	FLOOR FILL	
Copper or tin	0.05	Cinder concrete, per mm	0.017
Corrugated asbestos-cement roofing	0.19	Lightweight concrete, per mm	0.015
Deck, metal, 20 gage	0.12	Sand, per mm	0.015
Deck, metal, 18 gage	0.14	Stone concrete, per mm	0.023

(continued)

TABLE C3-1 — continued
MINIMUM DESIGN DEAD LOADS*

Component	Load (kN/m²)
FLOORS AND FLOOR FINISHES	
Asphalt block (51 mm), 13 mm mortar	1.44
Cement finish (25 mm) on stone-concrete fill	1.53
Ceramic or quarry tile (19 mm) on 13 mm mortar bed	0.77
Ceramic or quarry tile (19 mm) on 25 mm mortar bed	1.10
Concrete fill finish (per mm thickness)	0.023
Hardwood flooring, 22 mm	0.19
Linoleum or asphalt tile, 6 mm	0.05
Marble and mortar on stone-concrete fill	1.58
Slate (per mm thickness)	0.028
Solid flat tile on 25 mm mortar base	1.10
Subflooring, 19 mm	0.14
Terrazzo (38 mm) directly on slab	0.91
Terrazzo (25 mm) on stone-concrete fill	1.53
Terrazzo (25 mm), 51 mm stone concrete	1.53
Wood block (76 mm) on mastic, no fill	0.48
Wood block (76 mm) on 13 mm mortar base	0.77
FLOORS, WOOD-JOIST (NO PLASTER)	
DOUBLE WOOD FLOOR	

Joist sizes (mm):	305 mm spacing (kN/m²)	406 mm spacing (kN/m²)	610 mm spacing (kN/m²)
51 × 152	0.29	0.24	0.24
51 × 203	0.29	0.29	0.24
51 × 254	0.34	0.29	0.29
51 × 305	0.38	0.34	0.29

Component	Load (kN/m²)
FRAME PARTITIONS	
Movable steel partitions	0.19
Wood or steel studs, 13 mm gypsum board each side	0.38
Wood studs, 51 × 102, unplastered	0.19
Wood studs, 51 × 102, plastered one side	0.57
Wood studs, 51 × 102, plastered two sides	0.96
FRAME WALLS	
Exterior stud walls:	
51 mm × 102 mm @ 406 mm, 16 mm gypsum, insulated, 10 mm siding	0.53
51 mm × 152 mm @ 406 mm, 16 mm gypsum, insulated, 10 mm siding	0.57
Exterior stud walls with brick veneer	2.30
Windows, glass, frame and sash	0.38

Component					Load (kN/m²)
Clay brick wythes:					
102 mm					1.87
203 mm					3.78
305 mm					5.51
406 mm					7.42
Hollow concrete masonry unit wythes:					
Wythe thickness (in mm)	102	152	203	254	305
Density of unit (16.49 kN/m³)					
No grout	1.05	1.29	1.68	2.01	2.35
1219 mm		1.48	1.92	2.35	2.78
1016 mm grout		1.58	2.06	2.54	3.02
813 mm spacing		1.63	2.15	2.68	3.16
610 mm		1.77	2.35	2.92	3.45
406 mm		2.01	2.68	3.35	4.02
Full grout		2.73	3.69	4.69	5.70
Density of unit (125 pcf):					
No grout	1.25	1.34	1.72	2.11	2.39
1219 mm		1.58	2.11	2.59	2.97
1016 mm grout		1.63	2.15	2.68	3.11
813 mm spacing		1.72	2.25	2.78	3.26
610 mm		1.87	2.44	3.02	3.59
406 mm		2.11	2.78	3.50	4.17
Full grout		2.82	3.88	4.88	5.89
Density of unit (21.21 kN/m³)					
No grout	1.39	1.68	2.15	2.59	3.02
1219 mm		1.58	2.39	2.92	3.45
1016 mm grout		1.72	2.54	3.11	3.69
813 mm spacing		1.82	2.63	3.26	3.83
610 mm		1.96	2.82	3.50	4.12
406 mm		2.25	3.16	3.93	4.69
Full grout		3.06	4.17	5.27	6.37
Solid concrete masonry unit wythes (incl. concrete brick):					
Wythe thickness (in mm)	102	152	203	254	305
Density of unit (16.49 kN/m³):	1.53	2.35	3.21	4.02	4.88
Density of unit (19.64 kN/m³):	1.82	2.82	3.78	4.79	5.79
Density of unit (21.21 kN/m³):	1.96	3.02	4.12	5.17	6.27

* Weights of masonry include mortar but not plaster. For plaster, add 0.24 kN/m² for each face plastered. Values given represent averages. In some cases, there is a considerable range of weight for the same construction.

Material	Load (lb/ft³)	Material	Load (lb/ft³)
Aluminum	170	Earth (submerged)	
Bituminous products		Clay	80
Asphaltum	81	Soil	70
Graphite	135	River mud	90
Paraffin	56	Sand or gravel	60
Petroleum, crude	55	Sand or gravel and clay	65
Petroleum, refined	50	Glass	160
Petroleum, benzine	46	Gravel, dry	104
Petroleum, gasoline	42	Gypsum, loose	70
Pitch	69	Gypsum, wallboard	50
Tar	75	Ice	57
Brass	526	Iron	
Bronze	552	Cast	450
Cast-stone masonry (cement, stone, sand)	144	Wrought	48
Cement, portland, loose	90	Lead	710
Ceramic tile	150	Lime	
Charcoal	12	Hydrated, loose	32
Cinder fill	57	Hydrated, compacted	45
Cinders, dry, in bulk	45	Masonry, Ashlar stone	
Coal		Granite	165
Anthracite, piled	52	Limestone, crystalline	165
Bituminous, piled	47	Limestone, oolitic	135
Lignite, piled	47	Marble	173
Peat, dry, piled	23	Sandstone	144
Concrete, plain		Masonry, brick	
Cinder	108	Hard (low absorbtion)	130
Expanded-slag aggregate	100	Medium (medium absorbtion)	115
Haydite (burned-clay aggregate)	90	Soft (high absorbtion)	100
Slag	132	Masonry, concrete*	
Stone (including gravel)	144	Lightweight units	105
Vermiculite and perlite aggregate, non–load-bearing	25–50	Medium weight units	125
Other light aggregate, load-bearing	70–105	Normal weight units	135
Concrete, reinforced		Masonry grout	140
Cinder	111	Masonry, rubble stone	
Slag	138	Granite	153
Stone (including gravel)	150	Limestone, crystalline	147
Copper	556	Limestone, oolitic	138
Cork, compressed	14	Marble	156
Earth (not submerged)		Sandstone	137
Clay, dry	63	Mortar, cement or lime	130
Clay, damp	110	Particleboard	45
Clay and gravel, dry	100	Plywood	36
Silt, moist, loose	78	Riprap (Not submerged)	
Silt, moist, packed	96	Limestone	83
Silt, flowing	108	Sandstone	90
Sand and gravel, dry, loose	100	Sand	
Sand and gravel, dry, packed	110	Clean and dry	90
Sand and gravel, wet	120	River, dry	106

(continued)

Material	Load (lb/ft³)	Material	Load (lb/ft³)
Slag		Tin	459
Bank	70	Water	
Bank screenings	108	Fresh	62
Machine	96	Sea	64
Sand	52	Wood, seasoned	
Slate	172	Ash, commercial white	41
Steel, cold-drawn	492	Cypress, southern	34
Stone, quarried, piled		Fir, Douglas, coast region	34
Basalt, granite, gneiss	96	Hem fir	28
Limestone, marble, quartz	95	Oak, commercial reds and whites	47
Sandstone	82	Pine, southern yellow	37
Shale	92	Redwood	28
Greenstone, hornblende	107	Spruce, red, white, and Stika	29
Terra cotta, architectural		Western hemlock	32
Voids filled	120	Zinc, rolled sheet	449
Voids unfilled	72		

*Tabulated values apply to solid masonry and to the solid portion of hollow masonry.

Material	Load (kN/m³)	Material	Load (kN/m³)
Aluminum	170	Earth (submerged)	
Bituminous products		Clay	12.6
Asphaltum	12.7	Soil	11.0
Graphite	21.2	River mud	14.1
Paraffin	8.8	Sand or gravel	9.4
Petroleum, crude	8.6	Sand or gravel and clay	10.2
Petroleum, refined	7.9	Glass	25.1
Petroleum, benzine	7.2	Gravel, dry	16.3
Petroleum, gasoline	6.6	Gypsum, loose	11.0
Pitch	10.8	Gypsum, wallboard	7.9
Tar	11.8	Ice	9.0
Brass	82.6	Iron	
Bronze	86.7	Cast	70.7
Cast-stone masonry (cement, stone, sand)	22.6	Wrought	75.4
Cement, portland, loose	14.1	Lead	111.5
Ceramic tile	23.6	Lime	
Charcoal	1.9	Hydrated, loose	5.0
Cinder fill	9.0	Hydrated, compacted	7.1
Cinders, dry, in bulk	7.1	Masonry, Ashlar stone	
Coal		Granite	25.9
Anthracite, piled	8.2	Limestone, crystalline	25.9
Bituminous, piled	7.4	Limestone, oolitic	21.2
Lignite, piled	7.4	Marble	27.2
Peat, dry, piled	3.6	Sandstone	22.6
Concrete, plain		Masonry, brick	
Cinder	17.0	Hard (low absorbtion)	20.4
Expanded-slag aggregate	15.7	Medium (medium absorbtion)	18.1
Haydite (burned-clay aggregate)	14.1	Soft (high absorbtion)	15.7
Slag	20.7	Masonry, concrete*	
Stone (including gravel)	22.6	Lightweight units	16.5
Vermiculite and perlite aggregate, non–load-bearing	3.9–7.9	Medium weight units	19.6
Other light aggregate, load-bearing	11.0–16.5	Normal weight units	21.2
Concrete, reinforced		Masonry grout	22.0
Cinder	17.4	Masonry, rubble stone	
Slag	21.7	Granite	24.0
Stone (including gravel)	23.6	Limestone, crystalline	23.1
Copper	87.3	Limestone, oolitic	21.7
Cork, compressed	2.2	Marble	24.5
Earth (not submerged)		Sandstone	21.5
Clay, dry	9.9	Mortar, cement or lime	20.4
Clay, damp	17.3	Particleboard	7.1
Clay and gravel, dry	15.7	Plywood	5.7
Silt, moist, loose	12.3	Riprap (Not submerged)	
Silt, moist, packed	15.1	Limestone	13.0
Silt, flowing	17.0	Sandstone	14.1
Sand and gravel, dry, loose	15.7	Sand	
Sand and gravel, dry, packed	17.3	Clean and dry	14.1
Sand and gravel, wet	18.9	River, dry	16.7

(continued)

Material	Load (kN/m³)	Material	Load (kN/m³)
Slag		Tin	72.1
Bank	11.0	Water	
Bank screenings	17.0	Fresh	9.7
Machine	15.1	Sea	10.1
Sand	8.2	Wood, seasoned	
Slate	27.0	Ash, commercial white	6.4
Steel, cold-drawn	77.3	Cypress, southern	5.3
Stone, quarried, piled		Fir, Douglas, coast region	5.3
Basalt, granite, gneiss	15.1	Hem fir	4.4
Limestone, marble, quartz	14.9	Oak, commercial reds and whites	7.4
Sandstone	12.9	Pine, southern yellow	5.8
Shale	14.5	Redwood	4.4
Greenstone, hornblende	16.8	Spruce, red, white, and Stika	4.5
Terra cotta, architectural		Western hemlock	5.0
Voids filled	18.9	Zinc, rolled sheet	70.5
Voids unfilled	11.3		

*Tabulated values apply to solid masonry and to the solid portion of hollow masonry.

SECTION C4.0
LIVE LOADS

SECTION C4.2
UNIFORMLY DISTRIBUTED LOADS

C4.2.1 Required Live Loads. A selected list of loads for occupancies and uses more commonly encountered is given in Section 4.2.1, and the authority having jurisdiction should approve on occupancies not mentioned. Tables C4-1 and C4-2 are offered as a guide in the exercise of such authority.

In selecting the occupancy and use for the design of a building or a structure, the building owner should consider the possibility of later changes of occupancy involving loads heavier than originally contemplated. The lighter loading appropriate to the first occupancy should not necessarily be selected. The building owner should ensure that a live load greater than that for which a floor or roof is approved by the authority having jurisdiction is not placed, or caused, or permitted to be placed on any floor or roof of a building or other structure.

In order to solicit specific informed opinion regarding the design loads in Table 4-1, a panel of 25 distinguished structural engineers was selected. A Delphi [C4-1] was conducted with this panel in which design values and supporting reasons were requested for each occupancy type. The information was summarized and recirculated back to the panel members for a second round of responses; those occupancies for which previous design loads were reaffirmed, as well as those for which there was consensus for change, were included.

It is well known that the floor loads measured in a live-load survey usually are well below present design values [C4-2–C4-5]. However, buildings must be designed to resist the maximum loads they are likely to be subjected to during some reference period T, frequently taken as 50 years. Table C4-2 briefly summarizes how load survey data are combined with a theoretical analysis of the load process for some common occupancy types and illustrates how a design load might be selected for an occupancy not specified in Table 4-1 [C4-6]. The floor load normally present for the intended functions of a given occupancy is referred to as the sustained load. This load is modeled as constant until a change in tenant or occupancy type occurs. A live-load survey provides the statistics of the sustained load. Table C4-2 gives the mean, m_s, and standard deviation, σ_x, for particular reference areas. In addition to the sustained load, a building is likely to be subjected to a number of relatively short-duration, high-intensity, extraordinary, or transient loading events (due to crowding in special or emergency circumstances, concentrations

during remodeling, and the like). Limited survey information and theoretical considerations lead to the means, m_t, and standard deviations, σ_t, of single transient loads shown in Table C4-2.

Combination of the sustained load and transient load processes, with due regard for the probabilities of occurrence, leads to statistics of the maximum total load during a specified reference period T. The statistics of the maximum total load depend on the average duration of an individual tenancy, τ, the mean rate of occurrence of the transient load, v_e, and the reference period, T. Mean values are given in Table C4-2. The mean of the maximum load is similar, in most cases, to the Table 4-1 values of minimum uniformly distributed live loads and, in general, is a suitable design value.

For library stack rooms, the 150 psf (7.18 kN/m) uniform live load specified in Table 4-1 is intended to cover the range of ordinary library shelving. The most important variables that affect the floor loading are the book stack unit height and the ratio of the shelf depth to the aisle width. Common book stack units have a nominal height of 90 in. (2290 mm) or less, with shelf depths in the range of 8 in. (203 mm) to 12 in. (305 mm). Book weights vary, depending on their size and paper density, but there are practical limits to what can be stored in any given space. Book stack weights also vary, but not by enough to significantly affect the overall loading. Considering the practical combinations of the relevant dimensions, weights, and other parameters, if parallel rows of ordinary double-faced book stacks are separated by aisles that are at least 36 in. (914 mm) wide, then the average floor loading is unlikely to exceed the specified 150 psf (7.18 kN/m^2), even after allowing for a nominal aisle floor loading of 20 to 40 psf (0.96 to 1.92 kN/m^2).

The 150 psf floor loading is also applicable to typical file cabinet installations, provided that the 36 in. minimum aisle width is maintained. Five-drawer lateral or conventional file cabinets, even with two levels of book shelves stacked above them, are unlikely to exceed the 150 psf average floor loading unless all drawers and shelves are filled to capacity with maximum density paper. Such a condition is essentially an upper-bound for which the normal load factors and safety factors applied to the 150 psf criterion should still provide a safe design.

If a library shelving installation does not fall within the parameter limits that are specified in footnote 3 of Table 4-1, then the design should account for the actual conditions. For example, the floor loading for storage of medical X-ray film may easily exceed 200 psf (2.92 kN/m^2), mainly because of the increased depth of

the shelves. Mobile library shelving that rolls on rails should also be designed to meet the actual requirements of the specific installation, which may easily exceed 300 psf (14.4 kN/m^2). The rail support locations and deflection limits should be considered in the design, and the engineer should work closely with the system manufacturer in order to provide a serviceable structure.

SECTION C4.3
CONCENTRATED LOADS

C4.3.1 Accessible Roof-Supporting Members. The provision regarding concentrated loads supported by roof trusses or other primary roof members is intended to provide for a common situation for which specific requirements are generally lacking.

SECTION C4.4
LOADS ON HANDRAILS, GUARDRAIL SYSTEMS, GRAB BAR SYSTEMS, AND VEHICLE BARRIER SYSTEMS

C4.4.2 Loads.

a. Loads that can be expected to occur on handrail and guardrail systems are highly dependent on the use and occupancy of the protected area. For cases in which extreme loads can be anticipated, such as long straight runs of guardrail systems against which crowds can surge, appropriate increases in loading shall be considered.

b. When grab bars are provided for use by persons with physical disabilities, the design is governed by CABO A117, Accessible and Usable Buildings and Facilities.

c. Vehicle barrier systems may be subjected to horizontal loads from moving vehicles. These horizontal loads may be applied normal to the plane of the barrier system, parallel to the plane of the barrier system, or at any intermediate angle. Loads in garages accommodating trucks and buses may be obtained from the provisions contained in Standard Specifications for Highway Bridges, 1989, The American Association of State Highway and Transportation Officials.

d. This provision was introduced into the standard in 1998 and is consistent with the provisions for stairs.

e. Side rail extensions of fixed ladders are often flexible and weak in the lateral direction. OSHA (CFR 1910) requires side rail extensions with specific geometric requirements only. The load provided was introduced into the standard in 1998, and has been determined on the basis of a 250-lb person standing on a rung of the ladder, and accounting for reasonable angles of pull on the rail extension.

SECTION C4.6
PARTIAL LOADING

It is intended that the full intensity of the appropriately reduced live load over portions of the structure or member be considered, as well as a live load of the same intensity over the full length of the structure or member.

Partial-length loads on a simple beam or truss will produce higher shear on a portion of the span than a full-length load. "Checkerboard" loadings on multistoried, multipanel bents will produce higher positive moments than full loads, while loads on either side of a support will produce greater negative moments. Loads on the half span of arches and domes or on the two central quarters can be critical. For roofs, all probable load patterns should be considered. Cantilevers cannot rely on a possible live load on the anchor span for equilibrium.

SECTION C4.7
IMPACT LOADS

Grandstands, stadiums, and similar assembly structures may be subjected to loads caused by crowds swaying in unison, jumping to its feet, or stomping. Designers are cautioned that the possibility of such loads should be considered.

SECTION C4.8
REDUCTION IN LIVE LOADS

C4.8.1 General. The concept of, and methods for, determining member live load reductions as a function of a loaded member's influence area, A_I, was first introduced into this standard in 1982, and was the first such change since the concept of live load reduction was introduced over 40 years ago. The revised formula is a result of more extensive survey data and theoretical analysis [C4-7]. The change in format to a reduction multiplier results in a formula that is simple and more convenient to use. The use of influence area, now defined as a function of the tributary area, A_T, in a single equation has been shown to give more consistent reliability for the various structural effects. The influence area is defined as that floor area over which the influence surface for structural effects is significantly different from zero.

The factor K_{LL} is the ratio of the influence area (A_I) of a member to its tributary area (A_T), i.e.: $K_{LL} = A_I/A_T$, and is used to better define the influence area of a member as a function of its tributary area. Figure C4 illustrates typical influence areas and tributary areas for a structure with regular bay spacings. Table 4-2 has established K_{LL} values (derived from calculated K_{LL} values) to be used in Eq. 4-1 for a variety of structural members and configurations. Calculated K_{LL} values vary for column and beam members having adjacent cantilever construction, as is shown in Figure C4, and the Table 4-2 values have been set for these cases to result in live

load reductions that are slightly conservative. For unusual shapes, the concept of significant influence effect should be applied.

An example of a member without provisions for continuous shear transfer normal to its span would be a precast T-beam or double T beam, which may have an expansion joint along one or both flanges, or which may have only intermittent weld tabs along the edges of the flanges. Such members do not have the ability to share loads located within their tributary areas with adjacent members, thus resulting in K_{LL}, =1 for these types of members.

Reductions are permissible for two-way slabs and for beams, but care should be taken in defining the appropriate influence area. For multiple floors, areas for members supporting more than one floor are summed.

The formula provides a continuous transition from unreduced to reduced loads. The smallest allowed value of the reduction multiplier is 0.4 (providing a maximum 60% reduction), but there is a minimum of 0.5 (providing a 50% reduction) for members with a contributory load from just one floor.

C4.8.2 Heavy Live Loads. In the case of occupancies involving relatively heavy, basic live loads such as storage buildings, several adjacent floor panels may be fully loaded. However, data obtained in actual buildings indicate that rarely is any story loaded with an average actual live load of more than 80% of the average rated live load. It appears that the basic live load should not be reduced for the floor-and-beam design, but that it could be reduced a flat 20% for the design of members supporting more than one floor. Accordingly, this principle has been incorporated into the recommended requirement.

C4.8.3 Parking Garage Loads. Unlike live loads in office and residential buildings, which are generally spatially random, parking garage loads are due to vehicles parked in regular patterns and the garages are often full. The rationale behind the reduction according to area for other live loads therefore does not apply. A load survey of vehicle weights was conducted at nine commercial parking garages in four cities of different sizes [C4-13]. Statistical analyses of the maximum load effects on beams and columns due to vehicle loads over the garage's life were carried out using the survey results. Dynamic effects on the deck due to vehicle motions and on the ramp due to impact were investigated. The equivalent uniformly distributed loads (EUDL) that would produce the lifetime maximum column axial force and midspan beam bending moment are conservatively estimated at 34.8 psf. The EUDL is not sensitive to bay-size variation. In view of the possible impact of very heavy vehicles in the future such as sport-utility vehicles, however, a design load of 40 psf is recommended with no allowance for reduction according to bay area.

Compared with the design live load of 50 psf given in previous editions of the Standard, the design load contained herein represents a 20% reduction but is still 33% higher than the 30 psf one would obtain were an area-based reduction to be applied to the 50 psf value for large bays as allowed in most Standards. Also, the variability of the maximum parking garage load effect is found to be small with a coefficient of variation less than 5% in comparison with 20% to 30% for most other live loads. The implication is that when a live load factor of 1.6 is used in design, additional conservatism is built into it such that the recommended value would also be sufficiently conservative for special purpose parking (such as valet parking) where vehicles may be more densely parked causing a higher load effect. Therefore, the 50 psf design value was thought to be overly conservative and it can be reduced to 40 psf without sacrificing structural integrity.

In view of the large load effect produced by a single heavy vehicle (up to 10,000 lb), the current concentrated load of 2000 lb should be increased to 3000 lb acting on an area of 4.5 in. by 4.5 in., which represents the load caused by a jack in changing tires.

C4.8.5 Limitations on One-Way Slabs. One-way slabs behave in a manner similar to two-way slabs, but do not benefit from having a higher redundancy which results from two-way action. For this reason, it is appropriate to allow a live load reduction for one-way slabs, but restrict the tributary area, A_T, to an area which is the product of the slab span times a width normal to the span not greater than 1.5 times the span (thus resulting in an area with an aspect ratio of 1.5). For one-way slabs with aspect ratios greater than 1.5, the effect will be to give a somewhat higher live load (where a reduction has been allowed) than for two-way slabs with the same ratio.

Members such as hollow-core slabs, which have grouted continuous shear keys along their edges and that span in one direction only, are considered as one-way slabs for live load reduction even though they may have continuous shear transfer normal to their span.

SECTION C4.9
MINIMUM ROOF LIVE LOADS

C4.9.1 Flat, Pitched, and Curved Roofs. The values specified in Eq. 4-2 that act vertically upon the projected area have been selected as minimum roof live loads, even in localities where little or no snowfall occurs. This is because it is considered necessary to provide for occasional loading due to the presence of workers and materials during repair operations.

C4.9.2 Special Purpose Roofs. Designers should consider any additional dead loads that may be imposed by saturated landscaping materials. Special purpose or occupancy roof

live loads may be reduced in accordance with the requirements of Section 4.8.

SECTION C4.10
CRANE LOADS

All support components of moving bridge cranes and monorail cranes including runway beams, brackets, bracing, and connections shall be designed to support the maximum wheel load of the crane and the vertical impact, lateral, and longitudinal forces induced by the moving crane. Also, the runway beams shall be designed for crane stop forces. The methods for determining these loads vary depending on the type of crane system and support. See [C4-8 to C4-11], which describe types of bridge cranes and monorail cranes. Cranes described in these references include top running bridge cranes with top running trolley, underhung bridge cranes, and underhung monorail cranes. [C4-12] gives more stringent requirements for crane runway design that are more appropriate for higher capacity or higher speed crane systems.

REFERENCES

[C4-1] Corotis, R.B., Fox, R.R., and Harris, J.C. "Delphi methods: Theory and design load application." *J. Struct. Div.*, ASCE, 107(ST6), 1095–1105, 1981.

[C4-2] Peir, J.C. and Cornell, C.A. "Spatial and temporal variability of live loads." *J. Struct. Div.*, ASCE, 99(ST5), 903–922, 1973.

[C4-3] McGuire, R.K. and Cornell, C.A. "Live load effects in office buildings." *J. Struct. Div.*, ASCE, 100(ST7), 1351–1366, 1974.

[C4-4] Ellingwood, B.R. and Culver, C.G. "Analysis of live loads in office buildings." *J. Struct. Div.*, ASCE, 103(ST8), 1551–1560, 1977.

[C4-5] Sentler, L. A stochastic model for live loads on floors in buildings. Lund Institute of Technology, Division of Building Technology, Report 60, Lund, Sweden, 1975.

[C4-6] Chalk, P.L. and Corotis, R.B. "A probability model for design live loads." *J. Struct. Div.*, ASCE, 106(ST10), 2017–2030, 1980.

[C4-7] Harris, M.E., Corotis, R.B., and Bova, C.J. "Area-dependent processes for structural live loads." *J. Struct. Div.*, ASCE, 107(ST5), 857–872, 1981.

[C4-8] Material Handling Industry. "Specifications for Underhung Cranes and Monorail Systems." ANSI MH 27.1, Charlotte, N.C., 1981.

[C4-9] Material Handling Industry. "Specifications for Electric Overhead Traveling Cranes." No. 70, Charlotte, N.C., 1994.

[C4-10] Material Handling Industry. "Specifications for Top Running and Under Running Single Girder Electric Overhead Traveling Cranes." No. 74, Charlotte, N.C., 1994.

[C4-11] Metal Building Manufacturers Association (MBMA). *Low Rise Building Systems Manual*, Cleveland, Ohio, 1986.

[C4-12] Association of Iron and Steel Engineers. Technical Report No. 13, Pittsburgh, Penn., 1979.

[C4-13] Wen, Y. K. and Yeo, G. L. "Design live loads for passenger cars parking garages." *J. Struct. Engrg.*, ASCE, 127(3), 2001. (Based on "Design Live Loads for Parking Garages," ASCE, Reston, Va., 1999.)

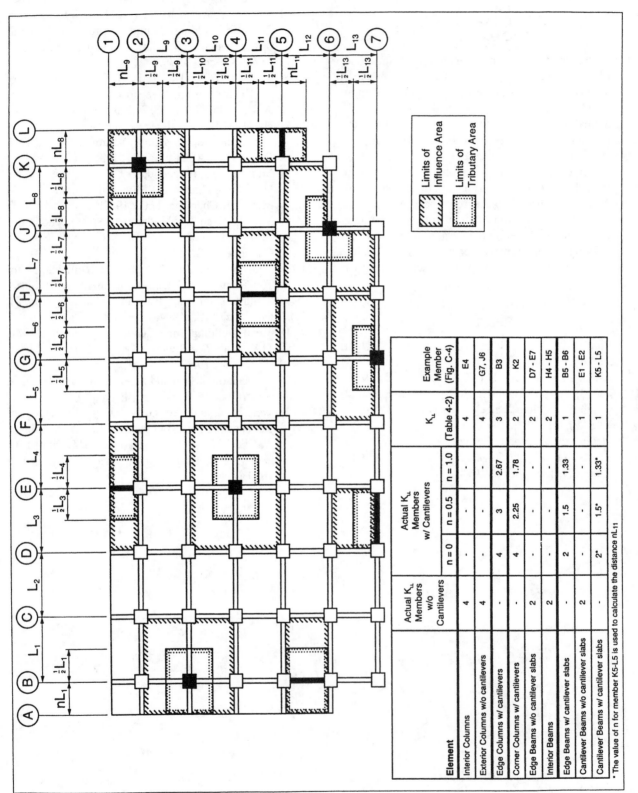

Element	Actual K_{LL} Members w/o Cantilevers	Actual K_{LL} Members w/ Cantilevers			K_{LL} (Table 4-2)	Example Member (Fig. C-4)
		n = 0	n = 0.5	n = 1.0		
Interior Columns	4	-	-	-	4	E4
Exterior Columns w/o cantilevers	4	-	-	-	4	G7, J6
Edge Columns w/ cantilevers	-	4	3	2.67	3	B3
Corner Columns w/ cantilevers	-	4	2.25	1.78	2	K2
Edge Beams w/o cantilever slabs	2	-	-	-	2	D7 - E7
Interior Beams	2	-	-	-	2	H4 - H5
Edge Beams w/ cantilever slabs	-	2	1.5	1.33	1	B5 - B6
Cantilever Beams w/o cantilever slabs	2	-	-	-	1	E1 - E2
Cantilever Beams w/ cantilever slabs	-	2*	1.5*	1.33*	1	K5 - L5

* The value of n for member K5-L5 is used to calculate the distance nL₁₁

FIGURE C4
TYPICAL TRIBUTARY AND INFLUENCE AREAS

Occupancy or use	Live Load lb/ft^2 (kN/m^2)	Occupancy or use	Live Load lb/ft^2 (kN/m^2)
Air-conditioning (machine space)	200* (9.58)	Laboratories, scientific	100 (4.79)
Amusement park structure	100* (4.79)	Laundries	150* (7.18)
Attic, Nonresidential		Libraries, corridors	80* (3.83)
Nonstorage	25 (1.20)	Manufacturing, ice	300 (14.36)
Storage	80* (3.83)	Morgue	125 (6.00)
Bakery	150 (7.18)	Office Buildings	
Exterior	100 (4.79)	Business machine equipment	100* (4.79)
Interior (fixed seats)	60 (2.87)	Files (see file room)	
Interior (movable seats)	100 (4.79)	Printing Plants	
Boathouse, floors	100* (4.79)	Composing rooms	100 (4.79)
Boiler room, framed	300* (14.36)	Linotype rooms	100 (4.79)
Broadcasting studio	100 (4.79)	Paper storage	**
Catwalks	25 (1.20)	Press rooms	150* (7.18)
Ceiling, accessible furred	10# (0.48)	Public rooms	100 (4.79)
Cold Storage		Railroad tracks	‡‡
No overhead system	250‡ (11.97)	Ramps	
Overhead system		Driveway (see garages)	
Floor	150 (7.18)	Pedestrian (see sidewalks and corridors in Table 2)	
Roof	250 (11.97)	Seaplane (see hangars)	
Computer equipment	150* (7.18)	Rest rooms	60 (2.87)
Courtrooms	50–100 (2.40–4.79)	Rinks	
Dormitories		Ice skating	250 (11.97)
Nonpartitioned	80 (3.83)	Roller skating	100 (4.79)
Partitioned	40 (1.92)	Storage, hay or grain	300* (14.36)
Elevator machine room	150* (7.18)	Telephone exchange	150* (7.18)
Fan room	150* (7.18)	Theaters	
File room		Dressing rooms	40 (1.92)
Duplicating equipment	150* (7.18)	Grid-iron floor or fly gallery:	
Card	125* (6.00)	Grating	60 (2.87)
Letter	80* (3.83)	Well beams, 250 lb/ft per pair	
Foundries	600* (28.73)	Header beams, 1000 lb/ft	
Fuel rooms, framed	400 (19.15)	Pin rail, 250 lb/ft	
Garages — trucks	§	Projection room	100 (4.79)
Greenhouses	150 (7.18)	Toilet rooms	60 (2.87)
Hangars	150§ (7.18)	Transformer rooms	200* (9.58)
Incinerator charging floor	100 (4.79)	Vaults, in offices	250* (11.97)
Kitchens, other than domestic	150* (7.18)		

* Use weight of actual equipment or stored material when greater.

‡ Plus 150 lb/ft^2 (7.18 kN/m^2) for trucks.

§ Use American Association of State Highway and Transportation Officials lane loads. Also subject to not less than 100% maximum axle load.

** Paper storage 50 lb/ft (2.40 kN/m^2) of clear story height.

‡‡ As required by railroad company.

Accessible ceilings normally are not designed to support persons. The value in this table is intended to account for occasional light storage or suspension of items. If it may be necessary to support the weight of maintenance personnel, this shall be provided for.

TABLE C4-2
TYPICAL LIVE LOAD STATISTICS

Occupancy or use	Survey Load		Transient Load		Temporal Constants			Mean maximum load*
	m_s lb/ft^2 (kN/m^2)	σ_s* lb/ft^2 (kN/m^2)	m_t* lb/ft^2 (kN/m^2)	σ_t* lb/ft^2 (kN/m^2)	τ_s† (years)	v_e‡ (per year)	T§ (year)	lb/ft^2 (kN/m^2)
Office buildings								
Offices	10.9 (0.52)	5.9 (0.28)	8.0 (0.38)	8.2 (0.39)	8	1	50	55 (2.63)
Residential								
Renter occupied	6.0 (0.29)	2.6 (0.12)	6.0 (0.29)	6.6 (0.32)	2	1	50	36 (1.72)
Owner occupied	6.0 (0.29)	2.6 (0.12)	6.0 (0.29)	6.6 (0.32)	10	1	50	38 (1.82)
Hotels								
Guest rooms	4.5 (0.22)	1.2 (0.06)	6.0 (0.29)	5.8 (0.28)	5	20	50	46 (2.2)
Schools								
Classrooms	12.0 (0.57)	2.7 (0.13)	6.9 (0.33)	3.4 (0.16)	1	1	100	34 (1.63)

* For 200-ft^2 (18.58 m^2) area, except 1000 ft^2 (92.9 m^2) for schools.

† Duration of average sustained load occupancy.

‡ Mean rate of occurrence of transient load.

§ Reference period.

SECTION C5.0
SOIL AND HYDROSTATIC PRESSURE AND FLOOD LOADS

SECTION C5.1
PRESSURE ON BASEMENT WALLS

Table 5-1 includes high earth pressures, 85 pcf (13.36 kN/m^2) or more, to show that certain soils are poor backfill material. In addition, when walls are unyielding, the earth pressure is increased from active pressure toward earth pressure at rest, resulting in 60 pcf (9.43 kN/m^2) for granular soils and 100 pcf (15.71 kN/m^2) for silt and clay type soils (see C5-4]). Examples of light floor systems supported on shallow basement walls mentioned in Table 5-1 are floor systems with wood joists and flooring, and cold-formed steel joists without cast-in-place concrete floor attached.

Expansive soils exist in many regions of the United States and may cause serious damage to basement walls unless special design considerations are provided. Expansive soils should not be used as backfill because they can exert very high pressures against walls. Special soil testing is required to determine the magnitude of these pressures. It is preferable to excavate expansive soil and backfill with nonexpansive freely draining sands or gravels. The excavated backslope adjacent to the wall should be no steeper than 45 degrees from the horizontal in order to minimize the transmission of swelling pressure from the expansive soil through the new backfill. Other special details are recommended, such as a cap of nonpervious soil on top of the backfill and provision of foundation drains. Refer to current reference books on geotechnical engineering for guidance.

SECTION C5.2
UPLIFT ON FLOORS AND FOUNDATIONS

If expansive soils are present under floors or footings, large pressures can be exerted and must be resisted by special design. Alternatively, the expansive soil can be excavated to a depth of at least 2 ft (0.60 m) and backfilled with nonexpansive freely draining sands or gravel. A geotechnical engineer should make recommendations in these situations.

SECTION C5.3
FLOOD LOADS

This section presents information for the design of buildings and other structures in areas prone to flooding. Design professionals should be aware that there are important differences between flood characteristics, flood loads, and flood effects in riverine and coastal areas (e.g., the potential for wave effects is much greater in coastal areas; the depth and duration of flooding can be much greater in riverine areas; the direction of flow in riverine areas tends to be more predictable; the nature and amount of flood-borne debris varies between riverine and coastal areas).

Much of the impetus for flood-resistant design has come about from the federal government-sponsored initiatives of flood-damage mitigation and flood insurance, both through the work of the U.S. Army Corps of Engineers and the National Flood Insurance Program (NFIP). The NFIP is based on an agreement between the federal government and participating communities that have been identified as being flood prone. The Federal Emergency Management Agency (FEMA) through the Federal Insurance and Mitigation Administration (FIMA) makes flood insurance available to the residents of communities provided that the community adopts and enforces adequate floodplain management regulations that meet the minimum requirements. Included in the NFIP requirements, found under Title 44 of the U.S. Code of Federal Regulations [C5-3], are minimum building design and construction standards for buildings and other structures located in Special Flood Hazard Areas (SFHA).

SFHA are those identified by FEMA as being subject to inundation during the 100-year flood. SFHA are shown on Flood Insurance Rate Maps (FIRM), which are produced for flood-prone communities. SFHA are identified on FIRM's as Zones A, A1-30, AE, AR, AO, AH, and coastal high hazard areas as V1-30, V and VE. The SFHA is the area in which communities must enforce NFIP-compliant, flood damage-resistant design and construction practices.

Prior to designing a structure in a flood-prone area, design professionals should contact the local building official to determine if the site in question is located in a SFHA or other flood-prone area that is regulated under the community's floodplain management regulations. If the proposed structure is located within the regulatory floodplain, local building officials can explain the regulatory requirements.

Answers to specific questions on flood-resistant design and construction practices may be directed to the Mitigation Division of each of FEMA's regional offices. FEMA has regional offices that are available to assist design professionals.

C5.3.1 Definitions. Three new concepts were added with ASCE 7-98. First, the concept of the Design Flood is introduced. The Design Flood will, at a minimum, be equivalent to the flood having a 1% chance of being equaled or exceeded in any given year (i.e., the Base Flood or

100-year flood, which served as the load basis in ASCE 7-95). In some instances, the Design Flood may exceed the Base Flood in elevation or spatial extent—this will occur where a community has designated a greater flood (lower frequency, higher return period) as the flood to which the community will regulate new construction.

Many communities have elected to regulate to a flood standard higher than the minimum requirements of the National Flood Insurance Program (NFIP). Those communities may do so in a number of ways. For example: a community may require new construction to be elevated a specific vertical distance above the Base Flood Elevation (this is referred to as freeboard); a community may select a lower frequency flood as its regulatory flood; a community may conduct hydrologic and hydraulic studies, upon which Flood Hazard Maps are based, in a manner different than the Flood Insurance Study prepared by the NFIP (the community may complete flood hazard studies based upon development conditions at build-out, rather than following the NFIP procedure, which uses conditions in existence at the time the studies are completed; the community may include watersheds smaller than 1 mi^2 in size in its analysis, rather than following the NFIP procedure, which neglects watersheds smaller than 1 mi^2).

Use of the Design Flood concept will ensure that the requirements of this Standard are not less restrictive than a community's requirements where that community has elected to exceed minimum NFIP requirements. In instances where a community has adopted the NFIP minimum requirements, the Design Flood described in this Standard will default to the Base Flood.

Second, this Standard also uses the terms Flood Hazard Area and Flood Hazard Map to correspond to and show the areas affected by the Design Flood. Again, in instances where a community has adopted the minimum requirements of the NFIP, the Flood Hazard Area defaults to the NFIP's Special Flood Hazard Area and the Flood Hazard Map defaults to the Flood Insurance Rate Map.

Third, the concept of a Coastal A Zone is used to facilitate application of load combinations contained in Section 2. Coastal A Zones lie landward of V Zones, or landward of an open coast shoreline where V Zones have not been mapped (e.g., the shorelines of the Great Lakes), Coastal A Zones are subject to the effects of waves, high-velocity flows, and erosion, although not to the extent that V Zones are. Like V Zones, flood forces in coastal A Zones will be highly correlated with coastal winds or coastal seismic activity.

C5.3.2 Design Requirements. Sections 5.3.4 (dealing with A-Zone design and construction) and 5.3.5 (dealing with V-Zone design and construction) of ASCE 7-98 have been deleted. These sections summarized basic principles of flood-resistant design and construction (building elevation, anchorage, foundation, below-DFE enclosures, breakaway walls, etc.). Some of the information contained in these deleted sections has been included in Section 5.3 of ASCE 7-02, and the design professional is also referred to SEI/ASCE 24-98 (*Standard for Flood Resistant Design and Construction*) for specific guidance.

C5.3.2.1 Design Loads. Wind loads and flood loads may act simultaneously at coastlines, particularly during hurricanes and coastal storms. This may also be true during severe storms at the shorelines of large lakes, and during riverine flooding of long duration.

C5.3.2.2 Erosion and Scour. The term "erosion" indicates a lowering of the ground surface in response to a flood event, or in response to the gradual recession of a shoreline. The term "scour" indicates a localized lowering of the ground surface during a flood due to the interaction of currents and/or waves with a structural element. Erosion and scour can affect the stability of foundations, and can increase the local flood depth and flood loads acting on buildings and other structures. For these reasons, erosion and scour should be considered during load calculations and the design process. Design professionals often increase the depth of foundation embedment to mitigate the effects of erosion and scour, and often site buildings away from receding shorelines (building set-backs).

C5.3.3.6 Loads on Breakaway Walls. Floodplain management regulations require buildings in coastal high hazard areas to be elevated to or above the design flood elevation by a pile or column foundation. Space below the DFE must be free of obstructions in order to allow the free passage of waves and high-velocity waters beneath the building [C5-6]. Floodplain management regulations typically allow space below the DFE to be enclosed by insect screening, open lattice, or breakaway walls. Local exceptions are made in certain instances for shear walls, firewalls, elevator shafts, and stairwells—check with the authority having jurisdiction for specific requirements related to obstructions, enclosures, and breakaway walls.

Where breakaway walls are used, they must meet the prescriptive requirements of NFIP regulations or be certified by a registered professional engineer or architect as having been designed to meet the NFIP performance requirements. The prescriptive requirements call for breakaway wall designs which are intended to collapse at loads not less than 10 psf ($0.48 kN/m^2$) and not more than 20 psf ($0.96 kN/m^2$). Inasmuch as wind or earthquake loads often exceed 20 psf ($0.96 kN/m^2$), breakaway walls may be designed for a higher load, provided the designer certifies that the walls have been designed to break away before base flood conditions are reached without damaging the elevated building or its foundation. A recent reference [C5-7] provides guidance on how to meet the performance requirements for certification.

C5.3.3.1 Load Basis. Water loads are the loads or pressures on surfaces of buildings and structures caused and induced by the presence of floodwaters. These loads are of two basic types: hydrostatic and hydrodynamic. Impact loads result from objects transported by floodwaters striking against buildings and structures or part thereof. Wave loads can be considered a special type of hydrodynamic load.

C5.3.3.2 Hydrostatic Loads. Hydrostatic loads are those caused by water either above or below the ground surface, free or confined, which is either stagnant or moves at velocities less than 5 ft/sec (1.52 m/s). These loads are equal to the product of the water pressure multiplied by the surface area on which the pressure acts.

Hydrostatic pressure at any point is equal in all directions and always acts perpendicular to the surface on which it is applied. Hydrostatic loads can be subdivided into vertical downward loads, lateral loads, and vertical upward loads (uplift or buoyancy). Hydrostatic loads acting on inclined, rounded, or irregular surfaces may be resolved into vertical downward or upward loads and lateral loads based on the geometry of the surfaces and the distribution of hydrostatic pressure.

C5.3.3.3 Hydrodynamic Loads. Hydrodynamic loads are those loads induced by the flow of water moving at moderate-to-high velocity above the ground level. They are usually lateral loads caused by the impact of the moving mass of water and the drag forces as the water flows around the obstruction. Hydrodynamic loads are computed by recognized engineering methods. In the coastal high hazard area, the loads from high-velocity currents due to storm surge and overtopping are of particular importance. Reference [C5-1] is one source of design information regarding hydrodynamic loadings.

Note that accurate estimates of flow velocities during flood conditions are very difficult to make, both in riverine and coastal flood events. Potential sources of information regarding velocities of floodwaters include local, state, and federal government agencies and consulting engineers specializing in coastal engineering, stream hydrology, or hydraulics.

As interim guidance for coastal areas, see Reference [C5-8], which gives a likely range of flood velocities as:

$$V = d_s/(1 \text{ sec}) \qquad \textbf{(Eq. C5-1)}$$

to

$$V = (gd_s)^{0.5} \qquad \textbf{(Eq. C5-2)}$$

where

V = average velocity of water in ft/sec (m/s)
d_s = local stillwater depth in ft (m)
g = acceleration due to gravity, 32.2 ft/sec (9.81 m/s^2)

Selection of the correct value of "a" in Eq. 5-1 will depend upon the shape and roughness of the object exposed to flood flow, as well as the flow condition itself. As a general rule, the smoother and more streamlined the object, the lower the drag coefficient (shape factor). Drag coefficients for elements common in buildings and structures (round or square piles, columns, and rectangular shapes) will range from approximately 1.0 to 2.0, depending upon flow conditions. However, given the uncertainty surrounding flow conditions at a particular site, ASCE 7-02 recommends a minimum value of 1.25 be used. Fluid mechanics texts should be consulted for more information on when to apply drag coefficients above 1.25.

C5.3.3.4 Wave Loads. The magnitude of wave forces (lbs/ft^2)(kN/m^2) acting against buildings or other structures can be 10 or more times higher than wind forces and other forces under design conditions. Thus, it should be readily apparent that elevating above the wave crest elevation is crucial to the survival of buildings and other structures. Even elevated structures, however, must be designed for large wave forces that can act over a relatively small surface area of the foundation and supporting structure.

Wave load calculation procedures in Section 5.3.3.4 are taken from References [C5-1] and [C5-5]. The analytical procedures described by Eqs. 5-2 through 5-9 should be used to calculate wave heights and wave loads unless more advanced numerical or laboratory procedures permitted by this Standard are used.

Wave load calculations using the analytical procedures described in this Standard all depend upon the initial computation of the wave height, which is determined using Eqs. 5-2 and 5-3. These equations result from the assumptions that the waves are depth-limited, and that waves propagating into shallow water break when the wave height equals 78% of the local stillwater depth and that 70% of the wave height lies above the local stillwater level. These assumptions are identical to those used by FEMA in its mapping of coastal flood hazard areas on FIRMs.

Designers should be aware that wave heights at a particular site can be less than depth-limited values in some cases (e.g., when the wind speed, wind duration, or fetch is insufficient to generate waves large enough to be limited in size by water depth, or when nearby objects dissipate wave energy and reduce wave heights). If conditions during the design flood yield wave heights at a site less than depth-limited heights, Eq. 5-2 may overestimate the wave height and Eq. 5-3 may underestimate the stillwater depth. Also, Eqs. 5-4 through 5-7 may overstate wave pressures and forces when wave heights are less than depth-limited heights. More advanced numerical or laboratory procedures permitted by this section may be used in such cases, in lieu of Eqs. 5-2 through 5-7.

It should be pointed out that present NFIP mapping procedures distinguish between A Zones and V Zones by the wave heights expected in each zone. Generally

speaking, A Zones are designated where wave heights less than 3 ft (0.91 m) in height are expected; V Zones are designated where wave heights equal to or greater than 3 ft (0.91 m) are expected. Designers should proceed cautiously, however. Large wave forces can be generated in some A Zones, and wave force calculations should not be restricted to V Zones. Present NFIP mapping procedures do not designate V Zones in all areas where wave heights greater than 3 ft (0.91 m) can occur during base flood conditions. Rather than rely exclusively on flood hazard maps, designers should investigate historical flood damages near a site to determine whether or not wave forces can be significant.

C5.3.3.4.2 Breaking Wave Loads on Vertical Walls. Equations used to calculate breaking wave loads on vertical walls contain a coefficient, C_p. Reference [C5-5] provides recommended values of the coefficient as a function of probability of exceedance. The probabilities given by Reference [C5-5] are not annual probabilities of exceedance, but probabilities associated with a distribution of breaking wave pressures measured during laboratory wave tank tests. Note that the distribution is independent of water depth. Thus, for any water depth, breaking wave pressures can be expected to follow the distribution described by the probabilities of exceedance in Table 5-2.

This Standard assigns values for C_p according to the building category with the most important buildings having the largest values of C_p. Category II buildings are assigned a value of C_p corresponding to a 1% probability of exceedance, which is consistent with wave analysis procedures used by FEMA in mapping coastal flood hazard areas and in establishing minimum floor elevations. Category I buildings are assigned a value of C_p corresponding to a 50% probability of exceedance, but designers may wish to choose a higher value of C_p. Category III buildings are assigned a value of C_p corresponding to a 0.2% probability of exceedance, while Category IV buildings are assigned a value of C_p corresponding to a 0.1% probability of exceedance.

Breaking wave loads on vertical walls reach a maximum when the waves are normally incident (direction of wave approach perpendicular to the face of the wall; wave crests are parallel to the face of the wall). As guidance for designers of coastal buildings or other structures on normally dry land (i.e., flooded only during coastal storm or flood events), it can be assumed that the direction of wave approach will be approximately perpendicular to the shoreline. Therefore, the direction of wave approach relative to a vertical wall will depend upon the orientation of the wall relative to the shoreline. Section 5.3.3.4.4 provides a method for reducing breaking wave loads on vertical walls for waves not normally incident.

C5.3.3.5 Impact Loads. Impact loads are those that result from logs, ice floes, and other objects striking buildings,

structures, or parts thereof. Reference [C5-2] divides impact loads into three categories: (1) normal impact loads, which result from the isolated impacts of normally encountered objects; (2) special impact loads that result from large objects, such as broken up ice floats and accumulations of debris, either striking or resting against a building, structure, or parts thereof; and (3) extreme impact loads that result from very large objects such as boats, barges, or collapsed buildings striking the building, structure, or component under consideration. Design for extreme impact loads is not practical for most buildings and structures. However, in cases where there is a high probability that a Category III or IV structure (see Table 1-1) will be exposed to extreme impact loads during the design flood, and where the resulting damages will be very severe, consideration of extreme impact loads may be justified. Unlike extreme impact loads, design for special and normal impact loads is practical for most buildings and structures.

The recommended method for calculating normal impact loads has been modified with ASCE 7-02. Previous editions of ASCE 7 used a procedure contained in Reference [C5-2] (the procedure, which had been unchanged since at least 1972, relied on an impulse-momentum approach with a 1000 lb (4.5 kN) object striking the structure at the velocity of the floodwater and coming to rest in 1.0 sec). Recent work [C5-9 and C5-10] has been conducted to evaluate this procedure, through a literature review and laboratory tests. The literature review considered riverine and coastal debris, ice floes and impacts, ship berthing and impact forces, and various methods for calculating debris loads (e.g., impulse-momentum, work-energy). The laboratory tests included log sizes ranging from 380 lbs (1.7 kN) to 730 lbs (3.3 kN) travelling at up to 4 ft/sec (1.2 m/s).

References [C5-9 and C5-10] conclude: (1) an impulse-momentum approach is appropriate; (2) the 1000 lb (4.5 kN) object is reasonable, although geographic and local conditions may affect the debris object size and weight; (3) the 1.0 second impact duration is not supported by the literature or by laboratory tests—a duration of impact of 0.03 seconds should be used instead; (4) a half-sine curve represents the applied load and resulting displacement well; and (5) setting the debris velocity equivalent to the flood velocity is reasonable for all but the largest objects in shallow water or obstructed conditions.

Given the short-duration, impulsive loads generated by flood-borne debris, a dynamic analysis of the affected building or structure may be appropriate. In some cases (e.g., when the natural period of the building is much greater than 0.03 seconds), design professionals may wish to treat the impact load as a static load applied to the building or structure (this approach is similar to that used by some following the procedure contained in Section C5.3.3.5 of ASCE 7-98).

In either type of analysis—dynamic or static— Eq. C5-3 provides a rational approach for calculating the

magnitude of the impact load.

$$F = \frac{\pi W V_b C_I C_O C_D C_B R_{max}}{2g \Delta t} \qquad \textbf{(Eq. C5-3)}$$

where

F = impact force, in lbs (N)

W = debris weight in lbs (N)

V_b = velocity of object (assume equal to velocity of water, V) in ft/sec (m/s)

g = acceleration due to gravity, = 32.2 ft/sec² (9.81 m/s²)

Δt = impact duration (time to reduce object velocity to zero), in seconds.

C_I = importance coefficient (see Table C5-3)

C_O = orientation coefficient, = 0.8.

C_D = depth coefficient (see Table C5-4, Figure C5-3)

C_B = blockage coefficient (see Table C5-5, Figure C5-4)

R_{max} = maximum response ratio for impulsive load (see Table C5-6)

The form of Eq. C5-3 and the parameters and coefficients are discussed below:

Basic Equation. The equation is similar to the equation used in ASCE 7-98, except for the $\pi/2$ factor (which results from the half-sine form of the applied impulse load) and the coefficients C_I, C_O, C_D, C_B, and R_{max}. With the coefficients set equal to 1.0, the equation reduces to $F = \pi W V_b / 2g \Delta t$, and calculates the maximum static load from a head-on impact of a debris object. The coefficients have been added to allow design professionals to "calibrate" the resulting force to local flood, debris, and building characteristics. The approach is similar to that employed by ASCE 7 in calculating wind, seismic, and other loads — a scientifically based equation is used to match the physics, and the results are modified by coefficients to calculate realistic load magnitudes. However, unlike wind, seismic, and other loads the body of work associated with flood-borne debris impact loads does not yet account for the probability of impact.

Debris Object Weight. A 1000-lb object can be considered a reasonable average for flood-borne debris (no change from ASCE 7-98). This represents a reasonable weight for trees, logs, and other large woody debris, which is the most common form of damaging debris nationwide. This weight corresponds to a log approximately 30 ft (9.1 m) long and just under 1 ft (0.3 m) in diameter. The 1000-lb object also represents a reasonable weight for other types of debris ranging from small ice floes, to boulders, to man-made objects.

However, design professionals may wish to consider regional or local conditions before the final debris weight is selected. The following text provides additional guidance. In riverine floodplains, large woody debris (trees and logs) predominates, with weights typically ranging from 1000 lbs (4.5 kN) to 2000 lbs (9.0 kN). In the Pacific Northwest, larger tree and log sizes suggest a typical 4000 lb (18.0 kN) debris weight. Debris weights in riverine areas subject to floating ice typically range from 1000 lbs (4.5 kN) to 4000 lbs (18.0 kN). In arid or semi-arid regions, typical woody debris may be less than 1000 lbs (4.5 kN). In alluvial fan areas, non-woody debris (stones and boulders) may present a much greater debris hazard.

Debris weights in coastal areas generally fall into three classes: in the Pacific Northwest, a 4000 lb (18.0 kN) debris weight due to large trees and logs can be considered typical; in other coastal areas where piers and large pilings are available locally, debris weights may range from 1000 lbs (4.5 kN) to 2000 lbs (9.0 kN); in other coastal areas where large logs and pilings are not expected, debris will likely be derived from failed decks, steps, and building components, and will likely average less than 500 lbs (2.3 kN) in weight.

Debris Velocity. The velocity that a piece of debris strikes a building or structure will depend upon the nature of the debris and the velocity of the floodwaters. Small pieces of floating debris, which are unlikely to cause damage to buildings or other structures, will typically travel at the velocity of the floodwaters, in both riverine and coastal flood situations. However, large debris, such as trees, logs, pier pilings, and other large debris capable of causing damage, will likely travel at something less than the velocity of the floodwaters. This reduced velocity of large debris objects is due in large part to debris dragging along the bottom and/or being slowed by prior collisions. Large riverine debris travelling along the floodway (the deepest part of the channel that conducts the majority of the flood flow) is most likely to travel at speeds approaching that of the floodwaters. Large riverine debris travelling in the floodplain (the shallower area outside the floodway) is more likely to be travelling at speeds less than that of the floodwaters, for those reasons stated above. Large coastal debris is also likely to be travelling at speeds less than that of the floodwaters. Eq. C5-2 should be used with the debris velocity equal to the flow velocity, since the equation allows for reductions in debris velocities through application of a depth coefficient, C_D and an upstream blockage coefficient, C_B.

Duration of Impact. A detailed review of the available literature [C5-9], supplemented by laboratory testing, concluded the previously suggested 1.0 second duration of impact is much too long and is not realistic. Laboratory tests showed that measured impact durations (from initial impact to time of maximum force Δt) varied from 0.01 sec to 0.05 sec [C5-9]. Results for one test, for example, produced a maximum impact load of 8300 lbs (37,000 N) for a log weighing 730 lbs (3250 N), moving at 4 ft/sec, and impacting with a duration of 0.016 sec. Over all the test conditions, the impact duration averaged about 0.026 sec. The recommended value for use in Eq. C5-2 is, therefore, 0.03 sec.

Coefficients C_I, C_O, C_D, and C_B. The coefficients are based in part on the results of laboratory testing and in part on engineering judgement. The values of the coefficients should be considered interim until more experience is gained with them.

The Importance Coefficient. C_I is generally used to adjust design loads for the structure category and hazard to human life following ASCE-7-98 convention in Table 1-1. Recommended values given in Table C5-3 are based on a probability distribution of impact loads obtained from laboratory tests in Reference [C5-10].

The Orientation Coefficient. C_O is used to reduce the load calculated by Eq. C5-3 for impacts that are oblique, not head-on. During laboratory tests [C5-10] it was observed that while some debris impacts occurred as direct or head-on impacts that produced maximum impact loads, most impacts occurred as eccentric or oblique impacts with reduced values of the impact force. Based on these measurements, an orientation coefficient of $C_O = 0.8$ has been adopted to reflect the general load reduction observed due to oblique impacts.

The Depth Coefficient. C_D is used to account for reduced debris velocity in shallow water due to debris dragging along the bottom. Recommended values of this coefficient are based on typical diameters of logs and trees, or on the anticipated diameter of the root mass from drifting trees that are likely to be encountered in a flood hazard zone. Reference [C5-9] suggests that trees with typical root mass diameters will drag the bottom in depths of less than 5 ft, while most logs of concern will drag the bottom in depths of less than 1 ft. The recommended values for the depth coefficient are given in Table C5-4 and Figure C5-3. No test data are available to fully

validate the recommended values of this coefficient. When better data are available, designers should use them in lieu of the values contained in Table C5-4 and Figure C5-3.

The Blockage Coefficient. C_B is used to account for the reductions in debris velocities expected due to screening and sheltering provided by trees or other structures within about 10 log-lengths (300 ft) upstream from the building or structure of interest. Reference [C5-9] quotes other studies in which dense trees have been shown to act as a screen to remove debris and shelter downstream structures. The effectiveness of the screening depends primarily on the spacing of the upstream obstructions relative to the design log length of interest. For a 1000-lb log, having a length of about 30 ft, it is therefore assumed that any blockage narrower than 30 ft would trap some or all of the transported debris. Likewise, typical root mass diameters are on the order of 3 to 5 ft, and it is therefore assumed that blockages of this width would fully trap any trees or long logs. Recommended values for the blockage coefficient are given in Table C5-5 and Figure C5-4 based on interpolation between these limits. No test data are available to fully validate the recommended values of this coefficient.

The Maximum Response Ratio. R_{max} is used to increase or decrease the computed load, depending on the degree of compliance of the building or building component being struck by debris. Impact loads are impulsive in nature with the force rapidly increasing from zero to the maximum value in time Δt, then decreasing to zero as debris rebounds from structure. The actual load experienced by the structure or component will depend on the ratio of the impact duration Δt relative to the natural period of the structure or component, T_n. Stiff or rigid buildings and structures with natural periods similar to the impact duration will see an amplification of the impact load. More flexible buildings and structures with natural periods greater than approximately 4 times the impact duration will see a reduction of the impact load. Likewise, stiff or rigid components will see an amplification of the impact load; more flexible components will see a reduction of the impact load. Successful use of Eq. C5-3, then, depends on estimation of the natural period of the building or component being struck by flood-borne debris. Calculating the natural period can be carried out using established methods that take building mass, stiffness, and configuration into account. One useful reference is Appendix C of ANSI/ACI 349 [C5-12]. Design professionals are also referred to Section 9 of ASCE 7-02 for additional information.

Natural periods of buildings generally vary from approximately 0.05 sec to several seconds (for high-rise, moment frame structures). For flood-borne debris impact loads with a duration of 0.03 sec, the critical period (above which loads are reduced) is approximately 0.11 sec (see Table C5-6). Buildings and structures with natural periods above approximately 0.11 sec will see a reduction in the debris impact load, while those with natural periods below approximately 0.11 sec will see an increase. Recent shake table tests of conventional, one- to two-story wood-frame buildings have shown natural periods of ranging from approximately 0.14 sec (7 Hz) to 0.33 sec (3 Hz), averaging approximately 0.20 sec (5 Hz). Elevating these types of structures for flood-resistant design purposes will act to increase these natural periods. For the purposes of flood-borne debris impact load calculations, a natural period of 0.5 to 1.0 sec is recommended for one- to three-story buildings elevated on timber piles. For 1- to 3-story buildings elevated on masonry columns, a similar range of natural periods is recommended. For 1- to 3-story buildings elevated on concrete piles or columns, a natural period of 0.2 to 0.5 sec is recommended. Design professionals are referred to Section 9.5.5.3 of ASCE 7-02 (or 9.5.3.3 of ASCE 7-98) where an approximate natural period for 1- to 12-story buildings (story height equal to or greater than 10 ft [3 m]), with concrete and steel moment-resisting frames can be approximated as 0.1 times the number of stories.

Special Impact Loads. Reference [C5-2] states that, absent a detailed analysis, special impact loads can be estimated as a uniform load of 100 lbs/ft (1.48 kN/m), acting over a 1 ft (0.31 m) high horizontal strip at the design flood elevation or lower. However, Reference [C5-9] suggests that this load may be too small for some large accumulations of debris, and suggests an alternative approach involving application of the standard drag force expression

$$F = (1/2)C_D \rho A V^2 \qquad \textbf{(Eq. C5-4)}$$

where

F = drag force due to debris accumulation, in lbs (N)

V = flow velocity upstream of debris accumulation, in ft/sec (m/s)

A = projected area of the debris accumulation into the flow, approximated by depth of accumulation times width of accumulation perpendicular to flow, in ft^2 (m^2)

ρ = density of water in slugs/ft^3 (kg/m$_3$)

C_D = drag coefficient = 1

This expression produces loads similar to the 100 lb/ft guidance from Reference [C5-2] when the debris depth is assumed to be 1 ft and when the velocity of the floodwater is 10 ft/sec. Other guidance from References [C5-9] and [C5-10] suggest that the depth of debris accumulation is often much greater than 1 ft, and is only limited by the water depth at the structure. Observations of debris accumulations at bridge piers listed in these references show typical depths of 5 to 10 ft, with horizontal widths spanning between adjacent bridge piers whenever the spacing of the piers is less than the typical log length. If debris accumulation is of concern, the design professional should specify the projected area of the debris accumulation based on local observations and experience, and apply the above equation to predict the debris load on buildings or other structures.

REFERENCES

[C5-1] U.S. Army Corps of Engineers, Coastal Engineering Research Center, Waterways Experiment Station, Shore Protection Manual, 4th Ed., 1984.

[C5-2] U.S. Army Corps of Engineers, Office of the Chief of Engineers, Flood Proofing Regulations, EP 1165-2-314, December 1995.

[C5-3] Federal Emergency Management Agency, National Flood Insurance Program, 44 CFR Ch. 1 Parts 59 and 60, (10-1-99 Edition), 1990.

[C5-4] Terzaghi, K. and Peck, R.B., Soil Mechanics in Engineering Practice, Wiley, 2nd Ed., 1967.

[C5-5] Walton, T.L., Jr., J.P. Ahrens, C.L. Truitt, and R.G. Dean, Criteria for Evaluating Coastal Flood Protection Structures, Technical Report CERC 89-15, U.S. Army Corps of Engineers, Waterways Experiment Station, 1989.

[C5-6] Federal Emergency Management Agency, Free-of-Obstruction Requirements for Buildings Located in Coastal High Hazard Areas in accordance with the National Flood Insurance Program, Technical Bulletin 5-93. Mitigation Directorate, 1993.

[C5-7] Federal Emergency Management Agency, Design and Construction Guidance for Breakaway Walls Below Elevated Coastal Buildings in accordance with the National Flood Insurance Program, Technical Bulletin 9-99. Mitigation Directorate, 1999.

[C5-8] Federal Emergency Management Agency, Revised Coastal Construction Manual, FEMA-55. Mitigation Directorate, 2000.

[C5-9] Kriebel, D.L., Buss, L., and Rogers, S. Impact Loads from Flood-Borne Debris. Report to the American Society of Civil Engineers, 2000.

[C5-10] Haehnel, R. and Daly, S., "Debris Impact Tests," Report prepared for the American Society of Civil Engineers by the U.S. Army Cold Regions

Research and Engineering Laboratory, Hanover, NH., 2001.

[C5-11] Clough, R.W. and Penzien, J., *Dynamics of Structures*, 2nd Ed. McGraw-Hill, New York, 1975.

[C5-12] American Concrete Institute. ANSI/ACI 349, Code Requirements for Nuclear Safety Related Concrete Structures, 1985.

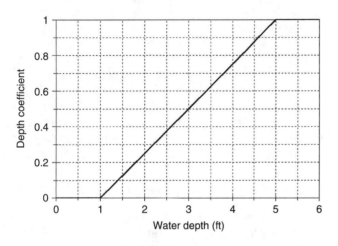

FIGURE C5-3
DEPTH COEFFICIENT, C_D

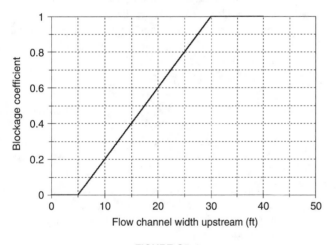

FIGURE C5-4
BLOCKAGE COEFFICIENT, C_B

TABLE C5-3
VALUES OF
IMPORTANCE
COEFFICIENT, C_I

Building Category	C_I
I	0.6
II	1.0
III	1.2
IV	1.3

TABLE C5-4
VALUES OF DEPTH COEFFICIENT, C_D

Building Location in Flood Hazard Zone and Water Depth	C_D
Floodway or V Zone	1.0
A Zone, stillwater depth > 5 ft	1.0
A Zone, stillwater depth = 4 ft	0.75
A Zone, stillwater depth = 3 ft	0.5
A Zone, stillwater depth = 2 ft	0.25
Any flood zone, stillwater depth < 1 ft	0.0

TABLE C5-5
VALUES OF BLOCKAGE COEFFICIENT, C_B

Degree of screening or sheltering within 100 ft upstream	C_B
No upstream screening, flow path wider than 30 ft	1.0
Limited upstream screening, flow path 20 ft wide	0.6
Moderate upstream screening, flow path 10 ft wide	0.2
Dense upstream screening, flow path <5 ft wide	0.0

TABLE C5-6
VALUES OF RESPONSE RATIO FOR IMPULSIVE LOADS, R_{max} (ADAPTED FROM [C5-11, FIGURE 5-6])

Ratio of Impact Duration to Natural Period of Structure	R_{max} (Response Ratio for Half-Sine Wave Impulsive Load)
0.00	0.0
0.10	0.4
0.20	0.8
0.30	1.1
0.40	1.4
0.50	1.5
0.60	1.7
0.70	1.8
0.80	1.8
0.90	1.8
1.00	1.7
1.10	1.7
1.20	1.6
1.30	1.6
≥ 1.40	1.5

SECTION C6.0
WIND LOADS

SECTION C6.1
GENERAL

The ASCE 7-02 version of the wind load standard provides three methods from which the designer can choose. An expanded "simplified method" (Method 1) for which the designer can select wind pressures directly without any calculation when the building meets all the requirements for application of the procedure; and two other methods (Analytical Method and Wind Tunnel Procedure), which are essentially the same methods as previously given in the standard except for changes that are noted.

Temporary bracing should be provided to resist wind loading on structural components and structural assemblages during erection and construction phases.

SECTION C6.2
DEFINITIONS

Several important definitions given in the standard are discussed further below. These terms are used throughout the standard and are provided to clarify application of the standard provisions.

MAIN WIND FORCE-RESISTING SYSTEM. Can consist of a structural frame or an assemblage of structural elements that work together to transfer wind loads acting on the entire structure to the ground. Structural elements such as cross-bracing, shear walls, roof trusses, and roof diaphragms are part of the main wind force-resisting system when they assist in transferring overall loads [C6-87].

BUILDING, ENCLOSED, OPEN, PARTIALLY EN-CLOSED. These definitions relate to the proper selection of internal pressure coefficients, GC_{pi}. Building, open and Building, partially enclosed are specifically defined. All other buildings are considered to be enclosed by definition, although there may be large openings in two or more walls. An example of this is a parking garage through which the wind can pass. The internal pressure coefficient for such a building would be ± 0.18, and the internal pressures would act on the solid areas of the walls and roof.

COMPONENTS AND CLADDING. Components receive wind loads directly or from cladding and transfer the load to the main wind force-resisting system. Cladding receives wind loads directly. Examples of components include fasteners, purlins, girts, studs, roof decking, and roof trusses. Components can be part of the main wind force-resisting system when they act as shear walls or roof diaphragms, but they may also be loaded as individual components. The engineer needs to use appropriate loadings for design of components, which may require certain components to be designed for more than one type of loading (e.g., long span roof trusses should be designed for loads associated with main wind force-resisting systems, and individual members of trusses should also be designed for component and cladding loads [C6-87]). Examples of cladding include wall coverings, curtain walls, roof coverings, exterior windows (fixed and operable) and doors, and overhead doors.

Effective wind area is the area of the building surface used to determine GC_p. This area does not necessarily correspond to the area of the building surface contributing to the force being considered. Two cases arise. In the usual case, the effective wind area does correspond to the area tributary to the force component being considered. For example, for a cladding panel, the effective wind area may be equal to the total area of the panel; for a cladding fastener, the effective wind area is the area of cladding secured by a single fastener. A mullion may receive wind from several cladding panels; in this case, the effective wind area is the area associated with the wind load that is transferred to the mullion.

The second case arises where components such as roofing panels, wall studs, or roof trusses are spaced closely together; the area served by the component may become long and narrow. To better approximate the actual load distribution in such cases, the width of the effective wind area used to evaluate GC_p need not be taken as less than one third the length of the area. This increase in effective wind area has the effect of reducing the average wind pressure acting on the component. Note, however, that this effective wind area should only be used in determining the GC_p in Figures 6-5 through 6-8. The induced wind load should be applied over the actual area tributary to the component being considered.

For membrane roof systems, the effective wind area is the area of an insulation board (or deck panel if insulation is not used) if the boards are fully adhered (or the membrane is adhered directly to the deck). If the insulation boards or membrane are mechanically attached or partially adhered, the effective wind area is the area of the board or membrane secured by a single fastener or individual spot or row of adhesive.

FLEXIBLE BUILDINGS AND OTHER STRUCTURES. A building or other structure is considered flexible if it contains a significant dynamic resonant response. Resonant response depends on the gust structure contained in the approaching wind, on wind loading pressures generated by the wind flow about the building, and on the dynamic properties of the building or structure. Gust energy in the wind is smaller at frequencies above about 1 Hz; therefore the resonant response of most buildings and structures with lowest natural frequency above 1 Hz will be sufficiently small that resonant response can often be ignored. When buildings or other structures have a height exceeding four times the least horizontal dimension or when there is reason to believe that the natural frequency is less than 1 Hz (natural period greater than 1 second), natural frequency for it should be investigated. A useful calculation procedure for natural frequency or period for various building types is contained in Section 9.

REGULAR SHAPED BUILDINGS AND OTHER STRUCTURES. Defining the limits of applicability of the analytical procedures within the standard is a difficult process, requiring a balance between the practical need to use the provisions past the range for which data has been obtained and restricting use of the provisions past the range of realistic application. Wind load provisions are based primarily on wind-tunnel tests on shapes shown in Figures 6-6 through 6-17. Extensive wind-tunnel tests on actual structures under design show that relatively large changes from these shapes can, in many cases, have minor changes in wind load, while in other cases seemingly small changes can have relatively large effects, particularly on cladding pressures. Wind loads on complicated shapes are frequently smaller than those on the simpler shapes of Figures 6-3 through 6-8, and so wind loads determined from these provisions reasonably envelope most structure shapes. Buildings which are clearly unusual should use the provisions of Section 6.4 for wind-tunnel tests.

RIGID BUILDINGS AND OTHER STRUCTURES. The defining criterion for rigid, in comparison to flexible, is that the natural frequency is greater than or equal to 1 Hz. A general guidance is that most rigid buildings and structures have height-to-minimum-width ratio of less than 4. Where there is concern about whether or not a building or structure meets this requirement, the provisions of Section 9 provide a method for calculating natural frequency (period = 1/natural frequency).

WIND-BORNE DEBRIS REGIONS. Some buildings located in a wind-borne debris region may not be vulnerable to wind-borne debris. For example, an isolated building located a substantial distance from natural and man-made debris sources would unlikely be impacted by debris, provided that building components from the building itself were not blown off, and provided that refuse containers, lawn furniture, and other similar items where not in the vicinity of the building. However, ASCE 7 does not allow an exception for such buildings to be excluded from the requirements applicable to buildings in a wind-borne debris region.

Although wind-borne debris can occur in just about any condition, the level of risk in comparison to the postulated debris regions and impact criteria may also be lower than that determined for the purpose of standardization. This possibility also applies to the "transition zone" as described above. For example, individual buildings may be sited away from likely debris sources that would generate significant risk of impacts similar in magnitude to pea gravel (i.e., as simulated by 2 g steel balls in impact tests) or butt-on 2×4 impacts as required in impact testing criteria. This situation describes a condition of low vulnerability only as a result of limited debris sources within the vicinity of the building. In other cases, potential sources of debris may be present, but extenuating conditions can lower the risk. These extenuating conditions include the type of materials and surrounding construction, the level of protection offered by surrounding exposure conditions, and the design wind speed. Therefore, the risk of impact may differ from those postulated as a result of the conditions specifically enumerated in the standard and the referenced impact standards. The committee recognizes that there are vastly differing opinions, even within the Standards Committee, regarding the significance of these parameters that are not fully considered in developing standardized debris regions or referenced impact criteria.

SECTION C6.3
SYMBOLS AND NOTATION

The following additional symbols and notation are used herein:

A_{ob} = average area of open ground surrounding each obstruction;

n = reference period, in years;

P_a = annual probability of wind speed exceeding a given magnitude (see Eq. C6-1);

P_n = probability of exceeding design wind speed during n years (see Eq. C6-1);

s_{ob} = average frontal area presented to the wind by each obstruction;

V_t = wind speed averaged over t seconds (see Figure C6-2), in mph (m/s);

V_{3600} = mean wind speed averaged over 1 hour (see Figure C6-2), in mph (m/s);

β = structural damping coefficient (percentage of critical damping).

SECTION C6.4
METHOD 1 — SIMPLIFIED PROCEDURE

Method 1 has been added to the Standard for a designer having the relatively common low-rise ($h \leq 60$ ft) regular shaped, simple diaphragm building case (see definitions for "simple diaphragm building" and "regular shaped building") where pressures for the roof and walls can be selected directly from a table. Two figures are provided: Figure 6-2 for the main wind force-resisting system and Figure 6-3 for components and cladding. Values are provided for enclosed buildings. Note that for the main wind force-resisting system in a diaphragm building, the internal pressure cancels for loads on the walls, but must be considered for the roof. This is true because when wind forces are transferred by horizontal diaphragms (such as floors and roofs) to the vertical elements of the main wind force-resisting system (such as shear walls, X-bracing, or moment frames), the collection of wind forces from windward and leeward sides of the building occurs in the horizontal diaphragms. Once transferred into the horizontal diaphragms by the wall systems, the wind forces become a net horizontal wind force that is delivered to the vertical elements. The equal and opposite internal pressures on the walls cancel in the horizontal diaphragm. Method 1 combines the windward and leeward pressures into a net horizontal wind pressure, with the internal pressures canceled. The user is cautioned to consider the precise application of windward and leeward wall loads to members of the roof diaphragm where openings may exist and where particular members such as drag struts are designed. The design of the roof members of the main wind force-resisting system is still influenced by internal pressures, but for simple diaphragm buildings with roof angles below 25 degrees, it can be assumed that the maximum uplift, produced by a positive internal pressure, is the controlling load case. From 25 to 45 degrees, both positive and negative internal pressure cases (Load Cases 1 and 2, respectively) must be checked for the roof, since the windward roof external pressure becomes positive at that point.

For the designer to use Method 1 for the design of the MWFRS, the building must conform to all eight requirements in 6.4.1.1; otherwise Method 2 or 3 must be used. Values are tabulated for Exposure B at $h = 30$ ft, and $I = 1.0$. Multiplying factors are provided for other exposures and heights. The following values have been used in preparation of the tables:

$h = 30$ ft Exposure B $K_z = 0.70$
$K_d = 0.85$ $K_{zt} = 1.0$ $I = 1.0$
$GC_{pi} = \pm 0.18$ (enclosed building)

Pressure coefficients are from Figures 6-10 and 6-11.

Wall elements resisting two or more simultaneous wind-induced structural actions (such as bending, uplift, or shear) should be designed for the interaction of the wind loads as part of the *main wind force-resisting system.*

The horizontal loads in Figure 6-2 are the sum of the windward and leeward pressures, and are therefore not applicable as individual wall pressures for the interaction load cases. Design wind pressures, p_s for Zones A and C should be multiplied by $+0.85$ for use on windward walls and by -0.70 for use on leeward walls (plus sign signifies pressures acting toward the wall surface). For side walls, p_s for zone C multiplied by -0.65 should be used. These wall elements must also be checked for the various separately acting (not simultaneous) *component and cladding* load cases.

Main wind force-resisting roof members spanning at least from the eave to the ridge or supporting members spanning at least from eave to ridge are not required to be designed for the higher end zone loads. The interior zone loads should be applied. This is due to the enveloped nature of the loads for roof members.

The component and cladding tables in Figure 6-3 are not related to the simple diaphragm methodology, but are a tabulation of the pressures on an enclosed, regular, 30-ft-high building with a roof as described. The pressures can be modified to a different exposure and height with the same adjustment factors as the MWFRS pressures. For the designer to use Method 1 for the design of the components and cladding, the building must conform to all five requirements in 6.4.1.2 otherwise Method 2 or 3 must be used. A building may qualify for use of Method 1 for its components and cladding only, in which case, its MWFRS should be designed using Method 2 or 3.

SECTION C6.5
METHOD 2 — ANALYTICAL PROCEDURE

C6.5.1 Scope. The analytical procedure provides wind pressures and forces for the design of main wind force-resisting systems and for the design of components and cladding of buildings and other structures. The procedure involves the determination of wind directionality and a velocity pressure, the selection or determination of an appropriate gust effect factor, and the selection of appropriate pressure or force coefficients. The procedure allows for the level of structural reliability required, the effects of differing wind exposures, the speed-up effects of certain topographic features such as hills and escarpments, and the size and geometry of the building or other structure under consideration. The procedure differentiates between rigid and flexible buildings and other structures, and the results generally envelope the most critical load conditions for the design of main wind force-resisting systems as well as components and cladding.

The Standard in Section 6.5.4 requires that a structure be designed for winds from all directions. A rational procedure to determine directional wind loads is as follows. Wind load for buildings using Section 6.5.12.2.1 and Figure 6-6 or Figure 6-7 are determined for eight wind directions

at 45-degree intervals, with four falling along primary building axes as shown in Figure C6-1. For each of the eight directions, upwind exposure is determined for each of two 45-degree sectors, one on each side of the wind direction axis. The sector with the exposure giving highest loads will be used to define wind loads for that direction. For example, for winds from the north, the exposure from sector one or eight whichever gives the highest load is used; for wind from the east, the exposure from sector two or three whichever gives the highest load is used; for wind coming from the northeast, the most exposed of sectors one or two is used to determine full x and y loading individually, and then 75% of these loads are to be applied in each direction at the same time according to the requirements of Section 6.5.12.3 and Figure 6-9. See Section C6.5.6 for further discussion of exposures. The procedure defined in this Section for determining wind loads in each design direction is not to be confused with the determination of the wind directionality factor k_d in Eq. 6-15. The k_d factor determined from Section 6.5.4 and Table 6-4 applies for all design wind directions (see Section C6.5.4.4).

Wind loads for cladding elements are determined using the upwind exposure for the single surface roughness in one of the eight sectors of Figure C6-1 which gives the highest cladding pressures.

C6.5.2 Limitations of Analytical Procedure. The provisions given under Section 6.5.2 apply to the majority of site locations and buildings and structures, but for some locations, these provisions may be inadequate. Examples of site locations and buildings and structures (or portions thereof) that require use of recognized literature for documentation pertaining to wind effects, or the use of the wind-tunnel procedure of Section 6.6 include:

1. Site locations that have channeling effects or wakes from upwind obstructions. Channeling effects can be caused by topographic features (e.g., mountain gorge) or buildings (e.g., a cluster of tall buildings). Wakes can be caused by hills or by buildings or other structures.

2. Buildings with unusual or irregular geometric shape, including barrel vaults, and other buildings whose shape (in plan or profile) differs significantly from a uniform or series of superimposed prisms similar to those indicated in Figures 6-3 through 6-8. Unusual or irregular geometric shapes include buildings with multiple setbacks, curved facades, irregular plan resulting from significant indentations or projections, openings through the building, or multitower buildings connected by bridges.

3. Buildings with unusual response characteristics, which result in across-wind and/or dynamic torsional loads, loads caused by vortex shedding, or loads resulting from instabilities such as flutter or galloping.

Examples of buildings and structures which may have unusual response characteristics include flexible buildings with natural frequencies normally below 1 Hz, tall slender buildings (building height-to-width ratio exceeds 4), and cylindrical buildings or structures. Note: Vortex shedding occurs when wind blows across a slender prismatic or cylindrical body. Vortices are alternately shed from one side of the body and then the other side, which results in a fluctuating force acting at right angles to the wind direction (across-wind) along the length of the body.

4. Bridges, cranes, electrical transmission lines, guyed masts, telecommunication towers, and flagpoles.

C6.5.2.1 Shielding. Due to the lack of reliable analytical procedures for predicting the effects of shielding provided by buildings and other structures or by topographic features, reductions in velocity pressure due to shielding are not permitted under the provisions of Section 6.5. However, this does not preclude the determination of shielding effects and the corresponding reductions in velocity pressure by means of the wind-tunnel procedure in Section 6.6.

C6.5.2.2 Air-Permeable Cladding. Air-permeable roof or wall claddings allow partial air pressure equalization between their exterior and interior surfaces. Examples include siding, pressure-equalized rain screen walls, shingles, tiles, concrete roof pavers, and aggregate roof surfacing.

The design wind pressures derived from Section 6.5 represent the pressure differential between the exterior and interior surfaces of the exterior envelope (wall or roof system). Because of partial air-pressure equalization provided by air-permeable claddings, the pressures derived from Section 6.5 can overestimate the load on air-permeable cladding elements. The designer may elect either to use the loads derived from Section 6.5 or to use loads derived by an approved alternative method. If the designer desires to determine the pressure differential across the air-permeable cladding element, appropriate full-scale pressure measurements should be made on the applicable cladding element, or reference be made to recognized literature [C6-9, C6-16, C6-37, C6-73] for documentation pertaining to wind loads.

C6.5.4 Basic Wind Speed. The ASCE 7 wind map included in the 1998 Standard and again for the 2002 edition has been updated from the map in ASCE 7-95 based on a new and more complete analysis of hurricane wind speeds [C6-78, C6-79]. This new hurricane analysis yields predictions of 50- and 100-year return period peak gust wind speeds along the coast, which are generally similar to those given in References [C6-88 and C6-89]. The decision within the Task Committee on Wind Loads to update the map relied on several factors important to an accurate wind specification:

1. The new hurricane results include many more predictions for sites away from the coast than have been available in the past. It is desirable to include the best available decrease in speeds with inland distance. Significant reductions in wind speeds occur in inland Florida for the new analysis.

2. The distance inland to which hurricanes can influence wind speed increases with return period. It is desirable to include this distance in the map for design of ultimate events (working stress multiplied by an appropriate load factor).

3. A hurricane coast importance factor of 1.05 acting on wind speed was included explicitly in past ASCE 7 Standards (1993 and earlier) to account for the more rapid increase of hurricane speeds with return period in comparison to non-hurricane winds. The hurricane coast importance factor actually varies in magnitude and position along the coast and with distance inland. In order to produce a more uniform risk of failure, it is desirable to include the effect of the importance factor in the map by first mapping an ultimate event and then reducing the event to a design basis.

The Task Committee on Wind Loads chose to use a map which includes the hurricane importance factor in the map contours. The map is specified so that the loads calculated from the Standard, after multiplication by the load factor, represent an ultimate load having approximately the same return period as loads for non-hurricane winds. (An alternative not selected was to use an ultimate wind speed map directly in the standard, with a load factor of 1.0).

The approach required selection of an ultimate return period. A return period of about 500 years has been used previously for earthquake loads. This return period can be derived from the non-hurricane speeds in ASCE 7-95. A factor of 0.85 is included in the load factor of ASCE 7-95 to account for wind and pressure coefficient directionality [C6-76]. Removing this from the load factor gives an effective load factor F of $1.30/0.85 = 1.529$ (round to 1.5). Of the uncertainties affecting the wind load factor, the variability in wind speed has the strongest influence [C6-77], such that changes in the coefficient of variation in all other factors by 25% gives less than a 5% change in load factor. The non-hurricane multiplier of 50-year wind speed for various return periods averages $Fc = 0.36 + 0.1 \ln (12T)$, with T in years [C6-74]. Setting $Fc = \sqrt{F} = \sqrt{1.5} = 1.225$ yields $T = 476$ years. On this basis, a 500-year speed can reasonably represent an approximate ultimate limit state event.

A set of design level hurricane speed contours, which include the hurricane importance factor, were obtained by dividing 500-year hurricane wind speed contours by $\sqrt{F} = 1.225$. The implied importance factor ranges from near 1.0 up to about 1.25 (the explicit value in ASCE 7-93 is 1.05).

The design level speed map has several advantages. First, a design using the map results in an ultimate load (loads inducing the design strength after use of the load factor) which has a more uniform risk for buildings than occurred with earlier versions of the map. Second, there is no need for a designer to use and interpolate a hurricane coast importance factor. It is not likely that the 500-year event is the actual speed at which engineered structures are expected to fail, due to resistance factors in materials, due to conservative design procedures which do not always analyze all load capacity, and due to a lack of a precise definition of "failure."

The wind speed map of Figure 6-1 presents basic wind speeds for the contiguous United States, Alaska, and other selected locations. The wind speeds correspond to 3-second gust speeds at 33 ft (10 m) above ground for Exposure Category C. Because the National Weather Service has phased out the measurement of fastest-mile wind speeds, the basic wind speed has been redefined as the peak gust which is recorded and archived for most NWS stations. Given the response characteristics of the instrumentation used, the peak gust is associated with an averaging time of approximately 3 seconds. Because the wind speeds of Figure 6-1 reflect conditions at airports and similar open-country exposures, they do not account for the effects of significant topographic features such as those described in Section 6.5.7. Note that the wind speeds shown in Figure 6-1 are not representative of speeds at which ultimate limit states are expected to occur. Allowable stresses or load factors used in the design equation(s) lead to structural resistances and corresponding wind loads and speeds that are substantially higher than the speeds shown in Figure 6-1.

The hurricane wind speeds given in Figure 6-1 replace those given in ASCE-7-95, which were based on a combination of the data given in References [C6-5, C6-15, C6-54, C6-88, and C6-20], supplemented with some judgment. The non-hurricane wind speeds of Figure 6-1 were prepared from peak gust data collected at 485 weather stations where at least 5 years of data were available [C6-29, C6-30, C6-74]. For non-hurricane regions, measured gust data were assembled from a number of stations in state-sized areas to decrease sampling error, and the assembled data were fit using a Fisher-Tippett Type I extreme value distribution. This procedure gives the same speed as does area-averaging the 50-year speeds from the set of stations. There was insufficient variation in 50-year speeds over the eastern 3/4 of the lower 48 states to justify contours. The division between the 90 and 85 mph (40.2 and 38.0 m/s) regions, which follows state lines, was sufficiently close to the 85 mph (38.0 m/s) contour that there was no statistical basis for placing the division off political boundaries. This data is expected to follow the gust factor curve of Figure C6-2 [C6-13].

Limited data were available on the Washington and Oregon coast; in this region, existing fastest-mile wind

speed data were converted to peak gust speeds using open-country gust factors [C6-13]. This limited data indicates that a speed of 100 mph is appropriate in some portions of the special coastal region in Washington and 90 mph in the special coastal region in Oregon; these speeds do not include that portion of the special wind region in the Columbia River Gorge, where higher speeds may be justified. Speeds in the Aleutian Islands and in the interior of Alaska were established from gust data. Contours in Alaska are modified slightly from ASCE 7-88 based on measured data, but insufficient data were available for a detailed coverage of the mountainous regions.

C6.5.4.1 Special Wind Regions. Although the wind-speed map of Figure 6-1 is valid for most regions of the country, there are special regions in which wind-speed anomalies are known to exist. Some of these special regions are noted in Figure 6-1. Winds blowing over mountain ranges or through gorges or river valleys in these special regions can develop speeds that are substantially higher than the values indicated on the map. When selecting basic wind speeds in these special regions, use of regional climatic data and consultation with a wind engineer or meteorologist is advised.

It is also possible that anomalies in wind speeds exist on a micrometeorological scale. For example, wind speed-up over hills and escarpments is addressed in Section 6.5.7. Wind speeds over complex terrain may be better determined by wind-tunnel studies as described in Section 6.6. Adjustments of wind speeds should be made at the micrometeorological scale on the basis of wind engineering or meteorological advice and used in accordance with the provisions of Section 6.5.4.2 when such adjustments are warranted.

C6.5.4.2 Estimation of Basic Wind Speeds from Regional Climatic Data. When using regional climatic data in accordance with the provisions of Section 6.5.4.2 and in lieu of the basic wind speeds given in Figure 6-1, the user is cautioned that the gust factors, velocity pressure exposure coefficients, gust effect factors, pressure coefficients, and force coefficients of this Standard are intended for use with the 3-second gust speed at 33 ft (10 m) above ground in open country. It is necessary, therefore, that regional climatic data based on a different averaging time, for example, hourly mean or fastest mile, be adjusted to reflect peak gust speeds at 33 ft (10 m) above ground in open country. The results of statistical studies of wind-speed records, reported by [C6-13] for extratropical winds and for hurricanes [C6-78], are given in Figure C6-2, which defines the relation between wind speed averaged over t seconds, V_t, and over one hour, V_{3600}. New research cited in [C6-78] indicates that the old Krayer-Marshall curve [C6-20] does not apply in hurricanes. Therefore it was removed in Figure C6-1 in ASCE 7-98. The gust factor adjustment to reflect peak gust speeds is not

always straightforward, and advice from a wind engineer or meteorologist may be needed.

In using local data, it should be emphasized that sampling errors can lead to large uncertainties in specification of the 50-year wind speed. Sampling errors are the errors associated with the limited size of the climatological data samples (years of record of annual extremes). It is possible to have a 20 mph (8.9 m/s) error in wind speed at an individual station with a record length of 30 years. It was this type of error that led to the large variations in speed in the non-hurricane areas of the ASCE 7-88 wind map. While local records of limited extent often must be used to define wind speeds in special wind areas, care and conservatism should be exercised in their use.

If meteorological data are used to justify a wind speed lower than 85 mph 50-year peak gust at 10 m, an analysis of sampling error is required to demonstrate that the wind record could not occur by chance. This can be accomplished by showing that the difference between predicted speed and 85 mph contains 2 to 3 standard deviations of sampling error [C6-67]. Other equivalent methods may be used.

C6.5.4.3 Limitation. In recent years, advances have been made in understanding the effects of tornadoes on buildings. This understanding has been gained through extensive documentation of building damage caused by tornadic storms and through analysis of collected data. It is recognized that tornadic wind speeds have a significantly lower probability of occurrence at a point than the probability for basic wind speeds. In addition, it is found that in approximately one-half of the recorded tornadoes, gust speeds are less than the gust speeds associated with basic wind speeds. In intense tornadoes, gust speeds near the ground are in the range of 150 to 200 mph (67–89 m/s). Sufficient information is available to implement tornado-resistant design for above-ground shelters and for buildings that house essential facilities for post-disaster recovery. This information is in the form of tornado risk probabilities, tornadic wind speeds, and associated forces. Several references provide guidance in developing wind load criteria for tornado-resistant design [C6-1, C6-2, C6-24–C6-28, C6-57].

Tornadic wind speeds, which are gust speeds, associated with an annual probability of occurrence of 1×10^{-5} (100,000 year mean recurrence interval) are shown in Figure C6-3. This map was developed by the American Nuclear Society committee ANS 2.3 in the early 1980s. Tornado occurrence data of the last 15 years can provide a more accurate tornado hazard wind speed for a specific site.

C6.5.4.4 Wind Directionality Factor. The existing wind load factor 1.3 in ASCE 7-95 includes a "wind directionality factor" of 0.85 [C6-76, C6-77]. This factor accounts for two effects: (1) The reduced probability of maximum winds coming from any given direction and (2) the reduced probability of the maximum pressure coefficient occurring

for any given wind direction. The wind directionality factor (identified as K_d in the Standard) has been hidden in previous editions of the Standard and has generated renewed interest in establishing the design values for wind forces determined by using the Standard. Accordingly, the Task Committee on Wind Loads, working with the Task Committee on Load Combinations, has decided to separate the wind directionality factor from the load factor and include its effect in the equation for velocity pressure. This has been done by developing a new factor, K_d, that is tabulated in Table 6-4 for different structure types. As new research becomes available, this factor can be directly modified without changing the wind load factor. Values for the factor were established from references in the literature and collective committee judgment. It is noted that the k_d value for round chimneys, tanks, and similar structures is given as 0.95 in recognition of the fact that the wind load resistance may not be exactly the same in all directions as implied by a value of 1.0. A value of 0.85 might be more appropriate if a triangular trussed frame is shrouded in a round cover. 1.0 might be more appropriate for a round chimney having a lateral load resistance equal in all directions. The designer is cautioned by the footnote to Table 6-4 and the statement in Section 6.5.4.4, where reference is made to the fact that this factor is only to be used in conjunction with the load combination factors specified in 2.3 and 2.4.

C6.5.5 Importance Factor. The importance factor is used to adjust the level of structural reliability of a building or other structure to be consistent with the building classifications indicated in Table 1-1. The importance factors given in Table 6-1 adjust the velocity pressure to different annual probabilities of being exceeded. Importance-factor values of 0.87 and 1.15 are, for the non-hurricane winds, associated, respectively, with annual probabilities of being exceeded by 0.04 and 0.01 (mean recurrence intervals of 25 and 100 years). In the case of hurricane winds, the annual exceedance probabilities implied by the use of the importance factors of 0.77 and 1.15 will vary along the coast; however, the resulting risk levels associated with the use of these importance factors when applied to hurricane winds will be approximately consistent with those applied to non-hurricane winds.

The probability P_n that the wind speed associated with a certain annual probability P_a will be equaled or exceeded at least once during an exposure period of n years is given by

$$P_n = 1 - (1 - P_a)^n \qquad \textbf{(Eq. C6-1)}$$

and values of P_n for various values of P_a and n are listed in Table C6-1. As an example, if a design wind speed is based upon $P_a = 0.02$ (50-year mean recurrence interval), there exists a probability of 0.40 that this speed will be equaled or exceeded during a 25-year period, and a 0.64 probability of being equaled or exceeded in a 50-year period.

For applications of serviceability, design using maximum likely events, or other applications, it may be desired to use wind speeds associated with mean recurrence intervals other than 50 years. To accomplish this, the 50-year speeds of Figure 6-1 are multiplied by the factors listed in Table C6-2. Table C6-2 is strictly valid for the non-hurricane winds only ($V < 100$ mph for continental United States and all speeds in Alaska), where the design wind speeds have a nominal annual exceedance probability of 0.02. Using the factors given in Table C6-2 to adjust the hurricane wind speeds will yield wind speeds and resulting wind loads that are approximately risk consistent with those derived for the non–hurricane-prone regions. The true return periods associated with the hurricane wind speeds cannot be determined using the information given in this Standard.

The difference in wind speed ratios between continental United States ($V < 100$ mph) and Alaska were determined by data analysis and probably represent a difference in climatology at different latitudes.

C6.5.6 Exposure Categories. A number of revisions have been made to the definitions of exposure categories in the ASCE 7-02 edition. In ASCE 7-98 the definitions of Exposures C and D were modified based on new research [C6-86]. Further changes in the new edition are:

1. Exposure A has been deleted. In the past, Exposure A was intended for heavily built-up city centers with tall buildings. However, the committee has concluded that in areas in close proximity to tall buildings the variability of the wind is too great, because of local channeling and wake buffeting effects, to allow a special category A to be defined. For projects where schedule and cost permit, in heavily built-up city centers, Method 3 is recommended since this will enable local channeling and wake buffeting effects to be properly accounted for. For all other projects, Exposure B can be used, subject to the limitations in 6.5.2.

2. A distinction has been made between surface roughness categories and exposure categories. This has enabled more precise definitions of Exposures B, C, and D to be obtained in terms of the extent and types of surface roughness that are upwind of the site. The requirements for the upwind fetch have been modified based on recent investigations into effects of roughness changes and transitions between them [C6-90].

3. The exposure for each wind direction is now defined as the worst case of the two 45-degree sectors either side of the wind direction being considered (see also Section C6.5.4).

4. Interpolation between exposure categories is now permitted. One acceptable method of interpolating between exposure categories is provided in C6.5.6.4.

The descriptions of the surface roughness categories and exposure categories in Section 6.5.6 have been expressed as far as is possible in easily understood verbal terms which are sufficiently precise for most practical applications. For cases where the designer wishes to make a more detailed assessment of the surface roughness category and exposure category, the following more mathematical description is offered for guidance. The ground surface roughness is best measured in terms of a roughness length parameter called z_0. Each of the surface roughness categories B through D corresponds to a range of values of this parameter, as does the even rougher category A used in earlier versions of the Standard in heavily built-up urban areas but removed in the present edition. The range of z_0 in meters (ft) for each terrain category is given in the boxed text below. Exposure A has been included in the table below as a reference which may be useful when using Method 3 (the wind tunnel procedure).

Exposure Category	Lower Limit of z_0, m (ft)	Typical Value of z_0, m (ft)	Upper Limit of z_0, m (ft)	z_0 Inherent in Tabulated K_z Values in Sec. 6.5.6.4, m (ft)
A	$0.7\ (2.3) \leq z_0$	2 (6.6)	—	—
B	$0.15\ (0.49) \leq z_0$	0.3 (0.98)	$z_0 < 0.7\ (2.3)$	0.15 (0.49)
C	$0.01\ (0.033) \leq z_0$	0.02 (0.066)	$z_0 < 0.15\ (0.49)$	0.02 (0.066)
D	—	0.005 (0.016)	$z_0 < 0.01\ (0.033)$	0.005 (0.016)

The value of z_0 for a particular terrain can be estimated from the typical dimensions of surface roughness elements and their spacing on the ground area using an empirical relationship, due to Lettau [C6-75], which is

$$z_0 = 0.5 H_{ob} \frac{S_{ob}}{A_{ob}} \qquad \textbf{(Eq. C6-2)}$$

where

H_{ob} = the average height of the roughness in the upwind terrain,

S_{ob} = the average vertical frontal area per obstruction presented to the wind,

A_{ob} = the average area of ground occupied by each obstruction, including the open area surrounding it.

Vertical frontal area is defined as the area of the projection of the obstruction onto a vertical plane normal to the wind direction. The area S_{ob} may be estimated by summing the approximate vertical frontal areas of all obstructions within a selected area of upwind fetch and dividing the sum by the number of obstructions in the area. The average height H may be estimated in a similar way by averaging the individual heights rather than using the frontal areas. Likewise A_{ob} may be estimated by dividing the size of the selected area of upwind fetch by the number of obstructions in it.

As an example, if the upwind fetch consists primarily of single-family homes with typical height $H = 6$ m, vertical frontal area 60 sq m, and ground area per home

of 1200 sq m, then z_0 is calculated to be $z_0 = 0.5 \times 6 \times 60/1200 = 0.15$ m, which falls into exposure category B according to the above table. Note that often a suburban area will have many trees which would add to the effective vertical frontal area per home, significantly increasing the ratio S_{ob}/A_{ob}.

Trees and bushes are porous and are deformed by strong winds, which reduces their effective frontal areas [C6-91]. For conifers and other evergreens no more than 50% of their gross frontal area can be taken to be effective in obstructing the wind. For deciduous trees and bushes no more than 15% of their gross frontal area can be taken to be effective in obstructing the wind. Gross frontal area is defined in this context as the projection onto a vertical plane (normal to the wind) of the area enclosed by the envelope of the tree or bush.

A recent study [C6-66] has estimated that the majority of buildings (perhaps as much as 60%–80%) have an exposure category corresponding to Exposure B. While the relatively simple definition in the Standard will normally suffice for most practical applications, oftentimes the designer is in need of additional information, particularly with regard to the effect of large openings or clearings (such as large parking lots, freeways, or tree clearings) in the otherwise "normal" ground surface roughness B. The following is offered as guidance for these situations:

1. The simple definition of Exposure B given in the body of the Standard, using the new surface roughness category definition, is shown pictorially in Figure C6-4. This definition applies for the surface roughness B condition prevailing 2630 ft (800 m) upward with insufficient "open patches" as defined below to disqualify the use of Exposure B.

2. An opening in the surface roughness B large enough to have a significant effect on the exposure category determination is defined as an "open patch." An open patch is defined as an opening greater than or equal to 164 ft (50 m) on each side (i.e., greater than 165 ft (50 m) by 164 ft (50 m)). Openings smaller than this need not be considered in the exposure category determination.

3. The effect of open patches of surface roughness C or D on the use of Exposure Category B is shown pictorially in Figures C6-5 and C6-6. Note that the plan location of any open patch may have a different effect for different wind directions.

Aerial photographs, representative of each exposure type, are included in the commentary to aid the user in establishing the proper exposure for a given site. Obviously, the proper assessment of exposure is a matter of good engineering judgment. This fact is particularly true in light of the possibility that the exposure could change in one or more wind directions due to future demolition and/or development.

EXPOSURE B
SUBURBAN RESIDENTIAL AREA WITH MOSTLY SINGLE-FAMILY DWELLINGS. STRUCTURES IN THE CENTER OF THE PHOTOGRAPH HAVE SITES DESIGNATED AS EXPOSURE B WITH SURFACE ROUGHNESS CATEGORY B TERRAIN AROUND THE SITE FOR A DISTANCE GREATER THAN 1500 FT OR TEN TIMES THE HEIGHT OF THE STRUCTURE, WHICHEVER IS GREATER, IN ANY WIND DIRECTION

EXPOSURE B
URBAN AREA WITH NUMEROUS CLOSELY SPACED OBSTRUCTIONS HAVING THE SIZE OF SINGLE-FAMILY DWELLINGS OR LARGER. FOR ALL STRUCTURES SHOWN, TERRAIN REPRESENTATIVE OF SURFACE ROUGHNESS CATEGORY B EXTENDS MORE THAN TEN TIMES THE HEIGHT OF THE STRUCTURE OR 800 M, WHICHEVER IS GREATER, IN THE UPWIND DIRECTION

EXPOSURE B
STRUCTURES IN THE FOREGROUND ARE LOCATED IN EXPOSURE B. STRUCTURES IN THE CENTER TOP OF THE PHOTOGRAPH
ADJACENT TO THE CLEARING TO THE LEFT, WHICH IS GREATER THAN 200 M IN LENGTH, ARE LOCATED IN EXPOSURE C WHEN
WIND COMES FROM THE LEFT OVER THE CLEARING (SEE FIGURE C6-5)

EXPOSURE C
FLAT OPEN GRASSLAND WITH SCATTERED OBSTRUCTIONS HAVING HEIGHTS GENERALLY LESS THAN 30 FT

EXPOSURE C
OPEN TERRAIN WITH SCATTERED OBSTRUCTIONS HAVING HEIGHTS GENERALLY LESS THAN 30 FT FOR MOST WIND DIRECTIONS, ALL 1-STORY STRUCTURES WITH A MEAN ROOF HEIGHT LESS THAN 30 FT IN THE PHOTOGRAPH ARE LESS THAN 1500 FT OR TEN TIMES THE HEIGHT OF THE STRUCTURE, WHICHEVER IS GREATER, FROM AN OPEN FIELD THAT PREVENTS THE USE OF EXPOSURE B

EXPOSURE D
A BUILDING AT THE SHORELINE (EXCLUDING SHORELINES IN HURRICANE-PRONE REGIONS) WITH WIND FLOWING OVER OPEN WATER FOR A DISTANCE OF AT LEAST 1 MILE. SHORELINES IN EXPOSURE D INCLUDE INLAND WATERWAYS, THE GREAT LAKES, AND COASTAL AREAS OF CALIFORNIA, OREGON, WASHINGTON, AND ALASKA

Minimum Design Loads for Buildings and Other Structures

C6.5.6.4 Velocity Pressure Exposure Coefficient. The velocity pressure exposure coefficient K_z can be obtained using the equation:

$$K_z = \begin{cases} 2.01 \left(\dfrac{z}{z_g}\right)^{2/\alpha} & \text{for } 15 \text{ ft} \leq z \leq z_g \quad \textbf{(Eq. C6-3a)} \\[2em] 2.01 \left(\dfrac{15}{z_g}\right)^{2/\alpha} & \text{for } z < 15 \text{ ft} \quad \textbf{(Eq. C6-3b)} \end{cases}$$

in which values of α and z_g are given in Table 6-2. These equations are now given in Table 6-3 to aid the user.

In ASCE 7-95, the values of α and the preceding formulae were adjusted to be consistent with the 3-second gust format introduced at that time. Other changes were implemented in ASCE 7-98 including truncation of K_z values for Exposure A and B below heights of 100 ft and 30 ft, respectively. Exposure A has been eliminated in the 2002 edition.

In the ASCE 7-02 standard, the K_z expressions are unchanged from ASCE 7-98. However, the possibility of interpolating between the standard exposures is recognized in the present edition. One rational method is provided below.

To a reasonable approximation, the empirical exponent α and gradient height z_g in Eqs. C6-3a and C6-3b for exposure coefficient K_z may be related to the roughness length z_0 by the relations

$$\alpha = c_1 z_0^{-0.157} \qquad \textbf{(Eq. C6-4)}$$

and

$$z_g = c_2 z_0^{0.125} \qquad \textbf{(Eq. C6-5)}$$

where

Units of z_0, z_g	c_1	c_2
m	5.14	450
ft	6.19	1273

The above relationships are based on matching the ESDU boundary layer model [C6-90 to C6-92] empirically with the power law relationship in Eq. C6-3a, b, the ESDU model being applied at latitude 45 degrees with a gradient wind of 50 m/s. If z_0 has been determined for a particular upwind fetch, Eq. C6-3, C6-4, and C6-5 can be used to evaluate K_z. The correspondence between z_0 and the parameters α and z_g implied by these relationships does not align exactly with that described in the commentary to ASCE 7-95 and 7-98. However, the differences are relatively small and not of practical consequence. The ESDU boundary layer model has also been used to derive the following simplified method of evaluating K_z following a transition from one surface roughness to another. For more precise estimates the reader is referred to the original ESDU model [C6-90 to C6-92].

In uniform terrain, the wind travels a sufficient distance over the terrain for the planetary boundary layer to reach an equilibrium state. The exposure coefficient values in Table 6-3 are intended for this condition. Suppose that at the site we have local terrain which, in equilibrium conditions, would have an exposure coefficient at height z of K_{z1} but there is, at a distance x km in the upwind direction, a change in the terrain. Upwind of the change the terrain is such that the normal equilibrium exposure coefficient would be K_{z2}. The effect of this change in terrain on the exposure coefficient at the site can be represented by adjusting K_{z1} by an increment ΔK, thus arriving at a corrected value K_z for the site.

$$K_z = K_{z1} + \Delta K \qquad \textbf{(Eq. C6-6)}$$

In this expression ΔK is calculated using

$$\Delta K = (K_{10,2} - K_{10,1}) \frac{K_{z1}}{K_{10,1}} F_{\Delta K}(x)$$

$$|\Delta K| \leq |K_{z2} - K_{z1}| \qquad \textbf{(Eq. C6-7)}$$

where $K_{10,1}$ and $K_{10,2}$ are, respectively, the local and upwind equilibrium values of exposure coefficient at 10 m height, and the function $F_{\Delta K}(x)$, with x measured in kilometers, is given by

$$F_{\Delta K}(x) = \log_{10}\left(\frac{x_1}{x}\right) / \log_{10}\left(\frac{x_1}{x_0}\right) \qquad \textbf{(Eq. C6-8)}$$

for $x_0 < x < x_1$

$$F_{\Delta k}(x) = 1 \quad \text{for } x < x_0$$
$$F_{\Delta k}(x) = 0 \quad \text{for } x_1 < x$$

The "starting" value of x is given by

$$x_0 = 10^{-(K_{10,1} - K_{10,2})^2 - 2.3} \qquad \textbf{(Eq. C6-9)}$$

The "finishing" value for x is given by $x_1 = 10$ km for $K_{10,1} < K_{10,2}$ (wind going from smooth terrain 2 to rough terrain 1) or $x_1 = 100$ km for $K_{10,1} > K_{10,2}$ (wind going from rough terrain 2 to smooth terrain 1).

As an example, suppose the building is 20 m high in suburban surroundings (Exposure B) but the edge of the suburban terrain is 0.6 km upwind with open country (Exposure C) beyond. We thus have $x = 0.6$ km and $K_{10,1}, K_{10,2}$ and $K_{20,1}$ are 0.72, 1.0 and 0.88, respectively, (from Table 6.3). From Equation C6-8, $F_{\Delta K}(x)$ is calculated to be 0.36 and ΔK is calculated from Eq. 6-7 to be $(1.00 - 0.72) \times (0.88/0.72 \times 0.36 = 0.124$. Thus, from Eq. C6-6 the value of K_z at 20 m height should be increased from its equilibrium value of 0.88 by an increment of 0.124, bringing it up to 1.004.

Note that in this case the simple rules in Section 6 would require Exposure C be applied since the extent of

Exposure B roughness is less than 800 m, resulting in $K_{20} = 1.16$. The above interpolation method would enable a value of 1.004 to be used instead. In some cases, the interpolation method will give a higher exposure coefficient than the simple rules. In these cases, the simple rules can be used if desired as the differences are typically within the range of uncertainty resulting from estimation of roughness lengths.

C6.5.7 Wind Speed-Up over Hills and Escarpments. As an aid to the designer, this Section was rewritten in ASCE 7-98 to specify when topographic effects need to be applied to a particular structure rather than when they do not as in the previous version. In addition, the upwind distance to consider has been lengthened from 50 times to 100 times the height of the topographic feature (100 H) and from 1 to 2 miles. In an effort to exclude situations where little or no topographic effect exists, condition (2) has been added to include the fact that the topographic feature should protrude significantly above (by a factor of 2 or more) upwind terrain features before it becomes a factor. For example, if a significant upwind terrain feature has a height of 35 feet above its base elevation and has a top elevation of 100 feet above mean sea level, then the topographic feature (hill, ridge, or escarpment) must have at least the H specified and extend to elevation 170 mean sea level (100 ft + 2 × 35 ft) within the 2-mile radius specified.

A recent wind-tunnel study [C6-81] and observation of actual wind damage has shown that the affected height H is less than previously specified. Accordingly, condition (5) was changed to 15 feet in Exposure C.

Buildings sited on the upper half of an isolated hill or escarpment may experience significantly higher wind speeds than buildings situated on level ground. To account for these higher wind speeds, the velocity pressure exposure coefficients in Table 6-3 are multiplied by a topographic factor, K_{zt}, defined in Eq. 6-15 of Section 6.5.10. The topographic feature (two-dimensional ridge or escarpment, or three-dimensional axisymmetrical hill) is described by two parameters, H and L_h. H is the height of the hill or difference in elevation between the crest and that of the upwind terrain. L_h is the distance upwind of the crest to where the ground elevation is equal to half the height of the hill. K_{zt} is determined from three multipliers, K_1, K_2, and K_3, which are obtained from Figure 6-4, respectively. K_1 is related to the shape of the topographic feature and the maximum speed-up near the crest, K_2 accounts for the reduction in speed-up with distance upwind or downwind of the crest, and K_3 accounts for the reduction in speed-up with height above the local ground surface.

The multipliers listed in Figure 6-4 are based on the assumption that the wind approaches the hill along the direction of maximum slope, causing the greatest speed-up near the crest. The average maximum upwind slope of the hill is approximately $H/2L_h$, and measurements

have shown that hills with slopes of less than about 0.10 ($H/L_h < 0.20$) are unlikely to produce significant speed-up of the wind. For values of $H/L_h > 0.5$, the speed-up effect is assumed to be independent of slope. The speed-up principally affects the mean wind speed rather than the amplitude of the turbulent fluctuations and this fact has been accounted for in the values of K_1, K_2, and K_3 given in Figure 6-4. Therefore, values of K_{zt} obtained from Figure 6-4 are intended for use with velocity pressure exposure coefficients, K_h and K_z, which are based on gust speeds.

It is not the intent of Section 6.5.7 to address the general case of wind flow over hilly or complex terrain for which engineering judgment, expert advice, or wind-tunnel tests as described in Section 6.6 may be required. Background material on topographic speed-up effects may be found in the literature [C6-18, C6-21, C6-56].

The designer is cautioned that, at present, the Standard contains no provision for vertical wind speed-up because of a topographic effect, even though this phenomenon is known to exist and can cause additional uplift on roofs. Additional research is required to quantify this effect before it can be incorporated into the Standard.

C6.5.8 Gust Effect Factors. ASCE 7-02 contains a single gust effect factor of 0.85 for rigid buildings. As an option, the designer can incorporate specific features of the wind environment and building size to more accurately calculate a gust effect factor. One such procedure, previously contained in the commentary, is now located in the body of the Standard [C6-63, C6-64]. A suggested procedure is also included for calculating the gust effect factor for flexible structures. The rigid structure gust factor is typically 0%–4% higher than in ASCE 7-95 and is 0%–10% lower than the simple, but conservative, value of 0.85 permitted in the Standard without calculation. The procedures for both rigid and flexible structures have been changed from the previous version to (1) keep the rigid gust factor calculation within a few percentages of the previous model, (2) provide a superior model for flexible structures which displays the peak factors g_Q and g_R, and (3) causes the flexible structure value to match the rigid structure as resonance is removed (an advantage not included in the previous version). A designer is free to use any other rational procedure in the approved literature, as stated in Section 6.5.8.3.

The gust effect factor accounts for the loading effects in the along-wind direction due to wind turbulence structure interaction. It also accounts for along-wind loading effects due to dynamic amplification for flexible buildings and structures. It does not include allowances for across-wind loading effects, vortex shedding, instability due to galloping or flutter, or dynamic torsional effects. For structures susceptible to loading effects that are not accounted for in the gust effect factor, information should be obtained from recognized literature [C6-60–C6-65] or from wind tunnel tests.

Along-Wind Response. Based on the preceding definition of the gust effect factor, predictions of along-wind response (e.g., maximum displacement, rms, and peak acceleration), can be made. These response components are needed for survivability and serviceability limit states. In the following, expressions for evaluating these along-wind response components are given.

Maximum Along-Wind Displacement. The maximum along-wind displacement $X_{max}(z)$ as a function of height above the ground surface is given by

$$X_{max}(z) = \frac{\phi(z)\rho B h C_{fx}\hat{V}_{\bar{z}}^2}{2m_1(2\pi n_1)^2}KG \qquad \textbf{(Eq. C6-10)}$$

where $\phi(z) =$ the fundamental model shape $\phi(z) = (z/h)^\xi$; $\xi =$ the mode exponent; $\rho =$ air density; $C_{fx} =$ mean along-wind force coefficient; $m_1 =$ modal mass $= \int_o^h \mu(z)\phi^2(z)dz$; $\mu(z) =$ mass per unit height: $K = (1.65)^{\hat{\alpha}}/(\hat{\alpha}+\xi+1)$; and $\hat{V}_{\bar{z}}$ is the 3-sec gust speed at height $\bar{z}$. This can be evaluated by $\hat{V}_{\bar{z}} = \hat{b}(\bar{z}/33)^{\hat{\alpha}}V$, where V is the 3-sec gust speed in Exposure C at the reference height (obtained from Figure 6-1); $\hat{b}$ and $\hat{a}$ are given in Table 6-2.

RMS Along-Wind Acceleration. The rms along-wind acceleration $\sigma_{\ddot{x}}(z)$ as a function of height above the ground surface is given by

$$\sigma_{\ddot{x}}(z) = \frac{0.85\phi(z)\rho B h C_{fx}\overline{V}_{\bar{z}}^2}{m_1}I_{\bar{z}}KR \qquad \textbf{(Eq. C6-11)}$$

where $\overline{V}_{\bar{z}}$ is the mean hourly wind speed at height $\bar{z}$, ft/sec

$$\overline{V}_{\bar{z}} = \bar{b}\left(\frac{\bar{z}}{33}\right)^{\bar{\alpha}}V$$

where $\bar{b}$ and $\bar{\alpha}$ are defined in Table 6-2.

Maximum Along-Wind Acceleration. The maximum along-wind acceleration as a function of height above the ground surface is given by

$$\ddot{X}_{max}(z) = g_{\ddot{x}}\sigma_{\ddot{x}}(z) \qquad \textbf{(Eq. C6-12)}$$

$$g_{\ddot{x}} = \sqrt{2\ln(n_1T)} + \frac{0.5772}{\sqrt{2\ln(n_1T)}}$$

where $T =$ the length of time over which the minimum acceleration is computed, usually taken to be 3600 sec to represent 1 hour.

Example. The following example is presented to illustrate the calculation of the gust effect factor. Table C6-3 uses the given information to obtain values from Table 6-2.

Table C6-4 presents the calculated values. Table C6-5 summarizes the calculated displacements and accelerations as a function of the height, z.

Given Values.

Basic wind speed at reference height in Exposure C = 90 mph
Type of Exposure = A (from ASCE 7-98)
Building height h = 600 ft
Building width B = 100 ft
Building depth L = 100 ft
Building natural frequency n_1 = 0.2 Hz
Damping ratio = 0.01
C_{fx} = 1.3
Mode exponent = 1.0
Building density = 12 lb/ft^3 = 0.3727 slugs/ft^3
Air density = 0.0024 slugs/ft^3

C6.5.9 Enclosure Classifications. The magnitude and sense of internal pressure is dependent upon the magnitude and location of openings around the building envelope with respect to a given wind direction. Accordingly, the Standard requires that a determination be made of the amount of openings in the envelope in order to assess enclosure classification (enclosed, partially enclosed, or open). "Openings" are specifically defined in this version of the Standard as "apertures or holes in the building envelope which allow air to flow through the building envelope and which are designed as "open" during design winds." Examples include doors, operable windows, air intake exhausts for air conditioning and/or ventilation systems, gaps around doors, deliberate gaps in cladding and flexible and operable louvers. Once the enclosure classification is known, the designer enters Figure 6-5 to select the appropriate internal pressure coefficient.

This version of the Standard has four definitions applicable to enclosure: "wind-borne debris regions," "glazing," "impact-resistant glazing," and "impact-resistant covering." "Wind-borne debris regions" are defined to alert the designer to areas requiring consideration of missile impact design and potential openings in the building envelope. "Glazing" is defined as "any glass or transparent or translucent plastic sheet used in windows, doors, skylights, or curtain walls." "Impact-resistant glazing" is specifically defined as "glazing which has been shown by testing in accordance with ASTM E 1886 [C6-70] and ASTM E 1996 [C6-71] or other approved test methods to withstand the impact of wind-borne missiles likely to be generated in wind-borne debris regions during design winds." "Impact-resistant covering" over glazing can be shutters or screens designed for wind-borne debris impact. Impact resistance can now be tested using the test method specified in ASTM E 1886 [C6-70], with missiles, impact speeds, and pass/fail criteria specified in ASTM E 1996 [C6-71]. Other

approved test methods are acceptable. Origins of missile impact provisions contained in these Standards are summarized in References [C6-72 and C6-80].

Attention is made to Section 6.5.9.3, which requires glazing in Category II, III, and IV buildings in wind-borne debris regions to be protected with an impact-resistant covering or be impact resistant. For Category II and III buildings (other than health care, jails and detention facilities, and power-generating and other public utility facilities), an exception allows unprotected glazing, provided the glazing is assumed to be openings in determining the building's exposure classification. The option of unprotected glazing for Category III health care, jails and detention facilities, power-generating and other public utility facilities, and Category IV buildings was eliminated in this edition of the Standard because it is important for these facilities to remain operational during and after design windstorm events.

Prior to this edition of the Standard, glazing in the lower 60 ft (18.3 m) of Category II, III, or IV buildings sited in wind-borne debris regions was required to be protected with an impact-resistant covering, or be impact-resistant, or the glazing was to be assumed to be openings. Recognizing that glazing higher than 60 ft (18.3 m) above grade may be broken by wind-borne debris when a debris source is present, a new provision was added. With this new provision, aggregate surface roofs on buildings within 1500 ft of the new building need to be evaluated. For example, loose roof aggregate that is not protected by an extremely high parapet should be considered as a debris source. Accordingly, the glazing in the new building, from 30 ft (9.2 m) above the source building to grade would need to be protected or assumed to be open. If loose roof aggregate is proposed for the new building, it too should be considered as a debris source because aggregate can be blown off the roof and be propelled into glazing on the leeward side of the building. Although other types of wind-borne debris can impact glazing higher than 60 ft above grade, at these high elevations, loose roof aggregate has been the predominant debris source in previous wind events. The requirement for protection 30 ft (9.1 m) above the debris source is to account for debris that can be lifted during flight. The following References provide further information regarding debris damage to glazing: [C6-72, C6-97–C6-100].

Levels of impact resistance specified in ASTM E 1996–1999*

Building Classification	Category II & III (Note 1)		Category III & IV (Note 2)	
Glazing Height	≤30 ft. (9.1 m)	>30 ft. (9.1 m)	≤30 ft. (9.1 m)	>30 ft. (9.1 m)
Wind Zone 1	Missile B	Missile A	Missile C	Missile C
Wind Zone 2	Missile B	Missile A	Missile C	Missile C
Wind Zone 3	Missile C	Missile A	Missile D	Missile C
*Reprinted with permission from ASTM.				

Wind Zone 1: Wind-borne debris region where basic wind speed is greater than or equal to 110 mph but less than 120 mph and Hawaii.

Wind Zone 2: Wind-borne debris region where basic wind speed is greater than or equal to 120 mph but less than 130 mph at greater than 1 mile (1.6 km) of the coastline (Note 3).

Wind Zone 3: Wind-borne debris region where basic wind speed is greater than or equal to 130 mph, or where the basic wind speed is greater than or equal to 120 mph and within 1 mile (1.6 km) of the coastline (Note 3).

Note 1. Category III other than health care, jails and detention facilities, power-generating and other public utility facilities.

Note 2. Category III health care, jails and detention facilities, power-generating and other public utility facilities only.

Note 3. The coastline shall be measured from the mean high waterline.

Note 4. For porous shutter assemblies that contain openings greater than 3/16 in. (5 mm) projected horizontally, missile A shall also be used where missile B, C, or D are specified.

Missile levels specified in ASTM E 1996–1999*		
Missile Level	**Missile**	**Impact Speed**
Missile A	2 g ± 5% steel ball	130 ft/sec (39.6 m/s)
Missile B	4.5 lb ± 0.25 lb (2050 g ± 100 g) 2 × 4 lumber 4′ − 0″ ± 4″ (1.2 m ± 100 mm) long	40 ft/sec (12.2 m/s)
Missile C	9.0 lb ± 0.25 lb (4100 g ± 100 g) 2 × 4 lumber 8′ − 0″ ± 4″ (2.4 m ± 100 mm) long	50 ft/sec (15.3 m/s)
Missile D	9.0 lb ± 0.25 lb (4100 g ± 100 g) 2 × 4 lumber 8′ − 0″ ± 4″ (2.4 m ± 100 mm) long	80 ft/sec (24.4 m/s)
*Reprinted with permission from ASTM.		

C6.5.10 Velocity Pressure. The basic wind speed is converted to a velocity pressure q_z in pounds per square foot (newtons/m^2) at height z by the use of Eq. 6-15.

The constant 0.00256 (or 0.613 in SI) reflects the mass density of air for the standard atmosphere, i.e., temperature of 59°F (15°C) and sea level pressure of 29.92 inches of mercury (101.325 kPa), and dimensions associated with wind speed in mph (m/s). The constant is obtained as follows:

$$\text{constant} = 1/2[(0.0765 \text{ lb/ft}^3)/(32.2 \text{ ft/s}^2)]$$
$$\times [(\text{mi/h})(5280 \text{ ft/mi})$$
$$\times (1 \text{ h}/3600 \text{ s})]^2$$
$$= 0.00256$$
$$\text{constant} = 1/2[(1.225 \text{ kg/m}^3)/(9.81 \text{ m/s}^2)]$$
$$\times [(\text{m/s})]^2[9.81 \text{ N/kg}] = 0.613$$

The numerical constant of 0.00256 should be used except where sufficient weather data are available to justify a different value of this constant for a specific design application. The mass density of air will vary as a function of altitude, latitude, temperature, weather, and season. Average and extreme values of air density are given in Table C6-6.

C6.5.11 Pressure and Force Coefficients. The pressure and force coefficients provided in Figures 6-6 through 6-22 have been assembled from the latest boundary-layer wind-tunnel and full-scale tests and from previously available literature. Since the boundary-layer wind-tunnel results were obtained for specific types of building such as low- or high-rise buildings and buildings having specific types of structural framing systems, the designer is cautioned against indiscriminate interchange of values among the figures and tables.

Loads on Main Wind Force-Resisting Systems.

Figures 6-6 and 6-10. The pressure coefficients for main wind force-resisting systems are separated into two categories:

1. Buildings of all heights (Figure 6-6); and
2. Low-rise buildings having a height less than or equal to 60 ft (18 m) (Figure 6-10).

In generating these coefficients, two distinctly different approaches were used. For the pressure coefficients given in Figure 6-6, the more traditional approach was followed and the pressure coefficients reflect the actual loading on each surface of the building as a function of wind direction; namely, winds perpendicular or parallel to the ridge line.

Observations in wind tunnel tests show that areas of very low negative pressure and even slightly positive pressure can occur in all roof structures, particularly as the distance from the windward edge increases and the wind streams reattach to the surface. These pressures can occur even for relatively flat or low-slope roof structures. Experience and judgment from wind-tunnel studies have been used to specify either zero or slightly negative pressures (-0.18) depending on the negative pressure coefficient. These new values require the designer to consider a zero or slightly positive net wind pressure in the load combinations of Section 2.

For low-rise buildings having a height less than or equal to 60 ft (18 m), however, the values of GC_{pf} represent "pseudo" loading conditions which, when applied to the building, envelope the desired structural actions (bending moment, shear, thrust) independent of wind direction. To capture all appropriate structural actions, the building must be designed for all wind directions by considering in turn each corner of the building as the reference corner shown in the sketches of Figure 6-10. In ASCE 7-02, these sketches were modified in an attempt to clarify the proper application of the patterns. At each corner, two load patterns are applied, one for each MWFRS direction. The proper orientation of the load pattern is with the end zone strip parallel to the MWFRS direction. The end zone creates the required structural actions in the end frame or bracing. Note also that for all roof slopes, all 8 load cases must be considered individually in order to determine the critical loading for a given structural assemblage or component thereof. Special attention should be given roof members such as trusses, which meet the definition of MWFRS, but are not part of the lateral resisting system. When such members span at least from the eave to the ridge or support members spanning at least from eave to ridge, they are not required to be designed for the higher end zone loads under MWFRS. The interior zone loads should be applied. This is due to the enveloped nature of the loads for roof members.

To develop the appropriate "pseudo" values of GC_{pf}, investigators at the University of Western Ontario [C6-11] used an approach which consisted essentially of permitting the building model to rotate in the wind tunnel through a full 360 degrees while simultaneously monitoring the loading conditions on each of the surfaces (see Figure C6-7). Both Exposures B and C were considered. Using influence coefficients for rigid frames, it was possible to spatially average and time average the surface pressures to ascertain the maximum induced external force components to be resisted. More specifically, the following structural actions were evaluated:

1. total uplift
2. total horizontal shear
3. bending moment at knees (2-hinged frame)
4. bending moment at knees (3-hinged frame), and
5. bending moment at ridge (2-hinged frame)

The next step involved developing sets of "pseudo" pressure coefficients to generate loading conditions which

would envelope the maximum induced force components to be resisted for all possible wind directions and exposures. Note, for example, that the wind azimuth producing the maximum bending moment at the knee would not necessarily produce the maximum total uplift. The maximum induced external force components determined for each of the above five categories were used to develop the coefficients. The end result was a set of coefficients which represent fictitious loading conditions, but which conservatively envelope the maximum induced force components (bending moment, shear, and thrust) to be resisted, independent of wind direction.

The original set of coefficients was generated for the framing of conventional pre-engineered buildings, i.e., single story moment-resisting frames in one of the principal directions and bracing in the other principal direction. The approach was later extended to single story moment-resisting frames with interior columns [C6-19].

Subsequent wind-tunnel studies [C6-69] have shown that the GC_{pf} values of Figure 6-10 are also applicable to low-rise buildings with structural systems other than moment-resisting frames. That work examined the instantaneous wind pressures on a low-rise building with a 4:12 pitched gable roof and the resulting wind-induced forces on its main wind force-resisting system. Two different main wind force-resisting systems were evaluated. One consisted of shear walls and roof trusses at different spacings. The other had moment-resisting frames in one direction, positioned at the same spacings as the roof trusses, and diagonal wind bracing in the other direction. Wind-tunnel tests were conducted for both Exposures B and C. The findings of this study showed that the GC_{pf} values of Figure 6-10 provided satisfactory estimates of the wind forces for both types of structural systems. This work confirms the validity of Figure 6-10, which reflects the combined action of wind pressures on different external surfaces of a building and thus take advantage of spatial averaging.

In the original wind-tunnel experiments, both B and C exposure terrains were checked. In these early experiments, B exposure did not include nearby buildings. In general, the force components, bending moments, and so on, were found comparable in both exposures, although GC_{pf} values associated with Exposure B terrain would be higher than that for Exposure C terrain because of reduced velocity pressure in Exposure B terrain. The GC_{pf} values given in Figures 6-10 through 6-15 are derived from wind-tunnel studies modeled with Exposure C terrain. However, they may also be used in other exposures when the velocity pressure representing the appropriate exposure is used.

In recent comprehensive wind-tunnel studies conducted by [C6-66] at the University of Western Ontario, it was determined that when low buildings ($h < 60$ ft) are embedded in suburban terrain (Exposure B, which included nearby buildings), the pressures in most cases are lower

than those currently used in existing standards and codes, although the values show a very large scatter because of high turbulence and many variables. The results seem to indicate that some reduction in pressures for buildings located in Exposure B is justified. The Task Committee on Wind Loads believes it is desirable to design buildings for the exposure conditions consistent with the exposure designations defined in the Standard. In the case of low buildings, the effect of the increased intensity of turbulence in rougher terrain (i.e., Exposure A or B versus C) increases the local pressure coefficients. In ASCE 7-95, this effect was accounted for by allowing the designer of a building situated in Exposure A or B to use the loads calculated as if the building were located in Exposure C, but to reduce the loads by 15%. In ASCE 7-98, the effect of the increased turbulence intensity on the loads is treated with the truncated profile. Using this approach, the actual building exposure is used and the profile truncation corrects for the underestimate in the loads that would be obtained otherwise. The resulting wind loads on components and cladding obtained using this approach are much closer to the true values than those obtained using Exposure C loads combined with a 15% reduction in the resulting pressures.

Figure 6-10 is most appropriate for low buildings with width greater than twice their height and a mean roof height that does not exceed 33 ft (10 m). The original database included low buildings with width no greater than 5 times their eave height, and eave height did not exceed 33 ft (10 m). In the absence of more appropriate data, Figure 6-10 may also be used for buildings with mean roof height which does not exceed the least horizontal dimension and is less than or equal to 60 ft (18 m). Beyond these extended limits, Figure 6-6 should be used.

All the research used to develop and refine the low-rise building method for MWFRS loads was done on gable roofed buildings. In the absence of research on hip roofed buildings, the committee has developed a rational method of applying Figure 6-10 to hip roofs based on its collective experience, intuition, and judgment. This suggested method is presented in Figure C6-8.

Recent research [C6-93, C6-94] indicates that in the past the low-rise method underestimated the amount of torsion caused by wind loads. In ASCE 7-02, Note 5 was added to Figure 6-10 to account for this torsional effect. The reduction in loading on only 50% of the building results in a torsional load case without an increase in the predicted base shear for the building. The provision will have little or no effect on the design of main wind force-resisting systems that have well-distributed resistance. However, it will impact the design of systems with centralized resistance, such as a single core in the center of the building. An illustration of the intent of the note on 2 of the 8 load patterns is shown in Figure 6-10. All 8 patterns should be modified in this way as a separate set of load conditions in addition to the 8 basic patterns.

Internal pressure coefficients (GC_{pi}) to be used for loads on main wind force-resisting systems are given in Figure 6-5. The internal pressure load can be critical in 1-story moment-resisting frames and in the top story of a building where the main wind force-resisting system consists of moment-resisting frames. Loading cases with positive and negative internal pressures should be considered. The internal pressure load cancels out in the determination of total lateral load and base shear. The designer can use judgment in the use of internal pressure loading for the main wind force-resisting system of high-rise buildings.

Figure 6-7. Frame loads on dome roofs are adapted from the proposed Eurocode [C6-102]. The loads are based on data obtained in a modeled atmospheric boundary-layer flow which does not fully comply with requirements for wind-tunnel testing specified in this Standard [C6-101]. Loads for three domes ($h_D/D = 0.5$, $f/D = 0.5$), ($h_D/D = 0$, $f/D = 0.5$), ($h_D/D = 0$, $f/D = 0.33$) are roughly consistent with data of Taylor [C6-103], who used an atmospheric boundary layer as required in this Standard. Two load cases are defined, one of which has a linear variation of pressure from A to B as in the Eurocode [C6-102] and one in which the pressure at A is held constant from 0 to 25 degrees; these two cases are based on comparison of the Eurocode provisions of Taylor [C6-103]. Case A (the Eurocode calculation) is necessary in many cases to define maximum uplift. Case B is necessary to properly define positive pressures for some cases (which cannot be isolated with current information) and which result in maximum base shear. For domes larger than 200 ft diameter, the designer should consider use of Method 3. Resonant response is not considered in these provisions; Method 3 should be used to consider resonant response. Local bending moments in the dome shell may be larger than predicted by this method due to the difference between instantaneous local pressure distributions and that predicted by Figure 6-7. If the dome is supported on vertical walls directly below, it is appropriate to consider the walls as a "chimney" using Figure 6-19.

Loads on Components and Cladding. In developing the set of pressure coefficients applicable for the design of components and cladding as given in Figures 6-11 through 6-15, an envelope approach was followed but using different methods than for the main wind force-resisting systems of Figure 6-10. Because of the small effective area which may be involved in the design of a particular component (consider, for example, the effective area associated with the design of a fastener), the point-wise pressure fluctuations may be highly correlated over the effective area of interest. Consider the local purlin loads shown in Figure C6-4. The approach involved spatial averaging and time averaging of the point pressures over the effective area transmitting loads to the purlin while the

building model was permitted to rotate in the wind-tunnel through 360 degrees. As the induced localized pressures may also vary widely as a function of the specific location on the building, height above ground level, exposure, and more importantly, local geometric discontinuities and location of the element relative to the boundaries in the building surfaces (walls, roof lines), these factors were also enveloped in the wind-tunnel tests. Thus, for the pressure coefficients given in Figures 6-11 through 6-15, the directionality of the wind and influence of exposure have been removed and the surfaces of the building "zoned" to reflect an envelope of the peak pressures possible for a given design application.

As indicated in the discussion for Figure 6-10, the wind-tunnel experiments checked both B and C exposure terrains. Basically, GC_p values associated with Exposure B terrain would be higher than those for Exposure C terrain because of reduced velocity pressure in Exposure B terrain. The GC_P values given in Figures 6-11 through 6-15 are associated with Exposure C terrain as obtained in the wind tunnel. However, they may be also used for any exposure when the correct velocity pressure representing the appropriate exposure is used. (See commentary discussion in Section C6.5.11 under Loads on Main Wind Force-Resisting Systems.)

The wind-tunnel studies conducted by [C6-66] determined that when low buildings ($h < 60$ ft) are embedded in suburban terrain (Exposure B), the pressures on components and cladding in most cases are lower than those currently used in the standards and codes, although the values show a very large scatter because of high turbulence and many variables. The results seem to indicate that some reduction in pressures for components and cladding of buildings located in Exposure B is justified.

The pressure coefficients given in Figure 6-17 for buildings with mean height greater than 60 ft were developed following a similar approach, but the influence of exposure was not enveloped [C6-42]. Therefore, Exposure Categories B, C, or D may be used with the values of GC_p in Figure 6-8 as appropriate.

Figure 6-11 The pressure coefficient values provided in this figure are to be used for buildings with a mean roof height of 60 ft (18 m) or less. The values were obtained from wind-tunnel tests conducted at the University of Western Ontario [C6-10, C6-11], at the James Cook University of North Queensland [C6-6], and at Concordia University [C6-40, C6-41, C6-44, C6-45, C6-47]. These coefficients have been refined to reflect results of full-scale tests conducted by the National Bureau of Standards [C6-22] and the Building Research Station, England [C6-14]. Pressure coefficients for hemispherical domes on ground or on cylindrical structures have been reported [C6-52]. Some of the characteristics of the values in the figure are as follows:

1. The values are combined values of GC_p; the gust effect factors from these values should not be separated.

2. The velocity pressure q_h evaluated at mean roof height should be used with all values of GC_p.

3. The values provided in the figure represent the upper bounds of the most severe values for any wind direction. The reduced probability that the design wind speed may not occur in the particular direction for which the worst pressure coefficient is recorded has not been included in the values shown in the figure.

4. The wind-tunnel values, as measured, were based on the mean hourly wind speed. The values provided in the figures are the measured values divided by $(1.53)^2$ (see Figure C6-2) to reflect the reduced pressure coefficient values associated with a 3-second gust speed.

Each component and cladding element should be designed for the maximum positive and negative pressures (including applicable internal pressures) acting on it. The pressure coefficient values should be determined for each component and cladding element on the basis of its location on the building and the effective area for the element. As recent research has shown [C6-41, C6-43], the pressure coefficients provided generally apply to facades with architectural features such as balconies, ribs, and various facade textures.

More recent studies [C6-104, C6-105, C6-106] have led to updating the roof slope range and the values of GC_p included in ASCE 7-02.

Figures 6-13 and 6-14A. These figures present values of GC_p for the design of roof components and cladding for buildings with multispan gable roofs and buildings with monoslope roofs. The coefficients are based on wind-tunnel studies reported by [C6-46, C6-47, C6-51].

Figure 6-14B. The values of GC_p in this figure are for the design of roof components and cladding for buildings with sawtooth roofs and mean roof height, h, less than or equal to 60 ft (18 m). Note that the coefficients for corner zones on segment A differ from those coefficients for corner zones on the segments designated as B, C, and D. Also, when the roof angle is less than or equal to 10 degrees, values of GC_p for regular gables roofs (Figure 6-11B) are to be used. The coefficients included in Figure 6-14B are based on wind-tunnel studies reported by [C6-35].

Figure 6-17. The pressure coefficients shown in this figure have been revised to reflect the results obtained from comprehensive wind-tunnel studies carried out by [C6-42]. In general, the loads resulting from these coefficients are lower than those required by ASCE 7-93. However,

the area averaging effect for roofs is less pronounced when compared with the requirements of ASCE 7-93. The availability of more comprehensive wind-tunnel data has also allowed a simplification of the zoning for pressure coefficients; flat roofs are now divided into three zones, and walls are represented by two zones.

The external pressure coefficients and zones given in Figure 6-17 were established by wind-tunnel tests on isolated "box-like" buildings [C6-2, C6-31]. Boundary-layer wind-tunnel tests on high-rise buildings (mostly in downtown city centers) show that variations in pressure coefficients and the distribution of pressure on the different building facades are obtained [C6-53]. These variations are due to building geometry, low attached buildings, non-rectangular cross sections, setbacks, and sloping surfaces. In addition, surrounding buildings contribute to the variations in pressure. Wind-tunnel tests indicate that pressure coefficients are not distributed symmetrically and can give rise to torsional wind loading on the building.

Boundary-layer wind-tunnel tests that include modeling of surrounding buildings permit the establishment of more exact magnitudes and distributions of GC_p for buildings that are not isolated or "box-like" in shape.

Figure 6-16. This figure for cladding pressures on dome roofs is based on Taylor [C6-103]. Negative pressures are to be applied to the entire surface, since they apply along the full arc, which is perpendicular to the wind direction and which passes through the top of the dome. Users are cautioned that only 3 shapes were available to define values in this figure ($h_D/D = 0.5$, $f/D = 0.5$; $h_D/D = 0.0$, $f/D = 0.5$; $h_D/D = 0.0$, $f/d = 0.33$).

Figure 6-8 and Figures 6-18 through 6-22. With the exception of Figure 6-22, the pressure and force coefficient values in these tables are unchanged from ANSI A58.1-1972 and 1982, and ASCE 7-88 and 7-93. The coefficients specified in these tables are based on wind-tunnel tests conducted under conditions of uniform flow and low turbulence, and their validity in turbulent boundary layer flows has yet to be completely established. Additional pressure coefficients for conditions not specified herein may be found in References [C6-3 and C6-36]. With regard to Figure 6-19, local maximum and minimum peak pressure coefficients for cylindrical structures with $h/D < 2$ are $GC_p = 1.1$ and $GC_p = -1.1$, respectively, for Reynolds numbers ranging from 1.1×10^5 to 3.1×10^5 [C6-23]. The latter results have been obtained under correctly simulated boundary layer flow conditions.

With regard to Figure 6-22, the force coefficients are a refinement of the coefficients specified in ANSI A58.1-1982 and in ASCE 7-93. The force coefficients specified are offered as a simplified procedure that may be used for trussed towers and are consistent with force coefficients given in ANSI/EIA/TIA-222-E-1991, Structural Standards

for Steel Antenna Towers and Antenna Supporting Structures, and force coefficients recommended by Working Group No. 4 (Recommendations for Guyed Masts), International Association for Shell and Spatial Structures (1981).

It is not the intent of the Standard to exclude the use of other recognized literature for the design of special structures such as transmission and telecommunications towers. Recommendations for wind loads on tower guys are not provided as in previous editions of the Standard. Recognized literature should be referenced for the design of these special structures as is noted in C6.4.2.1. For the design of flagpoles, see ANSI/NAAMM FP1001-97, 4th Ed., Guide Specifications for Design of Metal Flagpoles.

ASCE 7-02 has been modified to explicitly require the use of Figure 6-19 for the determination of the wind load on equipment located on a rooftop. Because there is a lack of research to provide better guidance for loads on rooftop equipment, this change was made based on the consensus opinion of the Committee. Because of the relatively small size of the equipment it is believed that the gust effect factor will be higher than 0.85; however, no research presently exists upon which to base a recommendation. Use of a gust effect factor of 1.1 or higher should be considered based on observations in a recent wind-tunnel study reported to the Task Committee on Wind Loads. Eq. 6-4 may be used as a guide in determining the gust effect factor. Caution should also be exercised in the positioning of the equipment on the roof. If rooftop equipment is located, either in whole or in part, in the higher pressure zones near a roof's edge, consideration should be given to increasing the wind load.

C6.5.11.1 Internal Pressure Coefficients. The internal pressure coefficient values in Figure 6-5 were obtained from wind-tunnel tests [C6-38] and full-scale data [C6-59]. Even though the wind-tunnel tests were conducted primarily for low-rise buildings, the internal pressure coefficient values are assumed to be valid for buildings of any height. The values $GC_{pi} = +0.18$ and -0.18 are for enclosed buildings. It is assumed that the building has no dominant opening or openings and that the small leakage paths that do exist are essentially uniformly distributed over the building's envelope. The internal pressure coefficient values for partially enclosed buildings assume that the building has a dominant opening or openings. For such a building, the internal pressure is dictated by the exterior pressure at the opening and is typically increased substantially as a result. Net loads, i.e., the combination of the internal and exterior pressures, are therefore also significantly increased on the building surfaces that do not contain the opening. Therefore, higher GC_{pi} values of $+0.55$ and -0.55 are applicable to this case. These values include a reduction factor to account for the lack of perfect correlation between the internal pressure and the external pressures on the building surfaces not containing the opening [C6-82] [C6-83]. Taken in isolation, the internal pressure coefficients

can reach values of ± 0.8, (or possibly even higher on the negative side).

For partially enclosed buildings containing a large unpartitioned space, the response time of the internal pressure is increased and this reduces the ability of the internal pressure to respond to rapid changes in pressure at an opening. The gust factor applicable to the internal pressure is therefore reduced. Eq. 6-14, which is based on References [C6-84 and C6-85] is provided as a means of adjusting the gust factor for this effect on structures with large internal spaces such as stadiums and arenas.

Glazing in the bottom 60 ft of buildings that are sited in hurricane-prone regions that is not impact-resistant glazing or is not protected by impact-resistant coverings should be treated as openings. Because of the nature of hurricane winds [C6-27], glazing in buildings sited in hurricane areas is very vulnerable to breakage from missiles, unless the glazing can withstand reasonable missile loads and subsequent wind loading, or the glazing is protected by suitable shutters. Glazing above 60 ft (18 m) is also somewhat vulnerable to missile damage, but because of the greater height, this glazing is typically significantly less vulnerable to damage than glazing at lower levels. When glazing is breached by missiles, development of high internal pressure results, which can overload the cladding or structure if the higher pressure was not accounted for in the design. Breaching of glazing can also result in a significant amount of water infiltration, which typically results in considerable damage to the building and its contents [C6-33, C6-49, C6-50].

If the option of designing for higher internal pressure (versus designing glazing protection) is selected, it should be realized that if glazing is breached, significant damage from overpressurization to interior partitions and ceilings is likely. The influence of compartmentation on the distribution of increased internal pressure has not been researched. If the space behind breached glazing is separated from the remainder of the building by a sufficiently strong and reasonably air-tight compartment, the increased internal pressure would likely be confined to that compartment. However, if the compartment is breached (e.g., by an open corridor door, or by collapse of the compartment wall), the increased internal pressure will spread beyond the initial compartment quite rapidly. The next compartment may contain the higher pressure, or it too could be breached, thereby allowing the high internal pressure to continue to propagate.

Because of the great amount of air leakage that often occurs at large hangar doors, designers of hangars should consider utilizing the internal pressure coefficients for partially enclosed buildings in Figure 6-5.

C6.5.11.5 Parapets. Previous versions of the Standard have had no provisions for the design of parapets, although the companion "Guide" [C6-87] has shown a methodology. The problem has been the lack of direct research in this area on which to base provisions. Research has yet to

be performed, but the Committee thought that a rational method based on its collective experience, intuition, and judgment was needed, considering the large number of buildings with parapets.

The methodology chosen assumes that parapet pressures are a combination of wall and roof pressures, depending on the location of the parapet, and the direction of the wind, see Figure C6-9. The windward parapet should receive the positive wall pressure to the front surface (exterior side of the building) and the negative roof edge zone pressure to the back surface (roof side). This concept is based on the idea that the zone of suction caused by the wind stream separation at the roof eave moves up to the top of the parapet when one is present. Thus the same suction which acts on the roof edge will also act on the back of the parapet.

The leeward parapet would receive the positive wall pressure on the back surface (roof side) with the negative wall pressure on the front surface (exterior side of building). There should be no reduction in the positive wall pressure to the leeward due to shielding by the windward parapet, since typically, they are too far apart to experience this effect. Since all parapets would be designed for all wind directions, each parapet would in turn be the windward and leeward parapet and be designed for both sets of pressures.

For the design of the main wind force-resisting system, the pressures used describe the contribution of the parapet to the overall wind loads on that system. A question arises whether to use the frame loads from MWFRS or the local loads from Components and Cladding. Both were studied using Figures 6-6 and 6-11. The use of Figure 6-10 coefficients would not have been appropriate since they are "pseudo-pressures" (see commentary on low-rise buildings). The results showed that the upper end of the range of frame loads match fairly closely with the lower end of the range of component loads. The judgment was made to use the upper end of the frame loads to determine the MWFRS coefficients used in the Standard. For simplicity, the front and back pressures on the parapet were combined into one coefficient for MWFRS design. The designer should not typically need the separate front and back pressures for MWFRS design. The internal pressures inside the parapet cancel out of the combined coefficient. The summation of these external and internal, front and back pressure coefficients is a new term GC_{pn}, the main wind force-resisting system parapet combined net pressure coefficient.

For the design of the components and cladding a similar approach was used. However, it is not possible to simplify the coefficients as much due to the increased complexity of the components and cladding pressure coefficients. In addition, the front and back pressures cannot necessarily be combined since the designer may be designing separate elements on each face of the parapet. The internal pressure is needed to account for net pressures on each surface. The new provisions guide the designer to the correct GC_p and

velocity pressure to use for each surface as illustrated in Figure C6-9.

Interior walls which protrude through the roof, such as party walls and fire walls, should be designed as windward parapets for both MWFRS and components and cladding.

The internal pressure that may be present inside a parapet is highly dependent on the porosity of the parapet envelope. In other words, it depends on the likelihood of the wall surface materials to leak air pressure into the internal cavities of the parapet. For solid parapets such as concrete or masonry, the internal pressure is zero, since there is no internal cavity. Certain wall materials may be impervious to air leakage, and as such have little or no internal pressure or suction, so using the value of GC_{pi} for an enclosed building may be appropriate. However, certain materials and systems used to construct parapets containing cavities are more porous, thus justifying the use of the GC_{pi} values for partially enclosed buildings, or higher. Another factor in the internal pressure determination is whether the parapet cavity connects to the internal space of the building, allowing the building's internal pressure to propagate into the parapet.

C6.5.12 Design Wind Loads on Buildings. The Standard provides specific wind pressure equations for both the main wind force-resisting systems and components and cladding.

In Eqs. 6-17, 6-19, and 6-23 a new velocity pressure term "q_i" appears that is defined as the "velocity pressure for internal pressure determination." The positive internal pressure is dictated by the positive exterior pressure on the windward face at the point where there is an opening. The positive exterior pressure at the opening is governed by the value of q at the level of the opening, not q_h. Therefore, the old provision which used q_h as the velocity pressure is not in accord with the physics of the situation. For low buildings this does not make much difference, but for the example of a 300-ft tall building in Exposure B with a highest opening at 60 ft, the difference between q_{300} and q_{60} represents a 59% increase in internal pressure. This is unrealistic and represents an unnecessary degree of conservatism. Accordingly, $q_i = q_z$ for positive internal pressure evaluation in partially enclosed buildings where height z is defined as the level of the highest opening in the building that could affect the positive internal pressure. For buildings sited in wind-borne debris regions, glazing that is not impact resistant or protected with an impact resistant covering, q_i should be treated as an opening. For positive internal pressure evaluation, q_i may conservatively be evaluated at height h ($q_i = q_h$).

C6.5.12.3 Design Wind Load Cases. Recent wind-tunnel research [C6-93–C6-96] has shown that torsional load requirements of Figure 9 in ASCE 7-98 often grossly underestimate the true torsion on a building under wind, including those that are symmetric in geometric form and stiffness. This torsion is caused by nonuniform pressure on

the different faces of the building from wind flow around the building, interference effects of nearby buildings and terrain and by dynamic effects on more flexible buildings. The revision to load cases two and four in Figure 6-9 increases the torsional loading to 15% eccentricity under 75% of the maximum wind shear for load case two (from ASCE 7-98 value of 3.625% eccentricity at 87.5% of maximum shear). Although this is more in line with wind-tunnel experience on square and rectangular buildings with aspect ratios up to about 2.5, it may not cover all cases, even for symmetric and common building shapes where larger torsions have been observed. For example, wind-tunnel studies often show an eccentricity of 5% or more under full (not reduced) base shear. The designer may wish to apply this level of eccentricity at full wind loading for certain more critical buildings even though it is not required by the Standard. The present more moderate torsional load requirements can in part be justified by the fact that the design wind forces tend to be an upperbound for most common building shapes.

In buildings with some structural systems, more severe loading can occur when the resultant wind load acts diagonally to the building. To account for this effect and the fact that many buildings exhibit maximum response in the across-wind direction (the Standard currently has no analytical procedure for this case), a structure should be capable of resisting 75% of the design wind load applied simultaneously along each principal axis as required by case three in Figure 6-9.

For flexible buildings, dynamic effects can increase torsional loading. Additional torsional loading can occur because of eccentricity between the elastic shear center and the center of mass at each level of the structure. The new Eq. 6-21 accounts for this effect.

It is important to note that significant torsion can occur on low-rise buildings also [C6-94] and therefore the wind loading requirements of Section 6.5.12.3 are now applicable to buildings of all heights.

As discussed in Section 6.6, the Wind-Tunnel Method 3 should always be considered for buildings with unusual shapes, rectangular buildings with larger aspect ratios, and dynamically sensitive buildings. The effects of torsion can more accurately be determined for these cases and for the more normal building shapes using the wind tunnel procedure.

SECTION C6.6
METHOD 3 — WIND-TUNNEL PROCEDURE

Wind-tunnel testing is specified when a structure contains any of the characteristics defined in Section 6.5.2 or when the designer wishes to more accurately determine the wind loads. For some building shapes, wind-tunnel testing can reduce the conservatism due to enveloping of wind loads inherent in Methods 1 and 2. Also, wind-tunnel testing accounts for shielding or channeling and can more accurately determine wind loads for a complex building shape than Methods 1 and 2. It is the intent of the Standard that any building or other structure be allowed to use the wind-tunnel testing method to determine wind loads. Requirements for proper testing are given in Section 6.6.2.

Wind-tunnel tests are recommended when the building or other structure under consideration satisfies one or more of the following conditions:

1. has a shape which differs significantly from a uniform rectangular prism or "box-like" shape,

2. is flexible with natural frequencies normally below 1 Hz,

3. is subject to buffeting by the wake of upwind buildings or other structures, or

4. is subject to accelerated flow caused by channeling or local topographic features.

It is common practice to resort to wind-tunnel tests when design data are required for the following wind-induced loads:

1. curtain wall pressures resulting from irregular geometry,

2. across-wind and/or torsional loads,

3. periodic loads caused by vortex shedding, and

4. loads resulting from instabilities such as flutter or galloping.

Boundary-layer wind tunnels capable of developing flows that meet the conditions stipulated in Section 6.4.3.1 typically have test-section dimensions in the following ranges: width of 6 to 12 ft (2–4 m), height of 6 to 10 ft (2–3 m), and length of 50 to 100 ft (15–30 m). Maximum wind speeds are ordinarily in the range of 25 to 100 mph (10–45 m/s). The wind tunnel may be either an open-circuit or closed-circuit type.

Three basic types of wind-tunnel test models are commonly used. These are designated as follows: (1) rigid pressure model (PM); (2) rigid high-frequency base balance model (H-FBBM); and (3) aeroelastic model (AM). One or more of the models may be employed to obtain design loads for a particular building or structure. The PM provides local peak pressures for design of elements such as cladding and mean pressures for the determination of overall mean loads. The H-FBBM measures overall fluctuating loads (aerodynamic admittance) for the determination of dynamic responses. When motion of a building or structure influences the wind loading, the AM is employed for direct measurement of overall loads, deflections, and accelerations. Each of these models, together with a model of the surroundings (proximity model), can provide information other than wind loads such as snow loads on complex roofs,

wind data to evaluate environmental impact on pedestrians, and concentrations of air-pollutant emissions for environmental impact determinations. Several references provide detailed information and guidance for the determination of wind loads and other types of design data by wind-tunnel tests [C6-4, C6-7, C6-8, C6-33].

Wind-tunnel tests frequently measure wind loads which are significantly lower than required by Section 6.5 due to the shape of the building, shielding in excess of that implied by exposure categories, and necessary conservatism in enveloping load coefficients in Section 6.5. In some cases, adjacent structures may shield the structure sufficiently that removal of one or two structures could significantly increase wind loads. Additional wind-tunnel testing without specific nearby buildings (or with additional buildings if they might cause increased loads through channeling or buffeting) is an effective method for determining the influence of adjacent buildings. It would be prudent for the designer to test any known conditions that change the test results and apply good engineering judgment in interpreting the test results. Discussion between the owner, designer, and wind-tunnel laboratory can be an important part of this decision. However, it is impossible to anticipate all possible changes to the surrounding environment that could significantly impact pressure for the main wind force-resisting system and for cladding pressures. Also, additional testing may not be cost effective. Suggestions written in mandatory language for users (e.g., code writers) desiring to place a lower limit on the results of wind tunnel testing are shown below:

Lower limit on pressures for main wind force-resisting system. Forces and pressures determined by wind-tunnel testing shall be limited to not less than 80% of the design forces and pressures which would be obtained in Section 6.5 for the structure unless specific testing is performed to show that it is the aerodynamic coefficient of the building itself, rather than shielding from nearby structures, that is responsible for the lower values. The 80% limit may be adjusted by the ratio of the frame load at critical wind directions as determined from wind-tunnel testing without specific adjacent buildings (but including appropriate upwind roughness), to that determined by Section 6.5.

Lower limit on pressures for components and cladding. The design pressures for components and cladding on walls or roofs shall be selected as the greater of the wind-tunnel test results or 80% of the pressure obtained for Zone 4 for walls and Zone 1 for roofs as determined in Section 6.5, unless specific testing is performed to show that it is the aerodynamic coefficient of the building itself, rather than shielding from nearby structures, that is responsible for the lower values. Alternatively, limited tests at a few wind directions without specific adjacent buildings, but in the presence of an appropriate upwind roughness, may be used to demonstrate that the lower pressures are due to the shape of the building and not to shielding.

REFERENCES

[C6-1] Abbey, R.F. "Risk probabilities associated with tornado wind speeds." *Proc., Symposium on Tornadoes: Assessment of Knowledge and Implications for Man*, R.E. Peterson, Ed., Institute for Disaster Research, Texas Tech University, Lubbock, Tex., 1976.

[C6-2] Akins, R.E. and Cermak, J.E. Wind pressures on buildings. Technical Report CER 7677REA-JEC15, Fluid Dynamics and Diffusion Lab, Colorado State University, Fort Collins, Colo., 1975.

[C6-3] ASCE "Wind forces on structures." *Trans. ASCE*, 126(2), 1124–1198, 1961.

[C6-4] Wind Tunnel Model Studies of Buildings and Structures. Manuals and Reports on Engineering Practice, No. 67, American Society of Civil Engineers, Reston, Va., 1997.

[C6-5] Batts, M.E., Cordes, M.R., Russell, L.R., Shaver, J.R., and Simiu, E. *Hurricane wind speeds in the United States*. NBS Building Science Series 124, National Bureau of Standards, Washington, D.C., 1980.

[C6-6] Best, R.J. and Holmes, J.D. Model study of wind pressures on an isolated single-story house. James Cook University of North Queensland, Australia, *Wind Engineering Rep*. 3/78, 1978.

[C6-7] Boggs, D.W. and Peterka, J.A. "Aerodynamic model tests of tall buildings." *J. Engrg. Mech.*, ASCE, 115(3), New York, 618–635, 1989.

[C6-8] Cermak, J.E. "Wind-tunnel testing of structures." *J. Engrg. Mech. Div.*, ASCE, 103(6), New York, 1125–1140, 1977.

[C6-9] Cheung, J.C.J. and Melbourne, W.H. "Wind loadings on porous cladding." *Proc., 9th Australian Conference on Fluid Mechanics*, 308, 1986.

[C6-10] Davenport, A.G., Surry, D., and Stathopoulos, T. Wind loads on low-rise buildings. "Final Report on Phases I and II, BLWT-SS8," University of Western Ontario, London, Ontario, Canada, 1977.

[C6-11] Davenport, A.G., Surry, D., and Stathopoulos, T. Wind loads on low-rise buildings. Final Report on Phase III, BLWT-SS4, University of Western Ontario, London, Ontario, Canada, 1978.

[C6-12] Interim guidelines for building occupants' protection from tornadoes and extreme winds. TR-83A, Defense Civil Preparedness Agency, Washington, D.C. 1975. [Available from Superintendent of Documents, U.S. Government Printing Office, Washington, D.C. 20402.]

[C6-13] Durst, C.S. "Wind speeds over short periods of time." *Meteor. Mag.*, 89, 181–187, 1960.

[C6-14] Eaton, K.J. and Mayne, J.R. "The measurement of wind pressures on two-story houses at Aylesbury." *J. Industrial Aerodynamics*, 1(1), 67–109, 1975.

[C6-15] Georgiou, P.N., Davenport, A.G., and Vickery, B.J. "Design wind speeds in regions dominated by tropical cyclones." *J. Wind Engrg. and Industrial Aerodynamics*, 13, 139–152, 1983.

[C6-16] Haig, J.R. *Wind Loads on Tiles for USA*, Redland Technology Limited, Horsham, West Sussex, England, June 1990.

[C6-17] Isyumov, N. "The aeroelastic modeling of tall buildings." *Proc, International Workshop on Wind Tunnel Modeling Criteria and Techniques in Civil Engineering Applications*, Cambridge University Press, Gaithersburg, Md., 373–407, 1982.

[C6-18] Jackson, P.S. and Hunt, J.C.R. "Turbulent wind flow over a low hill." *Quarterly J. Royal Meteorological Soc.*, Vol. 101, 929–955, 1975.

[C6-19] Kavanagh, K.T., Surry, D., Stathopoulos, T., and Davenport, A.G. Wind loads on low-rise buildings: Phase IV, BLWT-SS14, University of Western Ontario, London, Ontario, Canada, 1983.

[C6-20] Krayer, W.R. and Marshall, R.D. "Gust factors applied to hurricane winds." *Bull. AMS*, 73, 613–617, 1992.

[C6-21] Lemelin, D.R., Surry, D., and Davenport, A.G. "Simple approximations for wind speed-up over hills." *J. Wind Engrg. and Industrial Aerodynamics*, 28, 117–127, 1988.

[C6-22] Marshall, R.D. The measurement of wind loads on a full-scale mobile home. NBSIR 77-1289, National Bureau of Standards, U.S. Dept. of Commerce, Washington, D.C., 1977.

[C6-23] Macdonald, P.A., Kwok, K.C.S. and Holmes, J.H. Wind loads on isolated circular storage bins, silos and tanks: Point pressure measurements. Research Report No. R529, School of Civil and Mining Engineering, University of Sydney, Sydney, Australia, 1986.

[C6-24] McDonald, J.R. A methodology for tornado hazard probability assessment, NUREG/CR3058, U.S. Nuclear Regulatory Commission, Washington, D.C., 1983.

[C6-25] Mehta, K.C., Minor, J.E., and McDonald, J.R. "Wind speed analyses of April 3–4 Tornadoes." *J. Struct. Div.*, ASCE, 102(9), 1709–1724, 1976.

[C6-26] Minor, J.E. "Tornado technology and professional practice." *J. Struct. Div.*, ASCE, 108(11), 2411–2422, 1982.

[C6-27] Minor, J.E. and Behr, R.A. "Improving the performance of architectural glazing systems in hurricanes." *Proc., ASCE Conference on Hurricanes of 1992*, American Society of Civil Engineers (Dec. 1–3, 1993, Miami, Fla.), C1–11, 1993.

[C6-28] Minor, J.E., McDonald, J.R., and Mehta, K.C. The tornado: An engineering oriented perspective. TM ERL NSSL-82, National Oceanic and Atmospheric Administration, Environmental Research Laboratories, Boulder, Colo., 1977.

[C6-29] Peterka, J.A. "Improved extreme wind prediction for the United States." *J. Wind Engrg. and Industrial Aerodynamics*, 41, 533–541, 1992.

[C6-30] Peterka, J.A. and Shahid, S. "Extreme gust wind speeds in the U.S." *Proc., 7th U.S. National Conference on Wind Engineering*, UCLA, Los Angeles, Calif., 2, 503–512, 1993.

[C6-31] Peterka, J.A. and Cermak, J.E. "Wind pressures on buildings — Probability densities." *J. Struct. Div.*, ASCE, 101(6), 1255–1267, 1974.

[C6-32] Perry, D.C., Stubbs, N., and Graham, C.W. "Responsibility of architectural and engineering communities in reducing risks to life, property and economic loss from hurricanes." *Proc., ASCE Conference on Hurricanes of 1992*, Miami, Fla., 1993.

[C6-33] Reinhold, T.A. (Ed.). "Wind tunnel modeling for civil engineering applications." *Proc., International Workshop on Wind Tunnel Modeling Criteria and Techniques in Civil Engineering Applications*, Cambridge University Press, Gaithersburg, Md., 1982.

[C6-34] Australian Standard SAA Loading Code, Part 2: Wind Loads, published by Standards Australia, Standards House, 80 Arthur St., North Sydney, NSW, Australia, 1989.

[C6-35] Saathoff, P. and Stathopoulos, T. "Wind loads on buildings with sawtooth roofs." *J. Struct. Engrg.*, ASCE, 118(2) 429–446, 1992.

[C6-36] *Normen fur die Belastungsannahmen, die Inbetriebnahme und die Uberwachung der Bauten*, SIA Technische Normen Nr 160, Zurich, Switzerland, 1956.

[C6-37] *Standard Building Code*. Southern Building Code Congress International, 1994.

[C6-38] Stathopoulos, T., Surry, D., and Davenport, A.G. "Wind-induced internal pressures in low buildings." *Proc., 5th International Conference on Wind Engineering*, Colorado State University, Fort Collins, Colo., 1979.

[C6-39] Stathopoulos, T., Surry, D., and Davenport, A.G. "A simplified model of wind pressure coefficients for low-rise buildings." *4th Colloquium on Industrial Aerodynamics*, Aachen, West Germany, June 18–20, 1980.

[C6-40] Stathopoulos, T. "Wind loads on eaves of low buildings." *J. Struct. Div.*, ASCE, 107(10), 1921–1934, 1981.

[C6-41] Stathopoulos, T. and Zhu, X. "Wind pressures on buildings with appurtenances." *J. Wind Engrg. and Industrial Aerodynamics*, 31, 265–281, 1988.

[C6-42] Stathopoulos, T. and Dumitrescu-Brulotte, M. "Design recommendations for wind loading on buildings of intermediate height." *Can. J. Civ. Engrg.*, 16(6), 910–916, 1989.

[C6-43] Stathopoulos, T. and Zhu, X. "Wind pressures on buildings with mullions." *J. Struct. Engrg.*, ASCE, 116(8), 2272–2291, 1990.

[C6-44] Stathopoulos, T. and Luchian, H.D. "Wind pressures on building configurations with stepped roofs." *Can. J. Civ. Engrg.*, 17(4), 569–577, 1990.

[C6-45] Stathopoulos, T. and Luchian, H. "Wind-induced forces on eaves of low buildings." Wind Engineering Society Inaugural Conference, Cambridge, England, 1992.

[C6-46] Stathopoulos, T. and Mohammadian, A.R. "Wind loads on low buildings with mono-sloped roofs." *J. Wind Engrg. and Industrial Aerodynamics*, 23, 81–97, 1986.

[C6-47] Stathopoulos, T. and Saathoff, P. "Wind pressures on roofs of various geometries." *J. Wind Engrg. and Industrial Aerodynamics*, 38, 273–284, 1991.

[C6-48] Stubbs, N. and Boissonnade, A. "Damage simulation model or building contents in a hurricane environment." *Proc., 7th U.S. National Conference on Wind Engineering*, UCLA, Los Angeles, Calif., 2, 759–771, 1993.

[C6-49] Stubbs, N. and Perry, D.C. "Engineering of the building envelope." *Proc., ASCE Conference on Hurricanes of 1992*, Miami, Fla., 1993.

[C6-50] Surry, D., Kitchen, R.B., and Davenport, A.G. "Design effectiveness of wind tunnel studies for buildings of intermediate height." *Can. J. Civ. Engrg.*, 4(1), 96–116, 1977.

[C6-51] Surry, D. and Stathopoulos, T. The wind loading of buildings with monosloped roofs. Final Report, BLWT-SS38, University of Western Ontario, London, Ontario, Canada, 1988.

[C6-52] Taylor, T.J. "Wind pressures on a hemispherical dome." *J. Wind Engrg. and Industrial Aerodynamics*, 40(2), 199–213, 1992.

[C6-53] Templin, J.T. and Cermak, J.E. Wind pressures on buildings: Effect of mullions. Tech. Rep. CER76-77JTT-JEC24, Fluid Dynamics and Diffusion Lab, Colorado State University, Fort Collins, Colo., 1978.

[C6-54] Vickery, P.J. and Twisdale, L.A., "Wind field and filling models for hurricane wind speed predictions." *J. Struct. Engrg.*, ASCE, 121(11), 1700–1709, 1995.

[C6-55] Vickery, B.J., Davenport, A.G., and Surry, D. "Internal pressures on low-rise buildings." *4th Canadian Workshop on Wind Engineering*, Toronto, Ontario, 1984.

[C6-56] Walmsley, J.L., Taylor, P.A., and Keith, T. "A simple model of neutrally stratified boundary-layer flow over complex terrain with surface roughness modulations." *Boundary-Layer Metrology*, 36, 157–186, 1986.

[C6-57] Wen, Y.K. and Chu, S.L. "Tornado risks and design wind speed." *J. Struct. Div.*, ASCE, 99(12), 2409–2421, 1973.

[C6-58] Womble, J.A., Yeatts, B.B., and Mehta, K.C. "Internal wind pressures in a full and small scale building." *Proc., 9th International Conference on Wind Engineering*, New Delhi, India: Wiley Eastern Ltd., 1995.

[C6-59] Yeatts, B.B. and Mehta, K.C. "Field study of internal pressures." *Proc., 7th U.S. National Conference on Wind Engineering*, UCLA, Los Angeles, Calif., 2, 889–897, 1993.

[C6-60] Gurley, K. and Kareem, A. "Gust loading factors for tension leg platforms." *Applied Ocean Research*, 15(3), 1993.

[C6-61] Kareem, A. "Dynamic response of high-rise buildings to stochastic wind loads." *J. Wind Engrg. and Industrial Aerodynamics*, 41–44, 1992.

[C6-62] Kareem, A. "Lateral-torsional motion of tall buildings to wind loads." *J. Struct. Engrg.*, ASCE, 111(11), 1985.

[C6-63] Solari, G. "Gust buffeting I: Peak wind velocity and equivalent pressure." *J. Struct. Engrg.*, ASCE, 119(2), 1993.

[C6-64] Solari, G. "Gust buffeting II: Dynamic along-wind response." *J. Struct. Engrg.*, ASCE, 119(2), 1993.

[C6-65] Kareem, A. and Smith, C. "Performance of offshore platforms in hurricane Andrew." *Proc., ASCE Conference on Hurricanes of 1992*, ASCE, Miami, Fla., Dec., 1993.

[C6-66] Ho, E. "Variability of low building wind lands." Doctoral Dissertation, University of Western Ontario, London, Ontario, Canada, 1992.

[C6-67] Simiu, E. and Scanlan, R.H. *Wind effects on structures*, Third Edition, John Wiley & Sons, New York, 1996.

[C6-68] Cook N. *The designer's guide to wind loading of building structures, part I*, Butterworth's, 1985.

[C6-69] Isyumov, N. and Case P. *Evaluation of structural wind loads for low-rise buildings contained in ASCE standard 7-1995*. BLWT-SS17-1995, Univ. of Western Ontario, London, Ontario, Canada.

[C6-70] Standard Test Method for Performance of Exterior Windows, Curtain Walls, Doors and Storm Shutters Impacted by Missile(s) and Exposed to Cyclic Pressure Differentials. ASTM E1886-97, ASTM Inc., West Conshohocken, Pa., 1997.

[C6-71] Specification Standard for Performance of Exterior Windows, Glazed Curtain Walls, Doors ad Storm Shutters Impacted by Windborne Debris in Hurricanes. ASTM E 1996-99, ASTM Inc., West Conshohocken, Pa., 1999.

[C6-72] Minor J.E. "Windborne debris and the building envelope." *J. Wind Engrg. and Industrial Aerodynamics*, 53, 207–227, 1994.

[C6-73] Peterka, J.A., Cermak, J.E., Cochran, L.S., Cochran, B.C., Hosoya, N., Derickson, R.G., Harper, C., Jones, J., and Metz B. "Wind uplift model for asphalt shingles." *J. Arch. Engrg.*, December, 1997.

[C6-74] Peterka, J.A. and Shahid, S. "Design gust wind speeds for the United States." *J. Struct. Div.*, ASCE, February, 1998.

[C6-75] Lettau, H. "Note on aerodynamic roughness element description." *J. Applied Meteorology*, 8, 828–832, 1969.

[C6-76] Ellingwood, B.R., MacGregor, J.G., Galambos, T.V., and Cornell, C.A. "Probability-based load criteria: Load factors and load combinations." *J. Struct. Div.*, ASCE 108(5)978–997, 1981.

[C6-77] Ellingwood, B. "Wind and snow load statistics for probability design." *J. Struct. Div.*, ASCE 197(7)1345–1349, 1982.

[C6-78] Vickery, P.J., Skerlj, P.S., Steckley, A.C., and Twisdale, L.A. "Hurricane wind field and gust factor models for use in hurricane wind speed simulations." *J. Struct. Engrg.*, ASCE, 126:1203–1221, 2000.

[C6-79] Vickery, P.J., Skerlj, P.S., and Twisdale, L.A. "Simulation of hurricane risk in the United States using an empirical storm track modeling technique." *J. Struct. Engrg.*, ASCE, 126:1222–1227, 2000.

[C6-80] Twisdale, L.A., Vickery, P.J., and Steckley, A.C. "Analysis of hurricane windborne debris impact risk for residential structures." State Farm Mutual Automobile Insurance Companies, March, 1996.

[C6-81] Means, B., Reinhold, T.A., Perry, D.C. "Wind loads for low-rise buildings on escarpments." *Proc., ASCE Structures Congress 14*, Chicago, Ill., April, 1996.

[C6-82] Beste, F. and Cermak, J.E. "Correlation of internal and area-averaged wind pressures on low-rise buildings." *3rd International Colloquium on Bluff Body Aerodynamics and Applications*, Blacksburg, Va., July 28–August 1, 1996.

[C6-83] Irwin, P.A. "Pressure model techniques for cladding loads." *J. Wind Engrg. and Industrial Aerodynamics*, 29, 69–78, 1987.

[C6-84] Vickery, B.J. and Bloxham, C. "Internal pressure dynamics with a dominant opening." *J. Wind Engrg. and Industrial Aerodynamics*, 41–44, 193–204, 1992.

[C6-85] Irwin, P.A. and Dunn, G.E. "Review of internal pressures on low-rise buildings." RWDI Report 93-270 for Canadian Sheet Building Institute, February 23, 1994.

[C6-86] Vickery, P.J., Skerlj, P.F. "On the elimination of exposure D along the hurricane coastline in ASCE-7." Report for Andersen Corporation by Applied Research Associates, ARA Project 4667, March, 1998.

[C6-87] Wind Design Guide. Mehta, K.C., Marshall, R.D. *Guide to the Use of the Wind Load Provisions of ASCE 7-95*, ASCE Press, Reston, Va., 1998.

[C6-88] Vickery, P.J. and Twisdale, L.A. "Prediction of hurricane wind speeds in the United States." *J. Struct. Engrg.*, ASCE, 121(11) 1691–1699, 1995.

[C6-89] Georgiou, P.N. "Design wind speeds in tropical cyclone regions." Ph.D. Thesis, University of Western Ontario, London, Ontario, Canada, 1985.

[C6-90] Engineering Sciences Data Unit. Item Number 82026. With Amendments A to C. "Strong winds in the atmospheric boundary layer. Part 1: Mean hourly wind speeds." ESDU, March 1990.

[C6-91] Engineering Sciences Data Unit. Item Number 83045. With Amendments A and B, April 1993. "Strong winds in the atmospheric boundary Layer. Part 2: Discrete gust speeds issued." ESDU, November, 1983.

[C6-92] Harris, R.I. and Deaves, D.M. "The structure of strong winds." *Proc., CIRIA Conference on Wind Engineering in the Eighties*. CIRCIA, 6 Story's Gate, London, 1981.

[C6-93] Isyumov, N. "Wind induced torque on square and rectangular building shapes." *J. Wind Engrg. and Industrial Aerodynamics*, 13, 183–186, 1983.

[C6-94] Isyumov, N. and Case, P.C. "Wind-induced torsional loads and responses of buildings." *Proc., ASCE-SEI Structures Congress 2000*, ASCE, Philadelphia, Pa., May 8–10, 2000.

[C6-95] Boggs, D.W., Noriaku H., Cochran, L. "Sources of torsional wind loading on tall buildings: Lessons from the wind tunnel." *Proc., ASCE-SEI Structures Congress 2000*, Philadelphia, Pa., May 8–10, 2000.

[C6-96] Xie, J., Irwi, P.A. "Key factors for torsional wind response of tall buildings." *Proc., ASCE-SEI Structures Congress*, Philadelphia, Pa., May 8–10, 2000.

[C6-97] Behr, R.A. and Minor, J.E. "A survey of glazing behavior in multi-story buildings during hurricane Andrew." *The Structural Design of Tall Buildings*, Vol. 3, 143–161, 1994.

[C6-98] Minor, J.E. "Window glass performance and hurricane effects." *Proc., ASCE Specialty Conference on Hurricane Alicia: One Year Later, Galveston, Tex., August 16–17, 1984*, ASCE, New York, 151–167, 1985.

[C6-99] Beason, W.L., Meyers, G.E., and James, R.W. "Hurricane related window glass damage in Houston." *J. Struct. Engrg.*, 110:12, 2843–2857 December, 1984.

[C6-100] Kareem, A. "Performance of cladding in hurricane Alicia." *J. Struct. Engrg.*, ASCE, 112(21), 2679–2693, 1986.

[C6-101] Blessman, J. "Pressures on domes with several wind profiles." *Proc., 3rd International Conference on Wind Effects on Buildings and Structures*, Tokyo, Japan, 317–326, 1971.

[C6-102] Eurocode. "Eurocode 1: Basis of design and actions on structures — Part 2–4: Actions on structures — Wind actions." ENV 1991-2-4, European Committee for Standardization, Brussels, 1995.

[C6-103] Taylor, T.J. "Wind pressures on a hemispherical dome." *J. Wind Engrg. and Industrial Aerodynamics*, 40, 199–213, 1991.

[C6-104] Stathopoulos, T., Wang, K., and Wu, H. "Wind standard provisions for low building gable roofs revisited." *10th International Conference on Wind Engineering*, Copenhagen, Denmark, June 21–24, 1999.

[C6-105] Stathopoulos, T., Wang, K., and Wu, H. "Wind pressure provisions for gable roofs of intermediate roof slope." *Wind and Structures*, Vol. 4, No. 2, 2001.

[C6-106] Stathopoulos, T., Wang, K., and Wu, H. "Proposed new Canadian wind provisions for the design of gable roofs." *Can. J. Civ. Engrg.*, 27(5), 1059–1072, 2000.

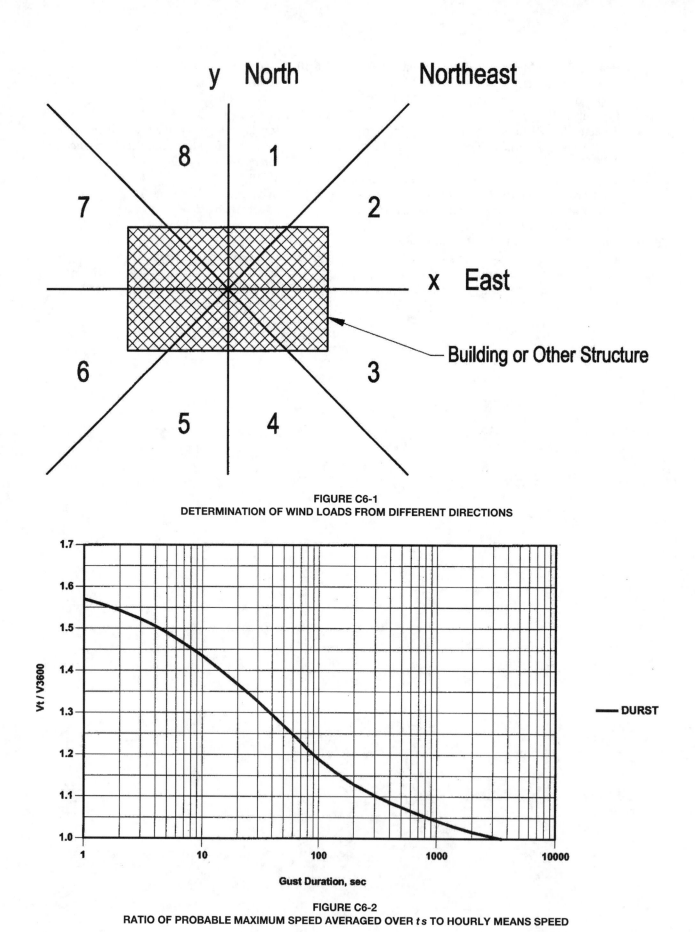

FIGURE C6-1
DETERMINATION OF WIND LOADS FROM DIFFERENT DIRECTIONS

FIGURE C6-2
RATIO OF PROBABLE MAXIMUM SPEED AVERAGED OVER *t s* TO HOURLY MEANS SPEED

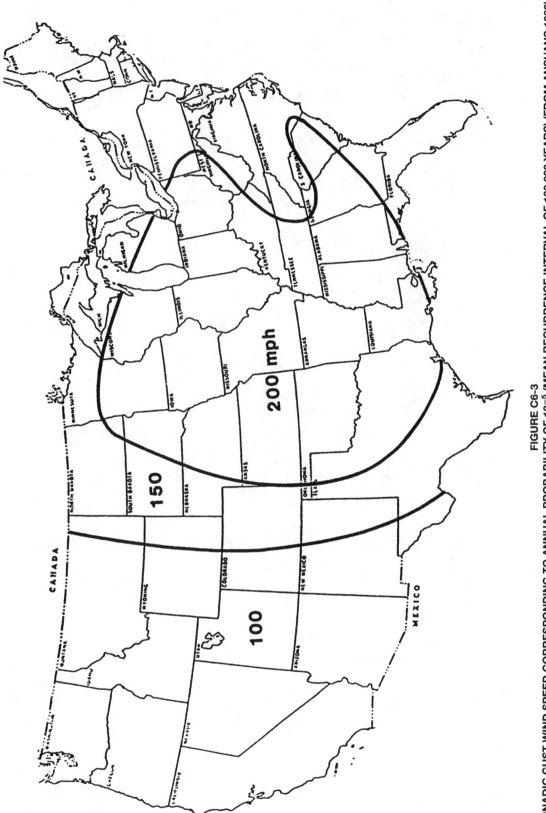

FIGURE C6-3
TORNADIC GUST WIND SPEED CORRESPONDING TO ANNUAL PROBABILITY OF 10⁻⁵ (MEAN RECURRENCE INTERVAL OF 100,000 YEARS) (FROM ANSI/ANS 1983)

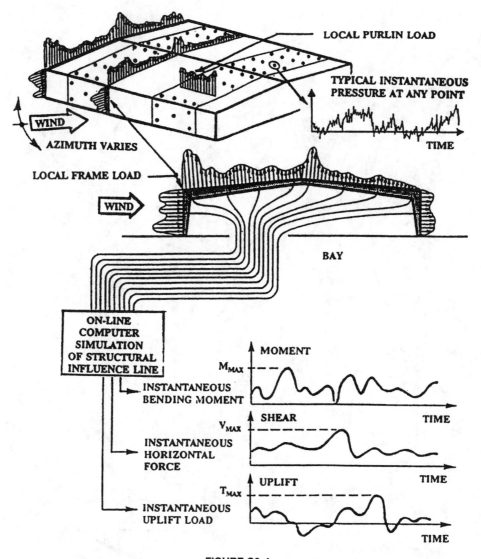

FIGURE C6-4
UNSTEADY WIND LOADS ON LOW BUILDINGS FOR GIVEN WIND DIRECTION (AFTER REFERENCE [C6-11])

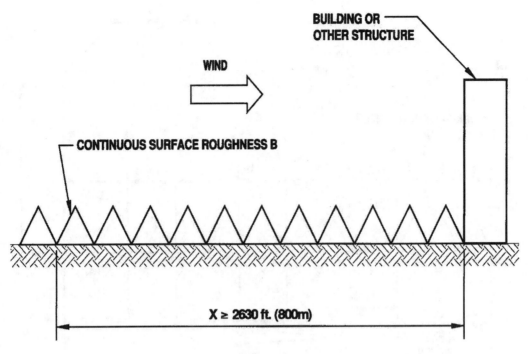

WIND

BUILDING OR OTHER STRUCTURE

CONTINUOUS SURFACE ROUGHNESS B

X ≥ 2630 ft. (800m)

**FIGURE C6-5
DEFINITION OF EXPOSURE B**

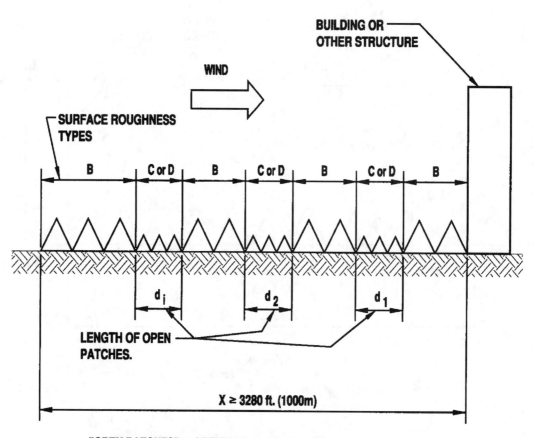

"OPEN PATCHES" - OPENINGS $\geq$ 164FT x 164FT (50m x 50m)

$d_1, d_2, ..., d_i \geq$ 164FT (50m)

$d_1 + d_2 + ... + d_i \leq$ 656FT (200m)

TOTAL LENGTH OF SURFACE ROUGHNESS B $\geq$ 2630FT (800m)
WITHIN 3280 FT (1000m) OF UPWIND FETCH DISTANCE.

FIGURE C6-6
EXPOSURE B WITH UPWIND OPEN PATCHES

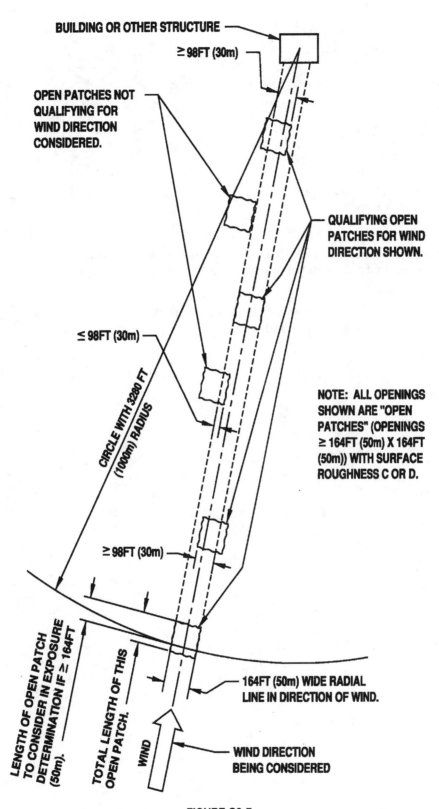

BUILDING OR OTHER STRUCTURE

≥ 98FT (30m)

OPEN PATCHES NOT QUALIFYING FOR WIND DIRECTION CONSIDERED.

QUALIFYING OPEN PATCHES FOR WIND DIRECTION SHOWN.

≤ 98FT (30m)

CIRCLE WITH 3280 FT (1000m) RADIUS

NOTE: ALL OPENINGS SHOWN ARE "OPEN PATCHES" (OPENINGS ≥ 164FT (50m) X 164FT (50m)) WITH SURFACE ROUGHNESS C OR D.

≥ 98FT (30m)

LENGTH OF OPEN PATCH TO CONSIDER IN EXPOSURE DETERMINATION IF ≥ 164FT (50m).

TOTAL LENGTH OF THIS OPEN PATCH.

164FT (50m) WIDE RADIAL LINE IN DIRECTION OF WIND.

WIND

WIND DIRECTION BEING CONSIDERED

FIGURE C6-7
EXPOSURE B WITH OPEN PATCHES

Minimum Design Loads for Buildings and Other Structures

303

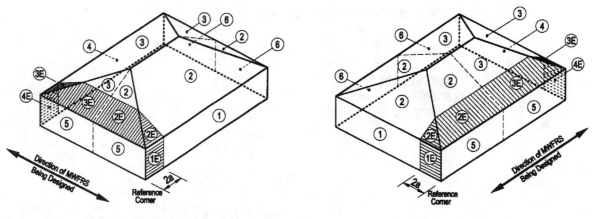

Notes:
1. Adapt the loadings shown in Figure 6-10 for hip roof buildings as shown above.
2. The total horizontal shear shall not be less than that determined by neglecting wind forces on roof surfaces.

FIGURE C6-8
HIP ROOFED LOW-RISE BUILDINGS

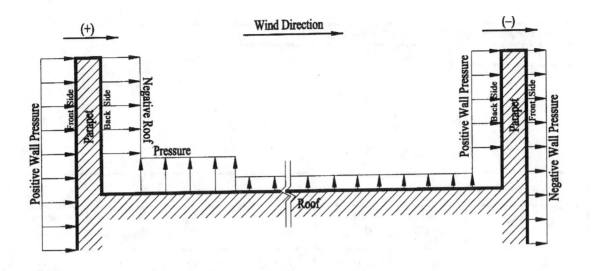

Methodology used to Develop External Parapet Pressures
(Main Wind Force Resisting Systems and Components and Cladding)

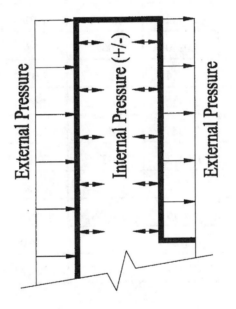

External and Internal Parapet Pressures
(Components and Cladding Only)

FIGURE C6-9
DESIGN WIND PRESSURES ON PARAPETS

Ambient Air Density Values for Various Altitudes

Table C6-1

Altitude		Ambient Air Density					
Feet	Meters	Minimum (lbm/ft^3)	Minimum (kg/m^3)	Average (lbm/ft^3)	Average (kg/m^3)	Maximum (lbm/ft^3)	Maximum (kg/m^3)
0	0	0.0712	1.1392	0.0765	1.2240	0.0822	1.3152
1000	305	0.0693	1.1088	0.0742	1.1872	0.0795	1.2720
2000	610	0.0675	1.0800	0.0720	1.1520	0.0768	1.2288
3000	914	0.0657	1.0512	0.0699	1.1184	0.0743	1.1888
3281	1000	0.0652	1.0432	0.0693	1.1088	0.0736	1.1776
4000	1219	0.0640	1.0240	0.0678	1.0848	0.0718	1.1488
5000	1524	0.0624	0.9984	0.0659	1.0544	0.0695	1.1120
6000	1829	0.0608	0.9728	0.0639	1.0224	0.0672	1.0752
6562	2000	0.0599	0.9584	0.0629	1.0064	0.0660	1.0560
7000	2134	0.0592	0.9472	0.0620	0.9920	0.0650	1.0400
8000	2438	0.0577	0.9232	0.0602	0.9632	0.0628	1.0048
9000	2743	0.0561	0.8976	0.0584	0.9344	0.0607	0.9712
9843	3000	0.0549	0.8784	0.0569	0.9104	0.0591	0.9456
10,000	3048	0.0547	0.8752	0.0567	0.9072	0.0588	0.9408

Probability of Exceeding Design Wind Speed During Reference Period

Table C6-2

Annual Probability P_a	Reference (Exposure) Period, n (years)					
	1	5	10	25	50	100
0.04 (1/25)	0.04	0.18	0.34	0.64	0.87	0.98
0.02 (1/50)	0.02	0.10	0.18	0.40	0.64	0.87
0.01 (1/100)	0.01	0.05	0.10	0.22	0.40	0.64
0.005 (1/200)	0.005	0.02	0.05	0.10	0.22	0.39

MRI (years)	Peak gust wind speed, V (mph) m/s		Alaska
	Continental U.S.		
	V = 85-100 (38-45 m/s)	V > 100 (hurricane) (44.7 m/s)	
500	1.23	1.23	1.18
200	1.14	1.14	1.12
100	1.07	1.07	1.06
50	1.00	1.00	1.00
25	0.93	0.88	0.94
10	0.84	0.74 (76 mph min.) (33.9 m/s)	0.87
5	0.78	0.66 (70 mph min.) (31.3 m/s)	0.81

Note: Conversion factors for the column "V > 100 (hurricane)" are approximate. For the MRI = 50 as shown, the actual return period, as represented by the design wind speed map in Fig. 6-1, varies from 50 to approximately 90 years. For an MRI = 500, the conversion factor is theoretically "exact" as shown.

Values Obtained From Table 6-4 of ASCE 7-98	
z_{min}	60 ft
$\bar{\epsilon}$	0.5
c	0.45
$\bar{b}$	0.3
$\bar{\alpha}$	0.33
$\hat{b}$	0.64
$\hat{\alpha}$	0.2
ℓ	180
C_{fx}	1.3
ξ	1
Height (h)	600 ft
Base (B)	100 ft
Depth (L)	100 ft

Calculated Values	
V	132 ft/s
$\bar{z}$	360 ft
$I_{\bar{z}}$	0.302
$L_{\bar{z}}$	594.52 ft
Q^2	0.589
$\bar{V}_{\bar{z}}$	87.83 ft/s
$\hat{V}_{\bar{z}}$	136.24 ft/s
N_1	1.354
R_n	0.111
η	1.047
R_B	0.555
η	6.285
R_h	0.146
η	3.507
R_L	0.245
R^2	0.580
G	1.01
K	0.502
m_1	745,400 slugs
g_R	3.787

z (ft)	φ(z)	X_{max} (z)	Rms.Acc. (ft/sec²)	Rms. Acc. (milli-g)	Max. Acc. (ft/sec²)	Max. Acc. (milli-g)
0	0.00	0.00	0.00	0.00	0.00	0.00
60	0.10	0.07	0.02	0.6	0.07	2.2
120	0.20	0.15	0.04	1.2	0.14	4.5
180	0.30	0.22	0.06	1.8	0.22	6.7
240	0.40	0.30	0.08	2.4	0.29	8.9
300	0.50	0.37	0.10	3.0	0.36	11.2
360	0.60	0.45	0.11	3.5	0.43	13.4
420	0.70	0.52	0.13	4.1	0.50	15.7
480	0.80	0.60	0.15	4.7	0.58	17.9
540	0.90	0.67	0.17	5.3	0.65	20.1
600	1.00	0.75	0.19	5.9	0.72	22.4

SECTION C7.0
SNOW LOADS

Methodology. The procedure established for determining design snow loads is as follows:

1. Determine the ground snow load for the geographic location (7.2 and the Commentary's C7.2).

2. Generate a flat roof snow load from the ground load with consideration given to: (a) roof exposure (7.3.1 and the Commentary's C7.3 and C7.3.1); (b) roof thermal condition (7.3.2 and the Commentary's C7.3 and C7.3.2); (c) occupancy and function of structure (7.3.3 and the Commentary's C7.3.3).

3. Consider roof slope (7.4 through 7.4.5 and the Commentary's C7.4).

4. Consider partial loading (7.5 and the Commentary's C7.5).

5. Consider unbalanced loads (7.6 through 7.6.4 and the Commentary's C7.6).

6. Consider snowdrifts: (a) on lower roofs (7.7 through 7.7.2 and the Commentary's C7.7) and (b) from projections (7.8 and the Commentary's C7.8).

7. Consider sliding snow (7.9 and the Commentary's C7.9).

8. Consider extra loads from rain-on-snow (7.10 and the Commentary's C7.10).

9. Consider ponding loads (7.11 and the Commentary's C7.11).

10. Consider existing roofs (7.12 and the Commentary's C7.12).

11. Consider other roofs and sites (the Commentary's C7.13)

12. Consider the consequences of loads in excess of the design value (immediately following).

Loads in Excess of the Design Value. The philosophy of the probabilistic approach used in this Standard is to establish a design value that reduces the risk of a snow load-induced failure to an acceptably low level. Since snow loads in excess of the design value may occur, the implications of such "excess" loads should be considered. For example, if a roof is deflected at the design snow load so that slope to drain is eliminated, "excess" snow load might cause ponding (as discussed in the Commentary's C7.11) and perhaps progressive failure.

The snow load/dead load ratio of a roof structure is an important consideration when assessing the implications of "excess" loads. If the design snow load is exceeded, the percentage increase in total load would be greater for a lightweight structure (that is, one with a high snow load/dead load ratio) than for a heavy structure (that is, one with a low snow load/dead load ratio). For example, if a 40-lb/ft² (1.92 kN/m²) roof snow load is exceeded by 20 lb/ft² (0.96 kN/m²) for a roof having a 25-lb/ft² (1.19 kN/m²) dead load, the total load increases by 31% from 65 to 85 lb/ft² (3.11 to 4.07 kN/m²). If the roof had a 60-lb/ft² (2.87 kN/m²) dead load, the total load would increase only by 20% from 100 to 120 lb/ft² (4.79 to 5.75 kN/m²).

SECTION C7.2
GROUND SNOW LOADS

The snow load provisions were developed from an extreme-value statistical analysis of weather records of snow on the ground [C7-1]. The log normal distribution was selected to estimate ground snow loads, which have a 2% annual probability of being exceeded (50-year mean recurrence interval).

Maximum measured ground snow loads and ground snow loads with a 2% annual probability of being exceeded are presented in Table C7-1 for 204 National Weather Service (NWS) "first-order" stations at which ground snow loads have been measured for at least 11 years during the period from 1952 to 1992.

Concurrent records of the depth and load of snow on the ground at the 204 locations in Table C7-1 were used to estimate the ground snow load and the ground snow depth having a 2% annual probability of being exceeded for each of these locations. The period of record for these 204 locations, where both snow depth and snow load have been measured, averages 33 years up through the winter of 1991 to 1992. A mathematical relationship was developed between the 2% depths and the 2% loads. The nonlinear best-fit relationship between these extreme values was used to estimate 2% (50-year mean recurrence interval) ground snow loads at about 9200 other locations at which only snow depths were measured. These loads, as well as the extreme-value loads developed directly from snow load measurements at 204 first-order locations, were used to construct the maps.

In general, loads from these two sources were in agreement. In areas where there were differences, loads from the 204 first-order locations were considered to be more valuable when the map was constructed. This procedure ensures that the map is referenced to the NWS observed loads and contains spatial detail provided by snow-depth measurements at about 9200 other locations.

The maps were generated from data current through the 1991 to 1992 winter. Where statistical studies using more recent information are available, they may be used to produce improved design guidance.

However, adding a big snow year to data developed from periods of record exceeding 20 years will usually not change 50-year values much. As examples, the databases for Boston and Chattanooga were updated to include the winters of 1992 to 1993 and 1993 to 1994 since record snows occurred there during that period. In Boston, 50-year loads based on water equivalent measurements only increased from 34 to 35 lb/ft² (1.63 to 1.68 kN/m²) and loads generated from snow depth measurements remained at 25 lb/ft² (1.20 kN/m²). In Chattanooga, loads generated from water equivalent measurements increased from 6 to 7 lb/ft² (0.29 to 0.34 kN/m²) and loads generated from snow depth measurements remained at 6 lb/ft² (0.29 kN/m²).

The following additional information was also considered when establishing the snow load zones on the map of the United States (Figure 7-1).

1. The number of years of record available at each location.

2. Additional meteorological information available from NWS, Soil Conservation Service (SCS) snow surveys, and other sources.

3. Maximum snow loads observed there.

4. Regional topography.

5. The elevation of each location.

The map is an updated version of those in the 1993 edition of this Standard and is the same as that in the 1995 and 1998 editions.

In much of the South, infrequent but severe snowstorms disrupted life in the area to the point that meteorological observations were missed. In these and similar circumstances, more value was given to the statistical values for stations with complete records. Year-by-year checks were made to verify the significance of data gaps.

The mapped snow loads cannot be expected to represent all the local differences that may occur within each zone. Because local differences exist, each zone has been positioned so as to encompass essentially all the statistical values associated with normal sites in that zone. Although the zones represent statistical values, not maximum observed values, the maximum observed values were helpful in establishing the position of each zone.

For sites not covered in Figure 7-1, design values should be established from meteorological information, with consideration given to the orientation, elevation, and records available at each location. The same method can also be used to improve upon the values presented in Figure 7-1. Detailed study of a specific site may generate a design value lower than that indicated by the generalized national map. It is appropriate in such a situation to use the lower value established by the detailed study. Occasionally a detailed study may indicate that a higher design value should be used than the national map indicates. Again, results of the detailed study should be followed.

Using the database used to establish the ground snow loads in Figure 7-1, additional meteorological data, and a methodology that meets the requirements of Section 7.2 [C7-61], ground snow loads have been determined for every town in New Hampshire [C7-62] [C7-63]. That information is presented in Table C7-4. Each tabulated value is for a specific elevation in each town. For lower elevations, the ground snow load should be decreased by 2.1 lb/ft² for every 100 ft of elevation difference (0.33 kN/m² for every 100 m of elevation difference). For higher elevations up to an elevation of 2500 ft (762 m), the ground snow load should be increased at the same rate. For example, in Hanover, the ground snow load from Table C7-4 is 75 lb/ft² at 1300 ft (3.6 kN/m² at 396 m). At an elevation of 900 ft (274 m) in Hanover, the ground snow load would be $75 - (2.1/100)(1300 - 900) = 75 - 8 = 67$ lb/ft² (in SI units: $3.6 - (0.33/100)(396 - 274) = 3.6 - 0.4 = 3.2$ kN/m²). In Hanover, at an elevation of 1700 ft (518 m), the ground snow load would be $75 + 8 = 83$ lb/ft² (in SI units: $3.6 + 0.4 = 4.0$ lb/m²). Above an elevation of 2500 ft (762 m), site-specific case studies are required. The rate of change of ground snow load with elevation 2.1 lb/ft² per 100 ft or 0.33 kN/m² per 100 m is applicable in New Hampshire, but is not necessarily appropriate for other case study areas.

The area covered by a site-specific case study will vary depending on local climate and topography. In some places, a single case study will suffice for an entire community, but in others, varying local conditions limit a "site" to a much smaller area. The area of applicability usually becomes clear as information in the vicinity is examined for the case study.

As suggested by the footnote, it is not appropriate to use only the site-specific information in Table C7-1 for design purposes. It lacks an appreciation for surrounding station information and, in a few cases, is based on rather short periods of record. The map or a site-specific case study provides more valuable information.

The importance of conducting detailed studies for locations not covered in Figure 7-1 is shown in Table C7-2.

For some locations within the CS areas of the northeast (Figure 7-1), ground snow loads exceed 100 lb/ft² (4.79 kN/m²). Even in the southern portion of the Appalachian Mountains, not far from sites where a 15 lb/ft² (0.72 kN/m²) ground snow load is appropriate, ground loads exceeding 50 lb/ft² (2.39 kN/m²) may be required. Lake-effect storms create requirements for ground loads in excess of 75 lb/ft² (3.59 kN/m²) along portions of the Great

Lakes. In some areas of the Rocky Mountains, ground snow loads exceed 200 lb/ft^2 (9.58 kN/m^2).

Local records and experience should also be considered when establishing design values.

The values in Table 7-1 are for specific Alaskan locations only and generally do not represent appropriate design values for other nearby locations. They are presented to illustrate the extreme variability of snow loads within Alaska. This variability precludes statewide mapping of ground snow loads there.

Valuable information on snow loads for the Rocky Mountain states is contained in References [C7-2 through C7-12].

Most of these references for the Rocky Mountain states use annual probabilities of being exceeded that are different from the 2% value (50-year mean recurrence interval) used in this Standard. Reasonable, but not exact, factors for converting from other annual probabilities of being exceeded to the value herein are presented in Table C7-3.

For example, a ground snow load based on a 3.3% annual probability of being exceeded (30-year mean recurrence interval) should be multiplied by 1.18 to generate a value of p_g for use in Eq. 7-1.

The snow load provisions of several editions of the National Building Code of Canada served as a guide in preparing the snow load provisions in this Standard. However, there are some important differences between the Canadian and the United States databases. They include:

1. The Canadian ground snow loads are based on a 3.3% annual probability of being exceeded (30-year mean recurrence interval) generated by using the extreme-value, Type-I (Gumbel) distribution, while the normal-risk values in this Standard are based on a 2% annual probability of being exceeded (50-year mean recurrence interval) generated by a log-normal distribution.

2. The Canadian loads are based on measured depths and regionalized densities based on four or less measurements per month. Because of the infrequency of density measurements, an additional weight of rain is added [C7-13]. In this Standard, the weight of the snow is based on many years of frequently measured weights obtained at 204 locations across the United States. Those measurements contain many rain-on-snow events and thus a separate rain-on-snow surcharge load is not needed except for some roofs with a slope less than 1/2 in./ft (2.38 degrees).

SECTION C7.3
FLAT ROOF SNOW LOADS, p_f

The live load reductions in Section 4.8 should not be applied to snow loads. The minimum allowable values of p_f presented in 7.3 acknowledge that in some areas, a single major storm can generate loads that exceed those developed from an analysis of weather records and snow load case studies.

The factors in this Standard that account for the thermal, aerodynamic, and geometric characteristics of the structure in its particular setting were developed using the National Building Code of Canada as a point of reference. The case study reports in References [C7-14 through C7-22] were examined in detail.

In addition to these published references, an extensive program of snow load case studies was conducted by eight universities in the United States, by the Corps of Engineers' Alaska District, and by the U.S. Army Cold Regions Research and Engineering Laboratory (CRREL) for the Corps of Engineers. The results of this program were used to modify the Canadian methodology to better fit United States conditions. Measurements obtained during the severe winters of 1976 to 1977 and 1977 to 1978 are included. A statistical analysis of some of that information is presented in Reference [C7-23]. The experience and perspective of many design professionals, including several with expertise in building failure analysis, have also been incorporated.

The minimum values of p_f account for a number of situations that develop on roofs. They are particularly important considerations where p_g is 20 lb/sq ft (0.96 kN/m^2) or less. In such areas, single storms result in loadings for which Eq. 7-1 and the C_e and C_t values in Tables 7-2 and 7-3, respectively, underestimate loads.

C7.3.1 Exposure Factor, C_e. Except in areas of "aerodynamic shade," where loads are often increased by snow drifting, less snow is present on most roofs than on the ground. Loads in unobstructed areas of conventional flat roofs average less than 50% of ground loads in some parts of the country. The values in this Standard are above-average values, chosen to reduce the risk of snow load-induced failures to an acceptably low level. Because of the variability of wind action, a conservative approach has been taken when considering load reductions by wind.

The effects of exposure are handled on two scales. First Eq. 7-1 contains a basic Exposure Factor of 0.7. Second, the type of terrain and the exposure of the roof are handled by Exposure Factor C_e. This two-step procedure generates ground-to-roof load reductions as a function of exposures that range from 0.49 to 0.91.

Table 7-2 has been changed from what appeared in a prior (1988) version of this Standard to separate regional wind issues associated with terrain from local wind issues associated with roof exposure. This was done to better define categories without significantly changing the values of C_e.

Although there is a single "regional" terrain category for a specific site, different roofs of a structure may have different exposure factors due to obstruction provided by

higher portions of the structure or by objects on the roof. For example in terrain Category C, an upper level roof could be fully exposed ($C_e = 0.9$) while a lower level roof would be partially exposed ($C_e = 1.0$) due to the presence of the upper level roof, as shown in Example 3 in the commentary.

The adjective "windswept" is used in the "mountainous areas" terrain category to preclude use of this category in those high mountain valleys that receive little wind.

The normal, combined exposure reduction in this standard is 0.70 as compared to a normal value of 0.80 for the ground-to-roof conversion factor in the 1990 National Building Code of Canada. The decrease from 0.80 to 0.70 does not represent decreased safety but arises due to increased choices of exposure and thermal classification of roofs (that is, six terrain categories, three roof exposure categories, and four thermal categories in this Standard versus three exposure categories and no thermal distinctions in the Canadian code).

It is virtually impossible to establish exposure definitions that clearly encompass all possible exposures that exist across the country. Because individuals may interpret exposure categories somewhat differently, the range in exposure has been divided into several categories rather than just two or three. A difference of opinion of one category results in about a 10% "error" using these several categories and an "error" of 25% or more if only three categories are used.

C7.3.2 Thermal Factor, C_t.

Usually, more snow will be present on cold roofs than on warm roofs. An exception to this is discussed below. The thermal condition selected from Table 7-3 should represent that which is likely to exist during the life of the structure. Although it is possible that a brief power interruption will cause temporary cooling of a heated structure, the joint probability of this event and a simultaneous peak snow load event is very small. Brief power interruptions and loss of heat are acknowledged in $C_t = 1.0$ category. Although it is possible that a heated structure will subsequently be used as an unheated structure, the probability of this is rather low. Consequently, heated structures need not be designed for this unlikely event.

Some dwellings are not used during the winter. Although their thermal factor may increase to 1.2 at that time, they are unoccupied, so their importance factor reduces to 0.8. The net effect is to require the same design load as for a heated, occupied dwelling.

Discontinuous heating of structures may cause thawing of snow on the roof and subsequent refreezing in lower areas. Drainage systems of such roofs have become clogged with ice, and extra loads associated with layers of ice several in. thick have built up in these undrained lower areas. The possibility of similar occurrences should be investigated for any intermittently heated structure.

Similar icings may build up on cold roofs subjected to meltwater from warmer roofs above. Exhaust fans and other mechanical equipment on roofs may also generate meltwater and icings.

Icicles and ice dams are a common occurrence on cold eaves of sloped roofs. They introduce problems related to leakage and to loads. Large ice dams that can prevent snow from sliding off roofs are generally produced by heat losses from within buildings. Icings associated with solar melting of snow during the day and refreezing along eaves at night are often small and transient. Although icings can occur on cold or warm roofs, roofs that are well insulated and ventilated are not commonly subjected to serious icings at their eaves. Methods of minimizing eave icings are discussed in References [C7-24 through C7-29]. Ventilation guidelines to prevent problematic icings at eaves have been developed for attics [C7-59] and for cathedral ceilings [C7-60].

Because ice dams can prevent load reductions by sliding on some warm ($C_t = \leq 1.0$) roofs, the "unobstructed slippery surface" curve in Figure 7-2a now only applies to unventilated roofs with a thermal resistance equal to or greater than 30 ft$^2 \cdot$ hr $\cdot$ F$^\circ$/Btu (5.3 K $\cdot$ m^2/W) and to ventilated roofs with a thermal resistance equal to or greater than 20 ft$^2 \cdot$ hr $\cdot$ F$^\circ$/Btu (3.5 K $\cdot$ m^2/W). For roofs that are well insulated and ventilated, $C_t = 1.1$ in Table 7-3.

Glass, plastic, and fabric roofs of continuously heated structures are seldom subjected to much snow load because their high heat losses cause snow melt and sliding. For such specialty roofs, knowledgeable manufacturers and designers should be consulted. The National Greenhouse Manufacturers Association [C7-30] recommends use of $C_t = 0.83$ for continuously heated greenhouses and $C_t = 1.00$ for unheated or intermittently heated greenhouses. They suggest a value of $I = 1.0$ for retail greenhouses and $I = 0.8$ for all other greenhouses. To qualify as a continuously heated greenhouse, a production or retail greenhouse must have a constantly maintained temperature of 50 °F (10 °C) or higher during winter months. In addition, it must also have a maintenance attendant on duty at all times or an adequate temperature alarm system to provide warning in the event of a heating system failure. Finally, the greenhouse roof material must have a thermal resistance, R-value, less than 2 ft$^2 \cdot$ hr $\cdot$ F$^\circ$/Btu (0.4 K $\cdot$ m^2/W). In this Standard, the C_t factor for such continuously heated greenhouses is set at 0.85. An unheated or intermittently heated greenhouse is any greenhouse that does not meet the requirements of a continuously heated single or double glazed greenhouse. Greenhouses should be designed so that the structural supporting members are stronger than the glazing. If this approach is used, any failure caused by heavy snow loads will be localized and in the glazing. This should avert progressive collapse of the structural frame. Higher design values should be used where drifting or sliding snow is expected.

Little snow accumulates on warm air-supported fabric roofs because of their geometry and slippery surface. However, the snow that does accumulate is a significant load for such structures and should be considered. Design methods for snow loads on air structures are discussed in References [C7-31 and C7-32].

The combined consideration of exposure and thermal conditions generates ground-to-roof factors that range from a low of 0.49 to a high of 1.09. The equivalent ground-to-roof factors in the 1990 National Building Code of Canada are 0.8 for sheltered roofs, 0.6 for exposed roofs, and 0.4 for exposed roofs in exposed areas north of the tree line, all regardless of their thermal condition.

Reference [C7-33] indicates that loads exceeding those calculated using this Standard can occur on roofs that receive little heat from below. Limited case histories for freezer buildings in which the air directly below the roof layer is intentionally kept cold suggest that the C_t factor may be larger than 1.2.

C7.3.3 Importance Factor, I. The importance factor I has been included to account for the need to relate design loads to the consequences of failure. Roofs of most structures having normal occupancies and functions are designed with an importance factor of 1.0, which corresponds to unmodified use of the statistically determined ground snow load for a 2% annual probability of being exceeded (50-year mean recurrence interval).

A study of the 204 locations in Table C7-1 showed that the ratio of the values for 4% and 2% annual probabilities of being exceeded (the ratio of the 25-year to 50-year mean recurrence interval values) averaged 0.80 and had a standard deviation of 0.06. The ratio of the values for 1% and 2% annual probabilities of being exceeded (the ratio of the 100-year to 50-year mean recurrence interval values) averaged 1.22 and had a standard deviation of 0.08. On the basis of the nationwide consistency of these values, it was decided that only one snow load map need be prepared for design purposes and that values for lower and higher risk situations could be generated using that map and constant factors.

Lower and higher risk situations are established using the importance factors for snow loads in Table 7-4. These factors range from 0.8 to 1.2. The factor 0.8 bases the average design value for that situation on an annual probability of being exceeded of about 4% (about a 25-year mean recurrence interval). The factor 1.2 is nearly that for a 1% annual probability of being exceeded (about a 100-year mean recurrence interval).

C7.3.4 Minimum Allowable Values of p_f for Low-Slope Roofs. These minimums account for a number of situations that develop on low-slope roofs. They are particularly important considerations where p_g is 20 lb/ft^2 (0.96 kN/m^2) or less. In such areas, single storms can result

in loading for which the basic Exposure Factor of 0.7 as well as the C_e and C_t factors do not apply.

SECTION C7.4
SLOPED ROOF SNOW LOADS, p_s

Snow loads decrease as the slopes of roofs increase. Generally, less snow accumulates on a sloped roof because of wind action. Also, such roofs may shed some of the snow that accumulates on them by sliding and improved drainage of meltwater. The ability of a sloped roof to shed snow load by sliding is related to the absence of obstructions not only on the roof but also below it, the temperature of the roof, and the slipperiness of its surface. It is difficult to define "slippery" in quantitative terms. For that reason, a list of roof surfaces that qualify as slippery and others that do not are presented in the Standard itself. Most common roof surfaces are on that list. The slipperiness of other surfaces is best determined by comparisons with those surfaces. Some tile roofs contain built-in protrusions or have a rough surface that prevents snow from sliding. However, snow will slide off other smooth-surfaced tile roofs. When a surface may or may not be slippery, the implications of treating it either as a slippery or non-slippery surface should be determined. Since valleys obstruct sliding on slippery surfaced roofs, the dashed lines in Figures 7-2a, b, and c should not be used in such roof areas.

Discontinuous heating of a building may reduce the ability of a sloped roof to shed snow by sliding, since meltwater created during heated periods may refreeze on the roof's surface during periods when the building is not heated, thereby "locking" the snow to the roof.

All these factors are considered in the slope reduction factors presented in Figure 7-2, which are supported by References [C7-33, C7-34, C7-35, and C7-36]. The thermal resistance requirements have been added to the "unobstructed slippery surfaces" curve in Figure 7-2a to prevent its use for roofs on which ice dams often form since ice dams prevent snow from sliding. Mathematically, the information in Figure 7-2 can be represented as follows:

1. Warm roofs ($C_t = 1.0$ or less):
   a. Unobstructed slippery surfaces with $R \geq 30$ ft$^2 \cdot$ hr $\cdot$ F°/Btu (5.3 K $\cdot$ m^2/W) if unventilated and $R \geq 20$ lb$^2 \cdot$ hr $\cdot$ F°/Btu (3.5 K $\cdot$ m^2/W) if ventilated:

0–5° slope	$C_s = 1.0$
5–70° slope	$C_s = 1.0 - (\text{slope} - 5°)/65°$
>70° slope	$C_s = 0$

   b. All other surfaces:

0–30° slope	$C_s = 1.0$
30–70° slope	$C_s = 1.0 - (\text{slope} - 30°)/40°$
>70° slope	$C_s = 0$

2. Cold Roofs with $C_t = 1.1$

   a. Unobstructed slippery surfaces:

0–10° slope	$C_s = 1.0$
10–70° slope	$C_s = 1.0 - (\text{slope} - 10°)/60°$
>70° slope	$C_s = 0$

   b. All other surfaces:

0–37.5° slope	$C_s = 1.0$
37.5–70° slope	$C_s = 1.0 - (\text{slope} - 37.5°)/32.5°$
>70° slope	$C_s = 0$

3. Cold Roofs ($C_t = 1.2$):

   a. Unobstructed slippery surfaces:

0–15° slope	$C_s = 1.0$
15–70° slope	$C_s = 1.0 - (\text{slope} - 15°)/55°$
>70° slope	$C_s = 0$

   b. All other surfaces:

0–45° slope	$C_s = 1.0$
45–70° slope	$C_s = 1.0 - (\text{slope} - 45°)/25°$
>70° slope	$C_s = 0$

If the ground (or another roof of less slope) exists near the eave of a sloped roof, snow may not be able to slide completely off the sloped roof. This may result in the elimination of snow loads on upper portions of the roof and their concentration on lower portions. Steep A-frame roofs that nearly reach the ground are subject to such conditions. Lateral as well as vertical loads induced by such snow should be considered for such roofs.

C7.4.3 Roof Slope Factor for Curved Roofs. These provisions have been changed from those in the 1993 edition of this Standard to cause the load to diminish along the roof as the slope increases.

C7.4.4 Roof Slope Factor for Multiple Folded Plate, Sawtooth, and Barrel Vault Roofs. Because these types of roofs collect extra snow in their valleys by wind drifting and snow creep and sliding, no reduction in snow load should be applied because of slope.

C7.4.5 Ice Dams and Icicles Along Eaves. The intent is to consider heavy loads from ice that forms along eaves only for structures where such loads are likely to form. It is also not considered necessary to analyze the entire structure for such loads, just the eaves themselves.

SECTION C7.5
UNLOADED PORTIONS

In many situations, a reduction in snow load on a portion of a roof by wind scour, melting, or snow-removal operations will simply reduce the stresses in the supporting members. However, in some cases, a reduction in snow load from an area will induce heavier stresses in the roof structure than occur when the entire roof is loaded. Cantilevered roof joists are a good example; removing half the snow load from the cantilevered portion will increase the bending stress and deflection of the adjacent continuous span. In other situations, adverse stress reversals may result.

The intent is not to require consideration of multiple "checkerboard" loadings.

Separate, simplified provisions have been added for continuous beams to provide specific partial loading requirements for that common structural system.

SECTION C7.6
UNBALANCED ROOF SNOW LOADS

Unbalanced snow loads may develop on sloped roofs because of sunlight and wind. Winds tend to reduce snow loads on windward portions and increase snow loads on leeward portions. Since it is not possible to define wind direction with assurance, winds from all directions should generally be considered when establishing unbalanced roof loads.

C7.6.1 Unbalanced Snow Loads on Hip and Gable Roofs. The unbalanced uniform snow load on the leeward side was $1.5\, p_s/C_e$ and $1.3\, p_s/C_e$ in the 1993 and 1995 editions of this Standard. In this edition, a 1993 approach is prescribed for roofs with small eave to ridge distances and the 15 degree cutoff is eliminated. For moderate to large roofs, the unbalanced snow load on the leeward side varies between $1.5\, p_s/C_e$ and $1.8\, p_s/C_e$ as a function of β. The gable roof drift parameter, β, is proportional to the percentage of the ground snow load that would drift across the ridge line.

The design snow load on the windward side for the unbalanced case, $0.3\, p_s$, is based upon case histories presented in References [C7-21] and [C7-56]. The lower limit of $\theta = 70/W + 0.5$ with W in ft (in SI: $21.3/W + 0.5$, with W in meters) is intended to exclude low-slope roofs, such as membrane roofs, on which significant unbalanced loads have not been observed [C7-57].

The provisions of the Standard and this Commentary correspond to the case where the gable or hip roof, in plan, is symmetric about the ridgeline, specifically roofs for which the eave to ridgeline distance, W, is the same for both sides. For asymmetric roofs, the unbalanced load for the side with the smaller W is expected to be somewhat larger as discussed in Reference [C7-56].

C7.6.2 Unbalanced Snow Loads for Curved Roofs. The method of determining roof slope is the same as the 1995 edition of this Standard. C_s is based on the actual slope, not an equivalent slope. These provisions do not apply to roofs that are concave upward. For such roofs, see Section 7.13.

C7.6.3 Unbalanced Snow Loads for Multiple Folded Plate, Sawtooth, and Barrel Vault Roofs. A minimum slope of 3/8 in./ft (1.79 degrees) has been established to preclude the need to determine unbalanced loads for most internally drained, membrane roofs which slope to internal drains. Case studies indicate that significant unbalanced loads can occur when the slope of multiple gable roofs is as low as 1/2 in./ft (2.38 degrees).

The unbalanced snow load in the valley is $2 \, p_f/C_e$ to create a total unbalanced load that does not exceed a uniformly distributed ground snow load in most situations.

Sawtooth roofs and other "up-and-down" roofs with significant slopes tend to be vulnerable in areas of heavy snowfall for the following reasons:

1. They accumulate heavy snow loads and are therefore expensive to build.

2. Windows and ventilation features on the steeply sloped faces of such roofs may become blocked with drifting snow and be rendered useless.

3. Meltwater infiltration is likely through gaps in the steeply sloped faces if they are built as walls, since slush may accumulate in the valley during warm weather. This can promote progressive deterioration of the structure.

4. Lateral pressure from snow drifted against clerestory windows may break the glass.

5. The requirement that snow above the valley not be at an elevation higher than the snow above the ridge may limit the unbalanced load to less than $2 \, p_f/C_e$.

C7.6.4 Unbalanced Snow Loads for Dome Roofs. This provision is based on a similar provision in the 1990 National Building Code of Canada.

SECTION C7.7
DRIFTS ON LOWER ROOFS
(AERODYNAMIC SHADE)

When a rash of snow-load failures occurs during a particularly severe winter, there is a natural tendency for concerned parties to initiate across-the-board increases in design snow loads. This is generally a technically ineffective and expensive way of attempting to solve such problems, since most failures associated with snow loads on roofs are caused not by moderate overloads on every ft^2 (m^2) of the roof, but rather by localized significant overloads caused by drifted snow.

It is extremely important to consider localized drift loads in designing roofs. Drifts will accumulate on roofs (even on sloped roofs) in the wind shadow of higher roofs or terrain features. Parapets have the same effect. The affected roof may be influenced by a higher portion of the same structure or by another structure or terrain feature nearby if the separation is 20 ft (6.1 m) or less. When a new structure is built within 20 ft (6.1 m) of an existing structure, drifting possibilities should also be investigated for the existing structure. The snow that forms drifts may come from the roof on which the drift forms, from higher or lower roofs, or, on occasion, from the ground.

The leeward drift load provisions are based on studies of snowdrifts on roofs [C7-37 through C7-40]. Drift size is related to the amount of driftable snow as quantified by the upwind roof length and the ground snow load. Drift loads are considered for ground snow loads as low as 5 lb/ft^2 (0.24 kN/m^2). Case studies show that, in regions with low ground snow loads, drifts 3 to 4 ft (0.9 to 1.2 m) high can be caused by a single storm accompanied by high winds.

A change from a prior (1988) edition of this Standard involves the width w when the drift height h_d from Figure 7-9, exceeds the clear height h_c. In this situation, the width of the drift is taken as $4 \, h_d^2/h_c$ with a maximum value of $8 \, h_c$. This drift width relation is based upon equating the cross-sectional area of this drift (i.e., $1/2 \, h_c \times w$) with the cross-sectional area of a triangular drift where the drift height is not limited by h_c (i.e., $1/2 \, h_d \times 4 \, h_d$). The upper limit of drift width is based on studies by Finney [C7-41] and Tabler [C7-42], which suggest that a "full" drift has a rise-to-run of about 1:6.5, and case studies [C7-43], which show observed drifts with a rise-to-run greater than 1:10.

The drift height relationship in Figure 7-9 is based on snow blowing off a high roof upwind of a lower roof. The change in elevation where the drift forms is called a "leeward step." Drifts can also form at "windward steps." An example is the drift that forms at the downwind end of a roof that abuts a higher structure there. Figure 7-7 shows "windward step" and "leeward step" drifts.

For situations having the same amount of available snow (i.e., upper and lower roofs of the same length), the drifts that form in leeward steps are larger than those that form in windward steps. In the previous version of the Standard, the windward drifts height was given as $1/2h_d$ from Figure 7-9 using the length of the lower roof for l_u. Based upon recent case history experience in combination with a prior study [C7-45], a value of 3/4 is now prescribed.

Depending on wind direction, any change in elevation between roofs can be either a windward or leeward step. Thus the height of a drift is determined for each wind direction as shown in Example 3, and the larger of the two heights is used as the design drift.

The drift load provisions cover most, but not all, situations. References [C7-41] and [C7-46] document a larger drift than would have been expected based on the length of the upper roof. The larger drift was caused when snow on a somewhat lower roof, upwind of the upper roof, formed a drift between those two roofs allowing

snow from the upwind lower roof to be carried up onto the upper roof then into the drift on its downwind side. It was suggested that the sum of the lengths of both roofs could be used to calculate the size of the leeward drift.

In another situation [C7-47], a long "spike" drift was created at the end of a long skylight with the wind about 30 degrees off the long axis of the skylight. The skylight acted as a guide or deflector that concentrated drifting snow. This caused a large drift to accumulate in the lee of the skylight. This drift was replicated in a wind tunnel.

As shown in Figure 7-8, the clear height, h_c, is determined based on the assumption that the upper roof is blown clear of snow in the vicinity of the drift. This is a reasonable assumption when the upper roof is nearly flat. However, sloped roofs often accumulate snow at eaves as illustrated in Figures 7-3 and 7-5. For such roofs, it is appropriate to assume that snow at the upper roof edge effectively increases the height difference between adjacent roofs. Using half the depth of the unbalanced snow load in the calculation of h_c produces more realistic estimates of drift loads.

Tests in wind tunnels [C7-48 and C7-49] and flumes [C7-44] have proven quite valuable in determining patterns of snow drifting and drift loads. For roofs of unusual shape or configuration, wind-tunnel or water-flume tests may be needed to help define drift loads.

SECTION C7.8
ROOF PROJECTIONS

Drifts around penthouses, roof obstructions, and parapet walls are also of the "windward step" type since the length of the upper roof is small or no upper roof exists. Solar panels, mechanical equipment, parapet walls, and penthouses are examples of roof projections that may cause "windward" drifts on the roof around them. The drift-load provisions in Sections 7.7 and 7.8 cover most of these situations adequately, but flat-plate solar collectors may warrant some additional attention. Roofs equipped with several rows of them are subjected to additional snow loads. Before the collectors were installed, these roofs may have sustained minimal snow loads, especially if they were windswept. Since a roof with collectors is apt to be somewhat "sheltered" by the collectors, it seems appropriate to assume the roof is partially exposed and calculate a uniform snow load for the entire area as though the collectors did not exist. In addition, the extra snow that might fall on the collectors and then slide onto the roof should be computed using the "cold roofs, all other surfaces" curve in Figure 7-2b. This value should be applied as a uniform load on the roof at the base of each collector over an area about 2 ft (0.6 m) wide along the length of the collector. The uniform load combined with the load at the base of each collector

probably represents a reasonable design load for such situations, except in very windy areas where extensive snow drifting is to be expected among the collectors. By elevating collectors several ft (1 m or more) above the roof on an open system of structural supports, the potential for drifting will be diminished significantly. Finally, the collectors themselves should be designed to sustain a load calculated by using the "unobstructed slippery surfaces" curve in Figure 7-2a. This last load should not be used in the design of the roof itself, since the heavier load of sliding snow from the collectors has already been considered. The influence of solar collectors on snow accumulation is discussed in [C7-50] and [C7-51].

SECTION C7.9
SLIDING SNOW

Situations that permit snow to slide onto lower roofs should be avoided [C7-52]. Where this is not possible, the extra load of the sliding snow should be considered. Roofs with little slope have been observed to shed snow loads by sliding. Consequently, it is prudent to assume that any upper roof sloped to an unobstructed eave is a potential source of sliding snow.

The final resting place of any snow that slides off a higher roof onto a lower roof will depend on the size, position, and orientation of each roof [C7-35]. Distribution of sliding loads might vary from a uniform load 5 ft (1.5 m) wide, if a significant vertical offset exists between the two roofs, to a 20 ft (6.1 m) wide uniform load, where a low-slope upper roof slides its load onto a second roof that is only a few ft (about 1 m) lower, or where snow drifts on the lower roof create a sloped surface that promotes lateral movement of the sliding snow.

In some instances, a portion of the sliding snow may be expected to slide clear of the lower roof. Nevertheless, it is prudent to design the lower roof for a substantial portion of the sliding load in order to account for any dynamic effects that might be associated with sliding snow.

Snow guards are needed on some roofs to prevent roof damage and eliminate hazards associated with sliding snow [C7-53]. When snow guards are added to a sloping roof, snow loads on the roof can be expected to increase. Thus, it may be necessary to strengthen a roof before adding snow guards. When designing a roof that will likely need snow guards in the future, it may be appropriate to use the "all other surfaces" curves in Figure 7-2 not the "unobstructed slippery surfaces" curves.

SECTION C7.10
RAIN-ON-SNOW SURCHARGE LOAD

The ground snow-load measurements on which this Standard is based contain the load effects of light rain-on-snow. However, since heavy rains percolate down

through snowpacks and may drain away, they might not be included in measured values. Where p_g is greater than 20 lb/ft^2 (0.96 kN/m^2), it is assumed that the full rain-on-snow effect has been measured and a separate rain-on-snow surcharge is not needed. The temporary roof load contributed by a heavy rain may be significant. Its magnitude will depend on the duration and intensity of the design rainstorm, the drainage characteristics of the snow on the roof, the geometry of the roof, and the type of drainage provided. Loads associated with rain-on-snow are discussed in References [C7-54], [C7-55], and [C7-58].

Water tends to remain in snow much longer on relatively flat roofs than on sloped roofs. Therefore, slope is quite beneficial since it decreases opportunities for drain blockages and for freezing of water in the snow.

The following example illustrates the evaluation of the rain-on-snow surcharge. For a roof with a 1/4 in./ft (1.19 degree) slope, where $p_g = 20$ lb/ft^2 (0.96 kN/m^2), $p_f = 18$ lb/ft^2 (0.86 kN/m^2), and the minimum allowable value of p_f is 20 lb/ft^2 (0.96 kN/m^2), the rain-on-snow surcharge of 5 lb/ft^2 (0.24 kN/m^2) would be added to the 18 lb/ft^2 (0.86 kN/m^2) flat roof snow load to generate a design load of 23 lb/ft^2 (1.10 kN/m^2). This rain-on-snow augmented design load corresponds to a uniform or balanced load case and hence need not be used in combination with drift, sliding, unbalanced, or partial loads.

SECTION C7.11
PONDING INSTABILITY

Where adequate slope to drain does not exist, or where drains are blocked by ice, snow meltwater and rain may pond in low areas. Intermittently heated structures in very cold regions are particularly susceptible to blockages of drains by ice. A roof designed without slope or one sloped with only 1/8 in./ft (0.6°) to internal drains probably contains low spots away from drains by the time it is constructed. When a heavy snow load is added to such a roof, it is even more likely that undrained low spots exist. As rainwater or snow meltwater flows to such low areas, these areas tend to deflect increasingly, allowing a deeper pond to form. If the structure does not possess enough stiffness to resist this progression, failure by localized overloading can result. This mechanism has been responsible for several roof failures under combined rain and snow loads.

It is very important to consider roof deflections caused by snow loads when determining the likelihood of ponding instability from rain-on-snow or snow meltwater.

Internally drained roofs should have a slope of at least 1/4 in./ft (1.19 degree) to provide positive drainage and to minimize the chance of ponding. Slopes of 1/4 in./ft (1.19 degree) or more are also effective in reducing peak loads generated by heavy spring rain-on-snow.

Further incentive to build positive drainage into roofs is provided by significant improvements in the performance of waterproofing membranes when they are sloped to drain.

Rain loads and ponding instability are discussed in detail in Section 8 of this Standard.

SECTION C7.12
EXISTING ROOFS

Numerous existing roofs have failed when additions or new buildings nearby caused snow loads to increase on the existing roof. A prior (1988) edition of this Standard mentioned this issue only in its Commentary where it was not a mandatory provision. The 1995 edition moved this issue to the Standard.

The addition of a gable roof alongside an existing gable roof as shown in Figure C7-1 most likely explains why some such metal buildings failed in the South during the winter of 1992 to 1993. The change from a simple gable roof to a multiple folded plate roof increased loads on the original roof as would be expected from Section 7.6.3. Unfortunately, the original roofs were not strengthened to account for these extra loads and they collapsed.

If the eaves of the new roof in Figure C7-1 had been somewhat higher than the eaves of the existing roof, the Exposure Factor C_e, for the original roof may have increased, thereby increasing snow loads on it. In addition, drift loads and loads from sliding snow would also have to be considered.

SECTION C7.13
OTHER ROOFS AND SITES

Wind-tunnel model studies, similar tests employing fluids other than air, for example, water flumes, and other special experimental and computational methods have been used with success to establish design snow loads for other roof geometries and complicated sites [C7-44], [C7-48], and [C7-49]. To be reliable, such methods must reproduce the mean and turbulent characteristics of the wind and the manner in which snow particles are deposited on roofs, then redistributed by wind action. Reliability should be demonstrated through comparisons with situations for which full-scale experience is available.

Examples. The following three examples illustrate the method used to establish design snow loads for most of the situations discussed in this Standard:

Example 1. Determine balanced and unbalanced design snow loads for an apartment complex in a suburb of Hartford, Connecticut. Each unit has an 8-on-12 slope

unventilated gable roof. The building length is 100 ft (30.5 m) and the eave to ridge distance, W, is 30 ft (9.1 m). Composition shingles clad the roofs. Trees will be planted among the buildings.

Flat Roof Snow Load.

$$p_f = 0.7C_eC_tIp_g$$

where:

$p_g = 30$ lb/ft^2 (1.44 kN/m^2) (from Figure 7-1)
$C_e = 1.0$ (from Table 7-2 for Terrain Category B and a partially exposed roof)
$C_t = 1.0$ (from Table 7-3); and $I = 1.0$ (from Table 7-4).

Thus

$$p_f = (0.7)(1.0)(1.0)(1.0)(30) = 21 \text{ lb/ft}^2$$
$$\text{(balanced load)}$$

in SI: $\quad p_f = (0.7)(1.0)(1.0)(1.0)(1.44) = 1.01 \text{ kN/m}^2$

Since the roof slope is greater than $70/W + 0.5 = 70/30 + 0.5 = 2.83°$ (in SI: $21.3/9.1 + 0.5 = 2.84°$), the minimum flat roof load does not apply (see Section 7.3).

Sloped-Roof Snow Load.

$$p_s = C_sp_f \text{ where } C_s = 0.91$$

[from solid line, Figure 7-2a].

Thus

$$p_s = 0.91(21) = 19 \text{ lb/ft}^2$$

in SI: $\quad p_s = 0.91(1.01) = 0.92 \text{ kN/m}^2$

Unbalanced Snow Load. Since the roof slope is greater than $70/W + 0.5 = 70/30 + 0.5 = 2.83°$ (in SI: $21.3/9.1 + 0.5 = 2.84°$), unbalanced loads must be considered. For $p_g = 30$ lb/ft^2 (1.44 kN/m^2), $\beta = 0.75$ by Eq. 7-3. Hence from Figure 7-5, the windward load is $0.3p_s = 6$ lb/ft^2 (0.29 kN/m^2) while the leeward unbalanced load is $1.2(1 + \beta/2)p_s/C_e = 31$ lb/ft^2 (1.5 kN/m^2).

Rain-on-Snow Surcharge. A rain-on-snow surcharge load need not be considered, since the slope is greater than 1/2 in./ft (2.38°) (see Section 7.10). See Figure C7-2 for both loading conditions.

Example 2. Determine the roof snow load for a vaulted theater, which can seat 450 people, planned for a suburb of Chicago, Illinois. The building is the tallest structure in a recreation-shopping complex surrounded by a parking lot. Two large deciduous trees are located in an area near the entrance. The building has an 80-ft (24.4-m) span and 15-ft (4.6-m) rise circular arc structural concrete roof covered with insulation and aggregate surfaced built-up roofing. The unventilated roofing system has a thermal resistance of 20 ft$^2 \cdot$ hr $\cdot$ F°/Btu (3.5 K $\cdot$ m^2/W). It is expected that the structure will be exposed to winds during its useful life.

Flat-Roof Snow Load.

$$p_f = 0.7C_eC_tIp_g$$

where

$p_g = 25$ lb/ft^2 (1.20 kN/m^2) (from Figure 7-1)
$C_e = 0.9$ (from Table 7-2 for Terrain Category B and a fully exposed roof)
$C_t = 1.0$ (from Table 7-3)
$I = 1.1$ (from Table 7-4)

Thus

$$p_f = (0.7)(0.9)(1.0)(1.1)(25) = 17 \text{ lb/ft}^2$$

In SI: $\quad pf = (0.7)(0.9)(1.0)(1.1)(1.19)$
$$= 0.83 \text{ kN/m}^2$$

Tangent of vertical angle from eaves to crown $= 15/40 = 0.375$ Angle $= 21°$.

Since the vertical angle exceeds 10°, the minimum allowable values of p_f do not apply. Use $p_f = 17$ lb/ft^2 (0.83 kN/m^2), see Section 7.3.4.

Sloped-Roof Snow Load.

$$p_s = C_sp_f$$

From Figure 7-2a, $C_s = 1.0$ until slope exceeds 30° which (by geometry) is 30 ft (9.1 m) from the centerline. In this area $p_s = 17(1) = 17$ lb/ft^2 (in SI $p_s = 0.83(1) = 0.83$ lb/m^2). At the eaves, where the slope is (by geometry) 41°, $C_s = 0.72$ and $p_s = 17(0.72) = 12$ lb/ft^2 (in SI $p_s = 0.83(0.72) = 0.60$ lb/m^2). Since slope at eaves is 41°, Case II loading applies.

Unbalanced Snow Load. Since the vertical angle from the eaves to the crown is greater than 10° and less than 60°, unbalanced snow loads must be considered.

Unbalanced load at crown
$$= 0.5p_f = 0.5(17) = 9 \text{ lb/ft}^2$$
$$\text{in SI} = 0.5(0.83) = 0.41 \text{ kN/m}^2$$

Unbalanced load at 30-degree point
$$= 2p_fC_s/C_e = 2(17)(1.0)/0.9 = 38 \text{ lb/ft}^2$$
$$\text{in SI} = 2(0.83)(1.0)/0.9 = 1.84 \text{ kN/m}^2$$

Unbalanced load at eaves
$$= 2(17)(0.72)/0.9 = 27 \text{ lb/ft}^2$$
$$\text{in SI:} = 2(0.83)(0.72)/0.9 = 1.33 \text{ kN/m}^2$$

Rain-on-Snow Surcharge. A rain-on-snow surcharge load need not be considered, since the slope is greater than 1/2 in./ft (2.38°) (see Section 7.10). See Figure C7-3 for both loading conditions.

Example 3. Determine design snow loads for the upper and lower flat roofs of a building located where $p_g = 40 \text{ lb/ft}^2$ (1.92 kN/m²). The elevation difference between the roofs is 10 ft (3 m). The 100-ft by 100-ft (30.5 m by 30.5 m) unventilated high portion is heated and the 170 ft wide (51.8 m), 100-ft (30.5-m) long low portion is an unheated storage area. The building is in an industrial park in flat open country with no trees or other structures offering shelter.

High Roof.

$$p_f = 0.7C_eC_tIp_g$$

where

$p_g = 40 \text{ lb/ft}^2$ (1.92 kN/m²) (given)
$C_e = 0.9$ (from Table 7-2)
$C_t = 1.0$ (from Table 7-3)
$I = 1.0$ (from Table 7-4).

Thus

$$p_f = 0.7(0.9)(1.0)(1.0)(40) = 25 \text{ lb/ft}^2$$

$$\text{in SI:} \quad p_f = 0.7(0.9)(1.0)(1.0)(1.92)$$

$$= 1.21 \text{ kN/m}^2$$

Since $p_g = 40 \text{ lb/ft}^2$ (1.92 kN/m²) and $I = 1.0$, the minimum value of $p_f = 20(1.0) = 20 \text{ lb/ft}^2$ (0.96 kN/m²) and hence does not control, see Section 7.3.

Low Roof.

$$p_f = 0.7C_eC_tIp_g$$

where

$p_g = 40 \text{ lb/ft}^2$ (1.92 kN/m²) (given)
$C_e = 1.0$ (from Table 7-2) partially exposed due to presence of high roof;

$C_t = 1.2$ (from Table 7-3)
$I = 0.8$ (from Table 7-4).

Thus

$$p_f = 0.7(1.0)(1.2)(0.8)(40) = 27 \text{ lb/ft}^2$$

$$\text{in SI:} \quad p_f = 0.7(1.0)(1.2)(0.8)(1.92) = 1.29 \text{ kN/m}^2$$

Since $p_g = 40 \text{ lb/ft}^2$ (1.92 kN/m²) and $I = 0.8$, the minimum value of $p_f = 20(0.8) = 16 \text{ lb/ft}^2$ (0.77 kN/m²) and hence does not control, see Section 7.3.

Drift Load Calculation.

$$\gamma = 0.13(40) + 14 = 19 \text{ lb/ft}^3 \text{ (Eq. 7-3)}$$
$$\text{in SI:} \quad \gamma = 0.426(1.92) + 2.2 = 3.02 \text{ kN/m}^3$$
$$h_b = p_f/19 = 27/19 = 1.4 \text{ ft}$$
$$\text{in SI:} \quad h_b = 1.29/3.02 = 0.43 \text{ meters}$$
$$h_c = 10 - 1.4 = 8.6 \text{ ft}$$
$$\text{in SI:} \quad h_c = 3.05 - 0.43 = 2.62 \text{ meters}$$
$$h_c/h_b = 8.6/1.4 = 6.1$$
$$\text{in SI:} \quad h_c/h_b = 2.62/0.43 = 6.1$$

Since $h_c/h_b \geq 0.2$ drift loads must be considered (see Section 7.7.1).

h_d (leeward step) = 3.8 ft (1.16 m)

[Figure 7-9 with $p_g = 40 \text{ lb/ft}^2$ (1.92 kN/m²)

and $l_u = 100$ ft (30.5 m)]

h_d (windward step) = 3/4 × 4.8 ft (1.5 m)

= 3.6 ft (1.1 m)4.8 ft (1.5 m)

from Figure 7-9 with $p_g = 40 \text{ lb/ft}^2$ (1.92 kN/m²)

and l_u = length of lower roof = 170 ft (52 m)

Leeward drift governs, use h_d = 3.8 ft (1.16 m)

Since $h_d < h_c$,

$$h_d = 3.8 \text{ ft } (1.16 \text{ m})$$

$$w = 4h_d = 15.2 \text{ ft } (4.64 \text{ m}),$$

$$\text{say 15 ft (4.6 m)}$$

$$p_d = h_d\gamma = 3.8(19) = 72 \text{ lb/ft}^2$$

$$\text{in SI:} \quad p_d = 1.16(3.02) = 3.50 \text{ kN/m}^2$$

Rain-on-Snow Surcharge. A rain-on-snow surcharge load need not be considered even though the slope is less than 1/2 in./ft (2.38°), since p_g is greater than

20 lb/ft² (0.96 kN/m²). See Figure C7-4 for snow loads on both roofs.

REFERENCES

[C7-1] Ellingwood, B., and Redfield, R. "Ground snow loads for structural design." *J. Struct. Engrg.*, ASCE, 109(4), 950–964, 1983.

[C7-2] MacKinlay, I., and Willis, W.E. 'Snow country design." National Endowment for the Arts, Washington, D.C., 1965.

[C7-3] Sack, R.L., and Sheikh-Taheri, A. *ground and roof snow loads for Idaho*. University of Idaho Press, Moscow, Idaho, 1986.

[C7-4] Structural Engineers Association of Arizona. "Snow load data for Arizona." Univ. of Arizona, Tempe, Ariz., 1973.

[C7-5] Structural Engineers Association of Colorado. "Snow load design data for Colorado." Denver, Colo., 1971. [Available from: Structural Engineers Association of Colorado, Denver, Colo., 1971.]

[C7-6] Structural Engineers Association of Oregon. "Snow load analysis for Oregon." Oregon Dept. of Commerce, Building Codes Division, Salem, Ore., 1971.

[C7-7] Structural Engineers Association of Washington. "Snow loads analysis for Washington." Seattle, Wash., 1981.

[C7-8] USDA Soil Conservation Service. "Lake Tahoe basin snow load zones." U.S. Dept. of Agriculture, Soil Conservation Service. Reno, Nev., 1970.

[C7-9] Videon, F.V., and Stenberg, P. "Recommended snow loads for Montana structures." Montana State Univ., Bozeman, Mont., 1978.

[C7-10] Structural Engineers Association of Northern California. "Snow load design data for the Lake Tahoe area." San Francisco, Calif., 1964.

[C7-11] Placer County Building Division. Snow Load Design, *Placer County Code*, Section 4.20(V). Auburn, Calif., 1985.

[C7-12] Brown, J. "An approach to snow load evaluation." *Proc., 38th Western Snow Conference*. 1970.

[C7-13] Newark, M. "A new look at ground snow loads in Canada." *Proc., 41st Eastern Snow Conference*. Washington, D.C., 37–48, 1984.

[C7-14] Elliott, M. "Snow load criteria for western United States, case histories and state-of-the-art." *Proc., First Western States Conference of Structural Engineer Associations*. Sun River, Ore., June 1975.

[C7-15] Lorenzen, R.T. "Observations of snow and wind loads precipitant to building failures in New York State, 1969–1970." American Society of Agricultural Engineers North Atlantic Region meeting, Newark, Del., Paper NA 70-305, August 1970. [Available from American Society of Agricultural Engineers, St. Joseph, Mo.]

[C7-16] Lutes, D.A., and Schriever, W.R. Snow accumulation in Canada: Case histories: II. National Research Council of Canada. DBR Tech. Paper 339, NRCC 11915. Ottawa, Ontario, Canada, 1971.

[C7-17] Meehan, J.F. "Snow loads and roof failures." *Proc., 1979 Structural Engineers Association of Calif.*, 1979. [Available from Structural Engineers Association of California, San Francisco, Calif.]

[C7-18] Mitchell, G.R. "Snow loads on roofs — An interim report on a survey." In *Wind and snow loading*. The Construction Press Ltd., Lancaster, England, 177–190, 1978.

[C7-19] Peter, B.G.W., Dalgliesh, W.A., and Schriever, W.R. "Variations of snow loads on roofs." *Trans. Eng. Inst. Can.* 6(A-1), 8, April 1963.

[C7-20] Schriever, W.R., Faucher, Y., and Lutes, D.A. Snow accumulation in Canada: Case histories: I. National Research Council of Canada, Division of Building Research. NRCC 9287. Ottawa, Ontario, Canada, 1967.

[C7-21] Taylor, D.A. "A survey of snow loads on roofs of arena-type buildings in Canada." *Can. J. Civ. Engrg.*, 6(1), 85–96, March 1979.

[C7-22] Taylor, D.A. "Roof snow loads in Canada." *Can. J. Civ. Engrg.* 7(1), 1–18, March 1980.

[C7-23] O'Rourke, M., Koch, P., and Redfield, R. "Analysis of roof snow load case studies: Uniform loads." U.S. Dept. of the Army, Cold Regions Research and Engineering Laboratory. CRREL Report 83-1, Hanover, N.H., 1983.

[C7-24] Grange, H.L., and Hendricks, L.T. Roof-snow behavior and ice-dam prevention in residential housing. St. Paul, Minn.: Univ. of Minnesota, Agricultural Extension Service. *Extension Bull.* 399, 1976.

[C7-25] Klinge, A.F. "Ice dams." *Popular Science*, 119–120, Nov. 1978.

[C7-26] Mackinlay, I. "Architectural design in regions of snow and ice." *Proc., First International Conference on Snow Engineering*, 441–455, Santa Barbara, Calif., July 1988.

[C7-27] Tobiasson, W. "Roof design in cold regions." *Proc., First International Conference on Snow Engineering*, 462–482, Santa Barbara, Calif., July 1988.

[C7-28] de Marne, H. "Field experience in control and prevention of leaking from ice dams in New England." *Proc., First International Conference on Snow Engineering*, 473–482, Santa Barbara, Calif., July 1988.

[C7-29] Tobiasson, W., and Buska, J. "Standing seam metal roofs in cold regions." *Proc., 10th Conference on Roofing Technology*, 34–44, Gaithersburg, Md., April 1993.

[C7-30] National Greenhouse Manufacturers Association. "Design loads in greenhouse structures." Taylors, S.C., 1988.

[C7-31] Air Structures Institute. *Design and standards manual.* ASI-77. [Available from the Industrial Fabrics Assn. International, St. Paul, Minn., 1977.]

[C7-32] American Society of Civil Engineers. *Air supported structures.* ASCE, New York, 1994.

[C7-33] Sack, R.L. "Snow loads on sloped roofs." *J. Struct. Engrg.*, ASCE, 114(3), 501–517, March 1988.

[C7-34] Sack, R., Arnholtz, D., and Haldeman, J. "Sloped roof snow loads using simulation." *J. Struct. Engrg.*, ASCE, 113(8), 1820–1833, Aug. 1987.

[C7-35] Taylor, D. "Sliding snow on sloping roofs." *Canadian Building Digest 228.* Ottawa, Ontario, Canada, National Research Council of Canada. Nov. 1983

[C7-36] Taylor, D. "Snow loads on sloping roofs: Two pilot studies in the Ottawa area." Division of Building Research Paper, *Can. J. Civ. Engrg.*, 1282(2), 334–343, June 1985.

[C7-37] O'Rourke, M., Tobiasson, W., and Wood, E. "Proposed code provisions for drifted snow loads." *J. Struct. Engrg.*, ASCE, 112(9), 2080–2092, Sept. 1986.

[C7-38] O'Rourke, M., Speck, R., and Stiefel, U. "Drift snow loads on multilevel roofs." *J. Struct. Engrg.*, ASCE, 111(2), 290–306, Feb. 1985.

[C7-39] Speck, R., Jr. "Analysis of snow loads due to drifting on multilevel roofs." Thesis, presented to the Department of Civil Engineering, at Rensselaer Polytechnic Institute, Troy, N.Y., in partial fulfillment of the requirements for the degree of Master of Science.

[C7-40] Taylor, D.A. "Snow loads on two-level flat roofs." *Proc., Eastern Snow Conference. 29, 41st Annual Meeting.* Washington, D.C., June 7–8, 1984.

[C7-41] Finney, E. "Snow drift control by highway design." Bulletin 86, Michigan State College Engineering Station, Lansing, Mich., 1939.

[C7-42] Tabler, R. "Predicting profiles of snow drifts in topographic catchments." *Western Snow Conf.*, Coronado, Calif., 1975.

[C7-43] Zallen, R. "Roof collapse under snow drift loading and snow drift design criteria." *J. Perf. of Constr. Fac.*, ASCE, 2(2), 80–98, May 1988.

[C7-44] O'Rourke, M., and Weitman, N. "Laboratory studies of snow drifts on multilevel roofs." *Proc., 2nd International Conf. Snow Engrg.*, Santa Barbara, Calif., June 1992.

[C7-45] O'Rourke, M., and El Hamadi, K. "Roof snow loads: Drifting against a higher wall." *Proc., 55th Western Snow Conf.*, pp. 124–132. Vancouver, B.C., April 1987.

[C7-46] O'Rourke, M. Discussion of "Roof collapse under snow drift loading and snow drift design criteria," *J. Perf. of Constr. Fac.*, ASCE, 266–268, 1989.

[C7-47] Kennedy, D., Isyumov, M., and Mikitiuk, M. "The effectiveness of code provisions for snow accumulations on stepped roofs." *Proc., 2nd International Conf. Snow Engrg.*, Santa Barbara, Calif., June 1992.

[C7-48] Isyumou, N., and Mikitiuk, M. "Wind tunnel modeling of snow accumulation on large roofs." *Proc., 2nd International Conf. Snow Engrg.*, Santa Barbara, Calif., June 1992.

[C7-49] Irwin, P., William, C., Gamle, S., and Retziaff, R. "Snow prediction in Toronto and the Andes Mountains: FAE simulation capabilities." *Proc., 2nd International Conf. Snow Engrg.*, Santa Barbara, Calif., June 1992.

[C7-50] O'Rourke, M.J. Snow and ice accumulation around solar collector installations. U.S. Dept. of Commerce, National Bureau of Standards. NBS-GCR-79 180. Washington, D.C., Aug. 1979.

[C7-51] Corotis, R.B., Dowding, C.H., and Rossow, E.C. "Snow and ice accumulation at solar collector installations in the Chicago metropolitan area." U.S. Dept. of Commerce, National Bureau of Standards. NBS-GCR-79 181. Washington, D.C., Aug. 1979.

[C7-52] Paine, J.C. "Building design for heavy snow areas." *Proc., First International Conference on Snow Engineering*, 483–492, Santa Barbara, Calif., July 1988.

[C7-53] Tobiasson, W., Buska, J., and Greatorex, A. "Snow guards for metal roofs." *Proc., 8th Conference on Cold Regions Engineering*, Fairbanks, Alaska, Aug. 1996.

[C7-54] Colbeck, S.C. "Snow loads resulting from rain-on-snow." U.S. Dept. of the Army, Cold Regions Research and Engineering Laboratory. CRREL Rep. 77-12. Hanover, N.H., 1977.

[C7-55] Colbeck, S.C. "Roof loads resulting from rain-on-snow: Results of a physical model." *Can. J. Civ. Engrg.* 4: 482–490, 1977.

[C7-56] O'Rourke, M., and Auren, M. "Snow loads on gable roofs." *J. Struct. Engrg.*, ASCE, 123(12) 1645–1651, Dec. 1997.

[C7-57] Tobiasson, W. discussion of [56] above, *J. Struct. Engrg.*, ASCE, 125(4) 470–471, 1999.

[C7-58] O'Rourke, M., and Downey, C. "Rain-on-snow surcharge for roof design." *J.Struct. Engrg.*, ASCE, 127(1), 74–79, 2001.

[C7-59] Tobiasson, W., Buska, J., and Greatorex, A. "Attic ventilation guidelines to minimize icings at eaves." *Interface*, XVI(1) Roof Consultants Institute, Raleigh, N.C., Jan. 1998.

[C7-60] Tobiasson, W., Tantillo, T., and Buska, J. "Ventilating cathedral ceilings to prevent problematic

icings at their eaves." *Proc., North American Conference on Roofing Technology*. National Roofing Contractors Association, Rosemont, Ill., September 1999.

[C7-61] Tobiasson, W., and Greatorex, A. "Database and methodology for conducting site specific snow load case studies for the United States." *Proc., 3ʳᵈ International Conference on Snow Engineering*, 249–256, Sendai, Japan, May 1996.

[C7-62] Tobiasson, W., Buska, J., Greatorex, A., Tirey, J., Fisher, J., and Johnson, S. "Developing ground snow loads for New Hampshire." *Proc., 4ᵗʰ International Conference on Snow Engineering*, Trondheim, Norway, 313–321, June 2000.

[C7-63] Tobiasson, W., Buska, J., Greatorex, A., Tirey, J., Fisher, J., and Johnson, S. "Ground snow loads for New Hampshire." U.S. Army Corps of Engineers, Engineer Research and Development Center (ERDC), Cold Regions Research and Engineering Laboratory (CRREL) Technical Report ERDC/CRREL TR-02-6. Hanover, N.H., 2002.

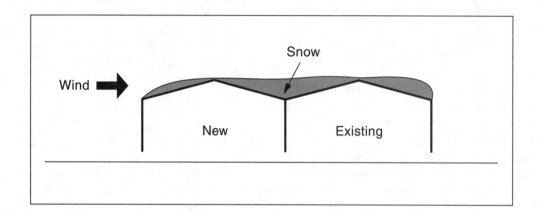

FIGURE C7-1
VALLEY IN WHICH SNOW WILL DRIFT IS CREATED WHEN NEW GABLE ROOF IS ADDED ALONGSIDE EXISTING GABLE ROOF

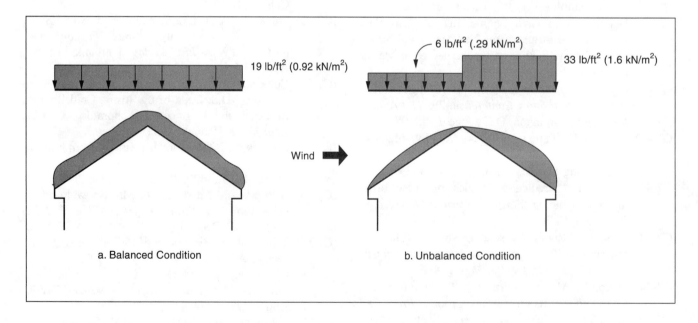

FIGURE C7-2
DESIGN SNOW LOADS FOR EXAMPLE 1

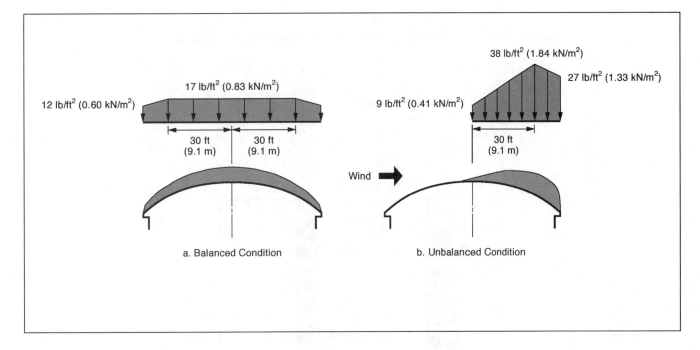

FIGURE C7-3
DESIGN SNOW LOADS FOR EXAMPLE 2

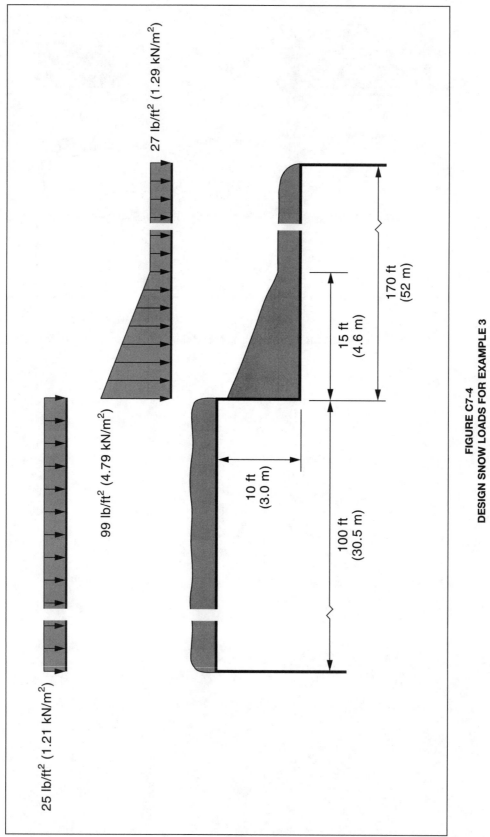

25 lb/ft² (1.21 kN/m²)

99 lb/ft² (4.79 kN/m²)

27 lb/ft² (1.29 kN/m²)

10 ft (3.0 m)

15 ft (4.6 m)

170 ft (52 m)

100 ft (30.5 m)

FIGURE C7-4
DESIGN SNOW LOADS FOR EXAMPLE 3

GROUND SNOW LOADS AT 204 NATIONAL WEATHER SERVICE LOCATIONS AT WHICH LOAD MEASUREMENTS ARE MADE (NOTE: TO CONVERT LB/FT² TO KN/M², MULTIPLY BY 0.0479)

Location	Ground Snow Load (lb/ft^2)			Location	Ground Snow Load (lb/ft^2)		
	Years of Record	Maximum Observed	2% Annual Probability*		Years of Record	Maximum Observed	2% Annual Probability*
ALABAMA				KANSAS			
Birmingham	40	4	3	Concordia	30	12	17
Huntsville	33	7	5	Dodge City	40	10	14
Mobile	40	1	1	Goodland	39	12	15
ARIZONA				Topeka	40	18	17
Flagstaff	38	88	48	Wichita	40	10	14
Tucson	40	3	3	KENTUCKY			
Winslow	39	12	7	Covington	40	22	13
ARKANSAS				Jackson	11	12	18
Fort Smith	37	6	5	Lexington	40	15	13
Little Rock	24	6	6	Louisville	39	11	12
CALIFORNIA				LOUISIANA			
Bishop	31	6	8	Alexandria	17	2	2
Blue Canyon	26	213	242	Shreveport	40	4	3
Mt. Shasta	32	62	62	MAINE			
Red Bluff	34	3	3	Caribou	34	68	95
COLORADO				Portland	39	51	60
Alamosa	40	14	14	MARYLAND			
Colorado Springs	39	16	14	Baltimore	40	20	22
Denver	40	22	18	MASSACHUSETTS			
Grand Junction	40	18	16	Boston	39	25	34
Pueblo	33	7	7	Nantucket	16	14	24
CONNECTICUT				Worcester	33	29	44
Bridgeport	39	21	24	MICHIGAN			
Hartford	40	23	33	Alpena	31	34	48
New Haven	17	11	15	Detroit City	14	6	10
DELAWARE				Detroit Airport	34	27	18
Wilmington	39	12	16	Detroit, Willow	12	11	22
GEORGIA				Flint	37	20	24
Athens	40	6	5	Grand Rapids	40	32	36
Atlanta	39	4	3	Houghton Lake	28	33	48
Augusta	40	8	7	Lansing	35	34	36
Columbus	39	1	1	Marquette	16	44	53
Macon	40	8	7	Muskegon	40	40	51
Rome	28	3	3	Sault Ste. Marie	40	68	77
IDAHO				MINNESOTA			
Boise	38	8	9	Duluth	40	55	63
Lewiston	37	6	9	International Falls	40	43	44
Pocatello	40	12	10	Minneapolis-St. Paul	40	34	51
ILLINOIS				Rochester	40	30	47
Chicago, O'Hare	32	25	17	St. Cloud	40	40	53
Chicago	26	37	22	MISSISSIPPI			
Moline	39	21	19	Jackson	40	3	3
Peoria	39	27	15	Meridian	39	2	2
Rockford	26	31	19	MISSOURI			
Springfield	40	20	21	Columbia	39	19	20
INDIANA				Kansas City	40	18	18
Evansville	40	12	17	St. Louis	37	28	21
Fort Wayne	40	23	20	Springfield	39	14	14
Indianapolis	40	19	22	MONTANA			
South Bend	39	58	41	Billings	40	21	15
IOWA				Glasgow	40	18	19
Burlington	11	15	17	Great Falls	40	22	15
Des Moines	40	22	22	Havre	26	22	24
Dubuque	39	34	32	Helena	40	15	17
Sioux City	38	28	28	Kalispell	29	27	45
Waterloo	33	25	32	Missoula	40	24	22

(continued)

GROUND SNOW LOADS AT 204 NATIONAL WEATHER SERVICE LOCATIONS AT WHICH LOAD MEASUREMENTS ARE MADE (NOTE: TO CONVERT LB/FT^2 TO KN/M^2, MULTIPLY BY 0.0479)

Location	Ground Snow Load (lb/ft^2)			Location	Ground Snow Load (lb/ft^2)		
	Years of Record	Maximum Observed	2% Annual Probability*		Years of Record	Maximum Observed	2% Annual Probability*
NEBRASKA				**OREGON**			
Grand Island	40	24	23	Astoria	26	2	3
Lincoln	20	15	22	Burns City	39	21	23
Norfolk	40	28	25	Eugene	37	22	10
North Platte	39	16	13	Medford	40	6	6
Omaha	25	23	20	Pendleton	40	9	13
Scottsbluff	40	10	12	Portland	39	10	8
Valentine	26	26	22	Salem	39	5	7
NEVADA				Sexton Summit	14	48	64
Elko	12	12	20	**PENNSYLVANIA**			
Ely	40	10	9	Allentown	40	16	23
Las Vegas	39	3	3	Erie	32	20	18
Reno	39	12	11	Harrisburg	19	21	23
Winnemucca	39	7	7	Philadelphia	39	13	14
NEW HAMPSHIRE				Pittsburgh	40	27	20
Concord	40	43	63	Scranton	37	13	18
NEW JERSEY				Williamsport	40	18	21
Atlantic City	35	12	15	**RHODE ISLAND**			
Newark	39	18	15	Providence	39	22	23
NEW MEXICO				**SOUTH CAROLINA**			
Albuquerque	40	6	4	Charleston	39	2	2
Clayton	34	8	10	Columbia	38	9	8
Roswell	22	6	8	Florence	23	3	3
NEW YORK				Greenville-Spartanburg	24	6	7
Albany	40	26	27	**SOUTH DAKOTA**			
Binghamton	40	30	35	Aberdeen	27	23	43
Buffalo	40	41	39	Huron	40	41	46
NYC, Kennedy	18	8	15	Rapid City	40	14	15
NYC, LaGuardia	40	23	16	Sioux Falls	39	40	40
Rochester	40	33	38	**TENNESSEE**			
Syracuse	40	32	32	Bristol	40	7	9
NORTH CAROLINA				Chattanooga	40	6	6
Asheville	28	7	14	Knoxville	40	10	9
Cape Hatteras	34	5	5	Memphis	40	7	6
Charlotte	40	8	11	Nashville	40	6	9
Greensboro	40	14	11	**TEXAS**			
Raleigh-Durham	36	13	14	Abilene	40	6	6
Wilmington	39	14	7	Amarillo	39	15	10
Winston-Salem	12	14	20	Austin	39	2	2
NORTH DAKOTA				Dallas	23	3	3
Bismark	40	27	27	El Paso	38	8	8
Fargo	39	27	41	Fort Worth	39	5	4
Williston	40	28	27	Lubbock	40	9	11
OHIO				Midland	38	4	4
Akron-Canton	40	16	14	San Angelo	40	3	3
Cleveland	40	27	19	San Antonio	40	9	4
Columbus	40	11	11	Waco	40	3	2
Dayton	40	18	11	Wichita Falls	40	4	5
Mansfield	30	31	17	**UTAH**			
Toledo Express	36	10	10	Milford	23	23	14
Youngstown	40	14	10	Salt Lake City	40	11	11
OKLAHOMA				Wendover	13	2	3
Oklahoma City	40	10	8				
Tulsa	40	5	8				

(*continued*)

GROUND SNOW LOADS AT 204 NATIONAL WEATHER SERVICE LOCATIONS AT WHICH LOAD MEASUREMENTS ARE MADE (NOTE: TO CONVERT LB/FT^2 TO KN/M^2, MULTIPLY BY 0.0479)

Location	Ground Snow Load (lb/ft^2)			Location	Ground Snow Load (lb/ft^2)		
	Years of Record	Maximum Observed	2% Annual Probability*		Years of Record	Maximum Observed	2% Annual Probability*
VERMONT				WEST VIRGINIA			
Burlington	40	43	36	Beckley	20	20	30
VIRGINIA				Charleston	38	21	18
Dulles Airport	29	15	23	Elkins	32	22	18
Lynchburg	40	13	18	Huntington	30	15	19
National Airport	40	16	22	WISCONSIN			
Norfolk	38	9	10	Green Bay	40	37	36
Richmond	40	11	16	La Crosse	16	23	32
Roanoke	40	14	20	Madison	40	32	35
WASHINGTON				Milwaukee	40	34	29
Olympia	40	23	22	WYOMING			
Quillayute	25	21	15	Casper	40	9	10
Seattle-Tacoma	40	15	18	Cheyenne	40	18	18
Spokane	40	36	42	Lander	39	26	24
Stampede Pass	36	483	516	Sheridan	40	20	23
Yakima	39	19	30				

*It is not appropriate to use only the specific information in this table for design purposes. Reasons are given in Commentary Section 7.2.

TABLE C7-2
COMPARISON OF SOME SITE-SPECIFIC VALUES AND ZONED VALUES IN FIGURE 7-1

State	Location	Elevation, ft (m)	Zoned Value lb/ft^2 (kN/m^2)	Case Study Value* lb/ft^2 (kN/m^2)
California	Mount Hamilton	4210 (1283)	0 to 2400' (732 m)	30 (1.44)
			0 to 3500' (1067 m)	
Arizona	Palisade Ranger Station	7950 (2423)	5 to 4600' (0.24 to 1402 m)	120 (5.75)
Tennessee	Monteagle	1940 (591)	10 to 5000' (0.48 to 1524 m)	15 (0.72)
Maine	Sunday River Ski Area	900 (274)	10 to 1800' (0.48 to 549 m)	100 (4.79)
			90 to 700' (4.31 to 213 m)	

*Based on a detailed study of information in the vicinity of each location.

TABLE C7-3
FACTORS FOR CONVERTING FROM OTHER ANNUAL PROBABILITIES OF BEING EXCEEDED AND OTHER MEAN RECURRENCE INTERVALS, TO THAT USED IN THIS STANDARD

Annual probability of being exceeded (%)	Mean recurrence interval (years)	Multiplication factor
10	10	1.82
4	25	1.20
3.3	30	1.15
1	100	0.82

TABLE C7-4
GROUND SNOW LOAD (P_g) AT A SPECIFIC ELEVATION FOR EACH TOWN IN NEW HAMPSHIRE

Town	P_g^* (lb/ft^2)**		Elevation (feet)***	Town	P_g^* (lb/ft^2)**		Elevation (feet)***
Acworth	90	@	1500	Danville	55	@	300
Albany	95	@	1300	Deerfield	70	@	700
Alexandria	85	@	1100	Deering	85	@	1200
Allenstown	70	@	700	Derry	65	@	600
Alstead	80	@	1300	Dixs Grant	90	@	1700
Alton	90	@	900	Dixville	90	@	1900
Amherst	70	@	600	Dorchester	80	@	1400
Andover	80	@	900	Dover	60	@	200
Antrim	80	@	1000	Dublin	90	@	1600
Ashland	75	@	800	Dummer	90	@	1400
Atkinson	55	@	400	Dunbarton	75	@	800
Atkinson & Gilmanton Academy Grant	85	@	1600	Durham	55	@	150
Auburn	65	@	500	East Kingston	50	@	200
Barnstead	85	@	900	Easton	85	@	1400
Barrington	70	@	500	Eaton	95	@	1000
Bartlett	100	@	1200	Effingham	85	@	600
Bath	65	@	1000	Ellsworth	90	@	1400
Beans Grant	105	@	1800	Enfield	85	@	1300
Beans Purchase	120	@	2000	Epping	55	@	300
Bedford	70	@	700	Epsom	75	@	800
Belmont	80	@	900	Errol	90	@	1600
Bennington	80	@	1000	Ervings Location	100	@	2100
Benton	90	@	1600	Exeter	50	@	200
Berlin	100	@	1600	Farmington	85	@	800
Bethlehem	105	@	1800	Fitzwilliam	75	@	1300
Boscawen	75	@	700	Francestown	80	@	1100
Bow	75	@	800	Franconia	95	@	1700
Bradford	85	@	1200	Franklin	75	@	700
Brentwood	50	@	250	Freedom	90	@	1000
Bridgewater	80	@	1000	Fremont	50	@	250
Bristol	80	@	1000	Gilford	90	@	1200
Brookfield	90	@	800	Gilmanton	90	@	1100
Brookline	60	@	500	Gilsum	80	@	1200
Cambridge	90	@	1300	Goffstown	75	@	800
Campton	85	@	1300	Gorham	100	@	1400
Canaan	80	@	1200	Goshen	90	@	1400
Candia	75	@	700	Grafton	90	@	1400
Canterbury	80	@	900	Grantham	90	@	1400
Carroll	95	@	1700	Greenfield	80	@	1100
Center Harbor	80	@	900	Greenland	50	@	100
Chandlers Purchase	120	@	2500	Greens Grant	105	@	1700
Charlestown	80	@	1100	Greenville	75	@	1000
Chatham	90	@	500	Groton	80	@	1200
Chester	65	@	500	Hadleys Purchase	100	@	1500
Chesterfield	70	@	1000	Hales Location	90	@	800
Chichester	75	@	700	Hampstead	55	@	300
Claremont	85	@	1100	Hampton	50	@	150
Clarksville	90	@	2000	Hampton Falls	50	@	150
Colebrook	80	@	1600	Hancock	85	@	1300
Columbia	80	@	1600	Hanover	75	@	1300
Concord	70	@	600	Harrisville	90	@	1500
Conway	95	@	900	Harts Location	100	@	1300
Cornish	85	@	1100	Haverhill	75	@	1200
Crawfords Purchase	110	@	1800	Hebron	80	@	900
Croydon	90	@	1200	Henniker	80	@	1000
Cutts Grant	110	@	1700	Hill	85	@	1100
Dalton	80	@	1300	Hillsborough	80	@	1000
Danbury	85	@	1000	Hinsdale	60	@	700

(continued)

Town	P_g^* (lb/ft^2)**		Elevation (feet)***	Town	P_g^* (lb/ft^2)**		Elevation (feet)***
Holderness	80	@	1000	Newport	85	@	1200
Hollis	60	@	500	Newton	50	@	250
Hooksett	70	@	600	North Hampton	50	@	100
Hopkinton	80	@	800	Northfield	75	@	800
Hudson	60	@	400	Northumberland	75	@	1200
Jackson	115	@	1800	Northwood	80	@	800
Jaffrey	80	@	1300	Nottingham	65	@	500
Jefferson	100	@	1700	Odell	90	@	1800
Keene	70	@	900	Orange	90	@	1500
Kensington	50	@	200	Orford	70	@	1100
Kilkenny	95	@	1700	Ossipee	85	@	1000
Kingston	50	@	200	Pelham	55	@	400
Laconia	80	@	900	Pembroke	70	@	700
Lancaster	75	@	1300	Peterborough	75	@	1000
Landaff	80	@	1300	Piermont	75	@	1400
Langdon	80	@	1000	Pinkhams Grant	115	@	2000
Lebanon	80	@	1200	Pittsburg	80	@	1700
Lee	55	@	200	Pittsfield	80	@	900
Lempster	95	@	1600	Plainfield	90	@	1300
Lincoln	95	@	1400	Plaistow	55	@	300
Lisbon	75	@	1100	Plymouth	75	@	900
Litchfield	60	@	250	Portsmouth	50	@	100
Littleton	75	@	1200	Randolph	110	@	1900
Livermore	100	@	1500	Raymond	60	@	500
Londonderry	65	@	500	Richmond	65	@	1100
Loudon	80	@	900	Rindge	80	@	1300
Low & Burbanks Grant	105	@	1800	Rochester	70	@	500
Lyman	75	@	1200	Rollinsford	60	@	200
Lyme	70	@	1100	Roxbury	80	@	1300
Lyndeborough	80	@	1000	Rumney	85	@	1300
Madbury	60	@	200	Rye	50	@	100
Madison	90	@	1100	Salem	55	@	300
Manchester	70	@	500	Salisbury	80	@	900
Marlborough	80	@	1300	Sanbornton	80	@	1000
Marlow	90	@	1600	Sandown	60	@	400
Martins Location	100	@	1300	Sandwich	85	@	1100
Mason	75	@	1000	Sargents Purchase	115	@	2000
Meredith	80	@	1000	Seabrook	50	@	100
Merrimack	60	@	400	Second College Grant	85	@	1500
Middleton	90	@	800	Sharon	80	@	1300
Milan	95	@	1500	Shelburne	90	@	800
Milford	70	@	600	Somersworth	60	@	250
Millsfield	90	@	1700	South Hampton	50	@	200
Milton	90	@	800	Springfield	95	@	1500
Monroe	65	@	1000	Stark	80	@	1200
Mont Vernon	75	@	900	Stewartstown	90	@	2000
Moultonborough	80	@	900	Stoddard	90	@	1600
Nashua	60	@	400	Strafford	80	@	800
Nelson	90	@	1500	Stratford	70	@	1100
New Boston	80	@	800	Stratham	50	@	150
New Castle	50	@	50	Success	100	@	1600
New Durham	90	@	900	Sugar Hill	90	@	1600
New Hampton	80	@	1000	Sullivan	90	@	1400
New Ipswich	80	@	1300	Sunapee	90	@	1400
New London	95	@	1400	Surry	80	@	1100
Newbury	90	@	1300	Sutton	85	@	1100
Newfields	50	@	150	Swanzey	65	@	800
Newington	50	@	100	Tamworth	85	@	1000
Newmarket	50	@	200	Temple	85	@	1300

(continued)

GROUND SNOW LOAD (P_g) AT A SPECIFIC ELEVATION FOR EACH TOWN IN NEW HAMPSHIRE

Town	P_g^* (lb/ft^2)**		Elevation (feet)***	Town	P_g^* (lb/ft^2)**		Elevation (feet)***
Thompson & Meserves Purchase	120	@	2500	Webster	75	@	700
Thornton	85	@	1200	Wentworth	80	@	1200
Tilton	80	@	900	Wentworth Location	80	@	1300
Troy	75	@	1300	Westmoreland	65	@	800
Tuftonboro	85	@	1100	Whitefield	80	@	1400
Unity	90	@	1500	Wilmot	90	@	1200
Wakefield	95	@	900	Wilton	75	@	900
Walpole	80	@	1200	Winchester	60	@	700
Warner	80	@	800	Windham	60	@	400
Warren	80	@	1300	Windsor	85	@	1200
Washington	95	@	1700	Wolfeboro	90	@	1000
Waterville Valley	105	@	1800	Woodstock	85	@	1200
Weare	80	@	900				

*These loads only apply at the elevations listed. For other elevations, P_g is adjusted according to the instructions and examples in Section C7-2.

**To convert lb/ft^2 to kN/m^2, multiply by 0.048.

***To convert feet to meters, multiply by 0.305.

SECTION C8.0
RAIN LOADS

SECTION C8.1
SYMBOLS AND NOTATION

A = roof area serviced by a single drainage system, in ft^2 (m^2)

i = design rainfall intensity as specified by the code having jurisdiction, in in./h (mm/h)

Q = flow rate out of a single drainage system, in gal/min (m^3/s)

SECTION C8.2
ROOF DRAINAGE

Roof drainage systems are designed to handle all the flow associated with intense, short-duration rainfall events. For example, the 1993 BOCA National Plumbing Code [C8-1], and Factory Mutual Loss Prevention Data 1–54, "Roof Loads for New Construction" [C8-2] use a 1-h duration event with a 100-year return period; the 1994 Standard Plumbing Code [C8-3] uses 1-h and 15-min duration events with 100-year return periods for the primary and secondary drainage systems, respectively, and the 1990 National Building Code [C8-4] of Canada uses a 15-minute event with a 10-year return period. A very severe local storm or thunderstorm may produce a deluge of such intensity and duration that properly designed primary drainage systems are temporarily overloaded. Such temporary loads are adequately covered in design when blocked drains (see Section 8.3) and ponding instability (see Section 8.4) are considered.

Roof drainage is a structural, architectural, and mechanical (plumbing) issue. The type and location of secondary drains and the hydraulic head above their inlets at the design flow must be known in order to determine rain loads. Design team coordination is particularly important when establishing rain loads.

SECTION C8.3
DESIGN RAIN LOADS

The amount of water that could accumulate on a roof from blockage of the primary drainage system is determined and the roof is designed to withstand the load created by that water plus the uniform load caused by water that rises above the inlet of the secondary drainage systems at its design flow. If parapet walls, cant strips, expansion joints, and other features create the potential for deep water in an area, it may be advisable to install in that area secondary (overflow) drains with separate drain lines rather than

overflow scuppers to reduce the magnitude of the design rain load. Where geometry permits, free discharge is the preferred form of emergency drainage.

When determining these water loads, it is assumed that the roof does not deflect. This eliminates complexities associated with determining the distribution of water loads within deflection depressions. However, it is quite important to consider this water when assessing ponding instability in Section 8.4.

The depth of water, d_h, above the inlet of the secondary drainage system (i.e., the hydraulic head) is a function of the rainfall intensity at the site, the area of roof serviced by that drainage system, and the size of the drainage system.

The flow rate through a single drainage system is as follows:

$$Q = 0.0104 \ Ai \ [\text{In SI: } Q = 0.278 \times 10^{-6} \ Ai]$$
$$\textbf{(Eq. C8-1)}$$

The hydraulic head, d_h, is related to flow rate, Q, for various drainage systems in Table C8-1. That table indicates that d_h can vary considerably depending on the type and size of each drainage system and the flow rate it must handle. For this reason the single value of 1 in. (25 mm) [i.e., 5 lb/ft^2 (0.24 kN/m^2)] used in ASCE 7–93 has been eliminated.

The hydraulic head, d_h, is zero when the secondary drainage system is simply overflow all along a roof edge.

SECTION C8.4
PONDING INSTABILITY

Water may accumulate as ponds on relatively flat roofs. As additional water flows to such areas, the roof tends to deflect more, allowing a deeper pond to form there. If the structure does not possess enough stiffness to resist this progression, failure by localized overloading may result. References [C8-1] through [C8-16] contain information on ponding and its importance in the design of flexible roofs. Rational design methods to preclude instability from ponding are presented in References [C8-5 and C8-6].

By providing roofs with a slope of 1/4 in./ft (1.19 degrees) or more, ponding instability can be avoided. If the slope is less than 1/4 in./ft, (1.19 degrees) the roof structure must be checked for ponding instability because construction tolerances and long-term deflections under dead load can result in flat portions susceptible to ponding.

SECTION C8.5
CONTROLLED DRAINAGE

In some areas of the country, ordinances are in effect that limit the rate of rainwater flow from roofs into storm drains. Controlled-flow drains are often used on such roofs. Those roofs must be capable of sustaining the storm water temporarily stored on them. Many roofs designed with controlled-flow drains have a design rain load of 30 lb/ft^2 (1.44 kN/m^2) and are equipped with a secondary drainage system (for example, scuppers) that prevents water depths ($d_s + d_h$) greater than 5 3/4 in. (145 mm) on the roof.

Examples. The following two examples illustrate the method used to establish design rain loads based on Section 8 of this Standard.

Example 1. Determine the design rain load, R, at the secondary drainage for the roof plan shown in Figure C8-1, located at a site in Birmingham, Alabama. The design rainfall intensity, i, specified by the plumbing code for a 100-yr, 1-hour rainfall is 3.75 in./hr (95 mm/hr). The inlet of the 4-in. diameter (102 mm) secondary roof drains are set 2 in. (51 mm) above the roof surface.

Flow rate, Q, for the secondary drainage 4 in diameter (102 mm) roof drain:

$$Q = 0.0104\ Ai \qquad \textbf{(Eq. C8-1)}$$
$$Q = 0.0104(2500)(3.75)$$
$$= 97.5 \text{ gal./min. } (0.0062 \text{ m}^3/\text{s})$$

Hydraulic head, d_h:

Using Table C8-1, for a 4 in. diameter (102 mm) roof drain with a flow rate of 97.5 gal./min. (0.0062 m^3/s) interpolate between a hydraulic head of 1 and 2 in. (25 and 51 mm) as follows:

$$d_h = 1 + [(97.5 - 80) \div (170 - 80)]$$
$$= 1.19 \text{ in. 30.2 mm)}$$

Static head $d_s = 2$ in. (51 mm); the water depth from drain inlet to the roof surface.

Design rain load, R, adjacent to the drains:

$$R = 5.2(d_s + d_h) \qquad \textbf{(Eq. 8-1)}$$
$$R = 5.2(2 + 1.19) = 16.6 \text{ psf } (0.80 \text{ kN/m}^2)$$

Example 2. Determine the design rain load, R, at the secondary drainage for the roof plan shown in Figure C8-2, located at a site in Los Angeles, California. The design rainfall intensity, i, specified by the plumbing code for a 100-yr, 1-hour rainfall is 1.5 in./hr (38 mm/hr). The inlet of the 12 in. (305 mm) secondary roof scuppers are set 2 in. (51 mm) above the roof surface.

Flow rate, Q, for the secondary drainage, 12 in. (305 mm) wide channel scupper:

$$Q = 0.0104\ A_i \qquad \textbf{(Eq. C8-1)}$$
$$Q = 0.0104(11,500)(1.5)$$
$$= 179 \text{ gal./min. } (0.0113 \text{ m}^3/\text{s})$$

Hydraulic head, d_h:

Using Table C8-1, by interpolation, the flow rate for a 12 in. (305 mm) wide channel scupper is twice that of a 6 in. (152 mm) wide channel scupper. Using Table C8-1, the hydraulic head, d_h, for one-half the flow rate, Q, or 90 gal./min. (0.0057 m^3/s), through a 6 in. (152 mm) wide channel scupper is 3 in. (76 mm).

$d_h = 3$ in. (76 mm) for a 12 in. wide (305 mm) channel scupper with a flow rate, Q, of 179 gal./min. (0.0113 m^3/s).

Static head, $d_s = 2$ in. (51 mm); depth of water from the scupper inlet to the roof surface.

Design rain load, R, adjacent to the scuppers:

$$R = 5.2(d_h + d_s) \qquad \textbf{(Eq. 8-1)}$$
$$R = 5.2(2 + 3) = 26 \text{ psf } (1.2 \text{ kN/m}^2)$$

REFERENCES

[C8-1] Building Officials and Code Administrators International. *The BOCA national plumbing code.* Country Club Hills, Ill., 1993.

[C8-2] Factory Mutual Engineering Corp. *Roof loads for new construction.* Loss Prevention Data 1-54, Norwood, Mass., 1991.

[C8-3] Southern Building Code Congress International (SBCCI). *Standard plumbing code.* Birmingham, Ala., 1991.

[C8-4] Associate Committee on the National Building Code. *National Building Code of Canada.* National Research Council of Canada. Ottawa, Ontario, Canada, 1990.

[C8-5] American Institute of Steel Construction (AISC). *Specification for structural steel for buildings, allowable stress design and plastic design.* AISC, New York, 1989.

[C8-6] AISC. *Load and resistance factor design specification for structural steel buildings.* AISC, New York, 1986.

[C8-7] American Institute of Timber Construction. "Roof slope and drainage for flat or nearly flat roofs." AITC Tech. Note No. 5, Englewood, Colo., 1978.

[C8-8] Burgett, L.B. "Fast check for ponding." *Eng. J. Am. Inst. Steel Construction,* 10(1), 26–28, 1973.

[C8-9] Chinn, J., Mansouri, A.H., and Adams, S.F. "Ponding of liquids on flat roofs." *J. Struct. Div.*, ASCE, 95(ST5), 797–808, 1969.

[C8-10] Chinn, J. "Failure of simply-supported flat roofs by ponding of rain." *Eng. J. Am. Inst. Steel Construction*, 3(2), 38–41, 1965.

[C8-11] Haussler, R.W. "Roof deflection caused by rain-water pools." *Civ. Engrg.* 32, 58–59, 1962.

[C8-12] Heinzerling, J.E. "Structural design of steel joist roofs to resist ponding loads." Steel Joist Institute, Tech. Dig. No. 3. Arlington, Va., 1971.

[C8-13] Marino, F.J. "Ponding of two-way roof systems." *Eng. J. Am. Inst. Steel Construction*, 3(3), 93–100, 1966.

[C8-14] Salama, A.E., and Moody, M.L. "Analysis of beams and plates for ponding loads." *J. Struct. Div.*, ASCE, 93(ST1), 109–126, 1967.

[C8-15] Sawyer, D.A. "Ponding of rainwater on flexible roof systems." *J. Struct. Div.*, ASCE, 93(ST1), 127–148, 1967.

[C8-16] Sawyer, D.A. "Roof-structural roof-drainage interactions." *J. Struct. Div.*, ASCE, 94(ST1), 175–198, 1969.

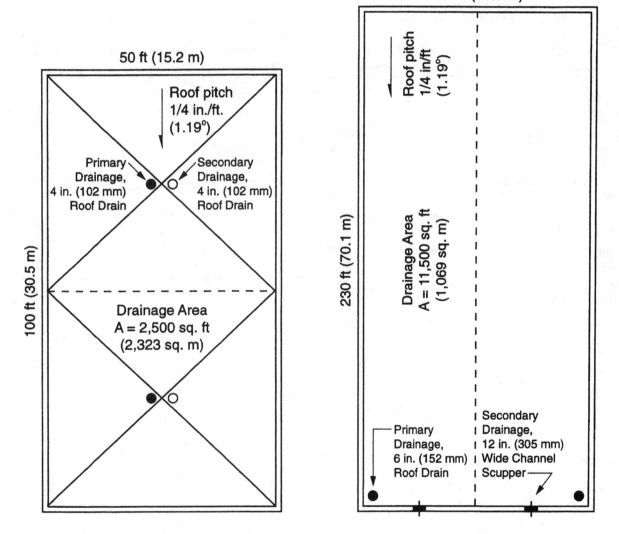

* Dashed lines indicate the boundary between separate drainage areas.

FIGURE C8-1
EXAMPLE 1 ROOF PLAN

FIGURE C8-2
EXAMPLE 2 ROOF PLAN

TABLE C8-1
FLOW RATE, Q, IN GALLONS PER MINUTE OF VARIOUS DRAINAGE SYSTEMS AT VARIOUS HYDRAULIC HEADS, d_h IN INCHES [C8-2]

Drainage System	Hydraulic Head d_h, Inches									
	1	2	2.5	3	3.5	4	4.5	5	7	8
4 in. diameter drain	80	170	180							
6 in. diameter drain	100	190	270	380	540					
8 in. diameter drain	125	230	340	560	850	1100	1170			
6 in. wide, channel scupper**	18	50	*	90	*	140	*	194	321	393
24 in. wide, channel scupper	72	200	*	360	*	560	*	776	1284	1572
6 in. wide, 4 in. high, closed scupper**	18	50	*	90	*	140	*	177	231	253
24 in. wide, 4 in. high, closed scupper	72	200	*	360	*	560	*	708	924	1012
6 in. wide, 6 in. high, closed scupper	18	50	*	90	*	140	*	194	303	343
24 in. wide, 6 in. high, closed scupper	72	200	*	360	*	560	*	776	1212	1372

*Interpolation is appropriate, including between widths of each scupper.
**Channel scuppers are open-topped (i.e., 3 sided). Closed scuppers are 4 sided.

IN SI, FLOW RATE, Q, IN CUBIC METERS PER SECOND OF VARIOUS DRAINAGE SYSTEMS AT VARIOUS HYDRAULIC HEADS, d_h IN MILLIMETERS [C8-2]

Drainage System	Hydraulic Head d_h, mm									
	25	51	64	76	89	102	114	127	178	203
102 mm diameter drain	.0051	.0107	.0114							
152 mm diameter drain	.0063	.0120	.0170	.0240	.0341					
203 mm diameter drain	.0079	.0145	.0214	.0353	.0536	.0694	.0738			
152 mm wide, channel scupper**	.0011	.0032	*	.0057	*	.0088	*	.0122	.0202	.0248
610 mm wide, channel scupper	.0045	.0126	*	.0227	*	.0353	*	.0490	.0810	.0992
152 mm wide, 102 mm high, closed scupper**	.0011	.0032	*	.0057	*	.0088	*	.0112	.0146	.0160
610 mm wide, 102 mm high, closed scupper	.0045	.0126	*	.0227	*	.0353	*	.0447	.0583	.0638
152 mm wide, 152 mm high, closed scupper	.0011	.0032	*	.0057	*	.0088	*	.0122	.0191	.0216
610 mm wide, 152 mm high, closed scupper	.0045	.0126	*	.0227	*	.0353	*	.0490	.0765	.0866

*Interpolation is appropriate, including between widths of each scupper.
**Channel scuppers are open-topped (i.e., 3-sided). Closed scuppers are 4-sided.

SECTION C9.0
EARTHQUAKE LOADS

Section 9 of the 2002 edition of ASCE 7 is primarily based on the 2000 edition of the *NEHRP Recommended Provisions for the Development of Seismic Regulations for New Buildings and Other Structures*, which is prepared by the Building Seismic Safety Council (BSSC) under sponsorship of the Federal Emergency Management Agency (FEMA). NEHRP stands for the National Earthquake Hazards Reduction Program, which is managed by FEMA. Since 1985, the NEHRP Provisions have been updated every 3 years. In most cases, revisions incorporated into the 2002 edition of ASCE 7 directly reflect changes made to the 1997 edition of the NEHRP Provisions, which was the basis of Section 9 of ASCE 7-98. Where ASCE 7 revisions vary significantly from those made to the NEHRP Provisions, new commentary has generally been provided.

The 2002 edition continues to utilize spectral response seismic design maps that reflect seismic hazards on the basis of contours. These maps were completed by the U.S. Geological Survey (USGS). Since these maps were developed, the USGS also created a companion CD-ROM which provides mapped spectral values for a specific site based on site's longitude, latitude, and site soil classification. Longitude and latitude for a given address can be found at Web sites such as *www.Geocode.com* (the EZ Locate option). The CD-ROM should be used for establishing spectral values for design since the maps found in ASCE 7 and at Web sites are at too large a scale to provide accurate spectral values for most sites. The CD-ROM may be purchased from BSSC. It is provided upon request with the purchase of the International Building Code.

The ATC and BSSC efforts had their origin in the 1971 San Fernando Valley earthquake, which demonstrated that design rules of that time for seismic resistance had some serious shortcomings. Each subsequent major earthquake has taught new lessons. ATC, BSSC, and ASCE have endeavored to work individually and collectively to improve each succeeding document so that it would be based upon the best earthquake engineering research as applicable to design and construction, and that they would have nationwide applicability. The natural evolution of this process has led BSSC's NEHRP Provisions into ASCE 7, some current model building codes, and was utilized to develop seismic provisions for The International Building Code (IBC 2000). The IBC 2000 provisions have also influenced the ASCE 7-98 provisions. There was considerable support in the ASCE 7 committee to review and adopt many changes IBC 2000 made to the original NEHRP 1997 provisions.

Content of Commentary. This Commentary does not attempt to explain the earthquake loading provisions in great detail. The reader is referred to two excellent resources:

Federal Emergency Management Agency (FEMA) *Part 2, Commentary*, of the *NEHRP Recommended Provisions for the Development of Seismic Regulations for New Buildings and Other Structures*, Building Seismic Safety Council, 2000.

Structural Engineers Association of California *Recommended Lateral Force Requirements and Commentary*, Seismology Committee, 1996.

Section 9 is organized such that the BSSC Commentary is easily used; simply remove the first "9" (or "A.9") from the section number in this Standard to find the corresponding section in the BSSC Commentary. Most of this Commentary is devoted to noting and explaining the differences of major substance between Section 9 of ASCE 7-02 and the 2000 edition of the *NEHRP Recommended Provisions*.

Nature of Earthquake "Loads." The 1988 edition of ASCE 7 and the 1982 edition of ANSI A58.1 contained seismic provisions based upon those in the *Uniform Building Code* (UBC) of 1985 and earlier. The UBC provisions for seismic safety have been based upon recommendations of the **Structural Engineers Association of California (SEAOC) and Predecessor Organizations** Until 1988, the UBC and SEAOC provisions had not yet been fully influenced by the ATC and BSSC efforts. The 1972 and 1955 editions of A58.1 contained seismic provisions based upon much earlier versions of SEAOC and UBC recommendations.

The two most far-reaching differences between the 1993, 1995, and 1998 editions and these prior editions are that the newer editions are based upon a strength level limit state rather than an equivalent loading for use with allowable stress design, and that it contains a much larger set of provisions that are not directly statements of loading. The intent is to provide a more reliable and consistent level of seismic safety in new building construction.

Earthquakes "Load" Structures Indirectly. As the ground displaces, a building will follow and vibrate. The vibration produces deformations with associated strains and stresses in the structure. Computation of dynamic response to earthquake ground shaking is complex. As

a simplification, this Standard is based upon the concept of a response spectrum. A response spectrum for a specific earthquake ground motion does not reflect the total time history of response, but only approximates the maximum value of response for simple structures to that ground motion. The design response spectrum is a smoothed and normalized approximation for many different ground motions, adjusted at the extremes for characteristics of larger structures. The BSSC *NEHRP Commentary*, Chapters 4 and 5, contains a much fuller description of the development of the design response spectrum and the maps that provide the background information for various levels of seismic hazard and various ground conditions.

The provisions of Section 9 are stated in terms of forces and loads; however, the user should always bear in mind that there are no external forces applied to the above-ground portion of a structure during an earthquake. The design forces are intended only as approximations to produce the same deformations, when multiplied by the deflection amplification factor C_d, as would occur in the same structure should an earthquake ground motion at the design level occur.

The design limit state for resistance to an earthquake is unlike that for any other load within the scope of ASCE 7. The earthquake limit state is based upon system performance, not member performance, and considerable energy dissipation through repeated cycles of inelastic straining is assumed. The reason is the large demand exerted by the earthquake and the associated high cost of providing enough strength to maintain linear elastic response in ordinary buildings. This unusual limit state means that several conveniences of elastic behavior, such as the principle of superposition, are not applicable, and makes it difficult to separate design provisions for loads from those for resistance. This is the reason the *NEHRP Provisions* contain so many provisions that modify customary requirements for proportioning and detailing structural members and systems. It is also the reason for the construction quality assurance requirements. All these "nonload" provisions are presented in the seismic safety appendix to Section 9.

Use of Allowable Stress Design Standards. The conventional design of nearly all masonry structures and many wood and steel structures has been accomplished using allowable stress design (ASD) standards. Although the fundamental basis for the earthquake loads in Section 9 is a strength limit state beyond first yield of the structure, the provisions are written such that the conventional ASD standards can be used by the design engineer. Conventional ASD standards may be used in one of two fashions:

1. The earthquake load as defined in Section 9 may be used directly in allowable stress load combinations of Section 2.4 and the resulting stresses compared directly with conventional allowable stresses, or

2. The earthquake load may be used in strength design load combinations and resulting stresses compared with amplified allowable stresses (for those materials for which the design standard gives the amplified allowable stresses, e.g., masonry).

Method 1 is changed somewhat since the 1995 edition of the Standard. The factor on E in the ASD combinations has been reduced to 0.7 from 1.0. This change was accomplished simultaneously with reducing the factor on D in the combination where dead load resists the effects of earthquake loads from 1.0 to 0.6.

The factor 0.7 was selected as somewhat of a compromise among the various materials for which ASD may still be used. The basic premise suggested herein is that for earthquake loadings ASD is an alternative to strength-based design, and that ASD should generally result in a member or cross-section with at least as much true capacity as would result in strength-based design. As this Commentary will explain, this is not always precisely the case.

There are two general load combinations, one where the effects of earthquake load and gravity load add, and a second where they counteract. In the second, the gravity load is part of the resistance, and therefore, only dead load is considered. These combinations can be expressed as follows, where α is the factor on E in the ASD combination, calibrated to meet the premise of the previous paragraph. Using the combinations from Sections 2.3 and 2.4:

Additive combinations:

Strength: $1.2D + 0.5L + 0.2S + 1.0E \leq \phi^*$ Strength

ASD: $1.0D + 0.75*(1.0L + 1.0S + \alpha E) \leq$ Allowable Stress

and $1.0D + \alpha E \leq$ Allowable Stress

Counteracting combinations:

Strength: $0.9D + 1.0E \leq \phi^*$ Strength

ASD: $0.6D + \alpha E \leq$ Allowable Stress

For any given material and limit state, the factor α depends on the central factor of safety between the strength and the allowable stress, the resistance factor ϕ, and ratios of the effects of the various loads. Table C9-1 summarizes several common cases of interest, including those where the designer opts to use the one-third increase in allowable stress permitted in various reference standards.

The bold entries indicate circumstances in which the 0.7 factor in the ASD equations will result in a structural capacity less than required by strength design. Given the current basis of the earthquake load provisions, such situations should be carefully considered in design. This review suggests that the one-third increase in allowable stress given in some reference standards is not appropriate in some situations such as the design of certain steel

members. The review also suggests that a load factor on E of 0.8 would be more appropriate for the design of wood structures.

The amplification for Method 2 is accomplished by the introduction of two sets of factors to amplify conventional allowable stresses to approximate the equivalent yield strength: one is a stress increase factor (1.7 for steel, 2.16 for wood, and 2.5 for masonry) and the second is a resistance or strength reduction factor (less than or equal to 1.0) that varies depending on the type of stress resultant and component. The 2.16 factor is selected for conformance with the new design standard for wood (*Load and Resistance Factor Standard for Engineered Wood Construction*, ASCE 16-95) and with an existing ASTM standard; it should not be taken to imply an accuracy level for earthquake engineering.

Although the modification factors just described accomplish a transformation of allowable stresses to the earthquake strength limit state, it is not conservative to ignore the provisions in the Standard as well as the supplementary provisions in the appendix that deal with design or construction issues that do not appear directly related to computation of equivalent loads, because **the specified loads are derived assuming certain levels of damping and ductile behavior**. In many instances, this behavior is not necessarily delivered by designs conforming to conventional standards, which is why there are so many seemingly "nonload" provisions in this standard and appendix.

Past design practices (the SEAOC and UBC requirements prior to 1997) for earthquake loads produce loads intended for use with allowable stress design methods. Such procedures generally appear very similar to this Standard, but a coefficient R_w is used in place of the response modification factor R. R_w is always larger than R, generally by a factor of about 1.5; thus the loads produced are smaller, much as allowable stresses are smaller that nominal strengths. However, the other procedures contain as many, if not more, seemingly "nonload" provisions for seismic design to ensure the assumed performance.

Story Above Grade. Figure C9-1 illustrates this definition:

Occupancy Importance Factor. The NEHRP 1997 Provisions introduced the Occupancy Importance Factor, I. It was a new factor in NEHRP provisions but not for ASCE 7 or UBC provisions. Editions of this Standard prior to 1995, as well as other current design procedures for earthquake loads, make use of an occupancy importance factor, I, in the computation of the total seismic force. This factor was removed from the 1995 edition of the Standard when it introduced the provisions consistent with the 1994 edition of NEHRP provisions. The 1995 edition did include a classification of buildings by occupancy, but this classification did not affect the total seismic force.

The *NEHRP Provisions* in the opening Section 1.1, Purpose, identify two purposes of the provisions, one of which specifically is to "improve the capability of essential facilities and structures containing substantial quantities of hazardous materials to function during and after design earthquakes." This is achieved by introducing the occupancy importance factor of 1.25 for Seismic Use Group II structures and 1.5 for Seismic Use Group III structures. The NEHRP Commentary Sections 1.4, 5.2, and 5.2.8 explain that the factor is intended to reduce the ductility demands and result in less damage. When combined with the more stringent drift limits for such essential or hazardous facilities, the result is improved performance of such facilities.

Federal Government Construction. The Interagency Committee on Seismic Safety in Construction has prepared an order executed by the President, Executive Order 12699, that all federally owned or leased building construction, as well as federally regulated and assisted construction, should be constructed to mitigate seismic hazards and that the *NEHRP Provisions* are deemed to be the suitable standard. The ICSSC has also issued a report that currently only the 1997 UBC and ASCE 7-95 have been deemed to be in "substantial compliance" with the executive order. It is expected that this Standard would also be deemed equivalent, but the reader should bear in mind that there are certain differences, which are summarized in this Commentary.

C9.1.1 Purpose. The purpose of Section 9.1.1 is to clarify that when the design load combinations involving the wind forces of Section 6 produce greater effects than the design load combinations involving the earthquake forces of Section 9 such that the wind design governs the basic strength of the lateral force-resisting system, the detailing requirements and limitations prescribed in this Section and referenced standards are still required to be followed.

C9.4.1.1 Maximum Considered Earthquake Ground Motions. The 1998 edition replaced the use in the 1995 edition of seismic ground acceleration maps to define ground motion with a general procedure for determining maximum considered earthquake and design spectral response accelerations based on new maps included in NEHRP 1997 Provisions. The reasons and procedures for the development of these new maps are explained in Appendix B of NEHRP 1997 Recommended Provisions and Commentary documents.

C9.5.2.4 Redundancy. This Standard introduces a new reliability factor for structures in Seismic Design Categories D, E, and F to quantify redundancy. The value of this factor varies from 1.0 to 1.5. This factor has an effect of reducing

the R factor for less redundant structures, thereby increasing the seismic demand. The factor is introduced in recognition of the need to address the issue of redundancy in the design. The NEHRP Commentary Section 5.2.4 explains that this new requirement is "intended to quantify the importance of redundancy." The Commentary points out that "many nonredundant structures have been designed in the past using values of R that were intended for use in designing structures with higher levels of redundancy." In other words, the use of R factor in the design has led to slant in design in the wrong direction. The NEHRP Commentary cites the source of this factor as the work of SEAOC for inclusion of this factor in UBC 1997.

C9.5.2.5.1 Seismic Design Category A. This edition includes a new provision of minimum lateral force for Seismic Design Category A structures. The minimum load is a structural integrity issue related to the load path. It is intended to specify design forces in excess of wind loads in heavy low-rise construction. The design calculation is simple and easily done to ascertain if it governs or the wind load governs. This provision requires a minimum lateral force of 1% of the total gravity load assigned to a story to ensure general structural integrity.

C9.5.2.6.2.5 Nonredundant Systems. The general philosophy embodied by this Section is covered by the General Structural Integrity requirements of Section 1.4 and the Commentary of Section 1.4. Authorities having jurisdiction are cautioned that the term "consider" should not necessarily mean that a rigorous analysis of any sort should be conducted unless special circumstances exist.

C9.5.2.6.2.11 Elements Supporting Discontinuous Walls or Frames. The purpose of the special load combinations is to protect the gravity load-carrying system against possible overloads caused by over-strength of the lateral force-resisting system. Either columns or beams may be subject to such failure, therefore, both should include this design requirement. Beams may be subject to failure due to overloads in either the downward or upward directions of force. Examples include reinforced concrete beams, the weaker top laminations of glulam beams, or unbraced flanges of steel beams or trusses. Hence, the provision has not been limited simply to downward force, but instead to the larger context of "vertical load." A remaining issue that has not been fully addressed in this edition is clarification of the appropriate load case for the design of the connections between the discontinuous walls or frames and the supporting elements.

C9.5.2.6.2.8 Anchorage of Concrete or Masonry Walls. Where roof framing is not perpendicular to anchored walls, provision needs to be made to transfer both the tension and sliding components of the anchorage force into the roof diaphragm.

C9.5.2.7.1 Special Seismic Loads. Some elements of properly detailed structures are not capable of safely resisting ground shaking demands through inelastic behavior. To ensure safety, these elements must be designed with sufficient strength to remain elastic. The Ω_0 coefficient approximates the inherent overstrength in typical structures having different seismic force-resisting systems. The special seismic loads, factored by the Ω_0 coefficient, are an approximation of the maximum force these elements are ever likely to experience. This Standard permits the special seismic loads to be taken as less than the amount computed by applying the Ω_0 coefficient to the design seismic forces when it can be shown that yielding of other elements in the structure will limit the amount of load that can be delivered to the element. As an example, the axial load in a column of a moment-resisting frame will derive from the shear forces in the beams that connect to this column. The axial loads due to lateral seismic action need never be taken greater than the sum of the shears in these beams at the development of a full structural mechanism, considering the probable strength of the materials and strain hardening effects (for frames controlled by beam hinge-type mechanisms, this would typically be $2M_p/L$, where for steel frames, M_p is the expected plastic moment capacity of the beam as defined in the AISC Seismic Specification and for concrete frames, Mp would be the probable flexural strength of the beam, where L is the clear span length). In other words, as used in this Section, the term "capacity" means the expected or median anticipated strength of the element, considering potential variation in material yield strength and strain hardening effects. When calculating the capacity of elements for this purpose, material strengths should not be reduced by capacity or resistance factors.

C9.5.5.5.5 Torsion. Where earthquake forces are applied concurrently in two orthogonal directions, the 5% displacement of the center of mass should be applied along a single orthogonal axis chosen to produce the greatest effect, but need not be applied simultaneously along two axis (i.e., in a diagonal direction).

Most diaphragms of light-framed construction are somewhere between rigid and flexible for analysis purposes, i.e., semi-rigid. Such diaphragm behavior is difficult to analyze when considering torsion of the structure. As a result, it is believed that consideration of the amplification of the torsional moment is a refinement that is not warranted for light-framed construction.

Historically, the intent of the A_x, term was not to amplify the natural torsion component, only the accidental torsion component. There does not appear reason to further increase design forces by amplifying both components together.

C9.5.4.3 Horizontal Distribution. Engineers have traditionally analyzed buildings employing diaphragms sheathed with wood structural panels or untopped steel deck using flexible diaphragm assumptions. In recent years, there has been a trend to recognize that in buildings with flexible vertical elements of the lateral force-resisting system and with relatively compact diaphragm aspect ratios, the diaphragms behave in a semi-rigid manner, taking on some of the characteristics of flexible diaphragms and some of rigid diaphragms. This Standard requires consideration of diaphragm stiffness effects when performing analyses using the equivalent lateral force technique or more sophisticated analytical techniques. However, under the Simplified Method, the specified design forces are increased relative to those for other methods of analysis and the resulting distribution of design seismic forces to individual structure elements is considered to be sufficiently conservative even when the rather inaccurate procedures of flexible diaphragm analysis are applied.

C9.5.9 Soil Structure Interaction. A dynamic modifier (a_θ) is included in the formulation of rocking stiffness (K_θ). When back-analyzed period lengthening and foundation damping values from stiff shear-wall structures are compared to predictions from code-type analyses, the predictions become significantly more accurate with the addition of the a_θ term.

For the calculation of impedance terms K_y and K_θ, there are no specific recommendations for when half-space versus finite soil layer over rigid base solutions should be used. Studies have shown the stiffness of a two-layer soil model approaches the stiffness of a finite soil layer over a rigid base when the underlying soil layer has a shear wave velocity greater than twice the velocity of the surface layer. The restrictions originally placed on the use of the finite soil layer over rigid base model still apply ($r/D_s < 0.5$, where r = foundation radius and D_s = depth of finite soil layer).

For the calculation of static impedance terms with the half-space solution, one key issue is over what depth the actual soil shear wave velocities should be averaged to provide a representative half-space velocity. Studies have shown that for a variety of velocity profiles, a depth of $0.75r_a$ was appropriate for translational stiffness, and $0.75r_m$ was appropriate for rocking stiffness.

The definitions of K_y and K_θ no longer contain the word "static," since dynamic effects will be considered subsequently for K_θ.

C9.6.1.4 Seismic Relative Displacements. The design of some nonstructural components that span vertically in the structure can be complicated when supports for the element do not occur at horizontal diaphragms. The language in Section 9.6.1.4 has been amended to clarify that story drift must be accommodated in the elements that will actually distort. For example, a glazing system supported by precast concrete spandrels must be designed to accommodate the full story drift, even though the height of the glazing system is only a fraction of the floor-to-floor height. This condition arises because the precast spandrels will behave as rigid bodies relative to the glazing system and therefore all the drift must be accommodated by anchorage of the glazing unit, the joint between the precast spandrel and the glazing unit, or some combination of the two.

C9.6.2.10 Glass in Glazed Curtain Walls, Glazed Storefronts, and Glazed Partitions. The 2000 NEHRP Provisions contain seismic design provisions for glazing systems. For ASCE 7, it was found that clarity of the provisions could be improved by reformatting the equations in Section 9.6.2.10.1.

C9.7.4.1 Investigation. Earthquake motion is only one factor in assessing liquefaction potential and slope instability. Lateral spreading includes surface rupture and lateral ground displacement, both of which can lead to adverse consequences. Finally, identification alone has little value unless mitigation options are also identified.

C9.7.5 Foundation Requirements for Seismic Design Categories D, E, and F. The exception allows detached 1- and 2-story dwellings of light-framed construction in Seismic Design Categories C through F to be designed in accordance with the seismic provisions for Seismic Design Category C. This is similar to exceptions for this type of construction in other Chapters of this Standard.

C9.7.5.3 Liquefaction Potential and Soil Strength Loss. The reference to Section 9.4.1.3.4 is to clarify that the designs should be based on the design level ground motions, not maximum considered earthquake ground motions.

In the 1997 NEHRP Commentary, liquefaction requires consideration of both peak ground acceleration and earthquake magnitude. This Standard will retain a reference to a peak ground acceleration of $S_S/2.5$ in lieu of a site-specific evaluation, in order to provide a conservative default value for the designer. The 2.5 factor is a nominal amplification from peak ground acceleration to short period response acceleration. The use of S_S will be conservative for the entire United States, whereas the default value in previous versions of this Standard may have been nonconservative for the central and eastern parts of the country.

The assessment of liquefaction potential may be based on the Summary Report and supporting documentation contained in NCEER-97-0022, Proceedings of the NCEER Workshop on Evaluation of Liquefaction Resistance of Soils, available from the Multidisciplinary Center for Earthquake Engineering Research, State University of New York at Buffalo, Red Jacket Quadrangle, Buffalo, NY 14261.

C9.7.5.4 Special Pile and Grade Beam Requirements.
Anchorage of the pile into the pile cap should be conservatively designed to allow energy-dissipating mechanisms, such as rocking, to occur in the soil without structural failure of the pile. Precast prestressed concrete piles are exempt from the concrete special moment frame column confinement requirements since these requirements were never intended for slender precast prestressed concrete elements and will result in unbuildable piles. Precast prestressed concrete piles have been proven through cyclic testing to have adequate performance with substantially less confinement reinforcing than required by ACI 318. A reduced transverse steel ratio from that required in columns is permitted in concrete piles, which reflects confinement provided by the soil.

Batter pile systems that are partially embedded have historically performed poorly under strong ground motions. Difficulties in examining fully embedded batter piles have led to uncertainties as to the extent of damage for this type of foundation. Batter piles are considered as limited ductile systems by their nature and should be designed using the special seismic load combinations.

SECTION C9.9
STRUCTURAL CONCRETE

This Section references ACI 318-02 for structural concrete design and construction. The R-, Ω_{0^-}, and C_d-values listed in Table 9.5.2.2 for structural systems of structural concrete anticipate conformance with this Standard.

SECTION C9.12
WOOD

Reference to the 1995 One- and Two-Family Dwelling Code (CABO) was removed and reference to the International Residential Code 2000 (IRC) added to maintain a single residential design code reflecting current advances in analysis of structures with regard to seismic forces. It is always prudent to determine from the authority having jurisdiction which building code is currently being enforced by that jurisdiction.

SECTION C9.14
NONBUILDING STRUCTURES

The NEHRP Provisions contain additional design requirements for nonbuilding structures in an Appendix to Chapter 14. The NEHRP Commentary contains, in addition to its Chapter 14 for nonbuilding structures, additional guidance in a separate chapter titled Appendix to Chapter 14. These additional resources should be referred to in designing nonbuilding structures for seismic loads.

C9.14.2 Reference Standards. The NEHRP Provisions contain additional references for the design and construction of nonbuilding structures that cannot be referenced directly by ASCE 7. The references are as follows:

Ref. C9.14-1 American Society of Civil Engineers (ASCE). "Guidelines for seismic evaluation and design of petrochemical facilities." Petrochemical Energy Committee Task Rep. Reston, Va., 1997.

Ref. C9.14-2 ASCE. "Design of secondary containment in petrochemical facilities." Petrochemical Energy Committee Task Rep., Reston, Va., 1997.

Ref. C9.14-3 Troitsky, M. S. "Tubular Steel Structures: Theory and Design," The James F. Lincoln Arc Welding Foundation, Cleveland, Ohio, 1990.

Ref. C9.14-4 Wozniak, R. S. and Mitchell, W. W. "Basis of seismic design provisions for welded steel oil storage tanks." *Proc., Refining Department*, American Petroleum Institute, Washington, D.C., 1978.

C9.14.3 Industry Design Standards and Recommended Practice. Table C9.14.3 is a cross-reference of references that cannot be referenced directly by ASCE 7 and the applicable nonbuilding structures.

References to industry standards on nonbuilding structures have been added to aid the design professional and the authority having jurisdiction in the design of nonbuilding structures. The addition of these references to ASCE 7 provides a controlled link between the requirements of ASCE 7 and industry design standards.

Some of the standards listed are not consensus documents. Following the example of Section 9.6, the standards have been divided into "Consensus Standards" and "Accepted Standards." Several "Industry References" are listed in the Commentary.

Many of the referenced standards contain either seismic design provisions and/or appropriate design stress levels for the particular nonbuilding structure. In many cases, the proposed revisions to Section 9.14 modify these requirements found in the industry standards. A summary of some of the changes are shown below:

API 620, API 650, AWWA D100, AWWA D103, AWWA D110, AWWA D115, and ACI 350.3 all have seismic requirements based on earlier editions of seismic codes (MRI = 475 years) and all are working stress design based. The proposed revisions to ASCE 7 provide working stress based substitute equations for each of these industry standards to bring the seismic design force level "up to speed" with that required in the NEHRP document.

NFPA 59A has seismic requirements considerably in excess of ASCE 7 (MRI = 4,975 years). ASCE 7 provides analysis methods that can augment this industry standard.

Other standards such as the ASME BPVC, while containing few seismic requirements, dictate the stress levels that the applicable structures are designed to.

Other standards such as NFPA 30 and API 2510 provide guidance on safety, plant layout, and so on. These documents have significant impact on the actual level of risk that the general public is exposed to.

All nonbuilding structures supported by other structures were contained in Section 9.6 of previous editions of ASCE 7. Significant nonbuilding structures (where the weight of nonbuilding structure equals or exceeds 25% of the combined weight of the nonbuilding structure and the supporting structure) cannot be analyzed or designed for seismic forces independent of the supporting structure. The requirements of Section 9.14 are more appropriate for the design of these combined systems.

C9.14.7.3 Tanks and Vessels. This Section brings in specific requirements for tanks and vessels not currently in ASCE 7 or in any other single document. Most (if not all) industry standards covering the design of tanks and vessels contain seismic design requirements based on earlier (lower force level) seismic codes. Many of the provisions of the Standard show how to modify existing industry standards to get to the same force levels as required by ASCE 7-98/NEHRP 2000. Eventually, the organizations responsible for maintaining these industry standards will adopt seismic provisions based on NEHRP. When this occurs, the specific requirements in ASCE 7 can be deleted and direct reference made to the industry standards.

A structure specific response spectrum is added for ground-supported liquid storage tanks (Figure 9.14.7.3.6-1). This was necessary for two reasons. First, the response spectrum that serves as the basis for the seismic requirements in NERHP is based on 5% damping. Sloshing liquid (convective mass) has a damping of 0.5%. Therefore, the existing response spectrum had to be amplified for this lower damping. Second, the period of a sloshing wave in most practical tank sizes is on the order of several seconds (much longer than most building structures). Therefore, the lower tail of the response spectrum needs to vary with $1/T^2$ instead of $1/T$.

C9.14.7.3.6.1.4 Internal Components. A recognized analysis method for determining the lateral loads due to the sloshing liquid can be found in Reference C9.14-4.

C9.14.7.3.8.2 Bolted Steel. As a temporary structure, it may be valid to design for no seismic loads or for reduced seismic loads based on a reduced return period. The actual force level must be based on the time period that this structure will be in place. This becomes a decision

between the authority having jurisdiction and the design professional.

C9.14.7.7 Secondary Containment Systems. This Section, as presented in ASCE 7, differs from the requirements in NEHRP 2000. The ASCE 7 committee felt that the NERHP 2000 requirements for designing all impoundment dikes for the maximum considered earthquake ground motion when full and to size all impoundment dikes for the sloshing wave was too conservative. Designing the impoundment dike full for the maximum considered earthquake assumes the failure of the primary containment and the occurrence of a significant aftershock. Significant (same magnitude as the maximum considered earthquake ground motion) aftershocks are rare and do not occur in all locations.

While designing for aftershocks has never been part of the design loading philosophy found in ASCE 7, secondary containment must be designed full for an aftershock to protect the general public. The use of two-thirds of the maximum considered ground motion as the magnitude of the design aftershock is supported by Bath's Law, according to which the maximum expected aftershock magnitude may be estimated as 1.2 scale units below that of the main shock magnitude.

The risk assessment and risk management plan as described in Section 1.5.2 should be used to determine when the secondary containment is to be designed for the full maximum considered earthquake seismic when full. The decision to design secondary containment for this more severe condition should be based on the likelihood of a significant aftershock occurring at the particular site and the risk posed to the general public by the release of the hazardous material from the secondary containment.

Secondary containment systems must be designed to contain the sloshing wave where the release of liquid would place the general public at risk by either exposing them to hazardous materials or by the scouring of foundations of adjacent structures or by causing other damage to the adjacent structures.

C9.14.7.8 Telecommunication Towers. This Section, as presented in ASCE 7, differs from the requirements in NEHRP 2000. Telecommunication towers are contained in the Appendix to NEHRP 2000. This Section was previously included in ASCE 7-98. The ASCE 7 committee felt that it benefited the design professional and building officials to leave these requirements in the Standard.

CA.9.7.4.4.1 Uncased Concrete Piles. The transverse reinforcing requirements in the potential plastic hinge zone of uncased concrete piles in Seismic Design Category C is a selective composite of two ACI 318 requirements. In the potential plastic hinge region of an intermediate moment-resisting concrete frame column, the transverse

reinforcement spacing is restricted to the least of: (1) 8 times the diameter of the smallest longitudinal bar, (2) 24 times the diameter of the tie bar, (3) one-half the smallest cross-sectional dimension of the column, and (4) 12 in. Outside of the potential plastic hinge region of a special moment-resisting frame column, the transverse reinforcement spacing is restricted to the smaller of 6 times the diameter of the longitudinal column bars and 6 in.

CA.9.7.4.4.4 Precast Nonprestressed Concrete Piles. Transverse reinforcement requirements in and outside of the plastic hinge zone of precast nonprestressed piles are clarified. The transverse reinforcement requirement in the potential plastic hinge zone is a composite of two ACI 318 requirements (see CA.9.7.4.4.1). Outside of the potential plastic hinge region the 8 longitudinal-bar-diameter spacing is doubled. The maximum 8 in. tie spacing comes from current building code provisions for precast concrete piles.

CA.9.7.4.4.5 Precast Prestressed Piles. The transverse and longitudinal reinforcing requirements given in ACI 318, Chapter 21, were never intended for slender precast prestressed concrete elements and will result in unbuildable piles. The requirements are based on the *Recommended Practice for Design, Manufacture and Installation of Prestressed Concrete Piling*, PCI Committee on Prestressed Concrete Piling, 1993.

Eq. A.9.7.4.4.5-1, originally from ACI 318, has always been intended to be a lower-bound spiral reinforcement ratio for larger diameter columns. It is independent of the member section properties and can therefore be applied to large or small diameter piles. For cast-in-place concrete piles and precast prestressed concrete piles, the resulting spiral reinforcing ratios from this formula are considered to be sufficient to provide moderate ductility capacities.

Full confinement per Eq. A.9.7.4.4.5-1 is required for the upper 20 ft of the pile length where curvatures are large. The amount is relaxed by 50% outside of that length in view of lower curvatures and in consideration of confinement provided by the soil.

CA.9.7.5.4.1 Uncased Concrete Piles. The reinforcement requirements for uncased concrete piles are taken from current building code requirements, and are intended to provide ductility in the potential plastic hinge zones.

CA.9.7.5.4.3 Precast Concrete Piles. The transverse reinforcement requirements for precast nonprestressed concrete piles are taken from current building code requirements and are intended to provide ductility in the potential plastic hinge zone.

CA.9.7.5.4.4 Precast-Prestressed Piles. The last paragraph provides minimum transverse reinforcement outside of the zone of prescribed ductile reinforcing.

SECTION CA.9.9
SUPPLEMENTARY PROVISIONS FOR CONCRETE

Section A.9.9.1 makes modifications to ACI 318-02 that are essentially adopted from the 2000 edition of the *NEHRP Recommended Provisions*. Sections A.9.9.2 through A.9.9.7 are needed to make the referenced standard compatible with the provisions of this document. Section A.9.9.6 of ASCE 7-98 contained a complete set of provisions on bolts and headed stud anchors in concrete. These provisions have been deleted because ACI 318-02 now contains similar provisions. Section A.9.9.7 of this document makes one additional modification to ACI 318-02. That Standard currently requires laboratory testing to establish the strength of anchor bolts greater than 2 in. in diameter or exceeding 25 in. in tensile embedment depth. This modification makes the ACI 318 equation giving the basic concrete breakout strength of a single anchor in tension in cracked concrete applicable irrespective of the anchor bolt diameter and tensile embedment depth.

SECTION CA.9.11
SUPPLEMENTARY PROVISIONS FOR MASONRY

The appendix for masonry design is primarily necessary because the reference standard itself refers to the 1993 edition of ASCE 7. Since that time, significant changes have been made in terminology to distinguish among various types of masonry walls based upon the method of engineering design (as reinforced or unreinforced) and on the minimum amount of reinforcement. This change in terminology was mainly needed to accommodate the change in philosophy that the detailing rules be driven by the R factor rather than the old Seismic Performance Category.

Most of the serious recent development in improved seismic-resistant design for masonry structures has been expressed in terms of strength design methodology, whereas the reference standard is written solely in terms of allowable stress design. This is not to say that the committee responsible for the reference standard has not considered strength design; the committee simply has not reached consensus on an overall strength design standard. Even though this consensus has not been reached, a few of the concepts are necessary when using the highest R factors for design of masonry systems. One very significant rule from the strength design provisions of the NEHRP Recommended Provisions and the proposed new International Building Code has been incorporated into this appendix for those masonry walls with relatively generous R factors: the upper limit on reinforcement in shear walls. This requirement intends to deliver ductile performance by requiring substantial yield of tension steel before crushing of masonry at the compressive end of the wall (Method A) or by providing a boundary member to control the spread of crushing (Method B).

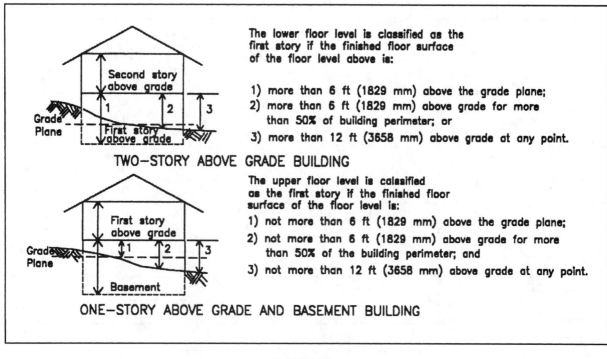

The lower floor level is classified as the first story if the finished floor surface of the floor level above is:

1) more than 6 ft (1829 mm) above the grade plane;
2) more than 6 ft (1829 mm) above grade for more than 50% of building perimeter; or
3) more than 12 ft (3658 mm) above grade at any point.

TWO–STORY ABOVE GRADE BUILDING

The upper floor level is calassified as the first story if the finished floor surface of the floor level is:

1) not more than 6 ft (1829 mm) above the grade plane;
2) not more than 6 ft (1829 mm) above grade for more than 50% of the building perimeter; and
3) not more than 12 ft (3658 mm) above grade at any point.

ONE–STORY ABOVE GRADE AND BASEMENT BUILDING

FIGURE C9-1
ILLUSTRATION OF DEFINITION OF STORY ABOVE GRADE

| Structural Element and Limit State | ASD Rules | Ratio of Load Effects (moment, axial load, etc.) in the load combination being considered | | Equivalency Factor |
		D/E	L/E	α
Steel girder; bending	Reduction per 2.4.3	0	0	0.67
		0.5	0.25	0.65
		1	0.5	0.48
	1/3 increase per reference standard	0	0	**0.89**
		0.5	0.25	**0.79**
		1	0.5	0.71
Steel brace, tension	Reduction per 2.4.3	0	0	0.67
		1	0	0.67
	1/3 increase per reference standard	0	0	**0.89**
		0.5	0	**0.80**
Steel brace, compression	Reduction per 2.4.3	0.25	0.25	0.66
		0.5	0.5	0.45
	1/3 increase per reference standard	0.25	0.25	0.67
Masonry wall, reinforcement for in-plane bending	Reduction per 2.4.3	0	0	0.50
		1.11	0	0.64
	1/3 increase per reference standard	0	0	0.67
		1.11	0	0.67
Masonry wall, in-plane shear (reinforced, with $M/Vd \geq 1$)	Reduction per 2.4.3	0	0	0.47
	1/3 increase per reference standard	0	0	0.63
Wood shear panel	Reduction per 2.4.3 (reference standard does not have explicit increase, although duration factors are not the same in LRFD and ASD)	0	0	**0.77**
Wood collector, tension		0	0	**0.77**
		1	0	**0.85**
Bolts in wood		0	0	**0.75**
		0.25	0.5	**0.76**

TABLE C9-14-3
STANDARDS, INDUSTRY STANDARDS, AND
REFERENCES

Application	Reference
Petrochemical structures	[Ref. C9.14-1]
Impoundment dikes and walls	[Ref. C9.14-2]
Guyed steel stacks and chimneys	[Ref. C9.14-3]
Welded steel tanks for petroleum and petrochemical storage	[Ref. C9.14-4]

SECTION C10.0
ICE LOADS — ATMOSPHERIC ICING

SECTION C10.1
GENERAL

In most of the contiguous United States, freezing rain is considered the cause of the most severe ice loads. Values for ice thicknesses due to in-cloud icing and snow suitable for inclusion in this Standard are not currently available.

Very few sources of direct information or observations of naturally occurring ice accretions (of any type) are available. Reference [C10-6] presents the geographical distribution of the occurrence of ice on utility wires from data compiled by various railroad, electric power, and telephone associations in the 9-year period from the winter of 1928 to 1929 to the winter of 1936 to 1937. The data includes measurements of all forms of ice accretion on wires including glaze ice, rime ice, and accreted snow, but does not differentiate between them. Ice thicknesses were measured on wires of various diameters, heights above ground, and exposures. No standardized technique was used in measuring the thickness. The maximum ice thickness observed during the 9-year period in each of 975 squares, 60 miles (97 km) on a side, in a grid covering the contiguous United States is reported. In every state except Florida, thickness measurements of accretions with unknown densities of approximately 1 radial in. were reported. Information on the geographical distribution of the number of storms in this 9-year period with ice accretions greater than specified thicknesses is also included.

Tattelman and Gringorten (1973) [C10-51] reviewed ice load data, storm descriptions, and damage estimates in several meteorological publications to estimate maximum ice thicknesses with a 50-year mean recurrence interval in each of seven regions in the United States.

Storm Data (NOAA 1959 to present) [C10-38] is a monthly publication that describes damage from storms of all sorts throughout the United States. The storms are sorted alphabetically by state within each month. The compilation of this qualitative information on storms causing damaging ice accretions in a particular region can be used to estimate the severity of ice and wind-on-ice loads. The Electric Power Research Institute has compiled a database of icing events from the reports in *Storm Data* (Shan and Marr 1996) [C10-50]. Damage severity maps were also prepared.

References [C10-7 and C10-48] provide information on freezing rain climatology for the 48 contiguous states based on recent meteorological data.

C10.1.1 Site-Specific Studies. In-cloud icing may cause significant loadings on ice-sensitive structures in mountainous regions and for very tall structures in other areas. Mulherin (1996) [C10-36] reports that of 120 communications tower failures in the United States due to atmospheric icing, 38 were due to in-cloud icing, and in-cloud icing combined with freezing rain caused an additional 26. In-cloud ice accretion is very sensitive to the degree of exposure to moisture-laden clouds, which is related to terrain, elevation, and wind direction and velocity. Large differences in accretion size can occur over a few hundred ft and cause severe load unbalances in overhead wire systems. Advice from a meteorologist familiar with the area is particularly valuable in these circumstances. In Arizona, New Mexico, and the panhandles of Texas and Oklahoma, the U.S. Forest Service specifies ice loads due to in-cloud icing for towers constructed at specific mountaintop sites (USFS 1994) [C10-52].

Snow accretions also can result in severe structural loads and may occur anywhere snow falls, even in localities that may experience only one or two snow events per year. Some examples of locations where snow accretion events resulted in significant damage to structures are Nebraska (NPPD 1976), Maryland (Mozer and West 1983), Pennsylvania (Goodwin et al. 1983), Georgia and North Carolina (Lott 1993), Colorado (McCormick and Pohlman 1993), and Alaska (Golden Valley Electric Association 1997) References [C10-16, C10-19, C10-31, C10-35, C10-39, and C10-34].

For Alaska, available information indicates that moderate to severe ice loads of all types can be expected. The measurements made by Golden Valley Electric Association referred to above are consistent in magnitude with visual observations across a broad area of central Alaska [C10-40]. Several meteorological studies using an ice accretion model to estimate ice loads have been performed for high voltage transmission lines in Alaska [C10-17, C10-18, C10-42, C10-45, C10-46, and C10-47]. Estimated 50-year mean recurrence interval accretion thicknesses from freezing rain range from 0.25 to 1.5 in. (6 to 38 mm), snow from 1.0 to 5.5 in. (25 to 140 mm), and in-cloud ice accretions from 0.5 to 6.0 in. (12 to 150 mm). The assumed accretion densities for snow and in-cloud ice accretions, respectively, were 5 to 31 lb/ft^3 (80 to 500 kg/m^3) and 25 lb/ft^3 (400 kg/m^3). These loads are valid only for the particular regions studied and are highly dependent on the elevation and local terrain features.

In Hawaii, for areas where snow and in-cloud icing are known to occur at higher elevations, site-specific meteorological investigations are needed.

Local records and experience should be considered when establishing the design ice thickness and concurrent wind speed. In determining equivalent radial ice thicknesses from historical weather data, the quality, completeness, and accuracy of the data should be considered along with the robustness of the ice accretion algorithm. Meteorological stations may be closed by ice storms because of power outages, anemometers may be iced over, and hourly precipitation data recorded only after the storm when the ice in the rain gauge melts. These problems are likely to be more severe at automatic weather stations where observers are not available to estimate the weather parameters or correct erroneous readings. Note also that (1) air temperatures are recorded only to the nearest 1 °F, at best, and may vary significantly from the recorded value in the region around the weather station, (2) the wind speed during freezing rain has a significant effect on the accreted ice load on objects oriented perpendicular to the wind direction, (3) wind speed and direction vary with terrain and exposure, (4) enhanced precipitation may occur on the windward side of mountainous terrain, and (5) ice may remain on the structure for days or weeks after freezing rain ends, subjecting the iced structure to wind speeds that may be significantly higher than those that accompanied the freezing rain. These factors should be considered both in estimating the accreted ice thickness at a weather station in past storms and in extrapolating those thicknesses to a specific site.

In using local data, it must also be emphasized that sampling errors can lead to large uncertainties in the specification of the 50-year ice thickness. Sampling errors are the errors associated with the limited size of the climatological data samples (years of record). When local records of limited extent are used to determine extreme ice thicknesses, care should be exercised in their use.

A robust ice accretion algorithm will not be sensitive to small changes in input variables. For example, because temperatures are normally recorded in whole degrees, the calculated amount of ice accreted should not be sensitive to temperature changes of fractions of a degree.

C10.1.2 Dynamic Loads. While design for dynamic loads is not specifically addressed in this edition of the Standard, the effects of dynamic loads are an important consideration for some ice-sensitive structures and should be considered in the design when they are anticipated to be significant. For example, large amplitude galloping of guys and overhead cable systems occurs in many areas. The motion of the cables can cause damage due to direct impact of the cables on other cables or structures and can also cause damage due to wear and fatigue of the cables and other components of the structure. Ice shedding from the guys on guyed masts can cause substantial dynamic loads in the mast.

C10.1.3 Exclusions. Additional guidance is available in References [C10-3, C10-9, and C10-10].

SECTION C10.2
DEFINITIONS

FREEZING RAIN. Freezing rain occurs when warm moist air is forced over a layer of subfreezing air at the earth's surface. The precipitation usually begins as snow that melts as it falls through the layer of warm air aloft. The drops then cool as they fall through the cold surface air layer and freeze on contact with structures or the ground. Upper air data indicates that the cold surface air layer is typically between 1000 and 3900 ft (300 and 1200 m) thick [C10-55], averaging 1600 ft (500 m) [C10-8]. The warm air layer aloft averages 5000 ft (1500 m) thick in freezing rain, but in freezing drizzle, the entire temperature profile may be below 32 °F (0 °C) [C10-8].

Precipitation rates and wind speeds are typically low to moderate in freezing rainstorms. In freezing rain, the water impingement rate is often greater than the freezing rate. The excess water drips off and may freeze as icicles, resulting in a variety of accretion shapes that range from a smooth cylindrical sheath, through a crescent on the windward side with icicles hanging on the bottom, to large irregular protuberances (see Figure C10-1). The shape of an accretion depends on a combination of varying meteorological factors and the cross-sectional shape of the structural member, its spatial orientation, and flexibility.

Note that the theoretical maximum density of ice (917 kg/m^3 or 57 lb/ft^3) is never reached in naturally formed accretions due to the presence of air bubbles.

ICE-SENSITIVE STRUCTURES. Ice-sensitive structures are structures for which the load effects from atmospheric icing control the design of part or all of the structural system. Many open structures are efficient ice collectors, so ice accretions can have a significant load effect. The sensitivity of an open structure to ice loads depends on the size and number of structural members, components, and appurtenances and also on the other loads for which the structure is designed. For example, the additional weight of ice that may accrete on a heavy wide-flange member will be smaller in proportion to the dead load than the same ice thickness on a light angle member. Also, the percentage increase in projected area for wind loads will be smaller for the wide-flange member than for the angle member. For some open structures, other design loads, for example, snow loads and live loads on a catwalk floor, may be larger than the design ice load.

IN-CLOUD ICING. This icing condition occurs when a cloud or fog (consisting of supercooled water droplets 100 μm or less in diameter) encounters a surface that is at or below freezing temperature. It occurs in mountainous areas where adiabatic cooling causes saturation of the atmosphere to occur at temperatures below freezing, in free air in supercooled clouds, and in supercooled fogs produced by a stable air mass with a strong temperature inversion. In-cloud ice accretions can reach thicknesses of 1 ft (0.30 m)

or more, since the icing conditions can include high winds and typically persist or recur episodically during long periods of subfreezing temperatures. Large concentrations of supercooled droplets are not common at air temperatures below about 0 °F (−18 °C).

In-cloud ice accretions have densities ranging from that of low-density rime to glaze. When convective and evaporative cooling removes the heat of fusion as fast as it is released by the freezing droplets, the drops freeze on impact. When the cooling rate is lower, the droplets do not completely freeze on impact. The unfrozen water then spreads out on the object and may flow completely around it and even drip off to form icicles. The degree to which the droplets spread as they collide with the structure and freeze governs how much air is incorporated in the accretion and thus its density. The density of ice accretions due to in-cloud icing varies over a wide range from 5 to 56 pcf (80 to 900 kg/m^3) [C10-13 and C10-23]. The resulting accretion can be either white or clear, possibly with attached icicles (see Figure C10-2).

The amount of ice accreted during in-cloud icing depends on the size of the accreting object, the duration of the icing condition, and the wind speed. If, as often occurs, wind speed increases and air temperature decreases with height above ground, larger amounts of ice will accrete on taller structures. The accretion shape depends on the flexibility of the structural member, component, or appurtenance. If it is free to rotate, such as a long guy or a long span of a single conductor or wire, the ice accretes with a roughly circular cross-section. On more rigid structural members, components, and appurtenances, the ice forms in irregular pennant shapes extending into the wind.

HOARFROST. Hoarfrost, which is often confused with rime, forms by a completely different process. Hoarfrost is an accumulation of ice crystals formed by direct deposition of water vapor from the air on an exposed object. Because it forms on surfaces whose temperatures have fallen below the frost point (a dew point temperature below freezing) of the surrounding air due to strong radiational cooling, hoarfrost is often found early in the morning after a clear, cold night. It is feathery in appearance and typically accretes up to about 1 in. (25 mm) in thickness with very little weight. Hoarfrost does not constitute a significant loading problem; however, it is a very good collector of supercooled fog droplets. In light winds, a hoarfrost coated wire may accrete rime faster than a bare wire [C10-43].

SNOW. Under certain conditions, snow falling on objects may adhere due to capillary forces, inter-particle freezing [C10-11] and/or sintering [C10-27]. On objects with circular cross-section such as a wire, cable, conductor or guy, sliding, deformation, and/or torsional rotation of the underlying cable may occur, resulting in the formation of a cylindrical sleeve, even around bundled conductors and wires (see Figure C10-3). Since accreting snow is often

accompanied by high winds, the density of accretions may be much higher than the density of the same snowfall on the ground.

Damaging snow accretions have been observed at surface air temperatures ranging from about 23 to 36 °F (−5 to 2 °C). Snow with high moisture content appears to stick more readily than drier snow. Snow falling at a surface air temperature above freezing may accrete even at wind speeds above 25 mph (10 m/s), producing dense 37 to 50 pcf (600 to 800 kg/m^3) accretions. Snow with lower moisture content is not as sticky, blowing off the structure in high winds. These accreted snow densities are typically between 2.5 and 16 pcf (40 and 250 kg/m^3) [C10-28].

Even apparently dry snow can accrete on structures [C10-14]. The cohesive strength of the dry snow is initially supplied by the interlocking of the flakes and ultimately by sintering, as molecular diffusion increases the bond area between adjacent snowflakes. These dry snow accretions appear to form only in very low winds and have densities estimated at between 5 and 10 pcf (80 and 150 kg/m^3) [C10-49 and C10-40].

SECTION C10.4
ICE LOADS DUE TO FREEZING RAIN

C10.4.1 Ice Weight. The ice thicknesses shown in Figures 10-2, 10-3, and 10-4 were determined for a horizontal cylinder oriented perpendicular to the wind. These ice thicknesses cannot be applied directly to cross-sections that are not round, such as channels and angles. However, the ice area from Eq. 10-1 is the same for all shapes for which the circumscribed circles have equal diameters. It is assumed that the maximum dimension of the cross-section is perpendicular to the trajectory of the raindrops. Similarly, the ice volume in Eq. 10-2 is for a flat plate perpendicular to the trajectory of the raindrops. For vertical cylinders and horizontal cylinders parallel to the wind direction, the ice area given by Eq. 10-1 is conservative.

C10.4.2 Nominal Ice Thickness. Three methods were used to determine the 50-year mean recurrence interval ice thicknesses shown in Figures 10-2, 10-3, and 10-4. In the Northeast, Mid-Atlantic, Midwest, and Plains regions, the mapped values are based on studies using an ice accretion model and local data. In the Pacific Northwest, values are based on icing studies performed for the Bonneville Power Administration. Ice thicknesses for the rest of the country, including the Southeast Atlantic and Gulf coasts and the region west of the Rocky Mountains, were extrapolated from the results of the other studies and from local data. The maps contain revisions and extrapolations not included in Reference [C10-5] based on information found in *Storm Data* (NOAA 1959 to Present) [C10-38], re-analysis of weather data for a longer period of record, and comments

from state climatologists. Specific regional information is given below.

Eastern Region. East of a line along the eastern slope of the Rocky Mountains from central Montana to West Texas in Figure 10-2 except for the extreme Southeast along the Atlantic and Gulf coasts, historical weather data from 330 National Weather Service (NWS), military, and Federal Aviation Administration (FAA) weather stations were used with the CRREL and Simple ice accretion models [C10-24 and C10-25] to estimate uniform radial glaze ice thicknesses in past freezing rainstorms. The station locations are shown in Figure C10-4. The period of record of the meteorological data at any station is typically 20 to 50 years. The ice accretion models use weather and precipitation data to simulate the accretion of ice on cylinders 33 ft (10 m) above the ground, oriented perpendicular to the wind direction in freezing rainstorms. Accreted ice is assumed to remain on the cylinder until after freezing rain ceases and the air temperature increases to at least 33 °F (0.6 °C). At each station, the maximum ice thickness and the maximum wind-on-ice load were determined for each storm. Severe storms, those with significant ice or wind-on-ice loads at one or more weather stations, were researched in *Storm Data*, newspapers, and utility reports to obtain corroborating qualitative information on the extent of and damage from the storm.

Extreme ice thicknesses were determined from an extreme value analysis using the peaks-over-threshold method and the generalized Pareto distribution [C10-1, C10-20, and C10-53]. To reduce sampling error, weather stations were grouped into superstations [C10-41] based on the incidence of severe storms, the frequency of freezing rainstorms, latitude, proximity to large bodies of water, elevation, and terrain. Concurrent wind-on-ice speeds were back calculated from the extreme wind-on-ice load and the extreme ice thickness. The analysis of the weather data and the calculation of extreme ice thicknesses are described in more detail in Jones [C10-26].

This portion of the map represents the most consistent and best available nationwide map for design ice loads. The icing model used to produce the map has not, however, been verified with a large set of collocated measurements of meteorological data and uniform radial ice thicknesses. Furthermore, the weather stations used to develop this map are almost all at airports. Structures in more exposed locations at higher elevations or in valleys or gorges (e.g., Signal and Lookout Mountains in Tennessee, the Ponatock Ridge and the edge of the Yazoo Basin in Mississippi, the Shenandoah Valley and Poor Mountain in Virginia, and Mt. Washington in New Hampshire), may be subject to larger ice thicknesses and higher concurrent wind speeds. In the plains and eastern slopes of the Rocky Mountains, loads from snow or in-cloud icing may be more severe than those from freezing rain. In-cloud icing, possibly combined with freezing drizzle, appears to be the most significant atmospheric icing process in Eastern Colorado and New Mexico.

Southeastern Region. Along the Atlantic and Gulf Coasts from North Carolina to Mississippi in Figure 10-2, ice thicknesses with a 50-year mean recurrence interval were extrapolated from the mapped region using historical weather data, summaries from the National Climatic Data Center, studies of a 1994 ice storm in the Southeast [C10-32] and information in a report by the Southeastern Regional Climate Center [C10-13]. This area of the map was revised from ASCE 7–98 using information from *Storm Data* and extending the analysis described above into the Eastern Piedmont in North Carolina. Based on the new analysis, the southern boundary of the 0.25-in. zone in Florida was moved northward and the 1-in. zone in the Eastern Piedmont in North Carolina was removed.

Western Region. For this revision of ASCE 7, ice thicknesses with concurrent wind speeds were extrapolated westward from the mapped zones in ASCE 7–98, excluding the Pacific Northwest. The mapped ice and wind-on-ice zones are based primarily on information from *Storm Data*. The first step in this process was separating freezing rainstorms from the snow and in-cloud icing events that can cause the same type of damage. The identification of freezing rainstorms from the storm descriptions was sometimes difficult, particularly along the eastern slopes of the Rockies in Colorado and New Mexico, where a mix of in-cloud icing and freezing rain occurs. From the freezing rainstorms that were identified, ice thicknesses were estimated based on the extent of damage to power and telephone lines and communication towers, the financial impact of ice and wind damage, duration of power outages, comparisons to previous ice storms, and reported ice thickness and wind speed. A region extending from New Mexico to Wyoming that was tentatively identified by the above analysis [C10-26] to be subject to minimal ice loads from freezing rain is included in the 0.25-in. zone in Figure 10-2. This portion of the map will be revised when the analysis of the weather data is extended to the Pacific coast.

Additional guidance for the extrapolations was provided by two papers on freezing rain in the United States [C10-7 and C10-48]. Both these publications provide information on the frequency and spatial distribution of freezing rain across the United States, but neither contains information on the duration, precipitation amounts, or accumulated ice thicknesses. Finally, climatologists and meteorologists from nine of the western states reviewed the mapped extrapolations and appropriate revisions were made based on their comments and discussions with them.

In some regions of the West, loads from accreted snow or in-cloud icing are more severe than those from freezing

rain. One region where in-cloud icing appears to be more severe is in Colorado and New Mexico along the eastern slopes of the Rockies.

Pacific Northwestern Region. The freezing rain map for the Pacific Northwest shown in Figure 10-4 is based on ice load maps produced by Meteorology Research Inc. (MRI) for the Bonneville Power Administration (BPA) [C10-44]. BPA commissioned the study in recognition of the need for more complete and detailed information regarding transmission line icing throughout its service area. Data were gathered from a variety of sources, including BPA's in-house records of observed ice accumulations, observed icing events, and/or outage reports from 93 power companies and 30 telephone companies, historical weather data for 42 weather stations from the National Climatic Data Center (NCDC), and personal interviews of pilots and forecasters to learn of local climatological effects. These data were processed and analyzed using methods developed by MRI. For this revision of the Standard, ice thicknesses in certain regions have been revised after consultation with Oregon's state climatologist. Ice thicknesses shown for the Pacific Northwest may not be consistent with those shown for the rest of the United States because of differences in methodology.

Special Icing Regions. Special icing regions are identified on the map. As described above, freezing rain occurs only under special conditions when a cold, relatively shallow layer of air at the surface is overrun by warm, moist air aloft. For this reason, severe freezing rainstorms at high elevations in mountainous terrain will typically not occur in the same weather systems that cause severe freezing rainstorms at the nearest airport with a weather station. Furthermore, in these regions, ice thicknesses and wind-on-ice loads may vary significantly over short distances because of local variations in elevation, topography, and exposure. In these mountainous regions, the values given in Figure 10-1 should be adjusted, based on local historical records and experience, to account for possibly higher ice loads from both freezing rain and in-cloud icing. See Section C10.1.1.

C10.4.4 Importance Factors. The importance factors for ice and concurrent wind adjust the nominal ice thickness and concurrent wind pressure for Category I structures from a 50-year mean recurrence interval to a 25-year mean recurrence interval. For category III and IV structures, they are adjusted to a 100-year mean recurrence interval. The concurrent wind speed used with the nominal ice thickness is based on both the winds that occur during the freezing rainstorm and those that occur between the time the freezing rain stops and the time the temperature rises to above freezing. When the temperature rises above freezing, it is

assumed that the ice melts enough to fall from the structure. In the colder northern regions, the ice will generally stay on structures for a longer period of time following the end of a storm that results in higher concurrent wind speeds. The results of the extreme value analysis show that the concurrent wind speed does not change significantly with mean recurrence interval. The lateral wind-on-ice load does, however, increase because the ice thickness increases. The importance factors differ from those used for wind loads in Section 6 because the extreme value distributions for the ice thickness and concurrent wind speed are different than the distributions used to determine the wind loads in Section 6. See also Table C10-1 and the discussion under C10.4.6.

C10.4.6 Design Ice Thickness for Freezing Rain. The design load on the structure is a product of the nominal design load and the load factors specified in Section 2. The load factors for Load Resistance Factored Design (LRFD) for atmospheric icing are 1.0. This is similar to the practice followed in Section 9 for seismic loads. Figures 10-2, 10-3, and 10-4 show the 50-year mean recurrence interval ice thickness due to freezing rain and the concurrent wind speeds. The probability of exceeding the 50-year event in 50 years of a structure's life is 64%. The design wind loads in this Standard (nominal loads times the load factors in Section 2) have a mean recurrence interval of approximately 500 years, which reduces the probability of being exceeded to approximately 10% in 50 years. Consistent with the design wind loads, the design level mean recurrence interval for atmospheric ice loads on ordinary structures, including load factors, is approximately 500 years. Table C10-1 shows the multipliers on the 50-year mean recurrence interval ice thickness and concurrent wind speed to adjust to other mean recurrence intervals.

The factor 2.0 in Eq. 10-5 is to adjust the design ice thickness from a 50-year mean recurrence interval to a 500-year mean recurrence interval. The multiplier is applied on the ice thickness rather than on the ice load because the ice load from Eq. 10-1 depends on the diameter of the circumscribing cylinder as well as the design ice thickness. The studies of ice accretion on which the maps are based indicate that the concurrent wind speed on ice does not increase with mean recurrence interval (see Section C10.4.4).

When the reliability of a system of structures or one interconnected structure of large extent is important, spatial effects should also be considered. All of the cellular telephone antenna structures that serve a state or a metropolitan area could be considered to be a system of structures. Long overhead electric transmission lines and communications lines are examples of large interconnected structures. Figures 10-2, 10-3, and 10-4 are for ice loads appropriate for a single structure of small areal extent. Large interconnected structures and systems of structures are hit by icing

storms more frequently than a single structure. The frequency of occurrence increases with the area encompassed and the linear extent. To obtain equal risks of exceeding the design load in the same icing climate, the individual structures forming the system or the large interconnected structure should be designed for a larger ice load than a single structure.

Several studies of the spatial effects of ice storms and windstorms have been published. Reference [C10-15] presents a simple method for determining the risk of ice storms to extended systems compared to single structures. Their results indicate that the mean recurrence interval of a given ice load for a transmission line decreases as the ratio of the line length to the ice storm width increases. For a line length to storm width ratio of 2, for example, the mean recurrence interval of a 50-year load as experienced by a single tower will be reduced to 17 years for the entire line. In another study, Laflamme and Periard [C10-30] analyzed the maximum annual ice thickness from triads of passive ice meters spaced about 50 km apart. The 50-year ice thicknesses obtained by extreme value analysis of the triad maxima averaged 10% higher than those for the single stations.

SECTION C10.5
WIND-ON-ICE-COVERED STRUCTURES

Ice accretions on structures change the structure's wind drag coefficients. The ice accretions tend to round sharp edges reducing the drag coefficient for such members as angles and bars. Natural ice accretions can be irregular in shape with an uneven distribution of ice around the object on which the ice has accreted. The shape varies from storm to storm and from place to place within a storm. The actual projected area of a glaze ice accretion may be larger than that obtained by assuming a uniform ice thickness.

SECTION C10.6
PARTIAL LOADING

Variations in ice thickness due to freezing rain on objects at a given elevation are small over distances of about 1000 ft (300 m). Therefore, partial loading of a structure from freezing rain is usually not significant [C10-12].

In-cloud icing is more strongly affected by wind speed, thus partial loading due to differences in exposure to in-cloud icing may be significant. Differences in ice thickness over several structures or components of a single structure are associated with differences in the exposure. The exposure is a function of shielding by other parts of the structure as well as by the upwind terrain.

Partial loading associated with ice shedding may be significant for snow or in-cloud ice accretions and for guyed structures when ice is shed from some guys before others.

REFERENCES

[C10-1] Abild, J., Andersen, E.Y., and Rosbjerg, L. "The climate of extreme winds at the Great Belt, Denmark." *J. Wind Eng. and Industrial Aerodynamics*, 41–44, 521–532, 1992.

[C10-2] American National Standards Institute/Electronic Industries Association/TIA. *Structural standards for steel towers and antenna supporting structures. EIA/TIA-222F*, Electronics Industries Association, Washington, D.C., 1996.

[C10-3] American Society of Civil Engineers (ASCE). "Loadings for electrical transmission structures by the committee on electrical transmission structures." *J. Struct. Div.*, ASCE, 108(5), 1088–1105, 1982.

[C10-4] ASCE. *Guidelines for electrical transmission line structural loading*. ASCE Manuals and Reps. on Engineering Practice No. 74, ASCE, New York, 1991.

[C10-5] ASCE. *Minimum design loads for buildings and other structures*. ASCE Standard 7–98, ASCE, Reston, Va., 2000.

[C10-6] Bennett, I. "Glaze: Its meteorology and climatology, geographical distribution and economic effects." Technical Rep. EP-105, Quartermaster Research and Engineering Center, Environmental Protection Research Div., 1959.

[C10-7] Bernstein, B.C., and Brown, B.G. "A climatology of supercooled large drop conditions based upon surface observations and pilot reports of icing." *Proc., 7th Conf. on Aviation, Range and Aerospace Meteorology*, Long Beach, Calif., 1997.

[C10-8] Bocchieri, J.R. "The objective use of upper air soundings to specify precipitation type." *Monthly Weather Rev.*, 108: 596–603, 1980.

[C10-9] Canadian Standards Association (CSA). "Overhead lines." CAN/CSA-C22.3 No. 1-01, Canadian Standards Association, Rexdale, Ontario, 1987.

[C10-10] CSA. "Antennas, towers, and antenna-supporting structures." CSA-S37-94, Canadian Standards Association, Rexdale, Ontario, 1994.

[C10-11] Colbeck, S.C., and Ackley, S.F. "Mechanisms for ice bonding in wet snow accretions on power lines." *Proc., 1st Int. Workshop on Atmospheric Icing of Structures*, U. S. Army CRREL Special Rep. 83-17, Hanover, N.H., 25–30, 1982.

[C10-12] Cluts, S., and Angelos, A. "Unbalanced forces on tangent transmission structures." *IEEE Winter Power Meeting*, Paper No. A77-220-7, 1977.

[C10-13] Davis, R.E., and Gay, D.A. *Freezing rain and sleet climatology of the Southeastern U.S.A.*, Research Paper No. 052593, Southeast Regional Climate Center, South Carolina Water Resources Commission, Columbia, S.C., 1993.

[C10-14] Gland, H., and Admirat, P. "Meteorological conditions for wet snow occurrence in France Calculated and measured results in a recent case study on 5 March 1985." *Proc., 3rd Int. Workshop on Atmospheric Icing of Structures*, Canadian Climate Program, Vancouver, Canada, 91–96, 1991.

[C10-15] Golikova, T.N., Golikov, B.F., and Savvaitov, D.S. "Methods of calculating icing loads on overhead lines as spatial constructions." *Proc., 1st Int. Workshop on Atmospheric Icing of Structures*, CRREL Special Rep. 83–17, 341–345, 1982.

[C10-16] Goodwin, E.J., Mozer, J.D., DiGioia, A.M. Jr., and Power, B.A. "Predicting ice and snow loads for transmission line design." *Proc., 3rd Int. Workshop on Atmospheric Icing of Structures*, Canadian Climate Program, Vancouver, Canada, 267–275, 1991.

[C10-17] Gouze, S.C., and Richmond, M.C. *Meteorological evaluation of the proposed Alaska transmission line routes*. Meteorology Research, Inc., Altadena, Calif., 1982.

[C10-18] Gouze, S.C., and Richmond, M.C. *Meteorological evaluation of the proposed Palmer to Glennallen transmission line route*. Meteorology Research, Inc., Altadena, Calif., 1982.

[C10-19] GVEA. Unpublished data of Golden Valley Electric Association, Fairbanks, Alaska, 1997.

[C10-20] Hoskings, J.R.M., and Wallis, J.R. "Parameter and quantile estimation for the generalized Pareto distribution." *Technometrics*, 29(3), 339–349, 1987.

[C10-21] International Standards Organization (ISO). "Atmospheric Icing of Structures." ISO 12494 Preliminary International Standard prepared by Technical Committee ISO/TC 98, Subcommittee SC 3, 1999.

[C10-22] International Electrotechnical Commission (IEC). "Loading and strength of overhead transmission lines International Standard 826", Technical Committee 11, 2nd Ed., ISO, Geneva, Switzerland, 1990.

[C10-23] Jones, K.F. "The density of natural ice accretions related to nondimensional icing parameters." *Quarterly J. Royal Meteorological Soc.*, 116, 477–496, 1990.

[C10-24] Jones, K.F. "Ice Accretion in Freezing Rain." CRREL Rep. 96–2, Cold Regions Research and Engineering Laboratory, Hanover, N.H., 1996.

[C10-25] Jones, K. "A simple model for freezing rain loads." *Atmospheric Res.*, 46, 87–97, 1998.

[C10-26] Jones, K.F. "Extreme ice loads from freezing rain." Report for the American Lifelines Alliance. http://www.americanlifelinesalliance.org., 2001.

[C10-27] Kuroiwa, D. "A study of ice sintering." Research Rep. 86, U.S. Army CRREL, Hanover, N.H., 1962.

[C10-28] Kuroiwa, D. "Icing and snow accretion on electric wires." Research Paper 123, U.S. Army CRREL, Hanover, N.H., 1965.

[C10-29] Langmuir, I., and Blodgett, K. "Mathematical investigation of water droplet trajectories." *The collected works of Irving Langmuir*, Pergamon Press, Elmsford, N.Y., 335–393, 1946.

[C10-30] Laflamme, J., and Periard, G. "The climate of freezing rain over the province of Quebec in Canada, a preliminary analysis." *Proc., 7th Int. Workshop on Atmospheric Icing of Structures*, Chicoutimi, Quebec, Canada, 19–24, 1996.

[C10-31] Lott, Neal. NCDC Technical Reps. 93-01 and 93-03: National Climatic Data Center, Asheville, N.C., 1993.

[C10-32] Lott, J.N., and Sittel, M.C. "The February 1994 Ice Storm in the Southeastern U.S." *Proc., 7th Int. Workshop on Atmospheric Icing of Structures*, Chicoutimi, Quebec, Canada, 259–264, 1996.

[C10-33] Macklin, W.C. "The density and structure of ice formed by accretion." *Quarterly J. Royal Meteorological Soc.*, 88, 30–50, 1962.

[C10-34] McCormick, T., and Pohlman, J.C. "Study of compact 220 kV line system indicates need for micro-scale meteorological information." *Proc., 6th Int. Workshop on Atmospheric Icing of Structures*, Budapest, Hungary, 1993.

[C10-35] Mozer, J.D., and West, R.J. "Analysis of 500 kV tower failures." Presented at the 1983 meeting of the Pennsylvania Electric Association, 1983.

[C10-36] Mulherin, N.D. "Atmospheric icing and tower collapse in the United States." *Presented at the 7th Int. Workshop on Atmospheric Icing of Structures*, Chicoutimi, Quebec, Canada, 1996.

[C10-37] National Electrical Safety Code (NESC). *National Electrical Safety Code, 2002 Ed*. The Institute of Electrical and Electronics Engineers, New York, 2002.

[C10-38] NOAA. (1959 to Present). *Storm Data*. National Oceanic and Atmospheric Administration, Washington, D.C.

[C10-39] NPPD. "The storm of March 29, 1976." Public Relations Dept., Nebraska Public Power District, 1976.

[C10-40] Peabody, A.B. "Snow loads on transmission and distribution lines in Alaska." *Proc., 6th Int. Workshop on Atmospheric Icing of Structures*, Budapest, Hungary, 1993.

[C10-41] Peterka, J.A. "Improved Extreme Wind Prediction for the United States." *J. Wind Engrg. and Industrial Aerodynamics*, 41–44, 533–541, New York, 1992.

[C10-42] Peterka, J.A., Finstad, K., and Pandy, A. K. *Snow and Wind Loads for Tyee Transmission Line*, Cermak Peterka Petersen, Fort Collins, Colo., 1996.

[C10-43] Power, B.A. "Estimation of climatic loads for transmission line design." CEA No. ST 198, Canadian Electric Association, Montreal, Quebec, Canada, 1983.

[C10-44] Richmond, M.C., Gouze, S.C., and Anderson, R.S. "Pacific Northwest icing study." Meteorology Research, Altadena, Calif., 1977.

[C10-45] Richmond, M.C. "Meteorological evaluation of Bradley Lake hydroelectric project 115kV transmission line route." M.C. Richmond Meteorological Consultant, Torrance, Calif., 1985.

[C10-46] Richmond, M.C. "Meteorological evaluation of Tyee Lake hydroelectric project transmission line route, Wrangell to Petersburg." Richmond Meteorological Consulting, Torrance, Calif., 1991.

[C10-47] Richmond, M.C. "Meteorological evaluation of Tyee Lake hydroelectric project transmission line route, Tyee power plant to Wrangell." Richmond Meteorological Consulting, Torrance, Calif., 1992.

[C10-48] Robbins, C.C., and Cortinas, J.V., Jr. "A climatology of freezing rain in the contiguous United States: Preliminary results." *Preprints, 15th AMS Conf. on Weather Analysis and Forecasting*, Norfolk, Va., 1996.

[C10-49] Sakamoto, Y., Mizushima, K., and Kawanishi, S. "Dry snow type accretion on overhead wires: Growing mechanism, meteorological conditions under which it occurs and effect on power lines." *Proc., 5th Int. Workshop on Atmospheric Icing of Structures*, Tokyo, Japan, 1990.

[C10-50] Shan, L., and Marr, L. "Ice storm data base and ice severity maps." Electric Power Research Institute, Palo Alto, Calif., 1996.

[C10-51] Tattelman, P. and Gringorten, I. "Estimated glaze ice and wind loads at the Earth's surface for the contiguous United States." Air Force Cambridge Research Laboratories Rep. AFCRL-TR-73-0646, 1973.

[C10-52] United States Forest Service (USFS). *Forest Service Handbook FSH6609.14 Telecommunications Handbook, R3 Supplement 6609.14-94-2 Effective 5/2/94*, United States Forest Service, Washington, D.C., 1994.

[C10-53] Wang, Q.J. "The POT model described by the generalized Pareto distribution with Poisson arrival rate." *J. Hydro.*, 129: 263–280, 1991.

[C10-54] Yip, T.C. "Estimating icing amounts caused by freezing precipitation in Canada." *Proc., 6th Int. Workshop on Atmospheric Icing of Structures*, Budapest, Hungary, 1993.

[C10-55] Young, W.R. (1978). "Freezing precipitation in the Southeastern United States." M.S. Thesis, Texas A&M Univ., 1978.

FIGURE C10-2
RIME ICE ACCRETION DUE TO IN-CLOUD ICING

**FIGURE C10-3
SNOW ACCRETION ON WIRES**

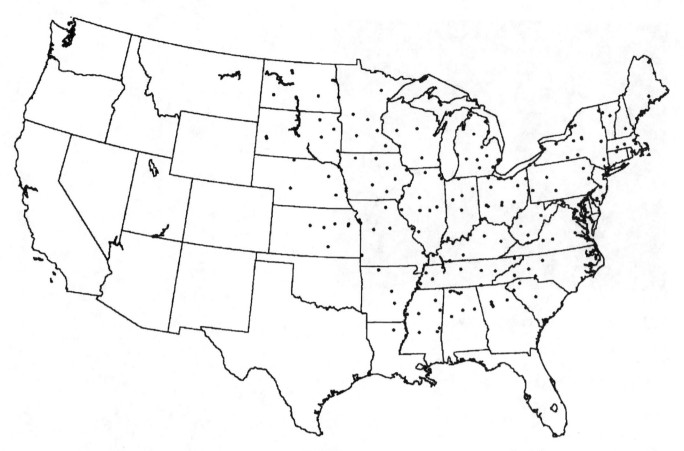

FIGURE C10-4
METEOROLOGICAL STATION LOCATIONS USED IN PREPARATION OF FIGURES 10-2 AND 10-3

Mean Recurrence Interval	Multiplier on Ice Thickness	Multiplier on Wind Pressure
25	0.80	1.0
50	1.00	1.0
100	1.25	1.0
200	1.5	1.0
250	1.6	1.0
300	1.7	1.0
400	1.8	1.0
500	2.0	1.0
1000	2.3	1.0
1400	2.5	1.0

SECTION CB
SERVICEABILITY CONSIDERATIONS

Serviceability limit states are conditions in which the functions of a building or other structure are impaired because of local damage, deterioration, or deformation of building components or because of occupant discomfort. While safety generally is not an issue with serviceability limit states, they may have severe economic consequences. The increasing use of the computer as a design tool, the use of stronger (but not stiffer) construction materials, the use of lighter architectural elements, and the uncoupling of the nonstructural elements from the structural frame may result in building systems that are relatively flexible and lightly damped. Limit states design emphasizes that serviceability criteria (as they always have been) are essential to ensure functional performance and economy of design for such building structural systems (Committee on Serviceability Research 1986; Commentary A 1990; Fisher and West 1990) [1] [10] [14].

There are three general types of unserviceability that may be experienced:

1. Excessive deflections or rotation that may affect the appearance, functional use, or drainage of the structure, or may cause damaging transfer of load to nonload supporting elements and attachments.

2. Excessive vibrations produced by the activities of building occupants, mechanical equipment, or the wind, which may cause occupant discomfort or malfunction of building service equipment.

3. Deterioration, including weathering, corrosion, rotting, and discoloration.

In checking serviceability, the designer is advised to consider appropriate service loads, the response of the structure, and the reaction of the building occupants.

Service loads that may require consideration include static loads from the occupants and their possessions, snow or rain on roofs, temperature fluctuations, and dynamic loads from human activities, wind-induced effects, or the operation of building service equipment. The service loads are those loads that act on the structure at an arbitrary point in time. (In contrast, the nominal loads have a small probability of being exceeded in any year; factored loads have a small probability of being exceeded in 50 years.) Appropriate service loads for checking serviceability limit states may be only a fraction of the nominal loads.

The response of the structure to service loads normally can be analyzed assuming linear elastic behavior. However, members that accumulate residual deformations under service loads may require examination with respect to this long-term behavior. Service loads used in analyzing creep or other long-terms effects may not be the same as those used to analyze elastic deflections or other short-term or reversible structural behavior.

Serviceability limits depend on the function of the building and on the perceptions of its occupants. In contrast to the ultimate limit states, it is difficult to specify general serviceability limits that are applicable to all building structures. The serviceability limits presented in Sections CB.1.1, CB.1.2, and CB.1.3 provide general guidance and have usually led to acceptable performance in the past. However, serviceability limits for a specific building should be determined only after a careful analysis by the engineer and architect of all functional and economic requirements and constraints in conjunction with the building owner. It should be recognized that building occupants are able to perceive structural deflections, motion, cracking, or other signs of possible distress at levels that are much lower than those that would indicate that structural failure was impending. Such signs of distress may be taken incorrectly as an indication that the building is unsafe and diminish its commercial value.

CB.1.1 Vertical Deflections and Misalignment. Excessive vertical deflections and misalignment arise primarily from three sources: (1) gravity loads such as dead, live, and snow loads; (2) effects of temperature, creep, and differential settlement; and (3) construction tolerances and errors. Such deformations may be visually objectionable, may cause separation, cracking, or leakage of exterior cladding, doors, windows and seals, and may cause damage to interior components and finishes. Appropriate limiting values of deformations depend on the type of structure, detailing, and intended use (Galambos and Ellingwood 1986) [16]. Historically, common deflection limits for horizontal members have been 1/360 of the span for floors subjected to full nominal live load and 1/240 of span for roof members. Deflections of about 1/300 of the span (for cantilevers, 1/150 of length) are visible and may lead to general architectural damage or cladding leakage. Deflections greater than 1/200 of the span may impair operation of moveable components such as doors, windows, and sliding partitions.

In certain long-span floor systems, it may be necessary to place a limit (independent of span) on the maximum deflection to minimize the possibility of damage of adjacent nonstructural elements (ISO 1977) [17]. For example, damage to nonload-bearing partitions may occur if vertical deflections exceed more than about 10 mm (3/8 in.) unless special provision is made for differential movement

(Cooney and King 1988) [11]; however, many components can and do accept larger deformations.

Load combinations for checking static deflections can be developed using first-order reliability analysis (Galambos and Ellingwood 1986) [16]. Current static deflection guidelines for floor and roof systems are adequate for limiting surficial damage in most buildings. A combined load with an annual probability of 0.05 of being exceeded would be appropriate in most instances. For serviceability limit states involving visually objectionable deformations, repairable cracking or other damage to interior finishes, and other short-term effects, the suggested load combinations are:

$$D + L \qquad \textbf{(Eq. CB-1a)}$$

$$D + 0.5S \qquad \textbf{(Eq. CB-1b)}$$

For serviceability limit states involving creep, settlement, or similar long-term or permanent effects, the suggested load combination is:

$$D + 0.5L \qquad \textbf{(Eq. CB-2)}$$

The dead load effect, D, used in applying Eqs. CB-1 and CB-2 may be that portion of dead load that occurs following attachment of nonstructural elements. Live load, L, is defined in Section 4. For example, in composite construction, the dead load effects frequently are taken as those imposed after the concrete has cured; in ceilings, the dead load effects may include only those loads placed after the ceiling structure is in place.

CB.1.2 Drift of Walls and Frames. Drifts (lateral deflections) of concern in serviceability checking arise primarily from the effects of wind. Drift limits in common usage for building design are on the order of 1/600 to 1/400 of the building or story height (ASCE Task Committee on Drift Control 1988) [8]. These limits generally are sufficient to minimize damage to cladding and nonstructural walls and partitions. Smaller drift limits may be appropriate if the cladding is brittle. An absolute limit on inter-story drift may also need to be imposed in light of evidence that damage to nonstructural partitions, cladding and glazing may occur if the interstory drift exceeds about 10 mm (3/8 in.) unless special detailing practices are made to tolerate movement (Freeman 1977; Cooney and King 1988) [15] [11]. Many components can accept deformations that are significantly larger.

Use of the factored wind load in checking serviceability is excessively conservative. The load combination with an annual probability of 0.05 of being exceeded, which can be used for checking short-term effects, is

$$D + 0.5\,L + 0.7\,W \qquad \textbf{(Eq. CB-3)}$$

obtained using a procedure similar to that used to derive Eqs. CB-1a and b. Wind load, W, is defined in Section 6.

Due to its transient nature, wind load need not be considered in analyzing the effects of creep or other long-term actions.

Deformation limits should apply to the structural assembly as a whole. The stiffening effect of nonstructural walls and partitions may be taken into account in the analysis of drift if substantiating information regarding their effect is available. Where load cycling occurs, consideration should be given to the possibility that increases in residual deformations may lead to incremental structural collapse.

CB.1.3 Vibrations. Structural motions of floors or of the building as a whole can cause the building occupants discomfort. In recent years, the number of complaints about building vibrations has been increasing. This increasing number of complaints is associated in part with the more flexible structures that result from modern construction practice. Traditional static deflection checks are not sufficient to ensure that annoying vibrations of building floor systems or buildings as a whole will not occur (Committee on Serviceability Research 1986) [1]. While control of stiffness is one aspect of serviceability, mass distribution and damping are also important in controlling vibrations. The use of new materials and building systems may require that the dynamic response of the system be considered explicitly. Simple dynamic models often are sufficient to determine whether there is a potential problem and to suggest possible remedial measurements (Bachmann and Ammann 1978; Ellingwood 1989) [9] [13].

Excessive structural motion is mitigated by measures that limit building or floor accelerations to levels that are not disturbing to the occupants or do not damage service equipment. Perception and tolerance of individuals to vibration is dependent on their expectation of building performance (related to building occupancy) and to their level of activity at the time the vibration occurs (ANSI 1983) [7]. Individuals find continuous vibrations more objectionable than transient vibrations. Continuous vibrations (over a period of minutes) with acceleration on the order of 0.005 g to 0.01 g are annoying to most people engaged in quiet activities, whereas those engaged in physical activities or spectator events may tolerate steady-state accelerations on the order of 0.02 g to 0.05 g. Thresholds of annoyance for transient vibrations (lasting only a few seconds) are considerably higher and depend on the amount of structural damping present (Murray 1991) [18]. For a finished floor with (typically) 5% damping or more, peak transient accelerations of 0.05 g to 0.1 g may be tolerated.

Many common human activities impart dynamic forces to a floor at frequencies (or harmonics) in the range of 2 to 6 Hz (Allen and Rainer 1976; Allen et al. 1985; Allen 1990a; Allen 1990b) [2] [3] [4] [5]. If the fundamental frequency of vibration of the floor system is in this range and if the activity is rhythmic in nature (e.g., dancing, aerobic exercise, cheering at spectator events), resonant amplification may occur. To prevent resonance from rhythmic activities, the floor system should be tuned

so that its natural frequency is well removed from the harmonics of the excitation frequency. As a general rule, the natural frequency of structural elements and assemblies should be greater than two times the frequency of any steady-state excitation to which they are exposed unless vibration isolation is provided. Damping is also an effective way of controlling annoying vibration from transient events, as studies have shown that individuals are more tolerant of vibrations that damp out quickly than those that persist (Murray 1991) [18].

Several recent studies have shown that a simple and relatively effective way to minimize objectionable vibrations to walking and other common human activities is to control the floor stiffness, as measured by the maximum deflection independent of span. Justification for limiting the deflection to an absolute value rather than to some fraction of span can be obtained by considering the dynamic characteristics of a floor system modeled as a uniformly loaded simple span. The fundamental frequency of vibration, f_o, of this system is given by,

$$f_o = \frac{\pi}{2\ell^2} \sqrt{\frac{EI}{\rho}} \qquad \textbf{(Eq. CB-4)}$$

in which EI = flexural rigidity of the floor, l = span, and $\rho = w/g$ = mass per unit length; g = acceleration due to gravity (9.81 m/s^2), and w = dead load plus participating live load. The maximum deflection due to w is,

$$\delta = (5/384)(w\, \ell^4/EI) \qquad \textbf{(Eq. CB-4)}$$

Substituting EI from this equation into Eq. CB-3, we obtain,

$$f_o \approx 18/\sqrt{\delta} \,(\delta \, in \, mm) \qquad \textbf{(Eq. CB-6)}$$

This frequency can be compared to minimum natural frequencies for mitigating walking vibrations in various occupancies (Allen and Murray 1993) [6]. For example, Eq. CB.6 indicates that the static deflection due to uniform load, w, must be limited to about 5 mm, independent of span, if the fundamental frequency of vibration of the floor system is to be kept above about 8 Hz. Many floors not meeting this guideline are perfectly serviceable; however, this guideline provides a simple means for identifying potentially troublesome situations where additional consideration in design may be warranted.

SECTION CB.2
DESIGN FOR LONG-TERM DEFLECTION

Under sustained loading, structural members may exhibit additional time-dependent deformations due to creep, which usually occur at a slow but persistent rate over long periods of time. In certain applications, it may be necessary to limit deflection under long-term loading to specified levels. This can be done by multiplying the immediate deflection by a creep factor, as provided in material standards, that ranges from about 1.5 to 2.0. This limit state should be checked using load combination in CB-2.

SECTION CB.3
CAMBER

Where required, camber should be built into horizontal structural members to give proper appearance and drainage and to counteract anticipated deflection from loading and potential ponding.

SECTION CB.4
EXPANSION AND CONTRACTION

Provision should be made in design so that if significant dimensional changes occur, the structure will move as a whole and differential movement of similar parts and members meeting at joints will be a minimum. Design of expansion joints to allow for dimensional changes in portions of a structure separated by such joints should take both reversible and irreversible movements into account. Structural distress in the form of wide cracks has been caused by restraint of thermal, shrinkage, and pre-stressing deformations. Designers are advised to provide for such effects through relief joints or by controlling crack widths.

SECTION CB.5
DURABILITY

Buildings and other structures may deteriorate in certain service environments. This deterioration may be visible upon inspection (weathering, corrosion, staining) or may result in undetected changes in the material. The designer should either provide a specific amount of damage tolerance in the design or should specify adequate protection systems and/or planned maintenance to minimize the likelihood that such problems will occur. Water infiltration through poorly constructed or maintained wall or roof cladding is considered beyond the realm of designing for damage tolerance. Waterproofing design is beyond the scope of this standard. For portions of buildings and other structures exposed to weather, the design should eliminate pockets in which moisture can accumulate.

REFERENCES

[CB-1] Ad Hoc Committee on Serviceability Research. "Structural serviceability: a critical appraisal and research needs." *J. Struct. Div., ASCE* 112(12), 2646–2664, 1986.

[CB-2] Allen, D.E., and Rainer, J.H. "Vibration criteria for long-span floors." *Can. J. Civil Engrg.,* 3(2), 165–173, 1976.

[CB-3] Allen, D.E., Rainer, J.H., and Pernica, G. "Vibration criteria for assembly occupancies." *Can. J. Civil Engrg.,* 12(3), 617–623, 1985.

[CB-4] Allen, D.E. "Floor vibrations from aerobics." *Can. J. Civil Engrg.,* 19(4), 771–779, 1990a.

[CB-5] Allen, D.E. (1990b). "Building vibrations from human activities." *Concrete Int.,* 12(6), 66–73, 1990b.

[CB-6] Allen, D.E., and Murray, T.M. "Design criterion for vibrations due to walking." *Engineering J., AISC* 30(4), 117–129, 1993.

[CB-7] American National Standards Institute (ANSI). "American national standard guide to the evaluation of human exposure to vibration in buildings." *ANSI S3.29-1983,* New York., 1983.

[CB-8] ASCE Task Committee on Drift Control of Steel Building Structures. "Wind drift design of steel-framed buildings: State of the art." *J. Struct. Div., ASCE* 114(9), 2085–2108, 1988.

[CB-9] Bachmann, H., and Ammann, W. "Vibrations in structures." *Structural Engineering, Doc. 3e,* International Assoc. for Bridge and Struct. Engrg., Zurich, 1987.

[CB-10] National Building Code of Canada. "*Commentary A, serviceability criteria for deflections and vibrations.*" National Research Council, Ottawa, Ontario, 1990.

[CB-11] Cooney, R.C., and King, A.B."Serviceability criteria for buildings." *BRANZ Report SR14,* Building Research Association of New Zealand, Porirua, New Zealand, 1988.

[CB-12] Ellingwood, B., and Tallin, A."Structural serviceability: Floor vibrations." *J. Struct. Div., ASCE* 110(2), 401–418, 1984.

[CB-13] Ellingwood, B. (1989). "Serviceability guidelines for steel structures." *Engrg. J., AISC,* 26(1), 1–8, 1989.

[CB-14] Fisher, J.M., and West, M.A. "Serviceability design considerations for low-rise buildings." *Steel Design Guide No. 3,* American Institute of Steel Construction, Chicago, 1990.

[CB-15] Freeman, S. "Racking tests of high rise building partitions." *J. Struct. Div., ASCE* 103(8), 1673–1685, 1977.

[CB-16] Galambos, T.V., and Ellingwood, B. (1986). "Serviceability limit states: Deflections." *J. Struct. Div., ASCE* 112(1), 67–84, 1986.

[CB-17] ISO. "Bases for the design of structures — Deformations of buildings at the serviceability limit states." ISO 4356, 1977.

[CB-18] Murray, T. "Building floor vibrations." *Engrg. J., AISC* 28(3), 102–109, 1991.

[CB-19] Ohlsson, S. "Ten years of floor vibration research — a review of aspects and some results." *Proc., Symposium on Serviceability of Buildings,* National Research Council of Canada, Ottawa, 435–450, 1988.

[CB-20] Tallin, A.G., and Ellingwood, B. "Serviceability limit states: Wind induced vibrations." *J. Struct. Eng., ASCE* 110(10), 2424–2437, 1984.

INDEX

accepted standards, 159
access floors, 164
accidents, 233
active fault, 97
additions, 3, 96, 97, 321, 326
adjacent structures, 80, 96
adjusted resistance, 97
adjustment factor, 46
aerodynamic shade, 79–80, 319–320
air density values, 306
air-permeable cladding, 274
Alaska, 90, 122–123, 207, 351
Alfred P. Murrah Federal Building, 233
allowable story drift, 145
allowable stress design, 1, 6, 132, 241, 340–341
along-wind displacement, 284
along-wind response, 284
alterations, 3, 97
alternate path method, 234
amusement structures, 206
analytical procedure, simplified, 146
analytical procedure, wind loads, 27–34
anchor bolts, 226
anchorage, 141, 142, 143, 196
anchorages, 225
   component, 160–161
   piles, 174–175
   walls, 342
anchorage systems, 201
appendage, 97
approval, 97
approximate period parameters, values of, 147
appurtenances, 207
arched roofs, 53
architectural components, 157, 188
   design, 162–164
   inspection, 219
   support, 97
ASME boilers, 203
ASME spheres, 204
atmospheric ice loads, 207, 241, 350
attachments, 97, 168
   boilers, 169
   components, 166–167
   electrical equipment, 170–171
   internal equipment, 203–204
   mechanical equipment, 169–170
   pressure vessels, 169–170
authority having jurisdiction, 1
axial load, 141
a-zones, 18

balanced snow loads, 85–88
balusters, 9
barrel vault roofs, 79
   roofs slope factor, 318
   unbalanced snow loads, 319
base, 97
base flood, 17
base flood elevation (BFE), 17
base shear, 97, 154, 151
basements, 97

basement walls, pressure on, 17, 261
basic wind speed, 23, 27, 28, 36–40, 274
beams, 142
bearing walls, 102
bearing wall systems, 133
bilateral load, 185
blockage coefficient, 266, 268
boilers, 169, 203–204
bolted steel structures, 201, 345
bolted steel tanks, 199
boundary elements, 97
boundary members, 97, 223
braced bay members, 223
braced frame, 99
breakaway walls, 17
   loads on, 18, 262
breaking wave pressures, 19–20, 21, 264
buckling failure, 202
building envelope, 23
building frame systems, 99, 133–134
building height, 43, 46
building height limit, 133–135, 136
buildings, 1, 97
   classification of, 2–3, 4, 235–236
   enclosed, 23
   large volume, 31
   low rise, 23
   new, 95–96
   open, 23
   partially enclosed, 23–24
   regular shaped, 25
   rigid, 25
   simple diaphragm, 25, 29
building separations, 145, 183

calculated period, upper limit on, 147
camber, 229, 367
cantilevered column systems, 97, 135
change of use, 95
chimneys, 69, 206, 209
cladding, 23, 25, 26, 27, 29, 31, 32, 33–34, 44, 271
   air-permeable, 27
   lower limit on pressures, 293
   wind loads on, 288
climatic data, regional, 28, 276
coastal A-zone, 17
coastal high-hazard area (V-zone), 17
cold-formed steel, seismic requirements for, 223
cold-formed steel framing, inspection, 219
cold-formed steel studs, 228
collapse, 233
   progressive, 232, 233–234
collector elements, 142, 143
columns, 19, 142
combination framing detailing requirements, 136
compliance procedures, 220
component certification, 167
component deformation, architectural, 161
component forces, architectural, 161
component importance factor, 158, 160
component supports, 167

components, 23, 25, 26, 27, 29, 31, 32, 33, 44, 207
   force transfer, 159
   lower limit on pressures, 293
   nonstructural, 182
   wind loads on, 288
   equipment, 97
   rigid, 97
composite structures, 174
concentrated live loads, 12–14, 255–256
concentrated loads, 9
concentrically braced frame (CBF), 99
concrete, reinforced, 97, plain, 97
concrete components, 174
concrete members, reinforcement, 224–225
concrete pedestal tanks, 203
concrete piles, 173–174
   metal-cased, 221, 222
   precast, 222
   precast nonprestressed, 221, 346
   uncased, 221, 345–346
concrete region, 97
concrete shear walls, 225
concrete walls, 141, 142, 143, 342
concrete-filled steel pipe piles, anchorage of, 221–222
connections, 143, 162, 223
construction documents, 97, 161
construction methods, alternate, 96
construction review, 184
containers, 97
continuous beams, 86
continuous load path, 132
contraction, 367
   structural, 229
contractor responsibility, 218
controlled drainage, 93, 336
counteracting structural actions, 2
coupling beam, 98
crane loads, 11–12, 257
cripple walls, 102
critical load condition, 32
critical load effects, 139
cross-sectional shapes, ice area, 210
curtain walls, 163, 165
curved roofs, 11, 85, 256
   slope factor, 78, 318
   unbalanced snow loads, 79, 318

damage propagation, 233
damping coefficient, 177
dead loads, 7, 100
   minimum design, 246–249
debris, 265
deflection limits, 192
deflections, 145, 151
deformability, 98
deformation, 98
deformational compatibility, 136
demolition, 233
depth coefficient, 266, 268
design acceleration parameters, 131
design displacement, effective period, 177
design earthquake, 98, 181
design earthquake ground motion, 98
design flood, 17, 18, 261–262
design flood elevation (DFE), 17
design force, 25
design ground motion, 131, 132

design ice thickness, 355
design lateral force, 181
design pressure, 25
design rain loads, 93, 335
design response spectrum, 130–131
design response spectrum curve, 129–130
design review, 154
design seismic forces, 132
design snow loads, 326, 327, 328
design story drift, 136, 145, 146
design strength, 1, 101
design values, 157
design wind load cases, 33, 54, 291–292
design wind loads, 291
design wind pressures, 33–34, 41–43
design wind speed, probability of exceeding, 307
designated seismic systems, 98
diagonal braces, 224
diaphragm boundary, 98
diaphragm chord, 98
diaphragm flexibility, 137
diaphragms, 142, 143, 144
   blocked, 98
direct shear, 148
discontinuous frames, 342
discontinuous members, 226, 342
discontinuous walls, 142, 342
displacement, 98, 181, 183
displacement restraint systems, 98
domed roofs, 52
dome roofs, 66
domed roofs, unbalanced snow loads, 79
drag strut, 98
drainage systems, roofs, 93, 335, 337, 338
   flow rate, 338
drift, structures, 229
drift, walls, 366
drift determination, p-delta effects, 149–150
drift height, snow, 89, 319
drift limitations, 145, 165, 179, 182, 192
drift loads. See snow drift
dual frame systems, 99, 132, 134
ductility, structural components, 195
ductwork systems, 167
durability, 229, 367
dynamic analysis, 176
dynamic lateral response, 179–182
dynamic loads, 207, 352
dynamic pressure coefficient, 22
dynamic response, 34
dynamically sensitive structures, 30

earthquake ground shaking, computation of, 339–340
earthquake loads, nature of, 339
earth-retaining structures, 194
eccentric loads, 143
eccentrically braced frame (EBF), 99
eccentricity, 33
effective building period, 154–155
effective damping, 155, 186–187
effective density factor, 201
effective height, 154
effective mass, 201, 204, 205
effective stiffness, 98, 185
effective wind area, 25, 57, 65
electrical component design, 165
electrical components, 157, 188

inspection, 219
  seismic coefficients, 166–167
electrical equipment, attachments, 170
  testing, 220–223
electrical power generating facilities, 194–195
elevated tanks, 201
elevators, 10
  design requirements, 171–172
enclosed buildings, 32, 41–43, 44–46, 49, 50
enclosure, 98
enclosure classifications, 30, 284–285
environmental effects, long term, 229
equipment, attached, 197–198
equipment support, 98
erosion, 18, 262
escarpments, 25, 29, 47, 283
essential facilities, 1, 98
existing structures, additions to, 96
expansion, 367
  structural, 229
expansive soils, 17
exposure categories, 27, 28, 29, 277–278
exposure factor, 77, 90, 315–316
exposures, 43, 46
external pressure coefficients, 31, 52
extraordinary events, 6, 242

facings, 101
factored loads, 1
  combining, 5
factored resistance, 98
Federal Emergency Management Agency (FEMA), 261
federal government construction, 341
fire protection, 168
fire resistance, 182
fixed ladder, 9, 10
fixed service equipment, weight of, 7
flat roofs, 11, 44, 252
  snow loads, 77, 315–316, 319
flexible buildings, 32, 271
flexible equipment connections, 99
flexible structures, 30
flexural members, 224
flood elevation, 18
flood hazard area, 17
flood hazard map, 17, 262
flood insurance rate map (FIRM), 17
flood loads, 5, 6, 17–20, 240, 242–243, 261–267
flood-resistant design, 261
floors, uplift on, 17
fluid load, 240
fluid-structure interaction, 205–206
footings, 226
force, horizontal distribution, 183
  vertical distribution, 179–180
force coefficients, 31, 31, 286
force distribution, 201
force-deflection characteristics, determination of, 185–186
foundation damping factor, 155, 156
foundation design, 132, 141
foundation requirements, 220, 343
foundation stiffness, 155
foundation ties, 173
foundations, 183–184
  design, 172–173
  inspection, 218
  uplift on, 17, 261

frame members, 227
frame systems, 99
frames, 142
  drift of, 229, 366
framing systems, combinations of, 132
freeboard, 197, 206
freezing rain, 207, 208, 352, 355–356, 359
  Lake Superior, 214
  map, 212–213
  Pacific Northwest, 215
fundamental period, 192

gable roofs, 44, 44, 50, 58, 87
  multispan, 62
  snow loads, 321–322, 326
  unbalanced snow loads, 77, 79, 85–88, 315–316, 318, 321–322
  annual probabilities, 329
gas spheres, 204–205
geologic reports, 172
glass, 163, 164, 165, 343
glass components, 165
glaze (ice), 207, 359
glazed curtain walls, 99, 343
glazed storefronts, 99, 343
glazing (glass), 25
  impact resistant, 25
gorges, 207
grab bar systems, 9, 10, 255
grade beam requirements, 344
grade plan, 99
grassland, 280
gravity load, 100
greenhouses, 11
ground motion, 152, 153, 180
ground motion accelerations, site-specific procedure, 130–131
ground snow loads, 77, 82–83, 90, 313–316, 329–331
  elevation, 332–334
guardrail systems, 9, 255
gust effect factor, 28, 30, 283, 309, 310
gust wind speeds, 299

handrails, 9, 255
Hawaii, 124–125
hazard assessment, 3
hazardous contents, 99
hazardous materials, 1, 2–3, 236–237
health hazard, 1
heavy live loads, 256
height factor, 208
height limitations, 132
high hazard exposure structures, 96
high roofs, snow loads, 322
high temperature energy source, 99
hills, 25, 29, 47, 283
hip roofs, 44, 50, 59, 79, 87, 304
  unbalanced snow loads, 318
hoarfrost, 207, 355
hoistway structural systems, 172–174
hold-down, 101
hoop reinforcement, 222
horizontal distribution, 146
horizontal shear distribution, 152, 343
hurricane wind speed, 274–275
hurricane-prone regions, 25, 240
HVAC ductwork, 167
hydraulic structures, 95, 206

hydrodynamic forces, distribution of, 197
hydrodynamic loads, 263
hydrostatic loads, 18, 18, 263

ice dams, 316, 318
ice load zones, map, 212–213
ice loads, 5, 6, 208, 241, 352
ice thickness, 208, 353–355
    Lake Superior, 214
    map, 212–213
    nominal, 208–209
    Pacific Northwest, 215
ice weight, 208, 353
ice-sensitive structures, 207, 352
icicles, 318
icing regions, 207, 353–355
impact loads, 10, 20, 255
    duration, 266
    dynamic analysis, 264–267
impact resistance, 285
impact-resistant covering, 25
importance coefficient, 266, 268
importance factor, 25, 27, 28, 73, 77, 91, 208, 215,
    277, 317, 355
impulsive loads, response ratio, 269
in-cloud icing, 207, 352–353, 360
index force analysis procedure, 145
industrial tanks, 200
industry design standards, 188, 344–345, 349
inelastic deformations, 153
inertia forces, distribution of, 197
inspection, 183, 218
    special, 99
inspector, special, 99–100
interaction effects, 136
intermediate moment frame (IMF), 99, 135
internal components, 346
    storage tanks, 198
internal equipment, attachments, 203–204
internal pressure, 199
internal pressure coefficient, 28, 31, 290
interstory drift, 153, 154, 179, 182
inverted pendulum-type structures, 142
isolation interface, 100
isolation systems, 100, 179, 180, 182
    deformation characteristics, 176–177
    design properties of, 185
    components, replacement, 183
    required tests, 184–185
    stability of, 175
isolator units, 100
    force deflection properties, 184
    spatial distribution, 180–181

joint, 100

knee walls. See cripple walls

lateral deflection, 229
lateral displacements, 177
    total, 178–179
lateral drift, 202
lateral force analysis, 140, 146
lateral force distribution, 200
lateral force procedure, 154
lateral forces, 11, 176, 178, 181
    transfer, 202–203

lateral loads, 182
lateral pressure, 201
lateral response, 176
lateral restoring force, 182
lateral sliding, 201
lateral soil loads, 17, 22
lateral stiffness, 155
lattice frameworks, 71, 209
library stack rooms, uniform live loads, 254
light-frame construction, 100, 223
light-framed walls, 102
light-framed wood shear wall, 102
limit deformation, 98
limit state, 1
limited local collapse, 233
linear response history analysis, 140, 152
liquefaction potential, 173, 343
liquefiable soils, 108
liquid spheres, 204
liquid storage tanks, refrigerated gas, 205
live load, 100
    minimum design, 10
live load element factor, 15
live loads, 9, 254–255
    heavy, 10
    reduction in, 10, 255
    typical, 260
    uniformly distributed, 259
load combinations, 5, 6, 145, 239–243
load effects, 1
    combinations of, 144–145
load factor, 1
load magnification effect, 27
load path connections, 141
load tests, 3, 236
loading, application of, 139
loads, 1
loads not specified, 10
local resistance method, 234
longitudinal, 12
long-term deflection, 367
lower roofs, 78, 79, 79–80, 89, 319–320
    snow loads, 322
low-rise buildings, 32, 33, 286–287, 300, 304
low-rise walls and roofs, 55–56
low-slope roofs, 77
    minimum allowable values, 317

machinery, 10
main wind force-resisting system, 25, 26, 27, 29, 31, 32, 41–43
    lower limit on pressures, 293
maintenance, 183
mansard roofs, 50
masonry, 175, 227
    structural, inspection, 218
masonry walls, 141, 142, 143, 342
mat foundations, 155
material strength, 204
materials, 96
    weight of, 7, 248
    minimum densities, 250–254
materials requirements, 192
mathematical models, 132, 150, 152, 153, 180–181
maximum considered earthquake, 152
maximum considered earthquake ground motion, 100, 107, 341
    map, 110–127
maximum considered earthquake spectrum, deterministic, 131

maximum displacement, effective period, 177–178
maximum response ratio, 266–267
mean recurrence intervals, 308, 363
mean roof height, 25, 29
mechanical components, 157, 188
   inspection, 219
   seismic coefficients, 166–167
mechanical equipment, 229
   attachments, 169
   testing, 220
member deformation, 153
member forces, 153
member strength, 132, 153
membrane roof systems, 271
metal decks, 143
meteorological station locations, 362
minimum wind pressure, 27
misalignment, 365–366
misuse, 233
modal analysis procedure, 150–152, 156–157
modal base shear, 150, 151, 156–157
modal effects, 157
modal forces, 151
modes of vibration, 150
moment frames, 99, 226, 227
moment resisting frames, 134, 136
monitoring, 183
monoslope roofs, 50, 63, 68–69, 77
mountainous terrain, 207
moving loads, 10
multiple folded plates, roof slope factor, 318
   sawtooth roofs, 78
   unbalanced snow loads, 79, 318
multiple use structures, 2, 96

National Flood Insurance Program (NFIP), 261
nominal loads, 1
   combining, 6
nominal strength, 1, 101
nonbearing walls, 102
nonbuilding structures, 100, 157, 183, 186, 188, 344
   importance factors, 191–192
   rigid, 192
   seismic coefficients, 190
   standards, 189–191
nonlinear response history analysis, 153
nonredundant systems, 142, 342
   anchorage of, 142
nonstructural walls, 102
nonstructural wall elements, exterior, 162–163
nonvertical walls, 20
nuclear reactors, 95

obliquely incident waves, 20
occupancy, 1
   nature of, 2, 4, 235
occupancy importance factor, 96, 100, 341
one story buildings, 33
one-way slabs, 11, 256–257
open buildings, 34, 49
open signs, 71
open terrain, 280, 281
openings, 25, 30, 141
ordinary concentrically braced frame (OCBF), 99
ordinary moment frame (OMF), 99
organic clays, 108
out-of-plane bending, 163

overturning, 148–149, 183
overturning moments, 156
owner, 100
owner's inspector, 99–100

panel fillers, 9
parapets, 32, 34, 290–291, 305
parking garage loads, 256
partial loading, 10, 86, 209, 255, 357
partially enclosed buildings, 32, 49, 50
partitions, 100, 164, 165
   provisions for, 9
passenger car garages, 11
p-delta effect, 1, 100, 141, 149–150, 152, 156, 202
peat, 108
pedestals, 227
pendulum systems, inverted, 135
pendulum-type structures, inverted, 100
period determination, 147
periods, 150
permitted analytical procedures, 140
petrochemical tanks, 200
physical hazard, 1
piers, 95, 194
pilasters, 143
pile requirements, 220, 344
piles, 173
pilings, vertical, 19
pipe, concrete filled, 221
pipe racks, 193
piping, attached, 197–198
piping attachments, flexibility of, 195
   minimum displacements, 196
piping systems, 168
   attachments, 168–169
pitched roofs, 11, 256
plan structural irregularities, 137
poles, 173
ponding instability, 80, 93, 321, 335
post-elastic energy dissipation, 95
posts, 173
post-supported spheres, 205–206
precast panel structures, 234–235
precast prestressed piles, 221, 222, 346
pressure coefficients, 286, 288–289
pressure piping, 168
pressure stability, 199
pressure vessels, 169, 203–204
prestressed concrete, inspection, 218
prestressed concrete structures, 201
prestressed concrete tanks, 200
prestressing steel, testing, 219
prestressing tendons, 224–225
probabilistic maximum considered earthquake, 130–131
probable maximum wind speed, 298
progressive collapse, 232
   resistance to, 234
prototype tests, isolator units, 184
public assembly occupancies, 11

quality assurance, 100, 131, 217–220
quality control, 183

rain loads, 93, 335
rain-on-snow surcharge load, 80, 320–321, 322–323, 327
rational analysis, 30
recurrence interval, snow loads, 331